PLINY THE ELDER
ON SCIENCE AND
TECHNOLOGY

PLINY THE ELDER ON SCIENCE AND TECHNOLOGY

John F. Healy

UNIVERSITY PRESS

OXFORD
UNIVERSITY PRESS

Great Clarendon Street, Oxford OX2 6DP

Oxford University Press is a department of the University of Oxford.
It furthers the University's objective of excellence in research, scholarship,
and education by publishing worldwide in

Oxford   New York
Athens   Auckland   Bangkok   Bogotá   Buenos Aires   Calcutta
Cape Town   Chennai   Dar es Salaam   Delhi   Florence   Hong Kong   Istanbul
Karachi   Kuala Lumpur   Madrid   Melbourne   Mexico City   Mumbai
Nairobi   Paris   São Paulo   Singapore   Taipei   Tokyo   Toronto   Warsaw

and associated companies in Berlin   Ibadan

Oxford is a registered trade mark of Oxford University Press
in the UK and certain other countries

Published in the United States
by Oxford University Press Inc., New York

© John F. Healy 1999

The moral rights of the author have been asserted
Database right Oxford University Press (maker)

First published 1999

All rights reserved. No part of this publication may be reproduced,
stored in a retrieval system, or transmitted, in any form or by any means,
without the prior permission in writing of Oxford University Press,
or as expressly permitted by law, or under terms agreed with the appropriate
reprographics rights organizations. Enquiries concerning reproduction
outside the scope of the above should be sent to the Rights Department,
Oxford University Press, at the address above

You must not circulate this book in any other binding or cover
and you must impose this same condition on any acquirer

British Library Cataloguing in Publication Data

Data available

Library of Congress Cataloging-in-Publication Data

Healy, John F.
Pliny the Elder on science and technology / John F. Healy.
p. cm.
Includes bibliographical references and indexes.
1. Pliny, the Elder—Knowledge—Science.   2. Pliny, the Elder
Naturalis historia.   3. Science, Ancient.   I. Title.
Q143.P64H4   1999   500—dc21   99-24316

ISBN 0-19-814687-6

1 3 5 7 9 10 8 6 4 2

Typeset by Newgen Imaging Systems (P) Ltd., Chennai, India
Printed in Great Britain
on acid-free paper by
Bookcraft (Bath) Ltd., Midsomer Norton

For Barbara

PREFACE

The highly successful colloquia held in Como in 1979, to mark the nineteenth centenary of the death of Pliny the Elder in the eruption of Vesuvius, led to an immediate and significant revival in interest in the *Natural History* as a work deserving serious consideration, especially by historians of science. The *Natural History* is, as the Younger Pliny aptly described his uncle's work, wide-ranging in its scope ('opus diffusum') and embraces the whole spectrum of human activity and achievement.

In the past decade important specialist studies have greatly increased our awareness and understanding of different aspects of Pliny's considerable talent ('acre ingenium'): among the most recent may be included, G. Isager, *Pliny on Art and Society: The Elder Pliny's Chapters on the History of Art* (London, 1991), Mary Beagon, *Roman Nature: The Thought of Pliny the Elder* (Oxford, 1992), Francisco de Oliviera, *Les Idées politiques et morales de Pline l'Ancien* (Coimbra, 1992), and R. French, *Ancient Natural History* (London, 1994), part of a series entitled 'Sciences in Antiquity'.

During this time, many errors due either to imperfect descriptions by Pliny, or faulty transmission of the text, have been corrected as a result of research, including simulated laboratory experiments. Hitherto, however, nearly all progress in this field has been presented in the form of papers of colloquia and conferences that have appeared in a wide range of 'Transactions' and journals.

Apart from the useful selection of papers from the Colloquium held at the Royal Institution Centre for the History of Science and Technology, London, in 1983, and published as *Science in the Early Roman Empire: Pliny the Elder, his Sources and Influence* (London and Sydney, 1986), edited by R. French and F. Greenaway, no definitive attempt has been made to re-examine Pliny's 'Science' since K. C. Bailey's, *Pliny's Chapters on Chemical Subjects*—published more than seventy years ago!

As Mary Beagon justly observes, in her perceptive study, 'The *Natural History* is an impressively huge and unwieldy work. It is impossible to give it a truly comprehensive examination in a book of ordinary length.' Similarly, the words of Pliny himself are not inappropriate: 'minime mirum est hominem genitum non omnia humana novisse' (*HN* 3. 1). The present treatment, therefore, is intended to provide a *preliminary* reassessment of some of Pliny's more interesting contributions, in the fields of 'Chemistry, Physics, and Earth Sciences'.

The text of the *Natural History* presents many problems of interpretation, complicated by the range and specialized character of its subject-matter. At the time of Pliny's death in AD 79 it was unrevised and, obviously, had not had the benefit of a copy-editor! Moreover, errors in the transmission of manuscripts by subsequent generations of copyists have often been compounded by emendations proposed by editors with inadequate scientific expertise, unable to understand what Pliny had originally written.

Such problems, however, are not new! As early as the fourteenth century, Petrarch complained about a codex of the *Natural History,* purchased in Mantua on July 1350. 'What', he wrote, 'would Cicero, or Livy, or the other great men of the past, Pliny above all, think if they could return to life and read their own books?' He answered his question with the surmise that they would scarcely recognize these corrupt and barbarous texts as theirs. Petrarch, therefore, began work on the emendation and annotation of the text of the *Natural History* in an attempt to sort out the disorder that had been introduced into a number of manuscripts of the ninth to the eleventh centuries.

The text of the *Natural History*, one of the first Latin books to be printed (Venice, 1469), had been seriously flawed in transmission over the preceding centuries. Hermalaos Barbarus, in his *Castigationes Plinianae* (1492–3), already claimed to have corrected five thousand textual errors! However, the first major challenge to Pliny came from Leoniceno's criticism in 1509 (see Ch. 20). The *Natural History* had reached its high-water mark in the Middle Ages but, thereafter, its popularity as an encyclopaedia began to decline, reaching its lowest ebb in the nineteenth century.

The understanding of Pliny's achievement has, until recently, also been seriously impeded by a lack of realization that Pliny—facing the same problem ('sermonis egestas') as Lucretius—had developed a 'language within a language', that is, as it were, a 'subset' of literary Latin in which words have a specialized meaning, different from the normal usage. In addition, numerous borrowings and adaptations from Greek, Celtic, and Spanish are to be found in the text. Study of this technical language has complemented evidence from laboratory experiments.

In an evaluation of Pliny's science there seems no point in avoiding the use of meaningful modern terminology, where appropriate, that is when Pliny describes materials, processes, and phenomena in the fields of Earth Sciences, 'Chemistry', and 'Physics', even if he does not understand the underlying principles. For example, although the science of crystallography was not fully established until the early nineteenth century, Pliny correctly describes the crystal system of the octahedral diamond, and the hexagonal systems of beryl and quartz. Elsewhere, he records processes—in nature—that are the equivalent of the precipitation of crystals in an evaporating, saturated solution. Similarly, he records properties of materials which come within the field of Physics. Finally, there seems no sustainable objection to examining the 'science' of the ancient world in the light of modern criteria, provided that the Greeks and Romans are not credited with technical knowledge that is the product of later centuries!

Some years ago H. Zehnacker well summarized the situation apropos Pliny's achievement when he wrote: 'Il est symptomatique de constater que nos progrès dans la connaissance de la technologie antique nous amènent toujours à apprécier positivement l'*Historie Naturelle*. Pline résiste à l'épreuve des faits; mieux: il en sort grandi.' This assessment is no less true today!

My thanks are offered to the many colleagues and friends mentioned in the text for their generous advice and help, and in particular, to Professor J. B. Hall, Mr Brian Kemp and Mrs Gillian Kemp, Mr John Kennedy, of Waterford Crystal Ltd., Mr M. McGarr of the Institution of Mining and Metallurgy, Mr R. Saunders of BBC Television, Dr P. Maxwell-Stuart, Dr R. Shepherd, Miss Emma Town, Mr D. Ward, and Professor H. Zehnacker.

I am indebted to members of the Oxford University Press for their care in the production of this book, to Miss Enid Barker and Ms Hilary O'Shea for help and encouragement throughout, and to Ms Georga Godwin for overseeing the final stages. Mr Julian Ward has made a much appreciated major contribution to this work and I have greatly benefited from his expertise, suggestions, and attention to detail in preparing the manuscript for publication.

Finally, I am once more grateful to my wife Barbara, on Pliny's behalf, for her patience throughout my often obsessive commitment to the study of his work and for her practical assistance in helping to resolve problems in translation and for reading through the final text.

Any errors and inconsistencies that remain are solely my own.

J.F.H.
Macclesfield
December 1997

CONTENTS

Abbreviations   xiii

1. Pliny's Life and Career   1
2. Pliny the Writer   31
3. The *Natural History*   36
4. Pliny's Sources   42
5. *Mirabilia*   63
6. Pliny and Research   71
7. Language and Style   79
8. Science in Antiquity   100
9. The *Natural History* and Technical Literature   106
10. Science in The *Natural History*   113
11. Chemistry   115

    APPENDICES   A: Niello   137
                                   B: Dyes and Dyeing   138

12. Physics   142

    APPENDIX   From Observer to Horizon   171

13. Earth Sciences   173
14. Minerals   190

    APPENDICES   A: Amber   250
                                   B: Hydrocarbons   254
                                   C: Meteorites   257

15. Minerals as Pigments	259
16. Gems and Precious Stones	263
17. Metals	271
18. Technology	347
19. Pliny and the Environment	371
20. The *Natural History* in the Middle Ages and After	380
Select Bibliography	393
Index Locorum	412
General Index	431

ABBREVIATIONS

Bailey, *Chemical Subjects*	K. C. Bailey, *The Elder Pliny's Chapters on Chemical Subjects*, 2 vols. (London, 1929–32)
Barton, *Roman Buildings*	I. M. Barton (ed.) *Roman Public Buildings* (Exeter Studies in History, 29; Exeter, 1989)
Beagon, *Roman Nature*	Mary Beagon, *Roman Nature: The Thought of Pliny the Elder* (Oxford, 1992)
CIL	*Corpus Inscriptionum Latinarum* (Berlin, 1863–)
Conophagos, *Laurium*	C. E. Conophagos, *Le Laurium antique et la technique grecque de la production de l' argent* (Athens, 1980)
Domergue, *Mineria y metalurgia*	C. Domergue (ed.), *Mineria y metalurgia en las antiguas civilizaciones mediterraneas y europeas*, 2 vols. (Madrid, 1989)
Eichholz. *Lap*	D. E. Eichholz (ed. and trans.), *Theophrastus: De Lapidibus* (Oxford, 1965)
Eichholz, *NH 36–37*	D. E. Eichholz (trans.), *Pliny the Elder, Natural History*, x: *Books 36–37* (Loeb: London and Cambridge, Mass., 1962)
FGrH	*Die Fragmente der griechischen Historiker*, ed. F. Jacoby, 15 vols. (Berlin and Leiden, 1923–58)
Forbes, *Ancient Technology*	R. J. Forbes, *Studies in Ancient Technology*, 9 vols. (Leiden, 1966–93)
French, *Ancient Natural History*	R. French, *Ancient Natural History: Histories of Nature* (London and New York, 1994)
French–Greenaway, *Roman Science*	R. French and F. Greenaway (eds.), *Science in the Early Roman Empire: Pliny the Elder, his Sources and Influence* (London and Sydney, 1986)
GGM	*Geographi Graeci Minores*, ed. C. Müller, 3 vols. (Paris, 1855–61)

Grant–Kitzinger, *Mediterranean Civilization*	M. Grant and Rachel Kitzinger (eds.), *Civilization of the Ancient Mediterranean: Greece and Rome*, 3 vols. (New York, 1988)
Hall–Metcalf, *Chemical and Metallurgical Investigation*	E. T. Hall and D. M. Metcalf (eds.), *Methods of Chemical and Metallurgical Investigation of Ancient Coinage* (Royal Numismatic Society, Special Publication, 8; London, 1972)
Healy, *Language and Style*	J. F. Healy, 'The Language and Style of Pliny the Elder', in *Filologia e Forme Letterarie: Studi offerti a Francesco della Corte* (Urbino, 1988), iv. 1–24
Healy, *Miniere*	J. F. Healy, *Miniere e metallurgia nel mondo greco e romano* (Rome, 1993): rev. edn. of Healy, *Mining* (see below)
Healy, *Mining*	J. F. Healy, *Mining and Metallurgy in the Greek and Roman World* (London, 1978)
Healy, *Pliny, NH: A Selection*	J. F. Healy (trans. and comm.), *Pliny the Elder, Natural History: A Selection* (London, 1991)
Hurlbut, *Mineralogy*	C. S. Hurlbut (ed.), *Dana's Manual of Mineralogy*, 18th edn. (London, 1971)
IG	*Inscriptiones Graecae* (1st edn. Berlin, 1873; 2nd edn. Berlin, 1913–)
Landels, *Ancient Engineering*	J. G. Landels, *Engineering in the Ancient World* (London, 1978)
LSJ	H. G. Liddell, R. Scott, and H. S. Jones, *A Greek–English Lexicon*, 9th edn. (Oxford, 1940)
Partington, *Chemistry*	J. R. Partington, *General and Inorganic Chemistry for University Students*, 4th edn. (London, 1996)
Rackham, *NH 33–35*	H. Rackham (trans.), *Pliny the Elder, Natural History*, ix: Books 33–35 (Loeb: London and Cambridge, Mass., 1952)
RE	*Real-Encyclopädie der classischen Altertumswissenschaft*, ed. A. Pauly and G. Wissowa (Stuttgart 1894–)
Tylecote, *Metallurgy*	R. F. Tylecote, *The Early History of Metallurgy in Europe* (London and New York, 1987)
Tylecote, *Metallurgy in Archaeology*	R. F. Tylecote, *Metallurgy in Archaeology: A Prehistory of Metallurgy in the British Isles* (London, 1962)

Whitten–Brooks, *Geology*	D. G. A. Whitten and J. R. V. Brooks, *A Dictionary of Geology* (London, 1974)
Zehnacker, *NH 33*	H. Zehnacker, *Pline l'Ancien: Histoire Naturelle, Livre 33* (Budé: Paris, 1983)

Standard abbreviations are used for ancient authors and their works. See especially the list of abbreviations in the *Oxford Classical Dictionary* (3rd edn.) and in LSJ.

Thesauros oportet esse non libros
Domitius Piso

1

Pliny's Life and Career

1.1 THE EARLY YEARS

Pliny the Elder,[1] Gaius Plinius Secundus, was born in late AD 23, or early 24, during the principate of Tiberius, a period of great political unrest, mutiny within the legions, and rivalries in the struggle for imperial power.

His birthplace, Novum Comum, a town of mixed population in Transpadane Gaul[2] (northern Italy), had twice received colonists in the first century BC. Comum was one of many communities that Pompeius Strabo[3] had reorganized in 89 BC. Subsequently, as a result of raids by Alpine tribes, it received a further five thousand colonists when Caesar, under the Lex Vatinia[4] in 59 BC, resettled it and renamed it Novum Comum.[5]

There is no record of Pliny's parents, but his father's name would indicate a family of native stock. What is more important, however, is that he came from a wealthy family that belonged to the municipal governing class. With this background, he became a member of the equestrian order (*ordo equester*), open to all Roman citizens who were at least 18 years

---

[1] Healy, *Pliny NH: a Selection*, pp. ix ff.
[2] See further, G. E. F. Chilver, *Cisalpine Gaul: Social and Economic History from 49 BC to the Death of Trajan* (Oxford, 1941), esp. 106 f.
[3] A Roman *legatus* who raised a considerable force on his private estates in Picenum during the Italian Wars (91–83 BC). As consul (89) he rewarded the semi-Celtic population of Transpadane Gaul by promotion to Latin status. He later suppressed the revolt of the Italian Confederacy in central Italy, while Sulla campaigned in the south.
[4] At the time of the First Triumvirate (59 BC), Caesar cleared the way for a wide programme of supplementary legislation which was submitted, partly in his own name, and partly in that of the tribune Vatinius who supported him.
[5] Suet. *Iul*. 28.

of age, of free birth, good character, and a property rating (*census*) of 400,000 sesterces. As a knight (*eques*) many opportunities were open to Pliny, and his status determined his career.

With the development of the empire, Augustus had organized a rudimentary civil service and to fill its ranks, had turned to the equestrian order, whose members had gained experience in tax-farming and other financial operations.[6] As the *equites* were likely to serve in administrative posts involving the command of troops, he restored the ancient connection between the order and military service. By the time of the emperor Claudius, *militiae equestres* were organized and the knight would normally serve for ten years. He would begin his career as the commander of an auxiliary cohort (*praefectus cohortis*), becoming, thereafter, a staff officer in a legion (*tribunus militum*). Finally he would rise to the rank of commander of an auxiliary squadron (*praefectus alae*). At the end of his period of military service, the knight was deemed qualified for appointment as a procurator.

Although, in origin, a personal agent of the emperor, the procurator soon became more clearly a state official. Procurators were fiscal agents in the provinces and could be governors of small, but not necessarily unimportant provinces. Their deployment throughout the empire gave them the opportunity to travel, to observe, and to collect information. Many of them turned to writing[7] to complement their political careers. Apart from Pliny the Elder,[8] one of the best known is, perhaps, Columella.[9]

After efficient and loyal service as a procurator, a knight could obtain one of various prefectures: of traffic, of the fleet,[10] of the fire brigade, of the corn supply, of Egypt, or, most powerful of all, of the Praetorian Guard.

Such, briefly, was the framework of Pliny the Elder's career. His own preoccupation with his status and public image, as

---

[6] Cf. Joyce Reynolds, 'The Elder Pliny and his Times', in French–Greenaway, *Roman Science*, 1.   [7] Beagon, *Roman Nature*, 6 ff.

[8] See Ch. 2 below.

[9] Columella was a native of Cadiz and a contemporary of the Younger Seneca (*c*.4 BC–AD 65) and Pliny the Elder. He wrote a *De Re Rustica*, (in 12 books). See further, K. D. White, *Roman Farming* (London, 1970), 26 ff.

[10] Pliny was *praefectus* of the *classis Misenensis*.

a knight, is evident from the nature of his digression on the history of the equestrian order (*NH* 33. 32 ff.). Tacitus (*Ann.* 16. 5), writing a generation later than Pliny (about Nero's behaviour), implies that men of municipal origin tended to favour stricter codes of behaviour than those characteristic of members of the imperial court and Roman aristocracy in the first century AD: 'People from remote country towns of austere, old-fashioned Italy, or visitors from distant provinces on official or private business had no experience of outrageous behaviour; they found the spectacle intolerable' ('sed qui remotis e municipiis severaque adhuc et antiqui moris retinente Italia, quique per longinquas provincias lascivia inexperti officio legationum aut privata utilitate advenerant, neque aspectum illum tolerare neque labori inhonesto sufficere'). The Younger Pliny (*Letters* 1. 14. 4) makes similar comments with reference to the inhabitants of Transpadane Gaul, although he is less trenchant.[11]

The evidence of material remains and artefacts excavated in the Campanian towns of Pompeii, Herculaneum, Stabiae, and Oplontis, buried by the eruption of Vesuvius in AD 79,[12] provides a picture of life in a Roman *municipium* in the first century AD. This is complemented by the commentary on social life and mores recorded by Pliny the Elder in his *Natural History* and by his nephew in his *Letters*.

Pompeii was, fundamentally, a commercial centre, as its public buildings confirm.[13] The private houses, with their peristyle gardens, suggest a life style that was comfortable, and to some extent, elegant, but not over ostentatious. Expensive materials such as coloured marble, imported woods, and other examples of luxury, appear to have been relatively rare in Pompeii. Herculaneum, by contrast, boasted many wealthy citizens in addition to artisans and fishermen.

1.2 ROME

The internal evidence of the *Natural History* itself suggests that Pliny was introduced to Rome during the principate of

---

[11] Reynolds, 'The Elder Pliny', 1.
[12] Vividly described by the Younger Pliny (*Letters* 6. 16. 4–20).
[13] See generally, Barton (ed.), *Roman Buildings*.

Caligula (AD 14–37), namely in the early 30s, and that his experience of Roman society, as, for example, when he attended the betrothal dinner for Lollia Paulina (*NH* 9. 117) made a lasting impression on him, as his vivid description of her jewellery shows: '...she was covered with emeralds and pearls which caught the light all over her head, hair, ears, neck, and fingers: these adornments were worth 40,000,000 sesterces!' ('sponsalium cena smaragdis margaritisque opertam alterno textu fulgentibus toto capite, crinibus, auribus, collo, digitis quae summa quadringenties sestertium colligebat.') His moral indignation at this ostentation was increased by the knowledge that these were not presents but items obtained with spoils from the provinces.

By Pliny's time, Rome had become a city of extremes. Nero's lavishly decorated Golden House[14] (Domus Aurea) and the town houses of the rich on the Palatine, provided a sharp contrast with the noisy, overcrowded tenement blocks (*insulae*) in which the poor lived: these were later, vividly described by Juvenal (*Satires* 3. 194 ff.) as poised like a house of cards... where fires are commonplace.

In the capital, Pliny was to find much to interest him in literary movements, the arts, oratory, philosophy, and science. His formal education began under the well-known soldier and tragic poet Pomponius Secundus (7. 80) 'consularis poeta' and (13. 83) 'civis clarissimus', whose biography he subsequently wrote as a debt to a friend ('munus debitum') and, possibly, to ensure his patronage.

Following the tradition of the time, an important part of Pliny's training was in rhetoric which greatly influenced his literary style,[15] for which he has been frequently criticized.[16]

The emperor Augustus had encouraged the idea that equestrian status should carry a certain obligation to seek at least one army posting. Not everyone, however, responded in the same way to this virtual challenge. Some found safe niches in administrative posts, ceremonial duties, supervisory, or non-combatant roles. This gave them the opportunity to indulge in

---

[14] R. Bandinelli, *Rome the Centre of Power: Roman Art to AD 200* (London, 1970), 132 ff.   [15] Healy, 'Language and Style' 13 ff.
[16] From the time of E. Norden, *Die antike Kunstprosa vom VI. Jhdt. v. Chr. bis in die Zeit der Renaissance*, 2 vols (Leipzig, 1898), i. 134.

leisure pursuits, writing, or whatever seemed conducive to the good life. Some indeed continued to live extravagant lifestyles, as did an officer of Pliny's acquaintance, who carried an expensive dinner service with him on active service. 'Pompeius Paulinus', he writes (33. 143), 'the son of a knight of Rome at Arles and, descended on his father's side from a tribe that went about in skins, had to my knowledge 12,000 pounds weight of silver plate with him while serving with an army that faced tribes of the greatest ferocity' ('at Hercules, Pompeium Paulinum, Arelatensis equitis Romani filium paternaque gente pellitum, XII pondo argenti habuisse apud exercitum ferocissimis gentibus oppositum scimus').

1.3 MILITIA EQUESTRIS

Pliny, however, took his military service more seriously and began his public career in the province of Germany (AD 47–57) rising to the command of a cavalry squadron (*praefectus alae*).

The long accepted model of his career, supported by Syme[17] and others,[18] assigns him three tours of duty, in which he campaigned against (*a*) the Chauci (AD 47), under Domitius Corbulo, (*b*) the Chatti (50), in Upper Germany during the governorship of Pomponius Secundus,[19] and (*c*), in Lower Germany, as a colleague of the future emperor Titus (son of Vespasian), to whom he dedicated the *Natural History*. The evidence linking Pliny with the legionary fortress of Xanten (Vetera), on the Rhine—an inscription on a horse-trapping[20] (*phalera*), PLINIO PRAEF. EQ.[21]—is now confirmed and dated to the later years of his service. However, Pliny's loyalty to the house of Germanicus earned him no reward.

---

[17] Cf. 'Pliny the Procurator', in *The Roman Papers of Sir Ronald Syme*, ed. A. R. Birley, 7 vols. (Oxford, 1979–91), ii. 742–73. Also 'Carrière et amis consulaires de Pline' *Helmantica*, 38 (1987), 223–31 and, with additions, 'The consular Friends of Pliny the Elder', in *Roman Papers*, vii. 496–511.

[18] For example, Reynolds, 'The Elder Pliny', 1–10.

[19] Commander of the army of the Upper Rhine *c*.50–1: cf. Tac. Ann. 12. 27. Patron of Pliny, Pomponius was also an amateur dramatist (Younger Pliny, *Letters* 7. 17. 11).

[20] *CIL* xiii. 10026. 2. J. F. Healy, 'Problems in Mineralogy and Metallurgy in Pliny the Elder's Natural History', in *Atti del Convegno di Como—Technologia, economia, e società nel mondo romano* (Como, 1980), 180.

[21] The revised reading EQ. removes any earlier doubts about Pliny's status.

During this period, Pliny wrote his *On the Use of the Javelin as a Cavalry Weapon* and began his first larger-scale work of the *History of Rome's Wars against the Germans*. The latter confirms his interest in the province.

In AD 59, Pliny returned to Rome with the intention of practising law. His activities, however, from then until the death of Nero (AD 68), of whom he was always unequivocally critical, are uncertain.[22] Attempts to explain his absence from public life and why no procuratorship was offered to him at that point in his career, tend to be inconclusive. Syme suggests that the reason was the loss of a patron (Pomponius Secundus had died), some minor indiscretion in the society of the capital, a sudden distaste for affairs, or a prescience of the dangers that lurked in the path of industry and integrity. In the light of the prevailing political climate, the last reason would seem to have been the most likely. Pliny (7. 45) writes that Nero, throughout his principate, was the enemy of mankind ('toto principatu suo hostem generis humani') and loses no opportunity to criticize his extravagance (35. 51; 37. 50). Nero's jealous and obsessive fear of all eminence of birth, or success in the military field, drove him to a policy of persecution. Added to this, his philhellenic outlook was resented by Romans generally. The Younger Pliny (*Letters* 3. 3. 5) sums up the situation, observing that Nero's attitude rendered dangerous every study of a free and elevated character ('cum omne studiorum genus paulo liberius et erectius periculosum servitus fecisset'). In such a climate, it is hardly surprising that Pliny apparently retired from public service and, keeping a low profile, devoted his talents to the safer pursuit of writing books on 'grammar'.[23]

The Neronian tyranny and catastrophes such as the Pisonian conspiracy against Nero (AD 65) involved a large number of knights as well as senators: this confirms Pliny's luck, or rather his sound judgement! On Nero's death, Pliny opportunely emerged as a partisan of the emperor Vespasian. Under him, he recommenced his public career which flourished in the new

---

[22] Beagon, *Roman Nature*, 3.
[23] See further A. della Casa, 'Plino grammatico', in *Atti del Convegno di Como* (1979–) *Plinio il Vecchio sotto il profilo storico e letterario* (Como, 1982), 109–15.

stability created by the Flavian dynasty. Pliny became a member of the emperor's council[24] (*amicus principis*) which meant that, when in Rome, he was expected to wait on Vespasian every morning (*salutatio*)[25] and that he might be called to attend business sessions as an adviser.[26]

Maxwell-Stuart, in a recent study of Pliny,[27] puts forward an alternative chronology from AD 53 onwards, in which he states that Pliny entered on the military *cursus* as *duovir quattuorvir*. He dates his first posting, as *praefectus cohortis* in Lower Germany under Pompeius Paulinus, to 58/9, Pliny was subsequently *tribunus militum* under Duvius Avitus and took part in a retaliatory campaign against the Frisii who had invaded and settled territory set aside for Roman veterans in the region south-east of the Isselmeer. In 61 he served with Titus (as *contubernalis*) and in 62/3 was promoted *praefectus alae*. Maxwell-Stuart suggests that he may have left the army in 65/6 but equally well could have stayed on until much nearer the start of his procuratorial career.

1.4 PROCURATORSHIPS

The evidence of Suetonius implies that Pliny held more than one procuratorship ('procurationes quoque splendidissimas et continuas summa integritate administravit').[28] Unfortunately, neither Suetonius, nor any other authority, except the Younger Pliny, who refers to his uncle's time in Hispania Tarraconensis (see below), mentions any further procuratorship.

Münzer's[29] list of four possible terms of office— in Gallia Narbonensis, Africa, Hispania Tarraconensis, and Gallia

---

[24] Cf. the Younger Pliny, *Letters* 3. 5. 7 ('amicitia principum').
[25] Suet. Vesp. 4. Vespasian consulted his cabinet in the dawn hours!
[26] The Younger Pliny, *Letters* 3. 5. 9.
[27] On the chronology of Pliny's career, see further P. Maxwell-Stuart's Ph.D. thesis, 'Studies in the Career of Pliny the Elder and the Composition of his *Naturalis Historia*' (St Andrews, 1996). I have benefited from the discussion of a number of critical issues with him.
[28] C. Roth (ed.), *Suetonius* (Teubner: Leipzig, 1862), p. 300. Cf. R. Syme, *Tacitus* (Oxford, 1979), i. 61 n. 2. Pliny ran through a sequence of procuratorships ending with Gallia Belgica which made him paymaster-general for the armies of the Rhine.
[29] 'Die Quelle des Tacitus für die Germanenkriege', *Bonner Jahrbücher*, 104 (1899), 103 ff.

Belgica—although open to challenge, has generally been accepted until recently, for example, by Syme,[30] Reynolds,[31] and, with reservations, myself.[32] Syme, however, observed that the procuratorship in Gallia Narbonensis seemed the least secure.[33]

As part of his alternative reconstruction of Pliny's career, Maxwell-Stuart rejects Münzer's findings. It is not inopportune, therefore, to re-examine in detail the internal evidence of the *Natural History*, although this is, admittedly, in the main circumstantial.

1.4.1 Hispania Tarraconensis[34]

The only certain and convincingly datable procuratorship is that of Hispania Tarraconensis (AD 72–4), which the Younger Pliny confirms in his correspondence with Baebius Macer (*Letters* 3. 5. 17), referring to an offer made to his uncle for his 'notes' ('cum procuraret in Hispania'). There, Pliny served as the financial agent of the *princeps*, being in charge of imperial revenue and expenditure in northern Spain when Larcius Licinius was governor of the province. Although his nephew's reference is quite explicit let us consider what other evidence there would have been to substantiate the claim for this procuratorship, if this confirmation had not been available.

Pliny (3. 28) records the census figures for this region,[35] but it is his interest in gold-mining and processing[36] that provides

---

[30] 'Pliny the procurator', in *Roman Papers*, ii. 742–73.

[31] 'The Elder Pliny', 8. On the basis of personal observations that he records, we may be sure he had visited Africa and Gallia Narbonensis (that is, in addition to Hispania Tarraconensis). He might perhaps have been to Narbonensis privately, conceivably diverging on the way to or from one of his postings in Germany.

[32] J. F. Healy, 'Pliny the Elder and Ancient Mineralogy', *Interdisciplinary Science Reviews*, 6/2 (June 1981), 166 ff.   [33] Syme, *Tacitus* 1, 61, n. 2.

[34] NH 3. 6. This province was renowned for its sources of gold: J. F. Healy, 'Greek and Roman Gold Sources', in Domergue, *Mineria y metalurgia*, ii. 12.

[35] The figures are 240, 000 (Astures), 161,000 (Lucus), and 285,000 (Braga).

[36] See P. R. Lewis and G. D. B. Jones, 'Roman Gold-mining in North-west Spain', *Journal of Roman Studies* (1970), 182ff.; C. Domergue, 'Introduction à l'étude des mines d'or du nord-ouest de la péninsule ibérique dans l' Antiquité: Legio VII Gemina', *Coloquio internacional* (León, 1970), 235–86,

the strongest possible circumstantial proof of his presence in Hispania Tarraconensis. The vividness of his narrative, when describing a mining operation (33. 72), suggests that this is an eyewitness account.

On completion of the work, the miners cut through the tops of the arches, beginning with the last. The opening fissure gives warning of the impending collapse, but this is only seen by a watchman perched on the top of the mountain. With a shout, or a wave, the look-out gives the order for the miners to be called off and, at the same time, rushes down from his vantage point. The ruptured mountain falls asunder with an unimaginable crash and is accompanied by an equally incredible blast of air. Like conquering heroes,[37] the miners contemplate their triumph over Nature.

Peracto opere cervices fornicum ab ultimo caedunt. dat signum rima eamque solus intellegit in cacumine eius montis vigil. hic voce, nutu evocari iubet operas pariterque ipse devolat. mons fractus cadit ab sese longe fragore qui concipi humana mente non possit, aeque et flatu incredibili. spectant victores ruinam naturae.

A further vivid picture is presented by Pliny (33. 75): 'The workmen cutting out the rock hang suspended by ropes, so that, viewed from a distance, the operation seems to involve, not so much a species of strange animals, as of birds. Most hang suspended as they take the levels and mark out the route—man leads rivers to run where there is no place for him to plant his own footsteps.' ('qui caedit, funibus pendet, ut procul intuenti species ne ferarum quidem, sed alitum fiat. pendentes maiore ex parte librant et lineas itineri praeducunt, quaque insistentis vestigiis hominis locus non est, amnes trahuntur ab homine.') These scenes must surely have been witnessed at first hand.

Perhaps the most important circumstantial evidence is Pliny's knowledge of the vocabulary and terminology[38] used

---

and C. Domergue, 'Apropos de Pline, *Naturalis Historia* 33. 70–8', *Archivo Español de Arte y Arquelogia*, 45–7 (1972–4). 499–548 and C. Domergue and G. Hérail, '*Mines d'or romaines d'Espagne: Le District de la Valduerna*, (León), (série B, vol. 4 Toulouse, 1978).

[37] Beagon, *Roman Nature*, 41: 'These *victores* are to be contrasted with the heroes who have conquered in *NH* 2. 54. Their probing of the heavens, unlike these men's probing of the earth, was righteous because it freed man by conquering his irrational fear.' Cf. also Beagon ch. II. 3.   [38] See Ch. 7 below.

by the miners—a mixed workforce—who spoke a hybrid language, a sort of lingua franca which included Greek, Spanish, and, sometimes, poorly understood Latin words.[39] The latter have often caused unnecessary emendations to the text and, for editors, needless traumas! Pliny (pref. 17) explains that his work is an update, or revision, and includes facts either ignored by his predecessors, or discovered as a result of his own experience.[40] This new material is additional to that obtained from secondary sources.

Although the evidence of book 33 suggests, in part, direct, on-the-spot experience, *nowhere* does Pliny specifically claim to have *seen* anything that he describes. Thus, judged solely by the criterion of any mention of *eyewitness* (that is, without the Younger Pliny's) evidence, his procuratorship in Hispania Tarraconensis might well have been open to question! We must conclude, therefore, that paucity, or lack, of eyewitness accounts does not preclude the possibility of Pliny having served as a procurator in any specific province. What is puzzling, however, is that, unlike Mucianus[41] who constantly refers to things he has personally encountered,[42] Pliny does not include more references to what *he* actually *saw*.

Pliny shows an interest in Hither Spain and in the Balearics, based on information from secondary sources. He describes flax at Tarraco, where cambrics were first invented (19. 10), and fabrics and other products manufactured from esparto, at Nova Carthago (19. 26 ff.). He also briefly mentions the moufflon and rabbit (8. 217): 'The animals in Spain called rabbit belong to the genus hare; their fertility is beyond counting and they bring famine to the Balearic Islands by ravaging the

---

[39] Pliny himself (pref. 13) highlights the problem when he apologizes for using rustic, or foreign terms—indeed barbarian words that have to be introduced with 'if you'll pardon the expression' ('aut rusticis vocabulis, aut externis, immo barbaris, etiam cum honoris praefatione ponendis'.)

[40] Especially in the field of earth sciences (mineralogy).

[41] Mucianus was consul on three occasions, namely in AD 52, 70, and 75. He was author of a work on 'Mirabilia'. See *NH* 12. 9.

[42] Among the many other things he saw when he was governor of Lycia (13. 88) was a letter of Sarpedon, written on paper, at Troy and, on another occasion, the beams of cedar of the roof of the temple of Diana at Ephesus (16. 213).

crops.... it is an established fact that the inhabitants of those islands petitioned the late emperor Augustus for military assistance against the spread of these animals.' ('leporis genus sunt et quos Hispania cuniculos appellat, fecunditatis innumerae famemque Baliarum insulis populatis messibus adferentes.... certum est Baliaricos adversus proventem eorum auxilium militare a divo Augusto petisse.') He confuses the etymology of *cuniculus*—which is derived from the Spanish word for a rabbit. From this was formed *cuniculum* meaning 'burrow', 'tunnel' and so, in a technical context, 'mine'.

1.4.2 Africa

Africa was, for the Greeks and Romans alike, a land of 'mirabilia'[43] and examples of the paranormal. Pliny and his contemporaries, no less than early writers, among whom Herodotus stands out, were fascinated by its legends and realities.[44] Many sources of secondary information were available, ranging from notes made by Hanno,[45] the Carthaginian explorer (5. 8), which were available in a Greek translation, the researches of King Juba[46] of Mauretania (5. 16 and elsewhere), and accounts brought back by Roman commanders, like Suetonius Paulinus[47] (5. 14).

Notwithstanding such sources, to which Pliny is widely indebted, his own knowledge of Africa, familiarity with African languages, and vividly descriptive accounts of many aspects of that province, its resources, and everyday life, support the

---

[43] Arist. *HA* 8. 28, 606^{b}20.

[44] (*HN* 8. 42) 'vulgare Graeciae dictum semper aliquid novi Africam adferre'. See also E. Bianchi, 'Teratologia e geografia', *Acme*, 34 (1981), 227–49.

[45] Of uncertain date. His *Periplus* contains an account of a voyage undertaken, beyond the Pillars of Hercules, to found settlements.

[46] In 30 BC Augustus reinstated Juba in his paternal kingdom of Numidia. He married Cleopatra (Selene), the daughter of Antony and Cleopatra. Later, in 25 BC, Augustus gave him Mauretania in exchange for Numidia which was reduced to a Roman province. Juba died in AD 19. Pliny (5. 16) writes: 'Juba was even more distinguished for his renown as a student than for his royal rule' ('Iuba...studiorum claritate memorabilior etiam quam regno').

[47] Propraetor in Mauretania in AD 42 and governor of Britain from 59 to 62, at the time of the revolt of Boudicca. He was consul in 66. After the death of Nero, in 68, he was one of Otho's generals in the war against Vitellius.

proposition that he may well have held a procuratorship there. Indeed, in many respects his treatment of Africa parallels that of Hispania Tarraconensis.

Pliny (5. 1 f.) begins by defining the boundaries of the province and describing the location of Antaeus' palace (5. 3) where the struggle with Hercules is said to have taken place. Many legends in the ancient world have their origin in fact:[48] interestingly, Pliny records such a rationalization of the story of the snake which was believed to guard the golden apples in the Garden of the Hesperides: 'A channel [5. 3] flows inland from the sea with a wandering course that, as people nowadays explain, looks like a snake guarding the place. It encompasses an island that is the only part not flooded by the tides, even though the neighbouring area is higher. On this island there is also an altar of Hercules, but nothing else, except wild olive-trees, remains of that famous grove which according to the legend bore golden apples' ('adfunditur autem aestuarium e mari flexuoso meatu, in quo draconis custodiae instar fuisse nunc interpretantur; amplectitur intra se insulam, quam solam e vicino tractu aliquanto excelsiore non tamen aestus maris inundant. exstat in ea et ara Herculis nec praeter oleastros aliud ex narrato illo aurifero nemore.')

Suetonius' report (*NH* 5. 14) confirmed earlier authorities' accounts of the region round and about the Atlas mountains:

the lower slopes are filled with dense forests of tall trees of an unknown species: they have very tall trunks notable for their sheen and freedom from knots. Their leaves, like those of the cypress, except for the heavy scent, are covered with a thin down, from which, with a suitable technique, clothing can be made just like that derived from the silkworm. The summit of Mt. Atlas is covered with deep snow, even in summer. ... Suetonius travelled beyond to the river Ger, across deserts of black dust, with projecting rocks in some places that looked as if they had been burnt—a place uninhabitable because of the heat, although it was winter when he experienced it. The Canarii live in the neighbouring forests which are full of every species of elephant, wild beast, and snake.

---

[48] Such as the Golden Fleece (Strabo 11. 2. 19), also guarded by a snake. Cf. Ap. Rhod. *Argon.* 4. 123–6. Healy, *Mining*, 75 f.

imas radices densis altisque repletas silvis incognito genere arborum, proceritatem spectabilem esse enodi nitore, frondes cupressi similes praeterquam gravitate odoris, tenui eas obduci lanugine, quibus addita arte posse quales e bombyce vestes confici. verticem altis etiam aestate operiri nivibus... ultra ad fluvium qui Ger vocatur per solitudines nigri pulveris, eminentibus interdum velut exustis cautibus, loca inhabitabilia fervore quamquam hiberno tempore experto. qui proximos inhabitent saltus refertos elephantorum ferarumque et serpentium omni genere Canarios appellari.

Pliny (5. 25) continues with a mention of Byzacium, in which district he claims (17. 41) to have 'seen a fertile plain which yields an increase of one hundred and fifty fold, land which, in dry weather, no bulls can plough, but which, after a spell of rain, *I have seen* being broken by a plough drawn by a wretched little donkey and an old woman at the other end of the yoke' ('contra in Byzacio Africae illum centena quinquagena fruge fertilem campum nullis, cum siccum est, arabilem tauris, post imbres vili asello et a parte altera iugi anu vomerem trahente *vidimus* scindi'). This is his only reference to having witnessed anything at first hand in Africa.

The country south of Cyrene (5. 33) is interesting: it is separated into three regions, namely (1) forest, (2) corn-growing, and (3) an area where only silphium[49] grows, forming a swathe some thirty miles wide and two hundred and fifty miles long. The corn is stored in pots called *siri*—a practice also found in Cappadocia, Thrace, and Spain (18. 306).

Pliny (18. 188) discusses the 'African city called Tacape. This is in the middle of the desert, on the way to the Syrtes and Leptis Magna, which has the exceptionally marvellous blessing of a well-watered soil. There is a spring that distributes water over a space of about three miles in every direction,

---

[49] See W. R. Philipson, 'Silphium: A Classical Example of Controlled Exploitation', *Illustrated London News*, 541 (8 Oct. 1949), 542. The umbelliferous silphium plant belonged to the genus *Asafoetida* and was used as medicine and as a flavouring in cooking. It appears as a type on the coinage of Cyrene. (see *British Museum Catalogue of Greek Coins: Cyrene*) from as early as the 6th cent. BC. When the province of Cyrenaica began to be ruled by short-term Roman governors they thought only of reaping the maximum profit during their office. Once the export of silphium was freed of strict control the over-exploitation of the plant led to its eventual disappearance.

giving a generous supply, although this is distributed among the population only at special fixed periods of the day.' ('civitas Africae in mediis harenis petentibus Syrtes Leptimque Magnam vocatur Tacape, felix super omne miraculum riguo solo. ternis fere milibus passuum in omnem partem fons abundat, largus quidem sed certis horarum spatiis dispensatur inter incolas.') He continues with a fascinating account of the intensive farming of the region: 'Here, underneath the palms of very great size, there are olives; under the olives are figs; under them, pomegranates; under the pomegranates are vines and, beneath the vines, corn is sown, then, later, leguminous plants and, finally, garden vegetables, all in the same year and nourished in the shade of some other crop.' ('palmae ibi praegrandi subditur olea, huic ficus, fico punica, illi vitis, sub vite seritur frumentum, mox legumen, deinde olus, omnia eodem anno, omniaque aliena umbra aluntur.')

A similar spring exists in Gallia Narbonensis which may be significant and Pliny (18. 190) adds: 'There is great difference in quality in the water supplied to irrigated places. In the province of Narbonne, there is a celebrated spring with the name Orga, in which grow plants so much sought after by oxen that they put their whole head under water to try to get them; but it is a well-known fact that those plants, though growing in water, only get their nutriment from showers of rain.' ('aquarum quoque differentia magna riguis. est in Narbonesi provincia nobilis fons Orgae nomine; in eo herbae nascuntur in tantum expetitae bubus ut mersis capitibus totis eas quaerant; sed illas in aqua nascentis certum est non nisi imbribus ali.')

Spelt grain (*alicia*) writes Pliny (8. 114) is enhanced by the addition of chalk in respect of its colour and fineness; and, when discussing Africa, (5. 115) he remarks that an ersatz form of *alicia* is made from an inferior kind of zea which grows there.

Communication by road is difficult in the province because of sabotage and Pliny (5. 38) cites the problem of the opening of the road to the Garamantes 'because brigands from that race fill up the wells with sand.... at the beginning of Vespasian's reign [after AD 69], however, a short route taking only four days to the Garamantes was discovered.'

Pliny (5. 34) records 'houses built of blocks of salt, quarried from the mountains like stone', and elsewhere (31. 81), that, 'in the vicinity of Utica,[50] heaps of salt occur like hills; when these have hardened under the sun and moon, they are not melted [he means dissolved] by any moisture and iron cuts them with difficulty'. The circumstances and context of both references, however, make it difficult to judge whether he had actually *seen* such salt deposits, or indeed houses built from this quarried salt. In the second passage a number of occurrences are listed which, like those in Crete, Egypt, and Babylon, Pliny is certainly not likely to have encountered in his travels.

Pliny often includes local Punic, or African words. In Numidia (5. 22) for example, the Nomads carry their homes (*mapalia*) about the country on wagons—possibly giving the appearance of Romany caravans: 'mapalia sua, hoc est domos, plaustris circumferentes'. This term is also used by Sallust,[51] Livy,[52] and Lucan,[53] whereas Virgil[54] uses the variant *magalia*, a Punic word describing 'shanties' rather than mobile homes. Pliny (8. 174) employs the local African word *lalisio* for a foal, an outstanding table delicacy: 'pullis eorum ceu praestantibus sapore Africa gloriatur, quos lalisiones appellat'. In addition the occasional Greek term appears, as in the place name Hippo Diarrhytus—Ἵππων Διαρρυτος[55]—(9.26) and *sirus*—σιρός—(18. 30), a hole in which grain is stored.

---

[50] Utica was second only in importance to Carthage. It was the site of the last stand made by the Pompeian party against Caesar, and where the Younger Cato committed suicide.
[51] *Iug.* 18. 8: 'aedifica Numidarum agrestium, quae mapalia illi vocant, oblonga, incurvis lateribus tecta, quasi navium carinae sunt.' Cf. also ibid. 46. 5.
[52] 29. 31: 'et raris habitata mapalia tectis'. Cf Verg. *G.* 3. 340 and Mart., *Epigr.* 10. 10. 8.
[53] *BC* 2. 89: 'vacuis mapalibus actus| nuda triumphatus iacuit per regna Iugurtae.'
[54] *Aen.* 1. 421: 'miratur molem Aeneas, magalia quondam.' Cf. *Aen.* 4. 259: 'ut primum alatis tetigit magalia plantis.'
[55] The name appears on a mosaic pavement in the Piazzale Corporazione at Ostia. The town (mod. Bizerta, Tunisia) is also featured in the well-known story of the friendly dolphin. See, for example, the Younger Pliny, *Letters* 9. 33.

In his discussion of the sources of the Nile, Pliny (5. 51) refers to the story that Juba brought a crocodile[56] from that river and placed it in the temple of Isis at Caesarea, where it was still on view in his day. ('crocodilus quoque inde ob argumentum hoc Caesareae in Iseo dicatus ab eo spectatur hodie.') The mention of Juba, the inundation of the Nile (5. 57), taken from Herodotus, and Timaeus' recondite account are reminders that Pliny is very much indebted to earlier authorities.

Fact and legend are often inextricably intertwined as in the description of Mt. Atlas and its peoples (5. 44 ff.). The following colourful account relies wholly on secondary sources.

Some authorities place the Atlas tribe in the middle of the desert and, next to them, the half-animal Goat-pans, the Blemmyae, Gamphasantes, Satyrs, and Strapfeet. The Atlas tribe is primitive and sub-human, *if we believe what we hear*; they do not call each other by names.... The Cave-dwellers hollow out caves which are their houses; their food is snake meat. They have no voice but make a shrill noise, thus lacking any communication by speech. The Garamantes do not marry but live promiscuously with their women. The Augilae worship only gods of the lower world. The Gamphasantes do not wear any clothes, are pacifists, and do not associate with any foreigner. The Blemmyae are reported as being without heads; their mouth and eyes are attached to their chest. The Satyrs have human characteristics except for their shape. The form of the Goat-pans is as commonly depicted. The Strapfeet are people with feet like thongs who naturally move by crawling.

quidam solitudinibus interposuerunt Atlantas eosque iuxta Aegipanas semiferos et Blemmyas et Gamphasantas et Satyros et Himantopodas. Atlantes degeneres sunt humani ritus, *si credimus*; nam neque nominum ullorum inter ipsos appellatio est.... Trogodytae specus excavant; hae illis domus, victus serpentium carnes, stridorque, non vox: adeo sermonis commercio carent. Garamantes matrimoniorum exortes passim cum feminis degunt. Augilae inferos tantum colunt. Gamphasantes nudi proeliorumque expertes nulli externo congregantur. Blemmyis traduntur capita abesse ore et oculis pectori adfixis. Satyris praeter figuram nihil moris humani. Aegipanum qualis vulgo

---

[56] The well-known Nilotic painting from Pompeii, however, illustrates alligators. Cf. D. Strong, *A History of Roman Painting* (Harmondsworth, 1976), 36.

pingitur forma. Himantopodes loripedes quidam quibus serpendo ingredi natura sit.

Building techniques provide a further interesting strand of evidence; 'In Africa and Spain [35. 169] there are earthen walls described as compacted [*formaceos*] because they are made by packing earth down between two sets of shuttering, so that the material is stuffed in rather than raised up. These last for ages, undamaged by rain, wind, or fire and are stronger than quarry stone. In Spain the watch-towers of Hannibal and turrets of earth placed on mountain ridges are still visible.' ('quid? non in Africa Hispaniaque e terra parietes, quos appellant formaceos, quoniam in forma circumdatis II utrimque tabulis inferciuntur verius quam struuntur, aevis durant, incorrupti imbribus, ventis, ignibus, omnique caemento firmiores? spectat etiam nunc speculas Hannibalis Hispania terrenasque turres iugis montium inpositas.') Pliny must surely have seen the latter during his procuratorship in Hispania Tarraconensis. He may even have encountered such walls in Africa—but this is only speculation.

There are several references to types of stones found in North Africa, among them a black variety (*anthracitis*,[57] 36. 148) and tufa, which, writes Pliny (36. 166), is the only stone in the vicinity of Carthage. However, he adds (36. 160) that selenite was recently discovered in Africa, but of an inferior quality to the Spanish variety.

Pliny (5. 37) mentions an inscription on Mt. Goriano indicating that precious stones were to be found in the vicinity: 'mons Gyri in quo gemmas nasci titulus praecessit.' Similarly, *carbunculi* (37. 95 f.), or 'red' stones, are also said to occur in Africa.

Pliny's account of Africa relies heavily on secondary sources, some of them contemporary; only once is there evidence of his being an eye-witness, but this is no decisive argument against his having held a procuratorship in Africa. However, it must be admitted that the supporting evidence is, in the main, circumstantial.

---

[57] This mineral is not coal (see Ch. 14, App. B), but ore which was partly magnetite (see Ch. 14, s.v.) and partly limonite (see Ch. 17.9.1–2).

1.4.3 Gallia Belgica

There are relatively few references to Gallia Belgica in the *Natural History*, and, although some of them are of great interest, they do not support the suggestion that Pliny was ever procurator in this province.

Pliny (4. 105) defines the region. 'The whole of Gaul included under the general name of Comata[58] divides into three races of people, which are chiefly separated by rivers: [1] from the Scheldt to the Seine is Gallia Belgica, [2] from the Seine to the Garonne is Gallia Celtica (also called Lyonese), and [3] from the Garonne to the projection of the Pyrenees is Gallia Aquitana—previously called Armorica.'

Pliny (8. 191 f.) mentions a type of wool employed for darning clothes and certain types of fleece for stuffing cushions.('[Lana] similis circa provinciae Narbonensis, similis et in Aegypto ex qua vestis detrita usu pinguitur rursusque aevo durat... aenis polientium extracta in tomenti usum veniunt. Galliarum, ut arbitror invento.')

The eunuch apple (*spadonia*,[59] 15. 51) is mentioned, and reeds which grow in the province have a number of uses. The reed (16. 158) 'is of slender appearance, jointed and divided with knots; it tapers gradually off to the top with a rather thick tuft of hair. This is not without value as it either serves instead of feathers to stuff the beds of innkeepers, or, in places where it grows very hard and woody in structure, as in Gallia Belgica, it is pounded up and inserted between the joints of ships to caulk the seams, holding better than glue and being more reliable for filling cracks than pitch.'('geniculata cetero gracilitas nodisque distincta, leni fastigio tenuatur in cacumina crassiore paniculae coma, neque hac supervacua—aut enim pro pluma strata cauponarum replet aut, ubi lignosiore induruit callo sicut in Belgis, contusa et interiecta navium commissuris feruminat textus glutino tenacior rimisque explendis fidelior pice.')

---

[58] C. J. Fordyce (ed. & comm.), *Catullus* (Oxford, 1961), 161 (on 29. 3): 'Gallia *comata* was the unofficial term for the Transalpine province of Gaul, where the natives wore their hair long... as *togata* was used for the Cisalpine province, where Roman dress had established itself; Caesar does not use it but, in Cicero, *Philippics* 8. 27, Antony states, "Galliam... togatam remitto, comatam postulo".'

[59] On apples generally, see White. *Roman Farming*, 258–9.

'In Gallia Belgica [36. 159] a white stone[60] is said to be cut with a saw, just like wood, only more easily so as to serve as ordinary roof tiles and as rain tiles or, if so desired, for a kind of roofing known as "peacock style"' ('in Belgica provincia candidum lapidem serra, qua lignum, faciliusque etiam secari aiunt ad tegularum et imbricum vicem vel, si libeat, quae vocant pavonacea tegendi genera').

Pliny (36. 164 f.) also describes whetstones found beyond the Alps and known by the Celtic name *passernices*: repertae sunt ... nec non et trans Alpis, quas passernices vocant.'

Pliny's (18. 296) reference to the use of a mechanical reaper,[61] is, perhaps, the most interesting: 'On the vast estates in the provinces of Gaul very large frames fitted with teeth at the edge, and carried on two wheels, are driven through the corn by a team of oxen pushing from behind; the ears thus torn off fall into the frame.' ('Galliarum latifundiis valli praegrandes dentibus in margine insertis duabus rotis per segetem inpelluntur iumento in contrarium iuncto; ita dereptae in vallum cadunt spicae'). The existence of such a mechanical device—originally known only from this reference in the *Natural History*—was disputed, but yet again Pliny was vindicated by the discovery of bas-relief sculptures illustrating such a device.[62] It made its appearance in north-east Gaul in the first century AD. Basically it was a 'comb' (*pecten*) mounted on wheels, with a 'grassbox'-type container.[63]

Finally, Pliny (31. 12) describes a remarkable spring the waters of which taste of iron rust—one is reminded of the Spa waters at Bath (tasting of flat irons) and of the springs at Treffiw[64] in Wales, both frequented by the Romans.

---

[60] Identified as selenite, see Ch. 14, s.v. 'Lapis Specularis'.

[61] White, *Roman Farming* 182 f. and *Agriculture Implements of the Roman World* (London, 1967), and 'The economics of the Gallo-Roman Harvesting Machines', in J. Bibauw (ed.), Hommages à Marcel Renard, iii: Histoire, histoire des religions, épigraphie (Collection *Latomus*, 102.), Brussels, 1969, 807–9. A similar machine, known as Ridley's Stripper, was in use in parts of Australia during the last century (ibid. 486, n. 46).

[62] White, *Roman Farming*, pls. 36–7.

[63] A later, heavier version is described by Palladius (7. 4), writing in the 4th cent. AD.

[64] The waters at Bath and Treffiw both contain iron salts which account for the flavour.

Pliny's cursory treatment of the province of Gallia Belgica is inconsistent with any extended official contact with the region.

1.4.4 Gallia Narbonensis

The possibility that Pliny may have visited Gallia Narbonensis privately has been considered, conceivably by way of a detour on his way to and from one of his postings to Germany,[65] and indeed this is more likely than that he held a procuratorship in that province. Pliny (3. 31 ff.) defines the province of Narbonne as 'that part of Gaul washed by the Mediterranean and previously known as Bracata'. It mainly covers the region known to us as Provence.

Only two references, namely Pliny's claim (2. 150) to have seen a meteorite[66] which had fallen to earth in the territory of the Vocontii ('*ego ipse vidi* in Vocontiorum agro paulo ante delapsum') and, possibly, the vivid description of men and dolphins catching fish together (9. 29), directly attest his presence in the province. The collaboration is particularly interesting.[67]

In the province of Gallia Narbonensis and the region of Nemausus there is a marsh called Latera where dolphins and men co-operate to catch fish. At a fixed season a huge number of mullet rushes through the narrow mouth of the marsh into the sea, after watching for a turn of the tide that prevents nets from being stretched across the channel. When the fishermen see this...their battle line appears and immediately takes up position where the fray is to commence. They put themselves between the open sea and the shore and drive the mullet into shallow water,[68] then the fishermen set their nets and lift the fish out of the water with two-pronged spears; the speed of some of the mullet enables them to leap over the barriers, but the dolphins still catch them. But satisfied for the moment with killing them,

---

[65] See Reynolds, 'The Elder Pliny', 5.   [66] See below, Ch. 14 App. C.

[67] Among the many stories of the relationship between dolphins and man, the best known is that recorded by the Younger Pliny (*Letters* 9. 33. 4ff). The god Dionysus was rescued from the pirates by dolphins and is the subject of the well-known Black-figure cup by Exekias (E. A. Lane, *Greek Pottery*, (London, 1948), pl. 14a).

[68] Local fishermen employ a similar technique when fishing for tuna off the coast of north-west Africa.

the dolphins put off their supper until complete victory has been achieved.

est provinciae Narbonensis et in Nemausiensi agro stagnum Latera appellatum ubi cum homine delphini societate piscantur. innumera vis mugilum stato tempore angustis faucibus stagni in mare erumpit observata aestus reciprocatione, qua de causa praetendi non queunt retia... quod ubi animadvertere piscantes... apparet acies quae protinus disponitur in loco ubi coniectus est pugnae; opponunt sese ab alto trepidosque in vada urgent. tum piscatores circumdant retia furcisque sublevant. mugilum nihilominus velocitas transilit; at illos excipiunt delphini et occidisse ad praesens contenti cibos in victoriam differunt.

Pliny (19. 3 ff.) makes some interesting observations on flax—a plant that brings Gallia Narbonensis within three, and Africa within four, days (of Italy). He continues (ibid. 8): 'The Cadurci, Caleti, Ruteni, Bituriges, and Morini who are believed to be the remotest of mankind, in fact, the whole of the Gallic provinces, weave sailcloth, and indeed by this time so do even our enemies across the Rhine and no dress material more fashionable is known to their women. This reminds me of the fact recorded by Varro that it is a family custom among the Serrani for the women not to wear linen dresses.'

The reference to Varro is significant since much of Pliny's information about the Gallic provinces, especially with regard to trees, viticulture, grain, wheat, and agricultural practices in general is derived from secondary sources.[69] Similarly, the account of the Druids and of the parasitic plant mistletoe (16. 245 ff.), although more detailed than some previous accounts, for example Caesar's (*BG* 6. 13 ff.), follows Theophrastus' (*HP* 3. 16. 1 and *CP* 2. 17. 1) description of mistletoe. Acorn-bearing trees produce another parasitic plant, agaric (16. 33), a species of

---

[69] Varro (RR. 1. 1. 7 ff.), to whom Pliny is much indebted, gives an impressive list of more than fifty Greek authorities who had written on agriculture, or whose works refer to topics related to the subject. Pliny rates Cato highly (*De agricultura*). Cf. *NH* 14. 14: 'by the admission of his contemporaries, a supremely competent and unrivalled agriculturalist' ('ille aevi confessione optimus ac sine aemulo agricola'). Columella (1. 1. 12) writes that it was Cato who taught agriculture to speak Latin. Pliny in book 17 relies heavily on Cato. Altogether there are 62 quotations from Cato but only 8 from Columella. See further, White, *Roman Farming*, 18 ff.

non-edible fungus, white in colour and with a strong odour: it grows on the tops of trees and is phosphorescent at night.

Pliny (34. 3) mentions copper-mining for which there is evidence in Gallia Narbonensis—in the Sallustian and Livian mines. Caesar (*BG* 3. 21. 3) had also referred to copper-mining in the region of the Sotiates. Pliny (34. 96) describes a different way of smelting copper:[70] the ore 'is smelted between stones heated red hot, as this roasting scorches it and renders it black and friable. Moreover, they only smelt it again once whereas if this process is repeated several times, the quality is greatly enhanced.' ('in Gallia... ubi lapides candefactos funditur; exurente enim coctura nigrum atque fragile conficitur, praeterea semel recoquunt quod saepius fecisse bonitati plurimum confert.')

Finally, the linguistic evidence relating to the provinces of Gaul is inconclusive. Pliny records a number of Celtic words and technical terms[71] but these are not specifically local to Gallia Belgica, or Gallia Narbonensis. For example, terms for red marl (17. 44) derive from the Celtic word for a stone (*agaunum*) and *eglecopala*, dove-coloured marl. Some words belong to Gallia Transpadana, as *padus* which Pliny (3. 122) explains is a pine-tree; and Eporedia, which is the name of a town which comes from the Gallic word for 'a man good at breaking horses' (*eporedias*). In Gaul certain types of gold bracelets (33. 39) are known as *viriolae*.

The holding of the office of procurator was an important landmark in Pliny's career. Apart, however, from his procuratorship in Hispania Tarraconensis, the internal evidence of the *Natural History* appears only to suggest, with any degree of likelihood, one further procuratorship, namely in Africa.

1.5 MISENUM

Pliny was in Italy about AD 76 and it has been suggested that he may have held the post of *praefectus vigilum*. The only certainty, however, is his appointment as admiral of the fleet stationed at Misenum, on the Bay of Naples (*praefectus classis Misenensis*). The fleet had no battle role in the Mediterranean

---

[70] See Healy, *Mining*, 158 ff.   [71] See below, Ch. 7. 3. 2 (end).

but acted as a transport and water-police force. The admiral's duties were mainly administrative involving responsibility for shipbuilding, repairs, victualling, and the provision of chandlery. The date of the appointment is uncertain, and, likewise, whether it involved being permanently stationed at Misenum.[72]

On 24 August 79, at the time of the eruption of Vesuvius, while going to the aid of survivors—among them Pomponianus, the son of Pomponius Secundus[73]—and no doubt spurred on by his scientific curiosity wishing to observe the volcano at close quarters, Pliny died. The dramatic manner of his death, whether from asphyxiation or from a heart attack,[74] is vividly described by his nephew (*Letters* 6. 16) in reply to an enquiry by Tacitus who was keen to hand down an accurate account to future generations. As he prophetically adds, his uncle's death would be renowned for all time if recorded by Tacitus.

1.6 PLINY THE MAN: HIS CHARACTER

Unfortunately, no statue, or portrait head, or other representation of Pliny survives from Roman times: nor is there any description of his physical appearance in extant literature.

---

[72] A. N. Sherwin-White (*Fifty Letters of Pliny*,[2] (Oxford, 1969), 102) suggests that his duties, in the later years of Vespasian, were partly performed at Rome, cf. *Letters* 6. 16. 4, 'classemque imperio praesens regebat'. However, the Younger Pliny's statement that his uncle called on Vespasian each morning (*salutatio*) does not provide evidence of what office required his presence in Rome. See further Syme, *Tacitus*, i. 61.

[73] The senator whose biography Pliny wrote (*NH* 24. 56), and who was probably his patron.

[74] The Younger Pliny (*Letters* 6. 16. 19) writes, 'ut ego colligo, crassiore caligine spiritu obstructo, clausoque stomacho qui illi natura invalidus et angustus et freqenter aestuans erat'. This suggests that Pliny suffered from a condition akin to asthma. The exact cause of Pliny's death, however, apart from the obvious, immediate conclusion that he was asphyxiated by toxic fumes, has led to much speculation. Among the numerous articles, there may be cited, as representative of the arguments advanced, M. D. Grmek, 'Les Circonstances de la mort de Pline: Commentaire médical d'une lettre destinée aux historiens *Helmantica*, 37 (1986), 25–43; Y. Grisé, 'L'Illustre Morte de Pline le Naturaliste', *Revue des Études Latines* 58 (1980), 338–43; L. Bessone, 'Sulla morte di Plinio il Vecchio', *Rivista di studi classici*, 17 (1969), 166–79; C. Zirkle, 'The Death of C. Plinius Secundus AD 23–79', *Isis* 58 (1967), 553–9; and R. M. Haywood, 'The Strange Death of Pliny the Elder', *Classical Weekly* 46 (1952), 1–3. Whether Pliny died of cardiac arrest, which is the commonly held view, or some other physical cause, cannot be determined on the evidence available.

The idealized genre statue of the fifteenth century, on the façade of Como Cathedral, by Tomasso Rodari, and the engaging, well-known miniature of Pliny in his study surveying the world of Nature, from an illumination in the Harley manuscript[75] of the same century, are products of artistic imagination.

Two main sources, however, throw considerable light on Pliny's character, temperament, and attitude to life and society under the early empire: these are the *Letters* of his nephew the Younger Pliny and, of course, the internal evidence of the *Natural History* itself.

Both give a vivid and revealing picture of his idiosyncratic life style. In *Letters* 3. 5. 7 ff. his nephew writes:

You may wonder how a busy man was able to complete so many volumes containing such detailed information; and wonder, even more, when you know that he practised as an advocate for some considerable time and died at the age of 55, and throughout the intervening years his time was much taken up with the important offices he held and his membership of the emperor's advisory council. But he combined a penetrating intellect with amazing powers of concentration and the capacity to manage with the minimum amount of sleep.... Admittedly he fell asleep very easily and would often doze and wake up again during his work. Before dawn he used to call on Vespasian [*salutatio*] (for he too worked at night) and return from the emperor to his appointed office. When he returned home he would devote the rest of his time to his literary pursuits. After a meal (which, during the day, was light and simple in the manner of people of earlier times), in summer, if he had any free time, he would often lie in the sun and have a book read aloud to him while he made notes and extracts from it. He did this for everything he read and always said that there was no book so bad that some good could not be derived from it ... When travelling he felt free from other responsibilities to devote every minute to work: he kept a secretary at his side with book and notebook and, in winter, saw that his hands were protected by long sleeves so that even bitter weather should not rob him of a working hour. For the same reason also he used to be carried about Rome in a sedan chair. I can remember how he scolded me for walking; according to him I need not have wasted those hours, for he thought any time wasted if it was not given over to work.... When you consider the extent of his reading and writing I wonder if you feel that he

---

[75] MS 2677, fo. 1, of the 15th cent., in the British Museum.

could never have been a public official, or member of the emperor's council, or, on the other hand, that he should have achieved more, now that you know his application? In fact his official duties put every possible obstacle in his path; and yet there was nothing that his energy could not surmount.

Miraris quod tot volumina multaque in his tam scrupulosa homo occupatus absolverit? Magis miraberis si scieris illum aliquamdiu causas actitasse, decessisse anno sexto et quinquagensimo, medium tempus distentum impeditumque qua officiis maximis qua amicitia principum egisse. Sed erat acre ingenium, incredibile studium, summa vigilantia. ... Erat sane somni paratissimi, non numquam etiam inter ipsa studia instantis et deserentis. Ante lucem ibat ad Vespasianum imperatorem (nam ille quoque noctibus utebatur), inde ad delegatum sibi officium. Reversus domum quod reliquum temporis studiis reddebat. Post cibum saepe (quem interdiu levem et facilem veterum more sumebat) aestate si quid otii iacebat in sole, liber legebatur, adnotabat excerpebatque. Nihil enim legit quod non excerperet; dicere etiam solebat nullum esse librum tam malum ut non aliqua parte prodesset. ... In itinere quasi solutus ceteris curis, huic uni vacabat: ad latus notarius cum libro et pugillaribus, cuius manus hieme manicis muniebantur, ut ne caeli quidem asperitas ullum studii tempus eriperet; qua ex causa Romae quoque sella vehebatur. Repeto me correptum ab eo, cur ambularem: 'poteras' inquit 'has horas non perdere'; nam perire omne tempus arbitrabatur, quod studiis non impenderetur. ... Nonne videtur tibi recordanti, quantum legerit, quantum scripserit, nec in officiis ullis nec in amicitia principis fuisse; rursus cum audis quid studiis laboris impenderit, nec scripsisse satis nec legisse? Quid est enim quod non aut illae occupationes impedire aut haec instantia non possit efficere?

Many fundamental aspects of Pliny's character appear in his nephew's thumbnail sketch. Noteworthy among these are his diligence, capacity for work, and utilization of every waking hour.[76] This is complemented by the more comprehensive picture which emerges from the pages of the *Natural History* itself.

Roman writers had an equivocal attitude towards leisure (*otium*). This, as Oliviera [77] writes, is often equated with idleness

---

[76] Pliny the Elder (pref. 19) writes, 'To be alive means to be awake' ('profecto enim vita vigilia est').
[77] Francisco de Oliviera, *Les Idées politiques et morales de Pline l'Ancient* (Coimbra, 1992), 264 ff. ('Vie de luxe et de loisir'), provides a comprehensive discussion of *otium*; see p. 301.

(*ignavia*) and seen as providing the opportunity for vices; it can also be a hindrance to the performance of one's duties (*officia/negotia*). A life of idleness, that is one opposite to the fulfilment of duties, merits the censure of Vespasian and Titus who appear as the guardians of good conduct (*boni mores*).

On the positive side, however, *otium* is part of a civilized way of life (*vitae lepos*).[78] So the Younger Pliny (Letters 2. 17. 1), like others, finds it is necessary, from time to time, to escape from the turmoil of Rome[79] to his Laurentine country villa, but only when he has fulfilled his duties (*peractis quae agenda fuerint*). His uncle has a positive attitude to rest and relaxation but, in keeping with his Stoic beliefs,[80] is concerned to explain how he has time to engage in literary pursuits without detriment to the fulfilment of his official duties. He writes (pref. 18), 'I am beset with duties, and pursue this sort of interest in my spare moments, that is at night—lest any of my readers should think that the night hours are given to idleness. The days I spend on you...('et occupati officiis, subsicivisque temporibus ista curamus, id est nocturnis, ne quis vestrum putet his cessatum horis. dies vobis inpendimus...').

Pliny's belief in the work ethic, not only moulds his character and outlook on life, but determines the overall emphasis of the *Natural History*. His is a typically Roman reaction to the aristocratic ideal of leisure—inherited from the Greeks—that led, by an easy decline, to the disdain of serious effort, and to the advertising of elegant accomplishments as a pretext for sloth and emptiness.

Similarly Cicero (*Planc.* 66) writes that he spent time composing speeches when others such as Cassius were at the Games,

---

[78] I have borrowed this expression from Pliny (*NH* 31. 88) where it is equated with *sales* ('wit') and implies amiability and charm.

[79] Cf. Juvenal (*Sat.* 3. 232 ff.), whose vivid descriptions of life in the capital, the noise, the broken sleep, the traffic gridlock, delapidated slum tenements, and constant conflagrations add up to a nightmare scenario.

[80] Panaetius (*c.*185–109 BC) sought to adapt Stoic ethics to the needs of active statesmen and soldiers and it was through him that Stoicism became so important an element in the life of the best representatives of Roman nobility. The later revisions of Posidonius (*c.*135–*c.*51 BC), who pursued a new natural philosophy which embraced all sciences, added further to the appeal of Stoicism. For a detailed analysis of the influence of Stoicism on Pliny's thought, see Beagon, *Roman Nature*, 26–54.

or celebrating festivals, that is, enjoying periods of leisure. (*cum otiosus sis, has ego scripsi ludis et feriis, ne omnino umquam essem otiosus.*)

Yet Pliny himself is well aware of the problems arising from official duties and writes (36. 27), 'The multitude of official functions and business must, after all, deter anyone from serious study, since the appreciation of works of art needs leisure and deep silence in our surroundings. Such leisure, however, must not be gained at the expense of one's duties.' ('officiorum negotiorumque acervi omnes a contemplatione tamen abducunt, quoniam otiosorum et in magno loci silentio talis admiratio est. ...')

Pliny's attitude to imperial rule and his political views are clear. Unusually for someone with Stoic beliefs,[81] he was close to the Flavians. He accepted the system as indispensable to Rome, being at the same time grateful for the security and stability afforded by the *pax Romana*. Even so, the latter, contrary to what might have been expected, could, in some respects, be counter-productive, as Pliny observes (2. 117 f.) in his criticism apropos the decline of arts and sciences, invention, and discovery in imperial times. 'Yet nowadays, in this happy time of peace, under an emperor who takes such pleasure in promoting literature and science, absolutely nothing is being added to the sum of knowledge as a result of original research; indeed not even the discoveries made by people long ago are thoroughly assimilated.' ('nunc vero pace tam festa, tam gaudente proventu litterarum artiumque principe, omnino nihil addisci nova inquisitione, immo ne veterum quidem inventa perdisci.') He continues, 'The rewards were not greater when the ample successes were spread over many students and in fact the majority of these made the discoveries in question with no other reward at all save the consciousness of benefiting posterity.' ('non erant maiora praemia in multos dispersa fortunae magnitudine, et ista plures sine praemio alio quam posteros iuvandi eruerunt.') In other words, peace and stability do not inevitably produce positive spin-offs.

---

[81] The Stoics were fundamentally in opposition to the Flavians. See Healy, *Pliny, NH: A Selection*, p. xii.

Pliny, essentially a 'laudator temporis acti'—and with justification—is conscious of the importance of ongoing research and of his debt to other authorities whom he is quick to acknowledge (pref. 21).

The place of Nature in Pliny's thoughts was central to his attitude towards religion.[82] Typically Roman, this was a blend of credulity and scepticism. He abhors the popular clamour over 'the gods' and includes as one of those popular quirks the ubiquitous Fortune. He writes (7. 130), 'There is always a fear that Fortune may grow tired, and, once this is entertained, happiness has no sure foundation' ('certe ne lassescat fortuna metus est, quo semel recepto solida felicitas non est'). The examples which follow (7. 133–46) are reminiscent of the 'loci de Fortuna' so popular in the Elder Seneca's *Suasoriae* and *Controversiae*: the vicissitudes of Fortune were a commonplace beloved by the rhetoricians and worked almost to death in school exercises. Nature, concludes Pliny, is what mortals call God.

Pliny appears as a man torn between respect of knowledge and his background of rhetorical training, which he could not totally forget in spite of his dedication to science. Like Lucretius,[83] Cicero, Seneca, and Juvenal, he ridicules the terrors found in Greek mythology and uses his powers to destroy the widespread Roman belief in life after death. He disposes unequivocally of man's claim to immortality (7. 188): 'All men are in the same state from their last day forward as they were before their first day, and neither body nor mind has any more sensation after death than it had before birth. But wishful thinking prolongs itself into the future and falsely invents for itself a life that continues beyond death, sometimes by giving the soul immortality, or a change of shape, sometimes by according feeling to those below, worshipping spirits and deifying one who has already ceased to be even a man.' ('omnibus a supremo die eadem quae ante primum, nec magis a morte sensus ullus aut corpori aut animae quam ante natalem—eadem enim, vanitas in futurum etiam se propagat et in mortis quoque tempora ipsa sibi vitam mentitur, alias immortalitatem animae, alias transfigurationem, alias sensum

---

[82] ibid p. xiv.   [83] Cf. Lucr. *DRN* 3. 37 ff. and elsewhere.

inferis dando et manes colendo deumque faciendo qui etiam homo esse desierit.')

Stoicism, which not only stressed the virtues of duty and the attainment of virtue through wise conduct, also encouraged the study of Nature, and this philosophy accorded well with Pliny's reaction to the mores and lifestyle of the early imperial period, as is seen again and again in his illuminating digressions on society.

Pliny's recurrent concern throughout the *Natural History* is for the environment.[84] His role, as an original 'Friend of the Earth', strikes a chord of immediate interest and relevance to our own society. The growth of luxury, which Stoicism defines as 'not living according to Nature',[85] the greed and materialism[86] of contemporary Romans, and their extravagant lifestyle are all anathema to him, since these are, in his judgement, the main causes of the decay of the moral standards which had formed the backbone of the Republic, as exemplified by the character of the Elder Cato.[87] The acquisitive attitude that accompanied the new lifestyle had many unwelcome repercussions. The rape of Nature, by mining and quarrying (33.1 ff.); the desire for pearls (9. 106 ff.) and precious stones to adorn the person (37. 54 ff.); 'purpurae insania', i.e. the demand for expensively dyed clothes (9. 127); the craving for exotic foods—birds for the table (10. 133 ff) and shellfish (9. 104 and 168–9); and the import of luxury goods from distant regions are a few of the many factors which, according to Pliny, precipitated the decline in Roman moral standards.

One further important facet of Pliny's character is revealed by his reaction to the Greeks and Greek culture. Like Cato, he has little time for either! His own feelings, however, appear more ambivalent, no doubt because he is often compelled to rely on Greek writers. Beagon[88] sums up as follows: 'Generally,

---

[84] See Ch. 19, below.
[85] Cf. Sen. *De otio* 5. 8. 1: 'ego secundum Naturam vivo si totum me illi dedi'.
[86] Pliny states (2. 118), 'The only pleasure consists of possession and the profit motive is uppermost in life... Men, in their blind obsession with avarice [mens et tantum avaritiae intenta], do not reflect that knowledge is a more reliable means even of making profit.'
[87] The Elder Marcus Porcius Cato (234–149 BC) is constantly put forward by Pliny as a role model.   [88] *Roman Nature*, 18 f.

the omnivorous learning of the Greeks was considered to be combined with a total lack of virtue and propriety. They were wanting in dignity, seriousness, and respect for themselves and for others. They were deceitful and even mendacious, unscrupulous, and above all highly boastful about their cultural superiority to the Romans.' A few brief examples will illustrate the basis of Pliny's judgement. The *vanitas*—in the sense of 'worthlessness'—of the Greeks, their tall stories, falsehoods (5. 4), exaggeration, gullibility (8. 82), the insidious effect of Greek medicine and medical literature (29. 14), and, worst of all, their unscrupulous and immoral use of blood and marrow from human bodies (28. 4) all contribute to the unacceptability of that race in Roman eyes.

In this widely held view Pliny is typical of his times.

2

Pliny the Writer

2.1 THE POLITICAL BACKGROUND

The deterioration in the political situation during the first century AD inevitably had far-reaching consequences on the content and character of the literature of the period. This resulted from opposition to imperial rule, the basis of which apparently lay in a sentimentalized view of the Republic, and in a concept less philosophical than political, namely that government should be in the hands of the best men, or best man, which, of course, meant the old oligarchy—naturally led by the dissidents themselves. Their Republican heroes were the Younger Cato,[1] Marcus Brutus,[2] and Gaius Cassius.[3]

The emperor Tiberius' mind was unhinged, and obsessed with a morbid fear of assassination. There followed a series of executions, notably of Paetus Thrasea,[4] Helvidius Priscus,[5]

---

[1] The Younger Cato (95–46 BC) was tribune-designate at the time of the Catilinarian conspiracy (63) and helped to secure the execution of the rebels.

[2] Marcus Iunius Brutus, the so-called tyrannicide, was not only pardoned by Caesar, after the battle of Pharsalia (48 BC), but made governor of Cisalpine Gaul (46) and praetor (44) Although promised the government of Macedonia, Brutus was persuaded by Cassius to murder Caesar under the delusive idea of again establishing the Republic.

[3] Brutus' co-conspirator, who, like Brutus, had been pardoned and, in 44 BC, made praetor. After the assassination of Caesar, he went to Syria which he claimed as his province; the senate, however, had given it to Dolabella and, instead, conferred Cyrene on Cassius.

[4] A high-minded senator and Stoic who fearlessly opposed Nero. He was condemned to death on a charge of treason in AD 66.

[5] Helvidius continued the opposition of his father-in-law Thrasea to imperial rule. Denounced in AD 74 by Mucianus, he was later put to death by Vespasian.

Rubellius Plautus,[6] and, in Domitian's reign, of Iunius Rusticus.[7]

Under Nero (AD 54–68) the political situation worsened.[8] His jealous fear of all eminence of birth, or success in the military field, drove him to a policy of persecution. Moreover, his philhellenic outlook was resented by Romans generally. The Younger Pliny (*Letters* 3. 5. 5) summarizes the situation: 'The slavery of the last years of Nero had rendered dangerous every study of a free and elevated character.' High-level opposition to the Flavians was mostly seen in a small, but loosely-knit, group of Stoics. Stoicism provided a philosophical foundation for the aristocratic opposition of those who tried to rule without, or against, the senate. Similarly a few Cynic philosophers joined in the disapprobation. By no means all thinking men and women, however, turned against the emperors. Agricola, the father-in-law of Tacitus, successfully pursued a military career and became governor of the province of Britain in 78.[9]

Pliny found it possible to serve the state and follow the path of scholarship at the same time. His political views are plain to see in the *Natural History*: he accepted the imperial system as indispensable to Rome, being at the same time grateful for the security afforded by the *pax Romana*. Italy he regards as the ruler and second mother of the world (37. 201). Close to the Flavians, as the tone of the Preface addressed to the emperor confirms, Pliny had easy access to Vespasian[10] and was a member of the emperor's advisory council.

In the ancient world, the dividing line between Arts and Sciences was not so rigidly drawn as it has been in more recent times. Poets and prose authors alike frequently crossed

---

[6] The son of Gaius Rubellius Blandus and a descendant of the emperor Augustus. His imperial connection made some regard him as a possible rival to Nero. In AD 60, on Nero's advice, he withdrew to Asia. Two years later he was forced to commit suicide by Nero's henchman Tigellinus.

[7] The writer of a Panegyric on his friend Paetus Thrasea, for which he was condemned to death by the emperor Domitian (? AD 93). He was a friend of the Younger Pliny and approved of Helvidius' opposition to Vespasian.

[8] See generally H. C. Boren, *Roman Society: A Social, Economic and Cultural History* (London, 1977), esp. 196 ff.   [9] Tac. Agr. 9.

[10] J. F. Healy, 'Pliny the Elder and Ancient Mineralogy', *Interdisciplinary Science Reviews*, 6/2 (1981), 166.

boundaries in their literary output. Thus Aristotle could write the *Poetics*, on the nature of drama, in addition to philosophical, political, and scientific works such as his *Historia Animalium*, *Physica*, and *Meteorologica*. Similarly, Virgil not only composed the *Aeneid*, a national epic glorifying Rome and the emperor Augustus, but also the *Georgics*, a didactic poem devoted to agriculture.[11]

2.2 THE WORKS

Pliny the Elder is an outstanding example of an interdisciplinary writer, whose works cover a wide range of topics, although all, except the *Natural History*, have been lost. The Younger Pliny, in his letter to Baebius Macer (5. 3 ff.), provides a list of his uncle's works in order of publication.[12] They are:

1. *De iaculatione equestri* (1 book)   AD 62–6
2. *De vita Pomponi Secundi* (2 books)   62–6
3. *Bella Germaniae* (20 books)   begun between 62 and 66
4. *Studiosus* (3 books)   66/7
5. *Dubius sermo* (8 books)   67/8
6. *A fine Aufidi Bassi* (31 books)   70–6
7. *Naturae historia* (37 books)   76–c.78

---

[11] J. F. Healy, 'Pliny the Elder and Ancient Mineralogy', *Interdisciplinary Science Reviews*, 6/2 (1981), 166.

[12] Cf. J. Reynolds, 'The Elder Pliny and his Times', in French–Greenaway, *Roman Science*, 5 ff. See the Younger Pliny, *Letters* 3. 5. 1–6: 'pergratum est mihi quod tam diligenter libros avunculi mei lectitas, ut habere omnes velis quaerasque qui sint omnes. Fungar indicis partibus, atque etiam quo sint ordine scripti notum tibi faciam; est enim haec quoque studiosis non iniucunda cognitio. "De iaculatione equestri unus"; hunc cum praefectus alae militaret, pari ingenio curaque composuit. "De vita Pomponi Secundi duo"; a quo singulariter amatus hoc memoriae amici quasi debitum munus exsolvit. "Bellorum Germaniae viginti"; quibus omnia quae cum Germanis gessimus bella collegit.... "Studiosi tres", in sex volumina propter amplitudinem divisi, quibus oratorem ab incunabulis instituit et perficit. "Dubii sermonis octo": scripsit sub Nerone novissimis annis, cum omne studiorum genus paulo liberius et erectius periculosum servitus fecisset. "A fine Aufidi Bassi triginta unus". "Naturae historiarum triginta septem", opus diffusum eruditum, nec minus varium quam ipsa natura.'

Military service in Germany, as a cavalry officer, must have encouraged Pliny to produce his monograph on *The Use of the Javelin as a Cavalry Weapon*, to which he makes a brief reference in the *Natural History* (8. 162). He was able to observe the people and countryside at first hand which, in turn, led to his interest in the history of Rome's campaigns against the Germans. Moreover, the success of Aufidius Bassus' *Bellum Germanicum* no doubt inspired him to write his comprehensive history, beginning where Bassus had ended his account. Tacitus refers to Pliny's authorship. The Younger Pliny (*Letters* 3. 5. 4) writes that his uncle's impulse to produce this definitive work came from a dream: 'Advised in a dream, he began the work [*Bella Germaniae*] when on military service in Germany. As he slept, the ghost of Nero Claudius Drusus, who died after his extensive conquests in the province, stood before him. Drusus committed his memory to my uncle's care and begged him to rescue him from the shame of being forgotten.' ('incohavit cum in Germania militaret, somnio monitus: adstitit ei quiescenti Drusi Neronis effigies, qui Germaniae latissime victor ibi periit, commendabat memoriam suam orabatque ut se ab iniuria oblivionis adsereret.') Nero Claudius Drusus was the brother of the future emperor Tiberius who directed the Germanic Wars of 12–9 BC, but died after reaching the river Elbe. Reynolds[13] observes that Pliny's dream should not be treated too dismissively, since it is fully in accord with the ethos of his times.

Pliny's major historical work, *A History of our Times* (*A fine Aufidi Bassi*), commenced with the accession of the emperor Claudius and was already completed when the *Natural History* was published (pref. 20): it must, therefore, have been composed between 70 and 76, when Pliny was absent from Rome fulfilling his assignments as procurator. He writes: 'This has long been finished and is in safe keeping and, in any case, it was my resolve to entrust it to my heir, to prevent it being thought that my life bestowed anything on ambition' ('iam pridem peracta sancitur; et alioqui statutum erat heredi mandare, ne quid ambitioni dedisse vita iudicaretur').

---

[13] Reynolds, 'The Elder Pliny', 5.

Tacitus (Hist. 2. 101. 1) regards Pliny the Elder as a historian and groups with him other Flavian writers against whom he issues a general anonymous warning: 'The historians who wrote their versions of this war, when the Flavians were in power, handed down to posterity their concern for peace and their love of the Republic—reasons twisted to suit their intention to flatter' ('scriptores temporum, qui potiente rerum Flavia domo monimenta belli huiusce composuerunt, curam pacis et amorem rei publicae, corruptas in adulationem causas tradidere'). He also (Ann. 13. 31. 1) criticizes writers who fill their works with what he considers irrelevant detail: Rome's histories should contain only important events.

The *Life of Pomponius Secundus* is a biography of Pliny's friend, literary mentor, and, possibly, patron (*NH* 14. 56). The *Studiosus*, a treatise on rhetorical training contains examples of declamation—and was the forerunner of Quintilian's *Institutio oratoria*. Pliny is quoted by Quintilian, but sometimes with disapproval (*Inst.* 3. 1. 21 and 11. 3. 143). The *Dubius sermo*, which Pliny refers to as *De grammatica* (pref. 28), treated phonology, inflexions, etymology, and parts of speech: it was often cited by other grammarians.

The surviving *Natural History*, Pliny's encyclopaedic account of the state of science, technology, and arts in the first century AD is treated separately.[14]

---

[14] See Ch. 3 below.

3

The *Natural History*

3.1 INTRODUCTION

In Greece, the Sophists had been the first to claim to impart to their pupils such knowledge as they might need in daily life.[1] Hippias of Elis (Pl. *Hp. mi.* 368b ff.) mastered all the subjects of instruction later referred to by Aristotle (*Pol.* 1337ᵇ15) as the branches of knowledge necessary for a freeman: these included astronomy, geometry, arithmetic, music, and grammar. Quintilian (*Inst.* 1. 10. 1) writes: 'orbis ille doctrinae quam Graeci ἐγκύκλιον παιδείαν vocant', he means a pupil's general studies before taking up his special subject.

Pliny (pref. 17) records the view of Domitius Piso that works of reference, not books, are needed: 'thesauros oportet esse non libros.' This, indeed, is the role of the encyclopaedia.[2] The Greeks never produced such a comprehensive work and it remained for the more practical Romans to establish the encyclopaedic genre.

Soon after 184 BC, Cato wrote a comprehensive account of agriculture, medical science, and rhetoric that may also have included military science and jurisprudence, but Varro's encyclopaedia—his *Disciplinae* in nine books—was the more important work. Celsus, in the principate of Tiberius (AD 14–54), composed an encyclopaedia—his *Artes*—similar in scope to that of Cato. The first seven books cover the

---

[1] Carolyn Dewald, 'Greek Education and Rhetoric', in Grant–Kitzinger, *Mediterranean Civilization*, ii. 1084–6.
[2] H. J. Mette, 'Enkyklios Paideia', *Gymnasium*, 67 (1960), 300–7. E. Norden, *Die antike Kunstprosa vom VI Jhdt v. Chr. bis in die Zeit der Renaissance*, 2 vols. (Leipzig, 1898; repr. Berlin, 1909), ii. 670, and *RE* Suppl. 6. 1256. N. P. Howe, 'In Defence of the Encyclopaedic Mode: On Pliny's Preface to the Natural History', *Latomus*, 44 (1985), 561–76.

so-called seven liberal arts, which as *trivium* (grammar, dialect, rhetoric) and as *quadrivium* (geometry, arithmetic, astronomy, music) continued to be practised into the Middle Ages.

Pliny's *Natural History*, published in AD 77, is the best surviving example of the genre. At the outset, he puts his work in context (pref. 14):

My path is not well worn by writers, nor the kind along which the mind aspires to wander. No Roman author has attempted the same project, nor has any Greek treated all these matters single-handed. Many of us seek pleasant fields for research, while others deal with matters of great complexity where one is overwhelmed and cannot see the wood for the trees. First and foremost I must deal with subjects that are part of what the Greeks term 'an all-round education', [ἐγκύκλιος παιδεία], but which are unknown or have been rendered obscure by scholarship.

praeterea iter est non trita auctoribus via nec qua peregrinari animus expetat: nemo apud nos qui idem temptaverit invenitur, nemo apud Graecos qui unus omnia ea tractaverit. magna pars studiorum amoenitates quaerimus quae vero tracta ab aliis dicuntur immensae subtilitatis obscuris rerum in tenebris premuntur, ante omnia attingenda quae Graeci τῆς ἐγκυκλίον παιδείας vocant; et tamen ignota aut incerta ingeniis facta.

The term 'encyclopaedia' has come to be synonymous with a work, in a number of volumes, which treats the whole range of human knowledge.

Pliny (36. 71 f.) discussing Egyptian obelisks, records that there were two in Rome;[3] one, in the Circus Maximus, was quarried for King Psemetnepserphreus, the other, in the Campus Martius, for Sesothis. He observes that they are inscribed with hieroglyphs comprising an account of natural science according to the theories of Egyptian philosophy

---

[3] There is some confusion here. Eichholz, *NH 36–37* (LCL), p. 56, note *a*, explains that the obelisk in the Circus Maximus (now the Piazza del Popolo) was that of Seti I (Dynasty 19) and his son Ramases II (1348–1282 BC) one of whose names was Sessura: hence 'Sesothis', or more frequently 'Sesostris'. The obelisk of the Campus Martius (now in Monte Citorio) is that of Psammetichus II (594–589, Dynasty 26) 'Psemetnepserphreus' is a corrupt form of two of his names Psamtik and Neferibre.

('inscripti ambo rerum naturae interpretationem Aegyptiorum philosophia continent'). Since the significance of hieroglyphs was unknown and they could not be translated, Pliny appears to have interpreted the pictograms of birds and animals as connected in some way with natural history—a logical, although incorrect deduction!

He introduces his work as the *Naturalis historia* (pref. 1) and (ibid. 13) defines its scope: 'My subject, the natural world, that is life (in its most basic aspects), is a barren one' ('sterilis materia, rerum natura, hoc est vita, narratur'). The Younger Pliny (*Letters* 3. 5. 6) similarly refers to it as *Naturae historia*— 'a learned and comprehensive work, as full of variety as nature herself' ('opus diffusum eruditum, nec minus varium quam ipsa natura').

The term ἱστορία which first occurs in Herodotus (1. 1) was used by him in two main senses as (1) inquiry,[4] or 'research',[5] and (2) the written account of his findings, that is narrative, or history. Plato[6] employs the term more specifically in the expression ἡ περὶ φύσεως ἱστορία, that is, the equivalent of natural science, which he further defines as the process of discovering the causes for which each thing comes and ceases and continues to be. The term φύσις is used, generally, to mean nature,[7] and, in Aristotle,[8] ἡ φυσικὴ ἐπιστήμη is the equivalent of natural science.

The expression 'rerum natura' occurs in Cicero and elsewhere, but, in its more specialized meaning, is first used by Lucretius (*DRN* 1. 24 f.): 'te sociam studeo scribendis versibus esse | quos ego de rerum natura pangere conor.' ('Yours is the partnership I seek in striving to compose these lines *On the Nature of the Universe*.' Translation here, and elsewhere, from R. E. Latham, *Lucretius: On the Nature of the Universe*

---

[4] Hdt. 1. 1. ἱστορίας ἀπόδειξις.   [5] Hdt. 2. 118 and cf. 2. 119.
[6] *Phdr.* 96a. This had been the title of a number of philosophical works during the 6th and 5th cents. BC, covering the topics embraced by 'natural science'. By φύσις those early philosophers meant the 'primary substance', later called ἀρχή. A different meaning, however, appears in the title of the poem by Empedocles (*c.*500 BC).
[7] As Xen. *Mem.* 1. 1. 11.
[8] Arist. *PA* 640^{a}2; *Ph.* 198^{a}22 and generally.

(London, 1997).) The adjective *naturalis*[9] is explicit and means, fundamentally, 'belonging to nature' (or less likely, 'to the nature of things'), as in the title of the Younger Seneca's *Quaestiones naturales*. Although we continue to follow the long-standing tradition of translating the title of Pliny's work as the *Natural History*, Beyet's version, *Recherches sur le monde*,[10] appears more accurately to reflect the scope of his treatment and well accords with his own definition of his theme.

In the presentation of his material, Pliny has much in common with Herodotus, whose history is full of anecdotes, digressions, and colourful description. There is, however, a major discrepancy between his stated intention and practice. In his Preface (12), he describes the *Natural History* as a somewhat lightweight work 'and one not abounding in talent—which, in my case, is exceedingly unremarkable. There is no place for digressions, speeches, discourses, miraculous happenings, or faits divers, although such matters might have been pleasant for me to recount and a source of entertainment for my reader. My subject *the Natural world, or life* (that is life in its most basic aspects), is a barren one...' ('nam nec ingenii sunt capaces, quod alioqui in nobis perquam mediocre erat, neque admittunt excessus aut orationes sermonesve aut casus mirabiles vel eventus varios, iucunda dictu aut legentibus blanda. sterilis materia, rerum natura, hoc est vita, narratur, et haec sordidissima sui parte.') In fact the *Natural History* is enlivened by the very embellishments he claims to omit.

The main reason for this apparent contradiction is that Pliny wishes to avoid others' mistakes, as he explains (pref. 15): 'Yet, other subjects have been so over-exposed by publication as to become boring. For it is difficult to give a new look to things that are old hat, an air of authority to what is novel, lustre to what is passé, to shed light on the obscure, bring acceptability to things that arouse aversion, credibility to

---

[9] As Cic. *Part. or.* 64.
[10] R. Schilling, 'La Place de Pline l'Ancien dans la littérature technique', *Revue de Philologie*, 52 (1978), 274.

matters open to question and indeed to give to all things Nature, and to Nature herself, all her intrinsic qualities. And so, even if I have not succeeded, willingness to make the attempt constitutes an excellent and high-minded endeavour.' ('res ardua vetustis novitatem dare, novis auctoritatem, obsoletis nitorem, obscuris lucem, fastiditis gratiam, dubiis fidem, omnibus vero naturam et naturae sua omnia. itaque nobis etiam non assecutis voluisse abunde pulchrum atque magnificum est.')

Both Pliny and Herodotus are avid cataloguers, with a more than passing interest in the unusual, whose approach, largely, I suspect, because of their enthusiasm, is often ingenuous and uncritical.

Although the *Natural History* is dedicated to the emperor Titus[11] (pref. 1), it is intended for a wide readership (ibid. 6): 'Why do you read this, my Emperor? It is written for the masses, for the great crowd [*turba*] of farmers and artisans, and, finally, for those who have time to devote to these pursuits.' ('quid ista legis, imperator? humili vulgo scripta sunt, agricolarum, opificum turbae, denique studiorum otiosis.')

3.2 CONTENTS

The *Natural History* consists, in essence, of a series of extended 'essays' on topics within the major fields of 'applied science'. The practical interests of the readership[12] envisaged by Pliny are certainly covered in its pages, but it was in no way intended as a 'handbook' on any individual subject, in the manner of the specialist publications of the authorities quoted by him. Nor was the work, as has sometimes been suggested,

---

[11] Similarly, Lucretius dedicates his *De rerum natura* to Gaius Memmius, an eminent Roman statesman, although his message is for mankind generally (*DRN* 1. 947).

[12] So Lucretius (*DRN* 1. 63 ff.) addresses all mankind, seeking to free men from superstitious fear: 'humana ante oculos foede cum vita iaceret | in terris, oppressa gravi sub religione' ('When human life lay prostrate in all men's sight, crushed under the weight of superstition').

a vade-mecum for some newly appointed provincial governor, or other imperial official!

Apart from factual material, Pliny includes much that clearly belongs to the world of fantasy, although even here a rational explanation can occasionally be advanced for what appears to be outside human experience.

4

Pliny's Sources

Although Pliny records information obtained[1] while on active service in Germany, and as procurator in Spain[2] (and, probably, North Africa[3]), most of the content of the *Natural History* is derived from secondary sources, and includes accounts of countries and peoples by geographers, historians, scholars, encyclopaedists, writers on special topics, local people, Roman commanders in the field (reports and *commentarii*), provincial governors, *equites,* and other officials. The accuracy and value of this information varies according to their criteria. In addition there are many facts from unidentified sources.

4.1 Authorities on Topography

Marcus Agrippa,[4] who served as general under Augustus and organized a geographical survey of the Roman empire, provides Pliny with dimensions and measurements of countries,

---

[1] Pliny, uniquely among ancient authors, lists his sources in book 1, but in the main body of the *Natural History*, accepts evidence from some which are not specifically identified. A complete account of these would be inappropriate here; some indication, however, of the nature of the evidence is included as a framework, within which an assessment of his use of available material—often from subsequently lost works—may be attempted.

[2] See Ch. 1.4.1.   [3] Ch. 1.4.2.

[4] Marcus Vipsanius Agrippa (63–12 BC), friend and supporter of Augustus, took an active part in military operations. Among his many political offices he held the consulship in 37, during which time he suppressed a rebellion in Aquitania and won two decisive naval engagements at Naulochus. Agrippa was sent on two missions to the East and wrote a geographical commentary (used by Strabo and, later, Pliny) that formed the basis of a map of the Roman empire which was displayed in the Porticus Vipsania, built after his death. Agrippa relied on Agatharcides for information about the eastern Mediterranean and Ethiopia, while Megasthenes and personnel attached to Alexander the Great's expedition supplied material relating to India.

seas, and continents. These are supplemented by topographical details from a number of other major authorities, including Artemidorus,[5] Ephorus,[6] Strabo,[7] and Isidorus:[8] the estimates, particularly of distances, are generally unreliable and differ significantly from each other and from the true figures.

In matters relating to Egypt (36. 64 ff.), a country with which Pliny is fascinated, he is indebted to Herodotus,[9] Dionysius,[10] and Alexander Polyhistor,[11] among other writers.

The accounts of India and Sri Lanka—which also refer to the social and cultural life of their peoples—are provided by

---

[5] Artemidorus, a Greek from Ephesus (*fl.* 104–101 BC), voyaged along the shores of Spain and, possibly, of Gaul. In Alexandria he wrote eleven books on geography. His records of distances, however, like those of other ancient writers, contain errors and misunderstandings.

[6] Ephorus of Cyme (*c.*405–330 BC) was the chief source of Diodorus Siculus 11–16. He consulted numerous authorities, including the Oxyrhyncus historian and Callisthenes. Ephorus was, with the exception of Xenophon, the most important historian of the 4th cent. BC.

[7] Strabo from Amaseia, Pontus (64/3 BC–AD 21), was of mixed descent, partly Greek and partly Asiatic. He was both historian and geographer and travelled widely in search of material for his *Geography*, of which seventeen books, covering the then known world, have survived. In a sense he brings Eratosthenes up to date, but does not consider mathematics and astronomy important to the study of geography. Similarly, he treats longitude, latitude, and 'climata' lightly. He holds Homer in high esteem but undervalues Herodotus. His work, a mine of information relating to aspects of physical and 'economic' geography, is both a history and philosophy of geography and an important source for Pliny.

[8] Isidorus, a Greek from Charax, near the mouth of the Tigris, wrote, *c.* AD 25, on Parthia and its pearl fisheries (Ath. 3. 93d), and a general geography, part of which is perhaps the Σταθμοὶ παρθικοί. The suggestion that he had been sent out by Augustus to collect facts about Parthia and the coasts of Arabia for Gaius Caesar (who was killed in Armenia in AD 4), is based on the assumption that Pliny (6. 141) misnames Isidorus as 'Dionysius'. See *GGM* i. 80 ff. and *FGrH* iii. 781.

[9] Pliny relies on Herodotus (book 2) for information about Egypt. Apart from his debt to Homer, Hesiod, Hecataeus, and other authorities, Herodotus derived much of his material orally from on-the-spot 'research' (ἱστορία). Unfortunately, he was often too ready, uncritically, to accept what he was told by priests and others.

[10] *NH* 36. 79. One of a list of unidentified writers about whom nothing further is known.

[11] Alexander Polyhistor, of Miletus (born *c.*105 BC), came to Rome as a prisoner of war and was freed by Sulla in 80 BC. He took the name Lucius Cornelius Alexander. His literary output, post 49 BC, was substantial and included many compilations relating, for example, to Delphi, Rome, the Jews, *mirabilia*, and literary criticism. He was industrious but lacked originality.

Megasthenes,[12] Eratosthenes,[13] Ctesias,[14] Onesicritus (and Nearchus),[15] Juba,[16] and the Younger Seneca,[17] who attempted to write a book about India, and are of particular interest. Pliny, however, uncritically includes some information—as, for example, from Megasthenes—that was already out of date

---

[12] Megasthenes (350–290 BC), an Ionian, served as ambassador for Seleucus I (302–291) to the Indian king Chandragupta, founder of the Mauryan empire in north India. He paid several visits to Chandragupta's capital, Pataliputra, and acquired a knowledge of the regions between the Indus and the Ganga. This firsthand experience of Indian topography and culture was embodied in a history (Ἰνδικά) in four books: (1) geography, peoples, and cities; (2–3) system of government, classification of citizens, and religious customs; (4) archaeology, history, and legends. Megasthenes' work coincided with a growing Western interest in India, because of the campaigns of Alexander the Great and subsequent trade. His work is important since, in spite of his credulous acceptance of fables, and inaccuracy of observation, it provided the most comprehensive account of India at that time and, for many centuries, remained the main source of the West's knowledge of India. Megasthenes was a major source for Strabo and for Arrian.

[13] Eratosthenes of Cyrene (c.275–194 BC) succeeded Apollonius Rhodius as head of the Library at Alexandria, at the invitation of Ptolemy Euergetes. His interests embraced mathematics, astronomy, and geography. Eratosthenes was the first systematic geographer and his *Geography* in three books, sketched the history of the subject, and dealt with physical, mathematical, and ethnographical geography. Strabo (n. 7 above), among others, did not approve of his work. One of his major contributions was his calculation of the circumference of the earth, with a high degree of accuracy, although he was less successful in estimating the distance of the sun and moon from the earth.

[14] Ctesias of Cnidus (late 5th cent. BC) was a Greek doctor at the Persian court and author of an unreliable history of Persia (Περσικά), in twenty-three books, a geographical treatise (Γεωγραφία), in three books, and a separate work on India (Ἰνδικά).

[15] Onesicritus of Astypalaea, together with Nearchus, served with Alexander in India. He steered Alexander's ship down the river Jhelum. Onesicritus' work was not a history but an historical romance, in the manner of Xenophon's *Cyropaedia*. Strabo and Pliny used it for its material relating to natural history.

[16] Juba II was taken as an infant to Rome in Caesar's triumph which celebrated his victory over the Pompeians at Utica in 46 BC. Educated in Italy, he received Roman citizenship form Octavian and in 25 was granted the kingdom of Mauretania. Juba was an erudite scholar who carried out a wide range of research and was keen to introduce Greek and Roman culture into his kingdom. He wrote many books in Greek—now lost—on Libya, Arabia, and Assyria, and also a history of Rome.

[17] Lucius Annaeus Seneca, the philosopher, born at Corduba between 4 BC and AD 1, studied grammar, rhetoric, and philosophy at Rome. A man of talents, Seneca's fame rests on his writings on moral and philosophical subjects. Pliny owes a limited debt to his *Questiones naturales*.

because of the subsequent exploration of India under the early principates.[18]

Ethiopia (6. 197) and the islands off the coast of Africa (6. 198 ff.) are described by Hanno,[19] Ephorus,[20] Eudoxus,[21] Timosthenes,[22] Clitarchus,[23] Polybius,[24] and Xenophon (of Lampsacus).[25]

Further random geographical references occur in a wide range of authors whose names Pliny records in his 'bibliography' in book 1: for example, Aristides,[26] Aglaosthenes,[27] Nepos,[28] Timaeus,[29] and many others.

---

[18] See A. Dihle, 'Plinius und die geographische Wissenschaft in der römischen Kaiserzeit', in *Atti del Convegno di Como—Technologia, economia e società nel mondo romano* (Como, 1980), 121 ff. See also generally W. W. Tarn and G. T. Griffith, *Hellenistic Civilization*³ (London, 1966), 239 ff. and 289 ff.

[19] Hanno, a Carthaginian explorer of the early 5th cent. BC, went to west Africa before 480 and founded cities there as far south as Sierra Leone. A version of his report on the region, originally written in the Punic language, survives in Greek. See *GGM* i. 1–14.   [20] See above, n. 6.

[21] Eudoxus of Cnidus (*c*.390–340 BC) was an outstanding mathematician who made significant contributions in the fields of astronomy and geography. In astronomy he was the first to construct a mathematical system to explain the apparent movement of the heavenly bodies—that of the 'homocentric spheres'.

[22] Timosthenes of Rhodes, an admiral under Ptolemy Philadelphus, *c*.280 BC, was an authority on harbours. (*NH* 5. 47, 129; 6. 18, 163, 183, 198) See Tarn and Griffith, *Hellenistic Civilization*, 250. Many cities in the 3rd cent. BC improved their ports and Timosthenes' *On Harbours* filled the place now held by the *Mediterranean Pilot*.

[23] Clitarchus compiled the history of Alexander's Asiatic expedition which he accompanied (*NH* 57; 6. 36, 198; 7. 129).

[24] Polybius, of Megalopolis (204–122 BC), wrote a history in forty books—only the first five of which survive in their entirety—covering the period *c*.220 BC–146 BC. Polybius undertook journeys to foreign countries in the interest of research for his history.

[25] Author of a *Periplus*, according to Pliny (*NH* 7. 155), who, unfortunately, does not give any further details. Cf. *NH* 4. 95; 6. 200.

[26] For other references to Aristides, see *NH* 4. 64, 70, and 7. 125, where he is identified as a painter.   [27] Otherwise unknown.

[28] Cornelius Nepos was the contemporary and friend of Cicero, Atticus, and Catullus. From his historical works only the *Vitae excellentium imperatorum* survive under his name. But, with the exception of the life of Atticus and the fragment of a life of Cato the censor, the rest are ascribed to an unknown author, Aemilius Probus, who lived under Theodosius at the end of the 4th cent. AD.

[29] Timaeus, of Tauromenium (*c*.356–260 BC) collated a history, primarily of Sicily, in thirty-eight books that was important for its standardization of previous accounts of that island's history and origins. It dealt mainly with events in Sicily, Italy, and Libya, ending either with Pyrrhus' death, or the beginning of the First Punic War.

4.2 HISTORIANS

Pliny has access to the major Greek and Roman historians and also to lesser known annalists such as Lucius Piso[30] and Cassius Hemina[31] (13. 84).

4.3 ENCYCLOPAEDISTS

Pliny, throughout the whole of his *Natural History*, derives much significant information from encyclopaedists.[32] He is, for example, indebted to the works of Varro[33] (116–28 BC), not only for geographical measurements but for a wide variety of general topics, ranging from outstanding physical characteristics of men and proposals for a bridge from Italy to Apollonia to unusual properties of waters. Pliny also quotes Verrius Flaccus,[34] who spans the period from the late first century BC to the reign of Tiberius and wrote on 'History and Antiquities'.

---

[30] Lucius Calpurnius Piso was in office in the middle years of the 2nd cent. BC. His *Annals* cover the period from the origins of Rome to his own times, in at least seven books (158 BC). Fragments of writings on antiquarian and mythological subjects are also attributed to him.

[31] Lucius Cassius Hemina, was also an annalist, whose *Annals* described events from early Italian times and the founding of Rome to the Second Punic War. A further title, *Bellum Punicum posterior* was written before the Third Punic War. Hemina's interests included the origins of Italy, etymology, religious and social customs, antiquities and synchronism. Fr. 8 (See H. Peter, *Historicorum Romanorum Reliquiae*, i²(1914)) reflects the influence of Cato's *Origines*.

[32] See generally Ch. 6 below, on the genre of the encyclopaedia at Rome.

[33] Marcus Terentius Varro, of Reate (116–27 BC), although a partisan of Pompey, regained Caesar's favour and was put in charge of the great library at Rome intended for public use. According to Gellius (3. 10. 17), he had already edited 490 books by the beginning of his seventy-eighth year. The titles known include: his *De lingua Latina*, *De re rustica* (see further, K. D. White, *Roman Farming* (London, 1970), 22 ff.), *Saturarum Menippearum libri*, *Antiquitatum rerum humanarum et divinarum libri,* and the *Logisticon libri*. His *Hebdomades vel De imaginibus* contained 700 portraits illustrating the text (*NH* 35. 11) and the *Disciplinarium libri* treated the liberal arts. Varro's writings covered nearly every aspect of science, history geography, rhetoric, jurisprudence, philosophy, music, medicine, architecture, and literary history, He was, possibly, the greatest scholar among then contemporary Romans, and, like the Elder Cato, a major source for Pliny, as the many references to his works confirm.

[34] Verrius Flaccus, a freedman, considered to be one of the most erudite of the Augustan scholars, and teacher of Augustus' grandsons. His works, now lost, include: the *Libri rerum memoria dignarum* (freely used by Pliny), *De obscuris Catonis, Libri rerum Etruscarum, De orthographia, Fasti Praenestini,* and the *Libri de significatu verborum,* in which he freely quotes from other republican authors.

In addition, occasional references are made to authorities such as Alexander Polyhistor[35] (9. 115) for his observations on the ageing of pearls and the mention of the so-called 'lion' tree used to build the *Argo* (13. 119).

4.4 MONOGRAPHS

Pliny derives specialist information, on a whole range of subjects, from experts like the Elder Cato[36] and Varro on agriculture—in particular wheat and grain crops, olive-growing, and viticulture—and, through Varro, much of his knowledge of art, originating in the works of Xenocrates[37] of Sicyon and Antignotus[38] of Carystus. Vitruvius[39] is a main source for architecture and building, and Dioscorides[40] and

---

[35] See above, n. 11.
[36] Marcus Porcius Cato, the elder Cato (234–149 BC), from Tusculum, was a military tribune in the Second Punic War. Pliny is greatly indebted to his *De agri cultura* (c.160), which deals with the development of viticulture, olive- and fruit-growing, and grazing in Latium. In spite of its shortcomings and archaic tone, this is an up-to-date account based on Cato's own knowledge and experience and related to the rise of capitalistic farming. See further White, *Roman Farming*, 19 f. The *Origines* (seven books) derive material from the historian Fabius Pictor, and from Hellenistic legends, local traditions, and inscriptions, supplemented by inserted speeches. The *Origines* played an important part in the establishment of Latin prose style and provided a model for subsequent works of a similar nature.
[37] According to Pliny (*NH* 34. 83), Xenocrates, a sculptor who had been a pupil of Tisicrates, or, by other accounts, of Euthycrates, produced a vast output of statues and books on sculpture.
[38] Antignotus, otherwise unmentioned, is credited by Pliny (*NH* 34. 86) with, among other statues, a version of the *Tyrannicides* (Harmodius and Aristogiton).
[39] Marcus Pollio Vitruvius, architect and military engineer under the Second Triumvirate through the early part of the principate of Augustus, based his *De architectura* on experience, supplemented by data drawn from similar works mainly by Greek architects. The ten books embrace the following topics: (1) town planning, architecture in general, and the qualifications appropriate for an architect; (2) building materials; (3–4) temples and the 'Orders' of architecture; (5) other civic buildings; (6) domestic buildings; (7) pavements and decorative plasterwork; (8) water supplies; (9) geometry, mensuration, astronomy etc.; (10) machines, civil and military.
[40] Pedanius Dioscorides, of Anazarbus (fl. 1st cent. AD). Well-versed in pharmacology, he wrote a Materia medica and works on plants. Vegetable, animal, and mineral remedies are described in careful categories. His Περὶ ὕλης ἰατρικῆς, bks. 1–5, contains almost 600 plants and nearly 1,000 drugs; Περὶ ἁπλῶν φαρμάκων, or Εὐπόριστα 1–2 may be spurious. See edn. of M. Wellmann, 3 vols. (Berlin 1907–14) for both works.

Celsus[41] for pharmacological and medical matters. A less well-known writer, Callixenus,[42] provides expertise on ships—especially Ptolemaic (36. 67).

4.5 MILITARY COMMANDERS

The expansion of the empire, during the first century AD, brought Rome into contact with a number of previously unexplored regions. Reports sent back by military commanders supplied useful information and encouraged further expeditions.

Pliny (5. 11) writes:

The first occasion Roman forces fought in Mauretania was during the principate of Claudius [41–54 AD]. King Ptolemy had been put to death by Caligula and his freedman Aedemon was seeking to avenge him; it is generally accepted that our soldiers went as far as Mt. Atlas and, at this point, the natives fled. And not only were the ex-consuls and generals drawn from the senate, who commanded in that campaign, able to boast of having penetrated the Atlas range, but this distinction was also shared by the Roman knights who subsequently governed the country.

Romana arma primum Claudio principe in Mauretania bellavere Ptolemaeum regem a Gaio Caesare interemptum ulciscente liberto Aedemone, refugientibusque barbaris ventum constat ad montem Atlantem. nec solum consulatu perfunctis atque e senatu ducibus qui tum res gessere sed equitibus quoque Romanis qui ex eo praefuere ibi Atlantem penetrasse in gloria fuit.

---

[41] Aulus Cornelius Celsus (AD 14–37) author of a comprehensive encyclopaedia on agriculture, medicine, military science, rhetoric, and, probably, philosophy and jurisprudence. Except for fragments, only the medical works—eight books—survive. On Celsus see further G. Sabbath and P. Mudry, *La Médicine de Celse: Aspects historiques, scientifiques et littéraires* Mémoires *XIII* du Centre Jean Palerne; (St-Étienne and Lyons, 1994).

[42] Callixenus, a Rhodian Greek (*fl. c.*155 BC) wrote περὶ Ἀλεξανδρείας. Athenaeus (5. 196a and 203e) quotes him on ships built by Ptolemy Philopator (*FGrH* iii. 627).

Later Pliny (5. 14) continues:

Suetonius Paulinus,[43] consul in my time, was the first Roman commander actually to cross the Atlas Mountains, and he went some miles further. His estimate of the height agrees with that of other authorities, but he adds that the lower slopes are filled with dense forests of tall trees of an unknown species: they have very tall trunks notable for their sheen and freedom from knots. Their leaves, like those of the cypress except for the heavy scent, are covered with a thin down, from which with a suitable technique, clothing can be made just like that derived from the silkworm. The summit of Mt. Atlas is covered with deep snow even in summer. Suetonius Paulinus reached there in ten days, and travelled beyond to the river Ger, across deserts of black dust, with projecting rocks in some places that looked as if they had been burnt—a place uninhabitable because of the heat, although it was winter when he experienced it. The Canarii live in the neighbouring forests, which are filled with every species of elephant, wild beast, and snake. The people are called Canarii because they share a common diet with dogs, living off the flesh of wild animals.

Suetonius Paulinus, quem consulem vidimus, primus Romanorum ducum transgressus quoque Atlantem aliquot milium spatio prodidit de excelsitate quidem eius quae ceteri, imas radices densis altisque repletas silvis incognito genere arborum, proceritatem spectabilem esse enodi nitore, frondes cupressi similes praeterquam gravitate odoris tenui eas obduci lanugine, quibus addita arte posse quales e bombyce vestes confici. verticem altis etiam aestate operiri nivibus, decumis se eo pervenisse castris et ultra ad fluvium qui Ger vocatur per solitudines nigri pulveris, eminentibus interdum velut exustis cautibus, loca inhabitabilia fervore quamquam hiberno tempore experto. quid proximos inhabitent saltus refertos elephantorum ferarumque et serpentium omni genere Canarios appellari, quippe victum eius animalis promiscuum his esse dividua ferarum viscera.'

---

[43] Gaius Suetonius Paulinus, propraetor in Mauretania in AD 42 and consul in 66, was governor of the province of Britain between 59 and 62. After the death of Nero (68) he was one of Otho's generals in the war against Vitellius. Pliny no doubt had access to reports brought from North Africa by Suetonius, on his return to Rome. The account reproduced here and that of Aelius Gallus (see following notes) gives an indication of the content and nature of the information likely to have been provided by military personnel.

Typical of the intelligence brought back by military commanders is the extended account of Aelius Gallus[44] (6. 160),

a knight, the only person to have carried arms into Arabia. (Gaius Caesar, son of Augustus, had only a brief view of that country.) He destroyed towns not named by authors who had written previously.... On his return, Gallus reported that the Nomads live on milk and wild animals' flesh, while the remaining tribes get wine from palm-trees—as the Indians do—and oil from sesame. The Homeritae are the most numerous people; the Minaei have land that grows palm groves and timber in abundance and is rich in flocks; the Cerbani, Aegraei and especially the Chatromotitae excel in fighting. The Sabaei are the richest because of the fertility of their forest in producing perfumes, their gold-mines, irrigated fields, and production of honey and wax. I shall speak of their perfumes in the appropriate book [*NH 12*]. The Arabs wear turbans, or leave their hair uncut; they shave their beards but leave a moustache; others, however, leave the beard also unshaven. Strange to relate, of these countless tribes, half live by trade, half by marauding. Overall they are the richest of peoples because very great wealth from Rome and Parthia settles in their coffers when they sell what they catch in the sea or forest and they buy nothing in return.

Romana arma solus in eam terram adhuc intulit Aelius Gallus ex equestri ordine; nam C. Caesar Augusti filius prospexit tantum Arabiam. Gallus oppida diruit non nominata auctoribus qui ante scripserunt.... Nomadas lacte et ferina carne vesci; reliquos vinum ut Indos palmis exprimere, oleum sesamae; numerosissimos esse Homeritas; Minaeis fertiles agros palmetis arbustoque, in pecore divitias; Cerbanos et Agraeos armis praestare, maxime Chatramotitas; Carreis latissimos et fertilissimos agros; Sabaeos ditissimos silvarum fertilitate odifera, auri metallis, agrorum riguis, mellis ceraque proventu; de odoribus suo dicemus volumine. Arabes mitrati degunt aut intonso crine, barba abraditur praeterquam in superiore labro; aliis et haec intonsa. mirumque dictu ex innumeris populis pars aequa in commerciis aut latrociniis degit; in universum gentes ditissimae, ut apud quas maximae opes Romanorum Parthorumque

---

[44] Aelius Gallus, a knight, was prefect of Egypt. He invaded Arabia in 24 BC. See above, n. 8.

subsidant, vendentibus quae e mari aut silvis capiunt, nihil invicem redimentibus.

Information about distances was also supplied by surveyors who accompanied expeditions, as for example, Diognetus and Baeton, in the case of Alexander the Great (6. 61).

Pliny, in his record (3. 136) of the inscription from a triumphal arch erected in the Alps celebrating Roman conquests, does not mention his source, but it may have come from some official. 'non alienum videtur hoc loco subicere inscriptionem e tropaeo Alpium, quae talis est. ...' Such a commemorative monument with inscription survives, in part, at La Turbie,[45] (French Riviera). There is also an arch carrying part of this inscription at Nicaea (Albania).

4.6 PROVINCIAL GOVERNORS

Governors of provinces often provided knowledge of local matters and curiosities. One of the most important, Mucianus,[46] who was three times consul, compiled a book of *Mirabilia* ('The Paranormal') based on his experiences and observations in Lycia and elsewhere. He refers to earthquakes on Delos (4. 66),[47] a sex change that he witnessed in which a woman became a man, after marriage (7. 36), and several animal feats—for example, an elephant which learnt to write the shapes of Greek letters and apes who played draughts and could tell real nuts from imitations made from wax. Mucianus describes a celebrated plane-tree in Lycia (*NH* 12. 9) 'like a dwelling-house, with a hollow cavity eighty-one feet across... so worthy to be deemed a marvel that Licinius Mucianus ... who was recently governor of that province thought it worth

---

[45] The Trophy of the Alps, located here on the Grand Corniche, was built by Augustus to celebrate victories over forty-four Gallic tribes, named in the inscription on its base.
[46] Licinius Mucianus was consul in AD 55, 70, and 75 and clearly had a fascination for the unusual (*mirabilia*).
[47] Although, as Pliny observes (ibid.), down to the time of Varro the island had never experienced an earth tremor.

handing down to posterity. He also records that he held a banquet with eighteen members of his staff inside the tree which itself provided couches of leaves on a lavish scale and that he had gone to bed in the same tree shielded from every breath of wind.' ('nunc est clara (platanus) in Lycia...domicilii modo, cava octaginta atque unius pedum specu, nemorosa vertice et se vastis protegens ramis arborum instar...tam digna miraculo ut Licinius Mucianus ter consul...prodendum etiam posteris putaverit epulatum intra eam se cum duodevicesimo comite, larga ipsa toros praebente frondis, ab omni adflatu securum.')

Mucianus claimed to have read, in a temple in Lycia, a letter of Sarpedon written on paper (*NH* 13. 88). Pliny, quite rightly, registers his disbelief. On a visit to Thrace (14. 54) Mucianus discovered that Maronean wine was mixed with water in the proportion of 1:8, that the wine was darker in colour, had a bouquet, and improved with age. In an interesting passage about oysters Pliny (32. 62 f.) writes: 'I shall quote the words of another authority, one who is the greatest expert of our time' ('dicemus aliena lingua quaeque peritissima huius censurae in nostro aevo fuit...'). 'Mucianus says that oysters from Cyzicus are larger than those from the Lucrine Lake, fresher than the British variety, sweeter than those from Medullae, sharper than oysters from Ephesus, fuller than those from Ilici [Spain], less slimy than those from Coryphas, softer than those from Istria, and whiter than those from Circeii. None are fresher, or softer, than the last named.' ('sunt ergo Muciani verba: Cyzicena maiora Lucrinis, dulciora Britannicis, suaviora Medullis, acriora Ephesis, pleniora Iliciensibus, sicciora Coryphantenis, teneriora Histricis, candidiora Cerceinsibus. sed his neque dulciora neque teneriora ulla compertum est.') Finally Mucianus comments on an earthenware dish which cost 1,000,000 sesterces (35. 164) and the extraordinary properties of the sarcophagus stone[48] (36. 131).

Occasionally members of the governor's entourage published information. So Trebius Niger[49] (*NH* 9. 90 ff.), on the staff of Lucius Lucullus, when the latter was governor of

---

[48] See Ch. 14, s.v.
[49] Trebius Niger, an otherwise undistinguished historian.

Baetica, vividly describes a large polyp found in the fishponds of Carteia; 'it was in the habit of getting into their uncovered tanks from the open sea and there foraging for salt fish—even the smell of which attracts all sea creatures in a surprising way, owing to which even fish-traps are smeared with them.' ('Carteiae in cetariis assuetus exire e mari in lacus eorum apertos atque ibi salsamenta populari,—mire omnibus marinis expetentibus odorem quoque eorum, qua de causa et nassis inlinuntur.') Trebius also describes a large cuttlefish.

Others, like Egnatius Calvinus, governor of the Alps, occasionally provide interesting facts, as, for example, when he 'states that the ibis, properly a native of Egypt, has been seen by him in the Alps' ('visam in Alpibus ab se peculiarem Aegypti et ibim Egnatius Calvinus praefectus earum prodidit').

4.7 COMMENTARII

Pliny had access to 'notes', as for example, 'those of the Carthaginian commander Hanno[50] (*NH* 5. 8) who, in the heyday of Carthage, was ordered to explore the circuit of Africa. The majority of Greek and Roman writers follow Hanno both in their legendary stories and in their accounts of the many settlements founded by him in Africa; neither memory nor trace of these settlements now exists.' ('fuere et Hannonis Carthaginiensium ducis commmentarii Punicis rebus florentissimis explorare ambitum Africae iussi, quem secuti plerique a Graecis nostrisque et alia quidem fabulosa et urbes multas ab eo conditas ibi prodidere, quarum nec memoria, nec vestigium extat.')

Similarly, Caesar's account of his campaigns of the Gallic War is styled 'Commentarii'. Pliny left 160 books of *commentarii* to his nephew (*Letters* 3. 5. 17); these, however, consisted of notes and selected excerpts from books. Larcius Licinus offered Pliny, who was procurator in Spain, 400,000 sesterces for these, which were considerably fewer in number at that time. ('hac intentione tot ista volumina peregit electorumque commentarios centum sexaginta mihi reliquit. ... referebat ipse

---

[50] See above, n. 19.

potuisse se, cum procuraret in Hispania, vendere hos commentarios Larcio Licino quadringentis milibus nummum; et tunc aliquanto pauciores erant.')

4.8 EXPLORERS

'While Scipio Aemilianus was commander in Africa, the historian Polybius[51] was given a fleet to explore that part of the world. Sailing round the coast, Polybius reported that west of Mt. Atlas are forests full of the wild animals that Africa produces. ... In the river Bambotus [Non] there are many crocodiles and hippopotamuses'(*NH* 5. 9 f.). ('Scipione Aemiliano res in Africa gerente Polybius annalium conditor ab eo accepta classe scrutandi illius orbis gratia circumvectus prodidit a monte eo ad occasum versus saltus plenos feris quas generat Africa. ... flumen Bambotum crocodilis et hippopotamis refertum.')

4.9 SPECIAL COMMISSIONS

'Ambassadors from Olisipo [*NH* 9. 9], sent on a mission with this purpose in view, reported to the emperor Tiberius that a Triton, whose appearance is well known, had been seen and heard playing on a shell in a certain cave.' ('Tiberio principi nuntiavit Olisiponensium legatio ob id missa visum auditumque in quodam specu concha canentem Tritonem qua noscitur forma.')

4.10 KNIGHTS

A further reference to Tritons is accepted by Pliny (9. 10) without critical comment: 'I have illustrious knights[52] as authority for the assertion that a Triton[53] has been seen by them in the Gulf of Cadiz, perfectly resembling a man in his

---

[51] See above, n. 24.

[52] Pliny accepts these stories about Tritons, although elsewhere (5. 13) he asserts that persons of rank (he has in mind members of the equestrian order) would be willing to say anything ... rather than admit ignorance. Such a statement surely implies that he questions their reliability in reporting on matters?

[53] In mythology Tritons were depicted as having bodies with a human upper half and that of a fish below. They carried conch shells which they blew, at Poseidon's bidding, to calm the restless waves.

appearance; they say that he climbs aboard ships during the night and the side of the ship on which he sits is weighed right down and if he should happen to stay unduly long, the ship is submerged.' ('auctores habeo in equestri ordine splendentes visum ab his in Gaditano oceano marinum hominem toto corpore absoluta similitudine; ascendere eum navigia nocturnis temporibus statimque degravari quas insederit partes, et si diutius permanent, etiam mergi.') 'The pistachio [15. 91]... is also a sort of nut, likewise brought first into Italy by Vitellius at the same time.... It was simultaneously introduced into Spain by Pompeius Flaccus, a knight of Rome, who was serving with Vitellius.' ('pistacium... et ipso nucum genere... et haec autem idem Vitellius in Italiam primus intulit eodem tempore, simulque in Hispaniam Flaccus Pompeius eques Romanus qui cum eo militabat.')

4.11 Local Authorities

Information often came from local people, as for example, Turranius Gracilis[54] (3. 4), who lived near Tarifa and was an authority on Spain. He no doubt supplied the vivid details of the giant 'tree' polypus in the Gulf of Cadiz (9. 8). During the principate of Tiberius (AD 14–37) no less than three hundred monsters were allegedly seen off the coast of southern France and a further three hundred off the coast of Saites. 'Turranius states [9. 11] that a monster was cast ashore on the coast of Cadiz that had 24 feet of tail-end between its two fins, and also a hundred and twenty teeth, the biggest 9 inches and the smallest 6 inches long' ('Turranius prodidit expulsam beluam in Gaditano litore cuius inter duas pinnas ultimae caudae cubita sedecim fuissent dentes eiusdem CXX, maximi dodrantium mensura, minimi semipedum'). Turranius also (18. 75) describes a smooth barley used for making barley water in Baetica and Africa.

4.12 Unnamed Sources

Much general information is attributed to unspecified Greeks, Romans, soldiers, people, and, even more vaguely, tradition.

---

[54] Turranius Gracilis is known only from these references.

Thus the story of the first accidental production of glass (36. 191 ff.) is unattributed, possibly because it was part of a long, ongoing traditional account and, in the circumstances, well known. In the case of gemstones, however, information is often lacking and identifications are correspondingly 'vague'.

4.13 Objects and Creatures Brought to Rome

'Claudius Caesar writes that a hippocentaur[55] was born in Thessaly but died on the same day. In his principate, I personally saw a hippocentaur brought here from Egypt, preserved in honey' (*NH* 7. 35) ('Claudius Caesar scribit hippocentaurum in Thessalia natum eodem die interisse, et nos principatu eius allatum illi ex Aegypto in melle *vidimus*').

Similarly, Pliny (9. 11) writes: 'The bones of the monster to which Andromeda[56] was said to have been exposed were brought by Marcus Scaurus from Jaffa, in Iudaea, during his aedileship and shown at Rome among the rest of the amazing items displayed. The monster was over forty feet long and the height of its ribs was greater than that of Indian elephants, while its spine was one and a half feet thick.' ('beluae cui dicebatur exposita fuisse Andromeda ossa Romae apportata ex oppido Iudaeae Ioppe ostendit inter reliqua miracula in aedilitate sua M. Scaurus longitudine pedum XL, altitudine costarum Indicos elephantos excedente, spinae crassitudine sesquipedali.')

4.14 Personal Observation

Pliny describes a number of things he saw when in Rome—for example, the hippocentaur already mentioned, Agrippina, the

---

[55] See also Cicero, *ND* 2. 2. 5 and *Tusc.* 1. 37, 90. The hippocentaur is a creature that is half man, half horse. While Claudius Caesar's assertion that a hippocentaur was born in Thessaly is, of course, in keeping with received legend—Mt. Pelion, in Thessaly, is the mythological home of the Centaurs—the actual origin of the specimen allegedly brought from Egypt is totally inexplicable.

[56] Cassiopeia boasted that the beauty of her daughter Andromeda surpassed that of the Nereids, so Poseidon sent a sea monster to lay waste the country. The oracle of Ammon promised deliverance if Andromeda was given up to the monster. Accordingly she was chained to a rock. Perseus, with the help of the Gorgon's head, destroyed the monster and married

wife of the emperor Claudius, in a cloak made entirely of cloth of gold[57] (33. 63), and Arellius Fuscus (who was expelled from the equestrian order on a singularly serious charge) wearing silver rings when he sought to acquire celebrity for his school for youths (33. 152). Likewise, his description of the forests of Germany appear to be from personal observation[58] (16. 5 ff.). He also tells us (13. 83): 'I have seen, at the house of Pomponius Secundus..., documents in the hands of Tiberius and Gaius Gracchus written nearly two hundred years ago; while as for autographs of Cicero and the late emperor Augustus and of Virgil, I see them constantly.' ('Tiberi Gaique Gracchorum manus apud Pomponium Secundum... *vidi* annos fere post ducentos; iam vero Ciceronis ac divi Augusti Vergilique saepenumero *videmus*.')

4.15 Archives

Pliny (35. 7) mentions a further source of information, namely archive rooms (in private houses) filled with books of records and with written memorials of official careers. ('tabulina codicibus implebantur et monimentis rerum in magistratu gestarum.')

Marcus Varro is said to have inserted in his volumes the portraits of seven hundred famous people (35. 11).

4.16 Mining and Metallurgy

As procurator in north-west Spain, Pliny had firsthand experience of gold-mining operations and subsequent processing.[59] Diodorus Siculus (5. 22. 2) gives a general description of tin collection and refining. Pliny's sources for his treatment of lead (which he confuses with tin), are twofold:[60] (1) a work of Cornelius Bocchus (33. 96 and 34. 159), and (2) a medical

---

Andromeda. A rocky island in the vicinity of the coastal town of Tartus (Syria) is one of the places which has been identified as the location of the legend. Cf. *NH* 6. 182.

[57] Gold, states Pliny, can be spun into thread and woven into a fabric like wool.   [58] On his military service in Germany, see Ch. 1.3 above.

[59] Ibid.

[60] See R. Halleux, 'Les Deux Métallurgies du plombe-argentifère dans l'Histoire Naturelle de Pline', *Revue de Philologie,* 49 (1975), 72–88, and J. P. Laskowski, 'La Métallurgie dans l'Histoire Naturelle de Pline l'Ancien', *Archaeologia,* 9 (1957), 99–122.

work by Sextus Niger (33. 106–8 and 34. 173). Polybius also states that there are two kinds of galena (34. 9–10). In his discussion of silver ores (33. 95 ff.), Pliny assimilates two metallurgical terminologies—the one probably Spanish, the other purely Greek. Both confuse, to some extent, galena (PbS) and litharge (PbO). Pliny appears to have understood his Spanish source better than his Greek, perhaps, as Halleux suggests, because the former wrote in Latin?

Pliny also had access to random information relating to aspects of mining[61] found in the major classical authors, especially Herodotus, Thucydides, Aristophanes (*Acharnians, Knights,* and *Wasps*), Xenophon (Πόροι), the Greek orators, Theophrastus, Diodorus Siculus, Strabo, and Lucretius, as well as in other sources, including specialist works, subsequently lost, such as, Theophrastus, *On Mines,* Straton (340–330 BC), *On Mining Tools and Machines,* and treatises of Posidonius (*c.*135–51 BC) and Philon.

4.17 AUTHORITIES ON EARTH SCIENCES, MINERALS, AND PIGMENTS

4.17.1 *General*

Theophrastus' *De lapidibus* is, for Pliny, an important sourcebook for minerals and pigments; and, for more fundamental matters relating to mineralization, Aristotle's *Meteorologica,* 1–3 and 4 (*NH* 36. 161). Pliny also appears to have access to the works of Posidonius[62] (*c.*135–51 BC), who had studied Spanish mines (cf. Strabo 3. 2. 4) and was keenly interested in the formation of stones, as is shown by a passage of Diodorus Siculus (2. 52. 1–4). Seneca (*QN* 2. 54. 1) confirms that the two exhalations played a part in the physical theories of Posidonius. Myrrhine (fluorspar) is not mentioned by Diodorus; this was probably still a rarity during the greater part of Posidonius' lifetime. Papirius Fabianus[63] (*NH* 36. 125) asserted that marble is self-regenerating.

---

[61] See generally, J. F. Healy, *Mining and Metallurgy in the Greek and Roman World* (London, 1978), and the revised edn., *Minicre e metallurgia nel mondo Greco e Romano* (Rome, 1993).

[62] Cf. Eichholz, *NH* 36–37, pp. xii ff.

[63] Described by Pliny as an outstanding natural scientist ('naturae rerum peritissimus'). The Younger Seneca was a pupil of his.

4.17.2 Specific Minerals

Theophrastus is Pliny's main source of information about cinnabar and mercury: whole sections are transferred verbatim from the *De lapidibus*[64] to the *Natural History*. Juba[65] and Timagenes[66] also give locations for cinnabar (33. 118) and Verrius Flaccus[67] (33. 111) adds an interesting comment on the use of cinnabar to colour the face of statues of Jupiter on special days.

Homer (*Il.* 11. 25, 18. 565, 574, 613) gives tin, which was highly prized even in the Trojan period, the name κασσίτερος.

Pliny (37. 23) describing rock-crystal, says that it was compacted by intense frost (reading 'concretum' for 'concreto'). This may owe something to Posidonius[68] but is a commonly held view. Both Diodorus (2. 52. 1–4) and Seneca (*QN* 3. 25. 10) agree that 'cold' is the effective cause of ordinary rock-crystal.

The term *lychnites*, used of Parian marble[69] (36. 14), is explained by Varro (in a lost work); it signifies that the marble is mined by the light of oil lamps.

Mucianus (*NH* 36. 131) confirms the petrifying effect of the sarcophagus stone,[70] in Lycia and the East, on objects associated with it.

Varro (*NH* 36. 135) provides evidence of the hardness of different kinds of black and white stone and the various means of cutting marble.

Pliny also (36. 146) makes a brief reference to Sotacus[71] who records five different kinds of haematite and believes that amber flows from crags (37. 35). He also names two varieties of gemstones called 'cerauniae' (37. 135).

---

[64] See, for example, 33. 113f. (the discovery of cinnabar) from sections 58–9. Zehnacker, *NH* 33, pp. 200f. compares the Greek and Latin texts in parallel.
[65] See above, n. 16.
[66] An otherwise unknown historian from Alexandria.
[67] See above, n. 34. [68] See above, n. 62.
[69] See Ch. 14, s.v. 'calcite'. [70] Ch. 14, s.v.
[71] Sotacus, who lived in the early 3rd cent. BC, the author of a book on stones. Pliny (*NH* 36. 146), when recording his statement that there are five kinds of haematite, refers to him as one of the earliest authorities ('e vetustissimis auctoribus').

Catullus (*NH* 36. 154) refers to the use of pumice in smoothing the edges of the book roll (Catull. 1. 2; 22. 8).

Cato (*NH* 36. 174) mentions lime, which is best prepared from white limestone. In the case of gypsum,[72] Pliny (36. 182) draws directly on Theophrastus (*Lap.* 64 and 69), who does not distinguish gypsum from dehydrated gypsum.

Juba asserts that rock-crystal (*NH* 37. 24) is found on an island called Necron, in the Red sea facing Arabia. Cornelius Bocchus[73] (ibid.) also mentions Portugal in the Ammaeensian mountains.

Xenocrates[74] provides further evidence (37. 27) and, later, refers to peridot (37. 108).

4.17.3 Precious Stones

In book 37 of the *Natural History* Theophrastus and Metrodorus[75] of Scepsis provide information about gemstones (61 ff.). Among other authorities cited by Pliny are: Apion[76] (37. 75), Demostratus[77] (37. 85), Sotacus[78] (37. 86), Zenothemis[79] (37. 87), Callistratus[80] (37. 94 f.), Bocchus[81] (37. 97), references in the plays of both Menander[82] and Philemon[83] (37. 106), and Democritus[84] (37. 146 and 149).

Pliny concludes the *Natural History* (37. 139 ff.) with a long list of less common precious stones, many of which are not able to be identified with certainty.

---

[72] See Ch. 14, s.v. 'calcium sulphate'.
[73] Cf. above n. 60. The works of Bocchus are lost.
[74] See above, n. 37.
[75] Metrodorus, a philosopher and statesman, under Mithridates Eupator, king of Pontus (120–63 BC), invented a system of *aides-mémoire* (*NH* 7.90).
[76] A scholar from Alexandria, who later lived in Rome during the principates of Tiberius and Claudius. He wrote on Egypt (cf. Joseph. *Ap*).
[77] A historian and Roman senator of the early 1st cent. AD.
[78] See above, n. 71.
[79] Known only as a writer of a Periplus. Cf. *NH* 37. 34 apropos the sardonyx here discussed.
[80] Callistratus refers to the stones generically described as 'carbunculi'. Cf. Theophr. *Lap*. 33; Eichholz, *Lap*. ad loc.
[81] See above, n. 60.   [82] *c*.342–290 BC.   [83] *c*.361–262 BC.
[84] An authority on gems, otherwise unknown.

Finally, the counterfeiting of gemstones[85] was widely practised in the ancient world and Bolos of Mende (37. 69) wrote on this subject as well as an account of *mirabilia*. Some of his works were wrongly attributed to Democritus.

14.17.4 Pigments

Pliny (34. 30 ff.) gives an account of the main pigments[86] used in the ancient world. These include cinnabar, minium, yellow ochre (oxides and hydroxides of iron), azurite (copper carbonate), realgar and orpiment (arsenic sulphides), malachite, calcium carbonate, magnesite, mixtures of lead and iron oxide. His main sources are Theophrastus, Vitruvius, and Celsus.

4.18 RESIN

For his discussion of resin, Pliny (37. 30) is indebted to Theophrastus.[87] Other authorities give varying explanations of the nature and origin of this resin, including Chares, Philemon (who believes it is a mineral), Sudines, Metrodorus,[88] Pytheas, Timaeus, Theochrestus, and Xenocrates.[89] The tragic poets Aeschylus, Sophocles, and Euripides,[90] give a highly fanciful explanation of the origin of amber. Pliny (37. 31) implies that Greek accounts are false ('occasio vanitatis'). A unique observation, however, by King Archelaus[91] of Cappadocia (37. 46) provided a vital clue as to its source, namely the resin of the pine-tree.

---

[85] The staining of rock-crystal in imitation of precious stones is well attested. See *NH* 37. 79, where Pliny states that the Indians produce counterfeit beryls in this manner. There were good forgeries of this precious stone in the ancient world. See further E. H. Warmington, *The Commerce Between the Roman Empire and India*² (Cambridge, 1974), 251.

[86] For a detailed discussion of the pigments see Ch. 15 below, under the names of individual compounds.

[87] *HP* 9. 18. 2. Pliny may well have seen amber while serving in Germany.

[88] See above, n. 75.

[89] Xenocrates of Aphrodisias was a physician—not the artist referred to in n. 37 above.

[90] *Hipp.* 741.

[91] The nominee of Mark Antony who, in 36 BC, received from him the kingdom of Cappadocia.

4.19 CONCLUSION

Pliny is often criticized for his gullibility in accepting the information provided by his sources, but, closer examination reveals that this is a sweeping and unjustified assessment. He records reservations on a number of occasions, and, as will be observed, exercises a degree of critical judgement[92] in assessing some of the more extravagant assertions of other writers.

---

[92] For an assessment of his attitude to research see Ch. 6 below.

5

Mirabilia

Words commonly used to categorize wonders, curiosities, monstrosities, extravagant fictions, the seemingly inexplicable, and things beyond normal, everyday experience include *mirabilia*, *miracula*, *prodigia*, and *portenta* (explained as 'praeter naturam hominum pecudumque').[1] Similarly, *mirum* and *mirabilis* are applied to the unusual. Such *mirabilia* are to be found in the shadowy realms of 'sub-science'.

Naturally occurring and man-made oddities, listed throughout the *Natural History*, are part of a long tradition which has its beginnings in Homer's *Odyssey*, where Odysseus encounters the Sirens,[2] Polyphemus,[3] Scylla and Charybdis,[4] and other legendary creatures. By the Hellenistic period,[5] writings specifically devoted to *mirabilia* were widespread, although many are known only from subsequently lost works: references and quotations preserved in the *Natural History* are an important source for their study.

Pliny's indexes to the *Natural History* include many authors of *mirabilia* listed in detail by Beagon,[6] among them Alexander Polyhistor (9. 115 and 13. 119),[7] Isigonus and Nymphodorus

---

[1] Cic. *ND* 2. 5. 14: this refers to monstrous births but may apply more generally. [2] *Od.* 12. 39 and 52.
[3] *Od.* 1. 70, 9. 371 ff. (Πολύφημος). *Od.* 1. 63 ff., 106 ff., 397, and 416. Cf. Hes. *Theog.* 144; Thuc. 6. 2 (Κύκλωψ).
[4] *Od.* 12. 255 (Σκύλλα), a legendary monster with six heads living in a cave opposite Charybdis, and *Od.* 12. 104 ff. and 441, 23. 327 (Χάρυβδις), a whirlpool in a narrow channel of the sea (the Straits of Messina).
[5] See, generally, W. W. Tarn and G. T. Griffith, *Hellenistic Civilization* 3 (London, 1966), 239 ff.   [6] Cf. Beagon, *Roman Nature*, 8.
[7] A frequent source for *mirabilia* in no less than twelve books of the *NH* (2–7, 9, 12–13, 16, 36–7).

(7. 16),[8] and Callimachus (31. 9).[9] In addition, numerous references to Licinius Mucianus[10] give an idea of the comprehensive coverage of such *mirabilia*. The tradition of cataloguing the bizarre continued into the second century AD.[11]

Curiosity for the unusual often underpins Pliny's 'science' and was one of the factors which led him to observe the eruption of Vesuvius at close quarters.[12] During the early principate, the Roman imagination was stimulated by exploration, military campaigns in far-off places, and journeys undertaken to expand commercial contacts. It would appear, however, that increased knowledge of foreign places and peoples may have enhanced rather than lessened reports of *mirabilia*.

Pliny recalls (8. 42) the saying of Aristotle (*HA* 8. 28, 606^{b}20) that Libya (Africa) is always producing some novelty,[13] 'vulgare Graeciae dictum semper aliquid novi Africam adferre'. Africa, it was believed, produced unusual creatures such as Goat-Pans and Satyrs (5. 7).

Not surprisingly, India and parts of Ethiopia (7. 21), at that time largely unexplored regions, are a major source of wonders ('miraculis scatent'). Tribes exhibit physical abnormalities and may possess intrinsic powers that are beyond those found in normal human beings. However, the incongruity of some of the stories brought back to Rome led Pliny to comment (11. 6), 'The more I observe Nature, the less prone I am to consider any statement about her to be impossible' ('nam mihi contuenti semper suasit rerum natura nihil incredibile existimare de ea').

---

[8] A. Giannini, *Paradoxographorum Graecorum Reliquiae* (Milan, 1966), fr. 15 (Isigonus) and fr. 8 (Nymphodorus). Both provide stories of African tribes with special powers.

[9] He records the curative powers of a Phrygian river in treating bladder stones. See further R. König, J. Hopp, and W. Glockner, *Plinius: Naturkunde, Buch 31* (Zurich, 1994), pp. 91 f., comm. ad loc.

[10] Mucianus was consul on three occasions and, as a provincial governor in the East, had the opportunity to witness what he considered to be *mirabilia*. [11] Cf. Beagon, *Roman Nature*, 9 n. 25.

[12] The Younger Pliny, *Letters* 6. 16. 4 ff. He refers to the eruption as a *miraculum*. 'poscit soleas. ascendit locum ex quo maxime miraculum illud conspici poterat.'

[13] See E. Bianchi, 'Teratologia e geographia', *Acme*, 34 (1981), 227–49.

Yet, elsewhere (5. 4) he expresses his unwillingness to accept Greek 'tall stories' ('fabulosa') that he considers to be an insult to the intelligence, 'minus profecto mirentur portentosa Graeciae mendacia'. His observations about human nature are often perspicacious, as when he writes, apropos of the knights (5. 12), 'Persons of rank, although unwilling to track down the truth, are not ashamed to tell falsehoods because they cannot bear to admit their ignorance.' But he adds, 'credulity never more readily falls flat on its face than when an authority of weight supports false assertions.' ('quia dignitates, cum indagare vera pigeat, ignorantiae pudore mentiri non piget, haut alio fidei proniore lapsu quam ubi falsae rei gravis auctor existit.')

Many *mirabilia* are found in the fields of zoology, especially in regard to man (the creature highest in the order of creation), botany, petrology, and mineralogy.

Book 7 is full of alleged examples of human freaks[14] and unusual anatomy, let alone antisocial behaviour—for example, cannibalism (7. 9): 'I have drawn attention to the fact that some Scythian tribes—indeed a percentage of them—feed on human bodies' ('esse Scytharum genera, et quidem plura, quae corporibus humanis vescerentur indicavimus'). He also quotes Isigonus as affirming (7. 12) that the Scythian cannibals (Anthropophagi) 'drink out of human skulls and use the scalps, with the hair attached, as napkins to cover their chests' ('ossibus humanorum capitum bibere cutibusque cum capillo pro mantelibus ante pectora').

The Arimaspi, like the Cyclops,[15] have one eye in the middle of their forehead (7. 10): 'uno oculo in fronte media insignes.' Some women have a double pupil in one eye and the image of a horse in the other (7. 17): 'notas tradit in altero oculo geminam pupillam in altero equi effigiem.'

There are tribes without necks, writes Ctesias, resulting in the eyes being displaced to the shoulders (7. 23): 'sine cervice oculos in umeris habentes.' Yet others (7. 25), among the

---

[14] Cf. *NH* 7. 32: 'These and similar varieties of the human race have been made by the ingenuity of Nature, as toys for herself and marvels for ourselves' ('haec atque talia ex hominum genere ludibria sibi, nobis miracula, ingeniosa fecit natura').   [15] See above, n. 3.

Indian Nomads, have nostrils like snakes, according to Megasthenes: 'gentem inter Nomadas Indos narium loco foramina tantum habentem anguium modo.' Near the source of the Ganga, live the Astomi, who, as their name indicates, have no mouth: they live on odours and scents which they inhale through their nostrils: 'circa fontem Gangis Astomorum gentem sine ore...halitu tantum viventem et odore quem naribus trahant.'

Legs and feet may be deformed and thong-like (5. 46), turned backwards (7. 11), or with eight toes on each foot (7. 23). The Monocoli, states Ctesias, have one leg and move forward in hops (ibid.). They are also called the Umbrella-foot tribe because in hotter weather they lie on their backs on the ground and protect themselves with the shadow of their feet: 'idem hominum genus qui Monocoli vocentur singulis cruribus mirae pernicitatis ad saltum, eosdem Sciapodas vocari, quod in maiori aestu humi iacentes resupini umbra se pedum protegant.'

Eudoxus claims that there are cavemen with feet eighteen inches long (7. 24). Overall heights vary from the very tall—Crates records men twelve feet tall in Ethiopia (7. 31)—to the abnormally short, as the Pygmies, who are about twenty-seven inches in height. It is fascinating that Pygmies are depicted in Nilotic paintings from Pompeii.[16]

Examples of longevity include centenarians and men who have lived 140, or even 200 years. Pliny (7. 30) records an interesting observation of Artemidorus about Sri Lankans who live to an advanced age, yet retain their physical powers undiminished: 'in Taprobane insula longissimam vitam sine ullo corporis languore traduci.' The Machlyes[17] perform the function of either sex alternately (7. 15). Aristotle adds that they have the left breast of a man and the right of a woman. Some Indians have union with wild animals and produce offspring of mixed race (7. 30). Other strange races have men with a dog's head and tail.

---

[16] A. Maiuri, *Roman Painting* (London, 1957); A. Stenico, *Roman and Etruscan Painting* (London, 1963); and generally, G. Isager, *Pliny on Art and Society* (London, 1991).

[17] The name of a legendary race which may derive from the Greek μαχλός, usually describing a lewd or lustful woman, or, very occasionally, man.

Pliny continues his account of the human race with details of unusual, multiple (7. 33), or premature births (7. 39)—including superfetation (7. 48), when twelve babies were stillborn at the same time.

There are also many instances of outstanding sight (7. 85), endurance (7. 87), strength (7. 81), and intellect (7. 91 ff.). A fascinating example was the claim by a man called Strabo that he could see, unaided, from Lilybaeum to Carthage. This can be tested. The results are interesting, since calculations[18] show that, even if such long sight had been physically possible, to offset the earth's curvature, he would have had to have stood at the top of Mt. Etna!

Examples of *mirabilia* from the animal kingdom, or from mythology, include a snake some 120 feet long seen in the river Bagradas (8. 37), a hippocentaur brought from Egypt to Rome (7. 35), and other legendary creatures such as the mantichora (8. 75) and werewolves (8. 80 ff.). Pliny unequivocally dismisses, as false, stories of men changed into wolves and restored to human form. As he explains, belief in such creatures would entail acceptance also of all other fantasies: 'homines in lupos verti rursusque restitui sibi falsum esse confidenter existimare debemus aut credere omnia quae fabulosa tot saeculis conperimus'.

Certain races possess unusual powers. For example, the Ophiogenes, are able, states Crates of Pergamum, to cure snakebite by touch and can draw off poison from the body by placing their hand on it (7. 13). Agatharcides records that the Psylli, an African tribe, have a poison in their bodies which is deadly to snakes (7. 14). Others have the evil eye and practise sorcery (7. 16).

The sea produces a rich harvest of bizarre creatures (*monstra*). Pliny (9. 9 f.) states that some knights had seen Nereids and Tritons near Lisbon and in the Gulf of Cadiz and refers (9. 11) to a creature washed ashore at Cadiz which had 24 feet of tail-end between its two fins and some hundred and twenty teeth. Marcus Scaurus brought to Rome the skeleton of a monster from which Andromeda is said to have been rescued

---

[18] See appendix to Ch. 12 below.

at Jaffa. This, and other marvels (*miracula*), were exhibited during his aedileship, in 58 BC. The skeleton was 40 feet long and the height of its ribs surpassed that of an Indian elephant: its spine was 18 inches thick. A giant octopus, which preyed on fishponds at Carteia (9. 92 f.), was eventually killed and preserved. Mucianus also saw a nauplius—a shell with a keel, not unlike a cuttlefish—in the Dardanelles (9. 94).

Cornelius Valerianus (10. 5) reported the appearance of a Phoenix[19] in Egypt (AD 36) that was later brought to Rome in 47.

Not a few portents connected with trees are found, such as those in the work of the Greek writer Aristander, and in the *Commentarii* of Gaius Epidius (17. 243), including trees that appeared to talk.[20] Pliny (17. 21 ff.) records exceptionally large trees, like the fig-tree beneath which squadrons of cavalry could shelter. Plane-trees (12. 9) are also particularly noteworthy, among them the one that grew in the Academy at Athens.[21] Yet another, which Licinius Mucianus considered a marvel (*miraculum*), was seen by him in Lycia. In this he held a banquet with eighteen of his retinue. Perhaps one of the best known of all was the plane-tree which had associations with the emperor Caligula (12. 10 ff.). 'When on a visit to an estate at Velitrae, Caligula was so impressed by the "flooring" of a single plane-tree and the horizontal branches which served as seats, that he too held a banquet in this tree. He constituted a considerable portion of the shadow[22]...this dining room he called his "nest".' ('aliud exemplum Gai principis in Veliterno rure mirati unius tabulata laxeque ramorum trabibus scamna patula, et in ea epulati, cum ipse pars esset umbrae...quam cenam appellavit ille nidum.') Elsewhere (13. 119), Alexander

---

[19] The Phoenix is said to live five hundred years and, from its ashes, a young Phoenix arises, cf. Ov. *Met.* 15. 393; Stat. *Silv.* 2. 4. 36; Sen. *Ep.* 42. 1; Tac. *Ann.* 6. 28. Pliny sensibly observes that, although 'this fact is attested by state records, nobody would doubt that this Phoenix was a fabrication' ('in comitio propositus, quod actis testatum est, sed quem falsum esse nemo dubitaret').

[20] H. Rackham (trans.) *Pliny the Elder: Natural History*, v (Loeb: London and Cambridge, Mass., 1950), p. 168 note *a*, suggests that noisy flocks of starlings roosting in the trees produced this impression.

[21] Cf. Pl. *Phdr.* 229a for a plane-tree by the Ilissus.

[22] A reference to Caligula's great height and obesity!

Polyhistor mentions a tree called the lion-tree, the timber of which, he says, was used to build the *Argo*, and Graecinus[23] claims that some vines lived for six hundred years.

In the fields of petrology and mineralogy, *mirabilia*, or *miracula*, are seen to impinge on the world of 'science'. Pliny (36. 125) follows Papirius Fabianus[24] in believing that marble and or minerals regenerate themselves, when they have been taken out of the earth. He also writes (36. 14) that a statue of Silenus was found in a block of marble, when split open. The origin of such a story is not difficult to imagine: it is little more than a fanciful conceit, based on the technique of carving statues from blocks of marble.

Among the mineralogical examples from Mucianus' book of *Mirabilia*, the sarcophagus stone's power of petrification[25] and its phenomenon of efflorescence are both included (36. 131). In Cyzicus (36. 99), the so-called 'runaway stone' ('lapis fugitivus'), which had been the *Argo*'s anchor, was fastened with a lead, because it had frequently wandered away from the President's House.

In his discussion of magnetite (36. 126 ff.) and its mysterious power of attracting iron, Pliny finds the concept of magnetism[26] difficult to grasp. 'For what is more strange than this stone? In what field has Nature displayed a greater wilfulness?' ('quid enim mirabilius aut qua in parte naturae maior improbitas?').

Finally, Pliny uses the term 'wonders' (*miracula*) to describe certain man-made structures and buildings.[27] So he refers (36. 82) to the pyramids ('pyramidum miracula'), to labyrinths (36. 84) as an abnormal achievement ('portentissimum humani...opus'), to the temple of Diana at Ephesus (36. 96 f.), with its architraves ('summa miraculi epistylia') and ornaments ('nihil ad specimen naturae pertinentia') as a wonder, and goes on to list the many amazing things to be seen in the city of Rome (36. 101 ff.).

---

[23] On Graecinus as a source, see further K. D. White, *Roman Farming* (London, 1970), 25 f.
[24] Described here by Pliny as an outstanding natural scientist ('rerum peritissimus'). [25] See Ch. 14 below, s.v. 'sarcophagus stone'.
[26] See Ch. 12. 3. [27] Isager, *Pliny on Art*, 199–205.

Not everyone approved of the widespread interest in *mirabilia* found throughout the ancient world. Seneca (*De brevitate vitae* 13. 3. 9 and 14. 1), for example, considers the recording of *mirabilia* without obvious moral purpose to be unworthy of a serious writer. Similarly, Plutarch (*On Curiosity* 517 f.), criticizes those who—in our society—have an unhealthy fascination for the abnormal, the sensational, and X-rated!

However, this interest in the unusual and sometimes bizarre, may, in a sense, constitute a logical, preliminary stage on the way to genuine 'scientific' curiosity, as has already been observed in Pliny's description of the sarcophagus stone and its strange properties.

6

Pliny and Research

Aristotle, in his *On Parts of Animals*, sets out some of his prerequisites for research. Although his observations refer to zoology, they also involve general principles in the context of which it is useful to examine Pliny's attitude to research. He writes (*PA* 639ª): 'There are, as it seems, two ways in which a person may be competent in respect of any study or investigation, whether it be a noble one, or humble: he may either have what can be called a scientific knowledge of the subject; or he may have what is roughly described as an educated man's competence and, therefore, be able to judge correctly which parts of any exposition are satisfactory and which are not. That, in fact, is the sort of person we take to be the "man of general education", whose education consists in the ability to do this.' Pliny (pref. 14) likewise claims that the topics included in the *Natural History* are those relevant to an all-round education.

Aristotle (ibid.) continues: 'In this case, however, we expect to find in the one individual the ability to judge almost all subjects, whereas, in the other case, the ability is confined to some special science; for, of course, it is possible to possess this ability for a limited field only. Hence it is clear that in the investigation of Nature, or Natural Science (φύσις), as in every other, there must first of all be certain defined rules by which the acceptability of the method of exposition may be tested, apart from whether the statements made represent the truth, or do not.'

Aristotle's *Meteorologica* treats astronomy, geology, seismology, and meteorology and is a work typical of a stage when natural sciences had not become fully differentiated from philosophy, but embraced a wide range of 'scientific' knowledge.

Although there were no comprehensive Greek works of reference, the appreciation of the problems to be solved and the ideas of matter, change, elements, and atoms, mostly despised by Plato, are abidingly positive contributions of Greek philosophy to science.

Pliny (8. 44) includes the oft-repeated account of the wide-ranging zoological research project, allegedly commissioned by Alexander the Great, the authenticity of which has been challenged.[1]

He delegated research in this field to Aristotle, a man of supreme authority in every branch of science. Orders were given to some thousands of people throughout the whole of Asia Minor and Greece—people who made their living by hunting, catching birds, and fishing, as well as those in charge of warrens, herds, apiaries, fish-ponds, and aviaries: they were to see that he [Aristotle] was informed about any creature born in any region. The result of his enquiries from such people led to the publication, in nearly fifty volumes, of his famous work *On Animals*. I ask my readers to be favourably disposed to my presentation of this information—together with facts of which Aristotle was unaware—while making their brief excursion, under my direction, into all the works of Nature.

---

[1] R. French, *Ancient Natural History*, p. 105, discusses this passage. The story has been rather sneered at by historians, for whom the conjunction of the great philosopher and the great king looks like an invention. Yet there are a number of reasons why we might accept some form of the story. Aristotle had, after all, been Alexander's tutor, and might well have taught him something about his beliefs about the natural world. He had also addressed a tract *On Kingship* to him and came to suffer banishment from Athens for his Macedonian sympathies; he also recommended to artists that they should paint Alexander's battles as being of lasting worth, so we may imagine that there was a special relationship between the two men. It would also have been natural for Alexander to take an interest in the nature of the countries he controlled. If nothing else, a general planning the logistics of an advance would need to know the nature of the country, its internal distances, and its peculiarities; a conqueror hoping to build a lasting empire would undertake a listing of resources, marvels worthy of *historiae*, and tales suitable for his historians. Victorious generals often enough sent strange animals and even trees back for the triumph, and although it would be rash to assert that any such specimen reached Aristotle, it is entirely possible that Alexander's lines of communication and administration, which must have amounted to a sort of civil service, could have handled some sort of questionnaire from Aristotle, on the lines here suggested by Pliny. See further W. Jaeger, *Aristotle: Fundamentals of the History of his Development*, trans. R. Robinson (Oxford, 1962). French, however, observes that the attribution of the *De mundo* to Aristotle, also addressed to Alexander the Great, is an invention.

Alexandro Magno rege inflammato cupidine animalium naturas noscendi delegataque hac commentatione Aristoteli, summo in omni doctrina viro, aliquot milia hominum in totius Asiae Graeciaeque tractu parere ei iussa, omnium quos venatus, aucupia piscatusque alebant quibusque vivaria, armenta, alvearia, piscinae, aviaria in cura erant, ne quid usquam genitum ignoraretur ab eo. quos percunctando quinquaginta ferme volumina illa praeclara de animalibus condidit. quae a me collecta in artum cum iis quae ignoraverat quaeso ut legentes boni consulant, in universis rerum naturae operibus medioque clarissimi regum omnium desiderio cura nostra breviter peregrinantes.

During the Hellenistic period there was a widening of learning. Alexandrian literature reflects a growing interest in aetiology, antiquarianism, and, above all, the collection of exotic and unusual information.[2] These trends are seen in a variety of works—among them Aratus'[3] *Phaenomena*, an account, in hexameters, of Eudoxus' old *Star Catalogue*, Nicander's[4] *Theriaca*, a scientific treatise on poisons and antidotes, and other definitive works on agriculture and bee-keeping. The success of the *Phaenomena*, a dry astronomical work, is puzzling, but was probably due, as has been suggested, to its illustration of the Stoic doctrine of Providence drawn from the utility of the stars to sailors and to farmers[5] alike. Other poems covered a variety of topics, including astronomy, geography, and fishing.

Research was conceived of as an imitative rather than a creative activity, and so scholarly works were little more than the results of reading and extensive note-taking, without critical, or scientific evaluation of the material excerpted.

In Rome there was an early interest in the compilation of facts. The Elder Cato[6] (234–149 BC) and Varro[7]

---

[2] W. W. Tarn and G. T. Griffith, *Hellenistic Civilization* 3(London, 1966), 268 ff.

[3] Aratus, of Soli in Cilicia (*fl. c.*270 BC), spent the latter part of his life at the court of Antigonus Gonatas, king of Macedonia. Cicero translated the *Phaenomena* into Latin.

[4] Nicander was born at Claros, near Colophon in Ionia (*fl. c.*185–135 BC). In addition to the *Theriaca* his *Alexipharmaca* is also extant.

[5] Tarn and Griffith, *Hellenistic Civilization*, 274.

[6] Cato wrote his *De agri cultura* in the middle of the 2nd cent. BC—the first work on agriculture in the Latin language. Columella (1. 1. 12) writes that he taught agriculture to speak Latin. See further K. D. White, *Roman Farming* (London, 1970), 19 ff.

[7] Seventy-four works are attributed to Varro, whose *De re rustica* was published in 37 BC. See White, *Roman Farming* 22 ff.

(116–27 BC) may be said to have domesticated the genre of the encyclopaedia.

Lucretius (*c*.94–*c*.55 BC) set out to expound the system of the Greek philosopher Epicurus and achieved a harmonious interplay of poetry and instruction never subsequently surpassed. Following Epicurus,[8] he offers a prescription for happiness, in particular asserting that the mental fears and forebodings that superstition (*religio* is equated with superstition, not with religion) engenders can be dispelled by a knowledge of the laws of 'physics'. These leave no place for divine intervention in life, or for human survival after death. So Lucretius writes (*DRN* 3. 866 ff.),

Rest assured that we have nothing to fear in death. One who no longer is, cannot suffer, or differ in any way from one who has never been born, when once this mortal life has been usurped by death the immortal.[9]

> scire licet nobis nil esse in morte timendum,
> nec miserum fieri qui non est posse, neque hilum
> differre an nullo fuerit iam tempore natus,
> mortalem vitam mors cum immortalis ademit.

So Pliny (7. 188) makes a similar assertion: 'There is some confusion concerning the spirits of the departed after burial. All men are in the same state from their last day forward as they were before their first day, and neither body nor mind has any more sensation after death than it had before birth.' ('post sepulturam variae manium ambages. omnibus a supremo die eadem quae ante primum, nec magis a morte sensus ullus aut corpori aut animae quam ante natalem.')

Lucretius exhibits the recognizable marks of a scientific spirit, namely a close and critical observation of the natural world, an unflagging insistence on the universality of natural

---

[8] Epicurus of Samos (341–270 BC) travelled in Lesbos and Troas and settled at Athens (307/6). The Epicureans believed that philosophy had the practical purpose of securing a happy life: the ideal state was *ataraxia* 'freedom from disturbance'. In order to attain this state they recommended, among other things, the avoidance of deep emotional attachments and abstention from the competitive world of politics. They also believed that the soul, composed of atoms like the body, died with it and that the gods did not interfere in the physical world, which owed its origin to natural causes.

[9] Translations of *DRN* here and below from R. E. Latham *Lucretius: On the Nature of the Universe* (London, 1997).

causes, and an overmastering (and, in the event overoptimistic) faith in reasoned arguments. He also had a concern for clear and systematic exposition, matched by a capacity to achieve this goal. His work *On the Nature of the Universe* thus paradoxically combines the passionate intensity of a proselytizing tract with qualities which place it among the forerunners of modern scientific publications.

Pliny and Lucretius share a number of themes.[10] Both have two main aims, namely (1) to explain the Universe and its phenomena in rational terms and (2) to free the minds of men from superstitious fear through a greater understanding of the world. Lucretius (1. 945 ff.) explains his aims as follows:

My object has been to engage your mind with my sweet-sounding verses... in the hope that I may be able to hold your attention, while you gain insight into the nature of the universe and how it has been shaped and framed.

volui tibi suaviloquenti
carmine Pierio rationem exponere nostram
. . . . .
si tibi forte animum tali ratione tenere
versibus in nostris possem, dum perspicis omnem
naturam rerum, qua constet compta figura.

With regard to his second aim, he writes (*DRN* 1. 146–58):

This dread and darkness of the mind cannot be dispelled by the sun's beams, the shining shafts of day, but only by an understanding of the outward form and inner workings of Nature. In tackling this theme, our starting point will be this principle: *Nothing can ever be created by divine power out of nothing.*[11] The reason why all mortals are so gripped by fear is that they see all sorts of things happening on earth and in the sky with no discernible cause, and these they attribute to the will of a god. Accordingly, when we have seen that nothing can be created out of nothing, we shall then have a clearer picture of the path ahead, the problem of how things are created and occasioned without the aid of the gods.

hunc igitur terrorem animi tenebrasque necessest
non radii solis nec lucida tela diei
discutiant, sed naturae species ratioque.
principium cuius hinc nobis exordia sumet,

---

[10] Healy, *Pliny, NH: A Selection*, p. xvii.
[11] Cf. the Law of the indestructibility of matter.

> nullam rem e nilo gigni divinitus umquam. 150
> quippe ita formido mortalis continet omnis,
> quod multa in terris fieri caeloque tuentur,
> quorum operum causas nulla ratione videre
> possunt ac fieri divino numine rentur.
> quas ob res ubi viderimus nil posse creari 155
> de nilo, tum quod sequimur iam rectius inde
> perspiciemus, et unde queat res quaeque creari,
> et quo quaeque modo fiant opera sine divum.

But Lucretius' and Pliny's treatments of the subject differ. Lucretius writes imaginatively as a poet and Epicurean, while Pliny reveals himself as a natural scientist with Stoic inclinations. He equates God with Nature and as a powerful, underlying theme criticizes man for his abuse of earth's resources. The most significant difference, however, is the wider scope of the *Natural History* and the fact that Pliny provides an in-depth commentary on the mores and aspirations of Romans of his time.

At the beginning of his discussion of vines and viticulture (14. 3), Pliny draws attention to the decline in research and knowledge of earlier achievements, by the beginning of the early imperial period. This is somewhat puzzling, since the establishment of the *pax Romana*, the improvement of living standards, and the civilizing influence of Roman culture should have facilitated physical communication between people and the exchange of knowledge and, as a spin-off, encouraged research. This did not happen. Other factors neutralized the benefits derived from the spread of empire, as Pliny (14. 3 f.) seeks to explain at the beginning of his account of vines and viticulture. 'Yet in all conscience, people who know much of what has been published by earlier writers cannot be found. The research of men of former times was even more productive, or their industry was more successful, a thousand years ago at the beginning of literature, when Hesiod[12] began to expound his principles for farmers. His research was followed by several writers, and this has resulted in more work for us, since now we have to investigate not only subsequent

---

[12] See generally M. L. West (ed.), *Hesiod: Works and Days*[2] (Oxford, 1997).

discoveries, but also those made by earlier authorities, because men's laziness has brought about a complete destruction of records. What cause for this shortcoming would there be other than the state of world affairs generally? The thing is that other customs have crept in; men's minds are preoccupied with other matters and the only arts practised are those of greed.'

Romans under the Republic had their talents circumscribed and were, in consequence, obliged to be self-sufficient;[13] moreover, in this context they set store by the arts. The extension of Rome's boundaries, however, afforded more opportunities for amassing a fortune and Pliny rightly observes that the materialistic society under the early empire, whose lifestyle was moulded by an acquisitive attitude, was easily distracted from higher pursuits. The new criterion of success had become wealth. He summarizes the situation (14. 5): 'The true prizes of life went to rack and ruin and all the arts that were called 'liberal', from liberty, expressing the greatest good, became quite the opposite' ('pessum iere vitae pretia, omnesque a maximo bono liberales dictae artes in contrarium cecidere').

The volume of research sources and authorities consulted by Pliny mirrors his diligence and commitment to research. He writes (pref. 17) that the *Natural History* contains 20,000 facts, from 2,000 works by 100 chosen authors. ('viginti milia rerum dignarum cura...lectione voluminum circiter duorum milium...ex exquisitis auctoribus centum inclusimus'.) These figures, however, represent a conservative estimate, since no fewer than 146 Roman and 327 foreign authors are quoted!

He shows a commendable attitude towards his sources whom he carefully acknowledges (pref. 21 ff.):

You will count as proof of my professionalism the fact that I have prefaced these books with the names of my authorities. In my opinion such acknowledgement of those who have contributed to one's success—unlike the practice of most of the authors I have mentioned—is a not ungracious gesture and abounds with honourable modesty. For, you ought to know that when I compared authorities, I found that writers of bygone times had been copied

---

[13] Pliny emphasizes this point especially in the fields of horticulture and medicine.

by the most reliable and modern authors, word for word, without acknowledgement.

argumentum huius stomachi mei habebis quod his voluminibus auctorum nomina praetexui. est enim benignum (ut arbitror) et plenum ingenui pudoris fateri per quos profeceris, non ut plerique ex his quos attigi fecerunt. scito enim conferentem auctores me deprehendisse a iuratissimis et proximis veteres transcriptos ad verbum neque nominatos.

Some of the authors, however, as for example King Juba, appear to have been included *honoris causa*.

To recapitulate: in the *Natural History*, Pliny rejects the liberal arts as an adequate framework for human knowledge: these subjects had become specialized and carried a high degree of abstraction. Although accessible to scholars, they were difficult to grasp for ordinary people with practical, rather than theoretical needs. Pliny (pref. 6) clearly defines the readership at whom he is aiming, namely 'the masses, the horde of farmers and artisans, and, finally, those who have time to devote to these studies'. Accordingly, he concentrates on topics of more immediate importance for human life in general. Disciplines such as arithmetic, geometry, and harmonics are discarded in favour of the study of topography—where man lives and works—and of animals, plants, food crops, minerals, and metals.

Pliny was the first to assimilate and survey all these topics within the compass of a single work. The *Natural History*, however, is not, as has been suggested, a vade-mecum for provincial governors, but essentially for the man in the street.

As to its form and presentation, there had been, in Rome, an early interest in the compilation of facts, and Pliny (pref. 17) records the assertion of Domitius Piso to the effect that 'works of reference not books were needed' ('thesauros oportet esse non libros').

7

Language and Style

A discussion of Pliny's use of language and of his style[1] may seem an unwarranted intrusion in a review of his contribution to early imperial science and technology. However, the problems that Pliny, like Lucretius and others, encountered because of the paucity of technical terms in Latin, and the style in which he presents his subject matter, have a critical bearing on the content and character of the *Natural History*. The importance also of nomenclature cannot be overstressed, as Fleury[2] writes in another context.

The present reassessment may help scientific colleagues with no specialist knowledge of Silver Latin to avoid the pitfall of subscribing to long-discredited views about Pliny's Latinity![3] Apart from other considerations, his style is essentially the product of the age in which he lived and not an idiosyncratic creation.

In a recent, wide-ranging treatment of the whole spectrum of attitudes to natural history in the ancient world, French includes, among a number of sweeping and not always appropriate generalizations: 'What we approve in Pliny, even when

---

[1] For an extended treatment see Healy, 'Language and Style', 1–24. This chapter contains many words and phrases found elsewhere in the text. Such duplication is deliberate, to provide the reader with easy reference in respect of this very important aspect of Pliny's work.

[2] *La Mécanique de Vitruve* (Caen, 1993), 328: 'Les dictionnaires et encyclopédies de l'Antiquité utilisés aujourd'hui contiennent un grand nombre d'erreurs sur l'identification des objects désinés par des mots techniques, particulièrement dans le domaine de la mécanique...Une première tâche pourrait donc être de proposer, à partir du texte de Vitruve, mais également à partir des autres mentions de ces mots dans l'ensemble de la littérature latine, de nouvelles définitions, fondées sur une connaissance précise des objects et de leur fonctionnement.'

[3] See also generally Healy, 'Language and Style'.

regretting his scientific credulity and appalling misuse of his own language, is that he appears, because he adopted so much from the Greeks, to fall into a history of Western thought that we like to derive from the Greeks.'[4] Such a conclusion is symptomatic of received attitudes to Pliny, which began with Leoniceno's[5] attempts to discredit Pliny's contribution to 'science'.

Errors in transmission of the text of the *Natural History*, from an early period, initiated many problems, which, in turn, have been compounded by the strenuous efforts of scientifically disadvantaged textual critics and commentators. Similarly, recent critics' unjustified and less than constructive observations about Pliny's language and style merely continue the prejudices of early scholars. According to Norden's[6] much quoted opinion, 'Sein Werk gehort stilistisch betractet, zu den schlechsten die wir haben'. Goodyear similarly subscribed to this assessment: 'Pliny is one of the prodigies of Latin literature, boundlessly energetic and catastrophically indiscriminate, wide-ranging and narrow-minded, a pedant who wanted to be a popularizer, a sceptic infected by traditional sentiment, and an aspirant to style who can hardly frame a coherent sentence.' Perhaps subconsciously parodying the style of his literary *bête noire*, Goodyear continued. 'Students of Latin language and style neglect Pliny at their peril. Here, better than in most places, we may see the contortions and obscurities, the odd combinations of preciosity and boldness, and the

---

[4] French, *Ancient Natural History*, 255. This is by any criterion a very sweeping generalization! Cf. also ibid., p. xxi, n. 12. The most important recent enquiry into the history of science that looks at the *nature* of science is D. C. Lindberg, *The Beginnings of Western Science* (Chicago, 1991). Lindberg allows that 'science' has changed in form, content, method, and function, which does not leave much by which we can recognize it in the past, or identify it as an enterprise. His principle of not looking for fragments of modern science in the past is of course sound... Such prejudgements and seemingly patronizing attitudes to Pliny, however, can only prevent an unbiased assessment of his contribution to our knowledge of early science and technology.

[5] In his *De Plinii et plurium aliorum medicorum, in medicina erroribus* (Ferrara, 1509).

[6] *Die antike Kunstprosa vom VI. Jhdt. v. Chr. bis in die Zeit der Renaissance*, 2 vols. (Leipzig 1898; repr. Berlin, 1909), i. 314.

pure vacuity to which rhetorical prose, handled by any but the most talented, could precipitously descend and would descend again.' Norden added: 'Man darf nicht sagen dass der Stoff daran schuld war.' In other words, the subject matter was not responsible for Pliny's 'bad' style.

Fortunately, any but the most superficial study of the *Natural History* soon reveals the wildly exaggerated nature of these generalizations. The fact that this unique encyclopaedic work, unlike others, has not only survived to the present day, but continues to excite the interest of scholars in different disciplines, is surely proof enough of the importance of its contribution to our knowledge of the early imperial period.

7.1 Influences

A number of factors helped to determine the character and style of Pliny's writing, among them his background,[7] political status, active service in the Roman army (*militia*), and experience as a procurator in at least two provinces,[8] his education in rhetoric, Stoic philosophy,[9] and, not least, his debt to other classical authors. Varro, Cato, Virgil, Columella, and the Younger Seneca were among the more important authorities whose works provided him with technical information and left their mark on the pages of the *Natural History*.

One of the first persons to exert a direct influence on Pliny was the celebrated soldier-poet Publius Pomponius Secundus[10] ('consularis poeta vatis civisque clarissimus') under whom he studied and whose biography[11] he wrote, as has been suggested, to discharge his debt of gratitude.

7.2 *Sermonis Egestas*

An ongoing problem that confronted many writers on philosophical or scientific topics, in prose almost to the same extent as

---

[7] See Ch. 1.1.   [8] See Ch. 1.3 and 1.4.
[9] Cf. French, *Ancient Natural History*, 198 ff.
[10] 13. 83 (all references in this form are from the *Natural History*)
[11] 14. 56, 'referentes vitam Pomponi Secundi vatis'.

in verse, was the inadequacy of the Latin language in respect of technical vocabulary. Latin has been described as a language of a semi-barbarous people who had displayed unequalled aptitude for the arts of government and war but had so far (that is until the time of Lucretius) devoted very little thought to the nature of the Universe. Although words like *mundus* and *vita* could translate the Greek κόσμος and βίος in some, at least, of their several senses, the almost total lack of suitable equivalents for philosophical terms hampered Lucretius in his attempts to expound the teachings of Epicurus: his problems were compounded by the restriction of writing in hexameters, a metre at an early stage of its development. Lucretius (*DRN* 1. 136–9) writes,

I am well aware that it is not easy to elucidate, in Latin verse, the obscure discoveries of the Greeks. The poverty of our language, and the novelty of the theme, compels me often to coin new words for the purpose.

> nec me animi fallit Graiorum obscura reperta
> difficile inlustrare Latinis versibus esse.
> multa novis verbis praesertim cum sit agendum
> propter egestatem linguae et rerum novitatem.

Lucretius also stresses the fact that philosophical questions (*res*) are unfamiliar (*novae*) to Romans of his day. It was left to Cicero to develop a vocabulary which would be a suitable vehicle for abstract thought.

At the simplest level, when confronted with difficulties, Lucretius renders keywords by periphrases—for example, the Greek term ἄτομοι becomes *rerum primordia*, or *principiorum corpora*, 'elements/first beginnings'; τὸ κένον becomes *natura inanis*, (or *inane*), 'vacuum, void'; and *rerum summa*, 'the Universe'.

When discussing Anaxagoras' explanation of the structure of matter, Lucretius (1. 830–3) again draws attention to a shortcoming in vocabulary.

Let us look into the theories of Anaxagoras that the Greeks call ὁμοιομέρεια (*homoeomeria*): the poverty of our own language will not let me translate the words, but the concept itself can be expressed readily enough;

> nunc et Anaxagorae scrutemus homoeomerian
> quam Grai memorant, nec nostra dicere lingua
> concedit nobis patrii sermonis egestas:
> sed tamen ipsam rem facilest exponere verbis.

What Lucretius means is that Latin does not have a single word to express ὁμοιομερεία.

Similarly Seneca also refers to the problems of deficiencies in Latin vocabulary. Pliny (pref. 13), however, is somewhat less specific in his observations, commenting that he has recourse to 'using... rustic terms, or foreign, indeed barbarian words which have to be introduced with the phrase "if you'll pardon the expression".' ('aut rusticis vocabulis aut externis, immo barbaris, etiam cum honoris praefatione ponendis'). In these remarks he is simply expressing impatience, as Lucretius had done: there is no suggestion that he is speaking with tongue in cheek.

The very title of Pliny's encyclopaedic *Naturalis historia*, is a hybrid, derived from *naturalis*, 'pertaining to, or concerning, Nature' and the Greek term ἱστορία,[12] which has a variety of meanings, including 'research', 'enquiry', or, at its simplest, a written account of one's investigations.

The influence of Pliny's military service is sometimes reflected in his use of imagery from army life and, uniquely, in the dedication of the *Natural History* to the future emperor Titus, under whom he had served ('castrense contubernium'), when he refers in 'barrack room language' ('castrense verbum'), to Catullus as his 'oppo': the reading 'conterraneum'[13] is more likely since both men were, as all Transpadanes, fiercely proud of their origin.[14]

7.3 TECHNICAL VOCABULARY

A number of Latin words, in general use, attracted specialized meanings, particularly in the fields of earth sciences, mineralogy, mining, and metallurgy[15] and formed a 'subset', as it were, of the literary language. Failure to understand this linguistic development has, in the past, been responsible for the creation of many non-existent problems and misjudgements in the interpretation of the text of the *Natural History*.

---

[12] See above, Ch. 3.1 and nn. 4–6.
[13] v.l. 'concerraneum, concerronem', or 'congerronem' in the sense of 'boon-companion'.
[14] C. J. Fordyce, *Catullus* (Oxford, 1961), p. x (ref. 39, line 13 and elsewhere).
[15] See below, Chs. 13–15 and 17 and, generally, Healy, *Mining*.

Some simple, straightforward examples include *marmor*,[16] which, in addition to its usual meaning of marble, is used of the mineral quartz; *glaesum*,[17] signifies amber; *strigiles*,[18] grains, or nuggets, of gold; *viscera*,[19] the soil between the earth's surface and bedrock; and *fibra*,[20] a vein of mineralization. Pliny also employs generic terms: for example *schistos*,[21] which appears to include any mineral with a lamellar structure; *adamas*,[22] any hard substance, including iron, diamonds, and other gemstones; and *chrysocolla*,[23] representing various forms of copper ore.

In detailing the outward appearance of minerals, however, Pliny, like his predecessors, is less precise. So *nitor* (and *fulgor*)[24] both refer to lustre, or brilliance, while the adjective *tralucidus*[25] may mean transparent, or translucent. *Pinguis*[26] has a range of meanings in addition to thick or rich—namely rich in colour, massive, or dull. *Crassitudo*[27] may denote opacity, or, more probably, thickness and bulk; *crassiores*[28] has the meaning more opaque and *crassius nitere*[29] means to shine with a rather dull lustre.

Physical characteristics that employ words in a technical sense include *fragilitas* (*fragilis*),[30] brittleness, and *facilitas* (*facilis*),[31] malleability.

In mineralogy, *sucus*[32] (usually the juice, or sap of a tree) refers to the streak of colour left after rubbing a mineral across a touchstone, or streak plate, on which it leaves a characteristic and identifiable trace. The term *tres partes*[33] in the context of volume, denotes 75 per cent.

---

[16] 33. 68. [17] 37. 42.
[18] 33. 62. For *strigilis* in its usual sense, as a scraper, see Cic. *Fin.* 4. 12. 30, Hor., *Sat.* 2. 7. 110, Pers. 5. 126, and elsewhere.
[19] 33. 2. [20] 33. 1. [21] 33. 84; cf. also 29. 124, 36. 144.
[22] J. F. Healy 'Pliny the Elder and Ancient Mineralogy', *Interdisciplinary Science Reviews*, 6/2 (1981), 172–4.
[23] 33. 86 ff. Zehnacker, *NH 33*, p. 184.
[24] Healy, 'Language and Style', 9. See also Healy 'Pliny on *Mineralogy and Metals*', in French–Greenaway *Roman Science*, 121 (ref. NH 37. 121 and 126).
[25] 37. 56 ('transparent'), 36. 163 ('translucent').
[26] 37. 66 ('rich in colour'), 37. 69f. ('massive'), and 37. 115 ('dull').
[27] 27. 71 (*cera*), 28. 187 (*mel*), and 32. 60 (*ostrea*).
[28] 37. 106. [29] Ibid. [30] 36. 141 (of jet) and generally.
[31] 33. 59 (of gold—here) and elsewhere. [32] 35. 192.
[33] R. C. A. Rottländer *et al.*, 'Glaserstellung bei Plinius dem Alteren', in *Glastechnische Berichte*, 52 (1979), 265–70.

Mining uses *piscina*[34] in the special sense of a 'reservoir' for producing a head of water; elsewhere, *piscina* can also mean a floodgate, or wooden vat. *Corrugi*,[35] derived by Pliny from *conrivatio*, are channels that bring water to mining sites. It is possible, therefore that *canalicius* and *canaliensis*[36] (cf. *canalis*) refer to gold obtained by hushing, or hydraulicking, above ground. *Fractaria*[37] is a machine for crushing flint or hard rock.

Some of the most interesting terms occur in Pliny's account of the refining of metals from their ores. Until preliminary research carried out by the Pliny Translation Group[38] was published, commentators were unaware that the periphrasis 'nucleus ferri'[39] means iron with a high carbon content and that 'acies' refers specifically to steel. Even the relatively straightforward term 'argentarium plumbum'[40]—lead, not tin—has, surprisingly, been misunderstood in spite of the conclusive evidence of ingots of lead stamped EXARGENT.[41]

But, perhaps the most significant advances have resulted from the understanding of the fundamental distinction between the meanings of *calefacere* and *coquere*,[42] where the former involves a simple heat process, as in boiling water, while the latter implies chemical change as in cooking and metallurgical process—namely cupellation, smelting, and similar processes. This discovery was one of the factors that made possible the certain identification of *myrrha*[43] as the mineral fluorspar, which undergoes modification of its colours as a result of chemical change, when subjected to heat. Other words with a technical meaning include *pecten*,[44] a mechanical reaper, or harvester; *machina*, a hoist;

---

[34] 33. 75.
[35] 33. 74. for *corrugi*, cf. arrugia. See also Zehnacker, *NH 33* pp. 176 f.
[36] 33. 68.   [37] 33. 71.
[38] See A. Locher, 'The Structure of Pliny the Elder's Natural History', in French–Greenaway, *Roman Science*, 20 ff.
[39] R. C. A. Rottländer, 'The Pliny Translation Group of Germany', in French–Greenaway, *Roman Science*, 14, and Tylecote, *Metallurgy*, 217 f. on the carburization of iron.   [40] Healy, *Mining*, 178 ff.
[41] Healy, 'Language and Style', 9 ff.   [42] Ibid. 9 f.
[43] 37. 18–22, and Ch. 14, s.v.
[44] 18. 296. *Pecten* is also a comb used for carding wool (11. 77).

symbolum,[45] a ring; and *cura*,[46] a term embracing all manner of 'industrial' operations.

7.3.1 Greek Words

From the time of Cicero's *Letters*, many Greek words, derived from the Koine, or universal dialect of the Hellenistic period, became current in Roman literature, much in the same way as French words have infiltrated our own language to fill gaps, or provide particular nuances. Among these are found ἀκηδία[47] (ennui); μετέωρος[48] (*distrait*); καχέκτης[49] (*mauvais sujet*), ἄπρακτος[50] (maladroit); παλινωδία[51] (volte-face); and ἀμφιλαφεία[52] (*embarras de richesse*).

Pliny's debt to Greek vocabulary is wide-ranging, but different in character. Greek words used by him (and common in the first century AD), include: *acontiae*[53] (ἀκοντίαι), meteor, shooting star; *adipsus*[54] (ἄδιψος), species of date that grew in Egypt; *aplysia*[55] (ἀπλύσιον), sponge that cannot be washed; *apudes*[56] (ἄποδες), swallows; *boletus*[57] (βωλίτης), a mushroom; *bolis*[58] (βόλις) a fiery meteor; *calliblepharum*[59] (καλλιβλέφαρον), a cosmetic for dilating the pupils to give a 'wide-eyed' look; *ceroma*[60] (κήρωμα), an ointment used by wrestlers, or the venue where the bouts took place; *cochlea*[61] (κοχλίας), snail (also used of the Archimedian screw); *coecas*[62] (κοίξ), a kind of date; *colaphus*[63] (κόλαφος), a blow with a fist; *colossus*[64] (κόλοσσος), a large statue; *cometes*[65] (κομήτης), a comet; *crotalia*[66] (κροτάλια), pendant earrings, or castanets; *crystallinum*[67] (κρύσταλλος), rock-crystal; *dasypus*[68] (δασύπους), a kind of rabbit; *diadumenos*[69]

---

[45] 33. 10, also known as *ungulus* by early Romans.
[46] 33. 1. Cf. Tac. *Ger* 31. [47] *Att.* 12. 45. 2.
[48] *Att.* 15. 14. 4 and 16. 5. 3. [49] *Att.* 1. 14. 6. [50] Ibid.
[51] *Att.* 4. 5. 1. [52] *Q Fr.* 2. 14. 3. [53] 2. 89.
[54] 12. 103. [55] 9. 150. [56] 10. 144. [57] 22. 92.
[58] 2. 96. [59] 21. 123; 23. 97; 33. 102. [60] 35. 168.
[61] Cf. Vitr. *De arch.* 5. 12, 10. 8, 10. 11. There is no reference to the Archimedian screw (*coclea*) or drainage-wheels in the mining sections of the *Natural History*. This is, however, perhaps not surprising, since Pliny was not particularly interested in mechanics, or mechanical devices.
[62] 13. 47. [63] 8. 130. [64] 35. 128 and 34. 41: cf. 34. 39.
[65] 2. 89ff. [66] 9. 114. [67] 37. 30. [68] 8. 219, 10. 179.
[69] 34. 55. Cf. also *doryphorus*, spear-carrying, and *astragalizontes*, [two boys] playing dice.

(διαδυμένος), wearing a diadem; *dibapha*[70] (δίβαφος), twice-dyed; *ectypus*[71] (ἔκτυπος), embossed, engraved in relief; *elenchus*[72] (ἔλεγχος), a scrutiny, disproof; *garum*[73] (γάρον), a fish sauce; *holosphyraton*[74] (ὁλοσφύρατος), beaten, solid; *hyphear*[75] (ὕφεαρ), mistletoe; *icas*[76] (εἰκάς), the twentieth of the month; *leucargillon*[77] (λευκάργιλλος), white clay; *lytta*[78] (λύττα), madness; *milax* (μίλαξ),[79] a yew-tree; *musaea*[80] (μουσεῖα), grottoes; *oenanthe*[81] (οἰνάνθη), the grape of an African wild vine; *paedagogium*[82] (παιδαγώγιον), a school where slave boys were educated; *pegma*[83] (πῆγμα), wooden stage-machinery; Phoenix[84] (Φοῖνιξ), the legendary phoenix; *plastice*[85] (πλασπκή sc. τέχνη), sculpture; *platyopthalmos*[86] (πλατυόφθαλμος), a cosmetic for dilating the eyes; *sirus*[87] (σίρος), a hole in the ground for storing grain *spondylus*[88] (σπόνδυλος), a vertebra of the spine; *stemma*[89] (στέμμα), a garland; *stomachichus*[90] (στομαχικός), having a stomach upset; *stomoma*[91] (στόμωμα), scales which fly off metal when forged; *struthocamelus*[92] (στρυθιοκάμηλος), an ostrich; *syce*[93] (σύκη), a fig-tree; *tragus*[94] (τράγος), kind of sponge; *xylina*[95] (ξύλινα), linen from a cotton-bush.

Pliny also incorporates Greek words directly into the text, untransliterated, as γνήσιον,[96] a true eagle; δοκοί,[97] meteoric lights; ἐγκύκλιος παιδεία,[98] an all-round education; ἀναπαυόμενος,[99] a Satyr reposing (in a painting by Protogenes); ἀπαθεῖς,[100] describing people incapable of showing normal emotion; and ἄρουρα,[101] used as the equivalent of a *iugerum*.

The nomenclature of minerals was largely adopted from the Greeks, especially those with terminations in *-ite*, *-itis*,

---

[70] 9. 137, 21. 45.   [71] 35. 152, 37. 173.   [72] 9. 113.
[73] 31. 93 ff. See R. I. Curtis, *Garum and Salsamenta: Production and Commerce in Materia Medica* (Studies in Ancient Medicine; Leiden, 1990).
[74] 33. 82.   [75] 16. 245 ff., 16. 120.
[76] 35. 5. See also Cic. *Fin.* 2. 32. 101.   [77] 17. 42.
[78] 29. 100. This refers to little worms on the tongue of dogs which the Greeks call λύττα.   [79] 16. 51.   [80] 36. 154.   [81] 12. 132.
[82] 33. 152. Cf. the Younger Pliny, *Letters* 7. 27. 13.   [83] 33. 53.
[84] 10. 3.   [85] 35. 151.   [86] 33. 102.   [87] 18. 306.
[88] 29. 67, 32. 116.   [89] 35. 6.
[90] 20. 100, 24. 123, 25. 60. Cf. Sen. *Ep.* 24. 14.
[91] 34. 108. Cf. Celsus 6. 6. 5.   [92] 10. 56, 11. 130, 29. 96.
[93] 27. 119.   [94] 9. 148, 31. 123, 32. 152.   [95] 19. 14.   [96] 10. 8.
[97] 2. 96.   [98] Preface 14.   [99] 35. 106.   [100] 7. 80.
[101] Originally a measure of land in Egypt: Hdt. 2. 168.

and -*ites*. Sometimes the Greek name indicates a colour, or constituent property, or locality where deposits were found. So haematite[102] derives its name from its reddish colour (αἷμα, blood). In other examples there may be a reference to a part of the human body reflected in the appearance of the surface, as hepatite[103] (ἧπαρ, liver), or its texture, *ammochrysus*[104] (ἀμμόχρυσος), sand, veined with gold, perhaps gold mica? Rock-crystal,[105] *crystallum*, is named after its likeness to ice (κρύσταλλος). White clay, *creta Cimolia*[106] (Κιμωλία γῆ), was found on the island of Kimolos. Ampelitis[107] (ἀμπελῖτις) is a kind of bituminous earth.

Gems (σφράγις),[108] or semiprecious stones, retain their Greek names in a transliterated form: *acaustoe*[109] (ἄκαυστοι), gems like rubies, or garnets; *acenteta*[110] (ἀκνέντητος), flawless rock-crystals; *adamas*[111] (ἄδαμας), diamond; *aërizusa*[112] (ἀερίζειν, 'to resemble air'), grey, or light-blue jasper; *alabastritis*[113] (ἀλαβαστρίτης), onyx marble; *anancites*[114] (ἀναγχίτης), diamond; *asteria*[115] (ἀστήριον), pale star sapphire; *astrion*[116] (ἄστριον), moonstone; *ceraunia*[117] (κέραυνος) onyx (?); *chalazias*[118] (χάλαζίας), granite with white spots; *iris*[119] (ἶρις), rainbow stone—quartz; *leptosephos*[120] (λεπτόψηφος), white-spotted porphyry; *lignyzon*[121] (λιγνύς), Indian carbuncle; *molochites*[122] (μολοχίτης), a stone of the colour of mallow; *paedaros*[123] (παιδέρως), used for two gem-stones—opal[124] and amethyst; *pyrropoecilus*[125] (πυρροποίκιλος), granite from Syene (Aswan), mottled red; *selenites*[126] (σεληνῖτις), moonstone; *tephrias*[127] (τέφρα), an unidentifiable, ash-coloured mineral, or gem; *topazus*[128] (τόπαζος), topaz, or chrysolite—possibly green jasper?

Some physical characteristics are also derived from Greek equivalents, as *sarcion*[129] (σάρκιον), an excrescence on a stone;

---

[102] 36. 129 and 146, cf. also 37. 169. Healy *Mining*, 39 ff.
[103] 36. 147, cf. 37. 186.   [104] 37. 188.   [105] 37. 23.
[106] 35. 195 ff. Cimolia is described as fuller's earth, also used in baths and barbers' shops (cf. Strabo 10. 5. 1). It is also found in medicine (Dioscorides 5. 156).   [107] 35. 194.   [108] 37. 117.   [109] 37. 92.
[110] 37. 28.   [111] 37. 55 ff.   [112] 37. 115.   [113] 37. 143.
[114] 37. 192.   [115] 37. 131.   [116] 37. 132.   [117] 37. 134 f.
[118] 36. 157, 37. 189.   [119] 37. 136.   [120] 36. 57.   [121] 37. 94.
[122] 37. 114.   [123] 37. 84.   [124] Ibid. (opal), 37. 123 (amethyst).
[125] 36. 63 and 157.   [126] 37. 181.   [127] 36. 56.   [128] 37. 107.
[129] 37. 73.

Language and Style 89

epipedos[130] (ἐπὶ πέδον), plain, or level; *monogrammos*[131] and *polygrammos* (μονόγραμμος, πολύγραμμος), single, or many-lined.

It is, however, in the field of medicine and the names for diseases and afflictions that Pliny, in common with other writers on these topics, owes his greatest debt to Greek terminology. This was largely due to the fact that medicine entered Roman life from Ionia through Alexandrian anatomy and physiology.[132] The earliest medical teacher in Rome was Asclepiades[133] of Bithynia (c.50–40 BC), and the Greeks had a virtual monopoly of the practice of medicine and surgery. There were very few Roman doctors and even these joined the 'brain-drain' to the Greeks ('paucissimi Quiritium medicina, attigere, et ipsi statim ad Graecos transfugae').[134] Furthermore, as Pliny (29. 17) explains, if medical treatises were written in a language other than Greek they had no prestige even among unlearned men ignorant of that language ('immo vero auctoritas aliter quam Graece eam tractantibus etiam apud imperitos expertesque lingua non est'). A small number of medical terms assimilated by Pliny are included, exempli gratia. *Clinice*[135] (κλινικὴ τέχνη), clinical medicine (that is, at the sick-bed); *iatraliptice*[136] (ἰατραλειπτική), healing by means of ointments; and *empirice*[137] (ἐμπειρική), experience, or practice. Drugs and medicines, likewise, retained their Greek designations, as *theriace*[138] (θηριακή), an antidote against snakebites; *acesis*[139] (ἄκεσις), borax. Finally, names of diseases—often difficult to identify precisely—are taken over from Greek. So, *lichen*[140] (λειχήν) a fungal infection of the skin; *angina*[141] (ἀγχόνη), quinsy, peritonsilar abscess, or, simply, sore throat; *parotis*[142] (παρῶτις), a swelling, or tumour of the glands by the ear (perhaps mumps?) *orthnopoea*[143] (ὀρθνοποιία),

---

[130] Contrast Pl. Criti. 112a and elsewhere—in the meaning 'level ground'.
[131] 37. 118.   [132] J. Scarborough, *Roman Medicine* (London, 1969), 26.
[133] 7. 124, 16. 15. See also Cic. *De or.* 1. 14. 62, and Celsus 3. 4.
[134] 29. 17.   [135] 29. 4, 30. 98 and sc. τέχνη (nn. 136–7 below).
[136] 29. 5.   [137] 29. 5.   [138] 20. 264, 29. 24.   [139] 33. 92.
[140] 26. 21.   [141] 23. 61.
[142] 20. 4, 20. 95, 20. 229, and numerous other references. Also Celsus 5. 18. 18, 6. 16. 142.
[143] 32. 37. Cf. *orthnopoicus* (adj.) 20. 193, 24. 145.

asthma; *zoster*[144] (ζώστηρ), shingles. Such terms, as other non-Roman technical words, are often introduced by 'vocant', the equivalent of 'are known as', or, in some contexts, 'is the local word for'.

Another sphere in which Greek words were readily assimilated was in building, or construction work. Three sizes of bricks were standard: *di-*, *tetra-*, and *pentadoron* (-δῶρον),[145] that is two, four, and five palms in length. *Isodomos*[146] (ἰσόδομος) is masonry with equal courses of varying thickness; *emplecton*[147] (ἔμπλεκτον) signifies that only the faces are dressed and that the rest of the material is laid at random, and *diatonicon*[148] (διατόνικον) that the core of the wall is packed with rubble, and single stones stretch from face to face. In addition, other terms include *pteron*[149] (πτέρον) the wing of a building; *buleuterium*[150] (βουλευτήριον) Council House; *echo*[151] (ἤχω) is used to describe the reverberation from a seven-towered gate; *heptaphonon*[152] (ἑπτάφωνος) is a portico in which the sound re-echoes seven times; *subdialia*[153] (ὑπαιθρία) are open galleries. *Lithostrata*[154] (λιθόστρατα) are mosaics, and *encaustus*[155] (ἔγκαυστος) refers to encaustic painting.

In short, Pliny frequently follows the exhortation of Horace (Ars P. 48 ff.) to Latinize Greek words:[156]

Should it happen to be necessary to indicate by new terms things before unknown, you may invent expressions... and words though new and lately invented, will gain credit, if derived from the Greek, and a little altered in form.

> si forte necesse est
> indiciis monstrare recentibus abdita rerum,
> fingere
> . . . . .
> et nova fictaque nuper habebunt verba fidem si
> Graeco fonte cadent parce detorta.

---

[144] 26. 121.   [145] δῶρον = palma and is equivalent to $4\frac{1}{2}$ inches in length.
[146] 36. 171; Vitr. *De arch.* 2. 8. 7.   [147] 36. 171.   [148] 36. 172.
[149] 36. 30, 36. 88.   [150] 36. 100, cf. Cic. Verr. 2. 2. 50.   [151] 36. 100.
[152] Ibid.
[153] 36. 186; *subdiales petrae*—of cadmea quarried above ground (34. 117).
[154] 36. 184. Varro, *R. R.* 3. 2. 4.   [155] 35. 149.
[156] It is interesting to observe that Seneca uses only 50 Greek words and these mainly in the *QN*.

7.3.2 Spanish and Greek Mining Terms

Pliny's assimilation of local words, in addition to his use of everyday vocabulary in a technical sense, sometimes causes problems in interpretation of his text. This is particularly so in his account of gold-mining and associated metallurgical processes in north-west Spain.[157] Commentators do not seem to have been sufficiently alerted by the well-known example from Cicero (Ad fam. 8.1.4), in which the phrase 'embaeneticam facere' was constantly challenged until the discovery of an inscription at Baiae that referred to 'embaenetarii trierum piscensium' (builders of boats used for fishing).[158] Ingenious, but unnecessary conjectures were sought for what was clearly an acceptable local phrase, used in the region of Campania and one with which Marcus Rufus, the writer of the letter would have been familiar.

As procurator in the province of Hispania Tarraconensis, an important gold-producing region throughout the first century AD, Pliny was able to observe operations and mingle with the workers. In his account Pliny records a mixture of local Iberian, Roman, and Greek words, no doubt reflecting the speech—a lingua franca—of the miners, who were a multinational workforce.

Words from Spanish include: *arrugia*[159] (*arrugia*, a hole dug while prospecting for gold—the term would appear to be connected with the ὀρύσσειν,[160] to dig, or mine); *segullum* (*segullo*, the first stratum of a gold working) and *talutium*,[161]

---

[157] See Ch. 17.4.3.
[158] R. G. Levens, *A Book of Latin Letters* (London, 1955), 91 n. 43.
[159] 33. 70, *arrugia*, and Zehnacker, *NH 33*, p. 176 refers to this as a 'borrowed' word ('emprunté').   [160] Cf. Hdt 1. 86; Thuc. 2. 76.
[161] 33. 67 and 77. O. Bloch (ed.), W. von Wartburg (rev.), *Dictionnaire étymologique de la langue française* (s.v. 'talus'), takes *talus* (*talutium*) as a Gallic word, derived from *talo*, 'front'. Zehnacker, *NH 33*, pp. 173 f. ad loc. disputes Bloch's interpretation: 'on peut objecter: (*a*) que rien dans le texte de Pline ne suggère une pente de terrain; (*b*) que talutium y désigne la couche de terre aurifère; (*c*) que le mot paraît hispanique, et non gaulois. See also C. Domergue, 'Apropos de Pline, *Naturalis Historia*, 33. 70–8', *Archivo Español de Arte y Arqueologia*, 45–7 (1972–4), 502. There is a further possibility that *talutium* is connected with *alutiae* (34. 157)— 'invenitur et in aurariis metallis quae alutias vocant'. See further H. Le Bonniec (ed. and trans.), and H. Gallet de Santerre (comm.), *Pline l'Ancien: Histoire Naturelle, Livre 34* (Budé: Paris, 1953), comm. ad loc.

possibly with a similar meaning, although its derivation is uncertain; *urium*,[162] earth in mines (perhaps derived from οὖρος, or ὄρος, mountain); *palaga, palacurna*[163] (*palacra, palacrana,* and πάλα)[164] all refer to fragments of native gold; *balux*,[165] or *baluca*, signifies an ingot or nugget of gold, as in Martial;[166] *obrussa*,[167] of gold (*obrizo -a*, pure, refined: χρυσίον ὄβρυζον);[168] *crudaria*, of silver (*crudo -a*, unrefined). *Tasconium*[169] (*tasconio*, clay), an argillaceous earth from which crucibles are made.

Other Iberian words occur in the Lex Vipascensis[170] of Hadrian's time (AD 117–38) discovered at Aljustrel, in Portugal. *Laurex*,[171] a rabbit, comes from the Balearic Islands.

Pliny's description of the metallurgical processes involved in lead refining, throws new light on the origin of some of his technical data, as Halleux[172] convincingly demonstrates. His account is the conflation of two lost sources: (1) a work of Cornelius Bocchus,[173] and (2) a medical treatise of Sextus Niger.[174] Two terminologies are used, the one from a Spanish source (*Stagnum, stannum,* and *galena*), the other, Greek (λιθάργυρος, ἀργυρῖτις, μολύβδαινα). As Halleux concludes, Pliny seems to have understood his Spanish source—probably written in Latin—better than the Greek.

Occasionally, in mining, Greek terms are transliterated, as, for example, *agogae*[175] (ἀγωγαί), channels, or cross-cuts

---

[162] 33. 75.   [163] 33. 77.
[164] Strabo 3. 2. 8. LSJ, inexplicably, regard πάλα as a Spanish word.
[165] 33. 77.   [166] *Epigr.* 12. 57.9.   [167] 33. 59, cf. Isid. *Orig.* 16. 18. 2.
[168] Scholiast to Thuc. 2. 13; also papyri of the 4th and 5th cent. AD. The mark 'OB' found is also on coins of the late empire.
[169] 33. 69. Cf. also Zehnacker, *NH 33*, p. 175 n. 4.
[170] Healy, *Mining*, 129 f. See further S. Riccobono, *Fontes Iuris Romani Ante Iustinani* (Florence, 1941), i. 104, translated, in full, by A. H. M. Jones, *A History of Rome through the Fifth century*, ii: *The Empire* (Oxford, 1965), 300 ff. See also A. H. M. Jones, *The Roman Economy* (Oxford, 1974), 68; C. Domergue, *La Mine antique d'Aljustrel (Portugal) et les tables de bronze de Vipasca* (Paris, 1983).   [171] 8. 217.
[172] 'Les Deux Métallurgies du plombe-argentifère dans l'Histoire Naturelle de Pline,' *Revue de Philologie*, 49 (1975), 77–88.   [173] 33. 96, 34. 159.
[174] 33. 106–9, 34. 173.
[175] Healy, *Mining*, 94. In Hdt. 6. 85 ἀγωγή has the basic meaning of 'carrying away'.

for drainage, and, elsewhere, whole passages are transferred verbatim from Greek authorities as in the case of Theophrastus' description of the discovery of cinnabar.

Relatively few Celtic words appear in the *Natural History* and not all of these relate specifically to Gallia Narbonensis, which tends to confirm the view that Pliny did not hold a procuratorship in that province, as has sometimes been suggested.[176] Much of his information on agriculture and viticulture relating to Gaul as a whole is derived from Cato, Varro, Virgil, and Columella who had already established a comprehensive vocabulary for these topics. Some Celtic terms do occur, however, among them *acaunamarga*[177] (*agaunum*, a stone), red marl: *alauda*[178] (*al*, high, or great, and *aud*, song), a lark; *chama*,[179] a lynx; *eglecopala*,[180] dove-coloured marl; *eporedias*,[181] men skilled at breaking in horses; *glastum*,[182] plantain; *padus*,[183] a pitch-pine; *passernix*,[184] a whetstone; *vettonica*,[185] a saw plant; *viriola*,[186] a small bracelet.

7.3.3 Assimilations from Other Languages

Occasional words are incorporated from a number of other languages.

Aethiopian: *nabun*,[187] a giraffe
African: *addax*,[188] a wild animal with crooked horns; *Astapus*[189] (*Astuapes* and *Astobores*), names of the Nile flowing through Aethiopia; *Celtis*,[190] a lotus-tree; *lalisio*,[191] a foal
Arabic: *cyna*,[192] a cotton-tree; *Hippalus*,[193] the west wind
Arimphaean: *Temarunda*,[194] the Palus Maeotis

---

[176] See above, Ch. 1.4.4.   [177] 17. 44.
[178] 11. 121, cf. Suet. *Iul*. 24. Cf. the French word *alouette*.   [179] 8. 70.
[180] 17. 46.   [181] 3. 123.   [182] 22. 2. This herb is used as a blue dye.
[183] 3. 117.   [184] 36. 165.   [185] 25. 184.
[186] 33. 40, cf. Isid. *Orig*. 19. 31.   [187] 8. 69 (also found in the form *nabis*).
[188] 11. 124.   [189] 5. 53.   [190] 13. 104.
[191] 8. 174. Also Mart. *Epigr*. 13. 97.   [192] 12. 39.   [193] 6. 104.
[194] 8. 69.

Carthaginian: *mapalia*,[195] caravans, cottages, or shanties (cf. *magalia*)[196]
Egyptian: *sacal*,[197] amber
German: *Morimarusa*,[198] the North Sea; *alces*[199] (*elaho*), and elk; *sapo*[200] (*Seife*), soap; *vibo*,[201] the herb called *Britannica*; *framea*,[202] a spear—found in Tacitus but not in *NH*; *ganta*[203] (*Gans*), goose; *glaesum*[204] (*Glas*, glass, and *glanzen*, to glimmer), amber
Indian: *Siptachora*,[205] a tree yielding amber
Persian: *tigris*,[206] an arrow, named from its swiftness (cf. river Tigris)
Phoenician: *gadir*,[207] a hedge, or fence (the name for Gades in Spain)
Raetian: *plaumoratum*,[208] a plough with two small wheels
Scythian: *sacrium*,[209] amber
Syrian: *harpax*,[210] amber
Trogodyte: *topazin*,[211] to seek; *zura*,[212] a wild thorn

7.4 PERIPHRASES

Lucretius employs periphrases to translate philosophical terms, such a *primordia rerum*[213] (ἄτομοι) and where a single

---

[195] On *mapalia*, see Sall. Iug. 18. 12; Livy 29. 31; Lucan, *BC* 2. 89; and see further above, Ch. 1.4.2, p. 11.
[196] Cf. Verg. *Aen.* 1. 421: *magalia* is a Punic word meaning shepherds' huts of some kind (only found in poetry, here and in *Aen.* 4. 259). Servius says that Cato (*Origines*) describes them as 'aedificia quasi cohortes rotundas', that is like round pens for cattle, or fowls. Others simply refer to them as 'casas Poenorum pastorales'. Servius further comments, 'debuit magaria dicere quia magar, non magal Poenorum lingua villam significat'. Plautus, *Poen.* 86, mentions a district of Carthage known as Magaria. The companion word *mapalia* (Verg. G. 3. 340) is used of Libyan shepherds' huts (Servius, on *Aen.* 4. 259, identifies *magalia* and *mapalia*). [197] 37. 36.
[198] 4. 95. [199] 8.39; Caes. *BG* 6. 27.
[200] 28. 191. This may, however, be a Celtic word. Cf. the French *savon*.
[201] 25. 21. [202] See Tac. *Ger.* 6, 11, 14, 18, 24. [203] 10. 54.
[204] 37. 42; Tac. *Ger.* 45. [205] 37. 39.
[206] 6. 127. Hor. *Odes* 4. 14. 46; Lucan, *BC* 3. 256, 3. 261, 8. 370.
[207] 4. 120. *gadir* = *saepes*. [208] 18. 172. [209] 37. 40.
[210] 37. 37. Likewise, sulphur (35. 176). [211] 37. 107.
[212] 24. 115. The seed of Christ's thorn.
[213] *DRN* 5. 419, *primordia rerum* translates the Greek term ἄτομος.

Language and Style

word presents a metrical problem. So we find *strata viarum*,[214] for 'viae'; *mortalia saecla*[215] ('mortales'); *prima virorum*[216] ('principes', 'duces'), and many similar expressions.

Pliny, however, rarely has recourse to this expedient, except in phrases such as *naturae numen*.[217]

On the other hand, he makes frequent use of abstract for concrete nouns, as *leonum feritas*[218] ('leones feri'); *universa mortalitas*[219] (also *mortalitas*). *Potestatum*,[220] signifies men of power, *servitus*[221] and *servitia*[222] slaves, and *ministeria*,[223] officials or attendants.

7.5 Neologisms

Pliny creates a number of abstract nouns from the neuter plural of adjectives: for example, *montuosa*,[224] mountainous regions (cf. *abrupta*,[225] rough, or dangerous ways); *austrina*,[226] southern, and *exortiva*,[227] eastern regions. Participles are similarly used, as *rigentia*,[228] frozen places.

In addition to sharing relatively rare technical terms with Varro, Columella, and the Younger Seneca, Pliny coins new words which are exclusively his. These include nouns ending in *-atio*, or *-io*, as, *decacumatio*,[229] pollarding trees; *exacutio*,[230] sharpening; and *nictatio*,[231] winking. Similarly, *-mentum*, as *duramentum*,[232] hardening; *incantamentum*,[233] charm; *nucamentum*[234] (pl.) fir cones; *piamentum*,[235] a means of expiation. He also adds new words in *-or*, as *anhelator*,[236] someone with breathing difficulties; *circumfossor*,[237] one who digs round;

---

[214] *DRN* 1. 315, 4. 415.   [215] Ibid. 5. 805, 982, 1199, 1238.
[216] Ibid. 1. 86, cf. Ov. *Am.* 1. 9. 37 ('summa ducum', the commander-in-chief). Cf. also 'saepta domorum' (*DRN* 1. 489).   [217] 2. 208.
[218] 7. 5. So Ov. *Fast.* 4. 217.   [219] 7. 147.   [220] 9. 26 (cf ἀρχαί).
[221] 14. 5.
[222] 33. 23 (cf. drones, 11. 27). See also Livy 2. 10. 8, 28. 11. 9, and elsewhere.
[223] 12. 10.   [224] 11. 280, 12. 48.   [225] Cf. Tac. *Agr.* 42.
[226] 6. 213 (Cyprus), 6. 214 (Sardinia), 6. 215 (Cappadocia).
[227] 6. 215.
[228] D. J. Campbell, *Natural History*, 2 (Aberdeen University Studies, 118; Aberdeen, 1936), 5 ff.
[229] 17. 236. Also *decorticatio*, removing bark.   [230] 17. 106.
[231] 11. 156.   [232] 17. 208.   [233] 28. 10.   [234] 16. 49.
[235] 25. 30, 25. 107.   [236] 21. 156, 22. 105.   [237] 17. 227.

infestator,[238] a trouble-maker, one who causes a disturbance; *-tas*, as *clauditas*,[239] lameness; *excelsitas*,[240] loftiness, high-level; *voracitas*,[241] greed, ravenousness; *-trix*, as *duratrix*,[242] making durable; *exulceratrix*,[243] causing soreness, or ulceration; *piscatrix*,[244] that fishes; *-ura*, as *divisura*,[245] division, or fork of a tree; *factura*,[246] manufacture; *indicatura*,[247] value, or price.

Nouns are also formed from the passive participle of verbs: *fluviatus*,[248] soaked, or steeped in a river; *mucronatus*,[249] pointed; and *papaveratus*,[250] made shining, or white, with poppies.

Further coinages include adjectives ending in *-alis*: *antegenialis*,[251] before birth; *auspicialis*,[252] suitable for auguries; *brachialis*,[253] belonging to the arm (shared with Plautus[254]); *-arius, -orius*, as *lutarius*,[255] living in mud; *crustarius*[256] (also a substantive), an engraver, or chaser of metal; *medicamentarius*,[257] here an apothecary; *clamatorius*,[258] screeching (as of a bird of illomen); *discussorius*,[259] dissolving; *excussorius*,[260] that serves for shaking out (as of a sieve); *mitigatorius*,[261] soothing; *vulnerarius*,[262] belonging to wounds; *-bilis* (a formation of which Seneca[263] disapproves, although he occasionally uses adjectives such as *infatigabilis*,[264] tireless): *computabilis*,[265] that may be accounted; *explicabilis*,[266] which can be explained; *inextirpabilis*,[267] unable to be rooted out; *-icius, -iceus, -acius*: *avenaceus*,[268] pertaining to oats; *cretaceus*,[269] chalk-like; *ericaceus*,[270] collected from the heather flower; *-ivus*, as *cadivus*,[271] falling of itself; *exortivus*,[272] ascendant; *impositivus*,[273] applied, or primitive; *-osus*, as *foliosus*,[274] leafy; *marmarosus*,[275] marble-like, hard as marble.

In addition, Pliny relies, to some extent, on adjectives ending in *-fer*, and *-ger*, many of which had been introduced by

---

[238] 6. 143. [239] 8. 169, 28. 35. [240] 2. 160, 16. 167.
[241] 2. 139 [242] 14. 17. [243] 27. 105. [244] 9. 143.
[245] 25. 167. [246] 34. 145. [247] 29. 21, 37. 18. [248] 16. 96.
[249] 25. 161, 32. 15. [250] 8. 195, 19. 21. [251] 7. 190.
[252] 32. 4. [253] 17. 123. [254] *Poen.* 1269. [255] 9. 65, 32. 32.
[256] 33. 157. [257] 19. 110. [258] 10. 37. [259] 30. 75.
[260] 18. 108. [261] 28. 63. [262] 23. 114. [263] *Ep.* 39, 1.
[264] Sen. *Vit. beat.* 7; *Ep.* 66; Pliny, *NH* 28, 257. [265] 19. 139.
[266] 4. 98. [267] 15. 84. [268] 22. 137, 30. 75. [269] 18. 86.
[270] 11. 41, describing wild honey. [271] 15. 59–60.
[272] 7. 160, 37. 39. [273] 28. 33. [274] 12. 40, 25. 161.
[275] 33. 158.

Lucretius: *glandifer*,[276] acorn-bearing; *corniger*,[277] horned; and others. He shares some with Varro, Virgil, and Columella: *bifer*,[278] bearing fruit twice a year; and with Ovid: *herbifer*,[279] grassy, or producing grass. Pliny invents similar formations, as: *annifer*,[280] bearing fruit the whole year; *plumiger*,[281] feathered; *silviger*,[282] wooded; *vitifer*,[283] supporting, or producing vines. He contributes adverbs ending in *-im*: *arcuatim*,[284] in the form of a bow; *fornicatim*,[285] in the form of an arch; *imbricatim*,[286] like a gutter-tile; *frustatim*,[287] piecemeal. Some comparatives and superlatives of adjectives and adverbs appear for the first time: *absolutissimus*,[288] supremely perfect; *scrupulosius*,[289] more detailed, more carefully; *tortuosius*,[290] more twisted; and *laxissime*,[291] very widely.

An important element of Pliny's vocabulary is his use of inceptive verbs, ending in *-escere*: these he shares with the poets, especially Lucretius and Virgil, and with the younger Seneca. However, only four new examples appear, namely *cornescere*,[292] to become like horn; *emarcescere*,[293] to wither, or dwindle away; *lapidescere*,[294] to petrify, or turn into stone; and *senescere*,[295] in the sense of to become worn out. *Tenerescere*[296] (rare), to grow soft, occurs in the form *tenerascere* in Lucretius.[297]

Finally, two further, unusual words deserve mention: these are *Marcipor* and *Lucipor*, meaning, respectively, slave of Marcus and slave of Lucius.[298]

7.6 STYLE

Both style and presentation in the *Natural History*, as in other Silver Latin works, inevitably owe a significant debt

---

[276] *DRN* 5. 939; Cic. *Leg.* 1. 1. 2.
[277] 11. 212. Cf. also *DRN* 2. 368, 3. 751, and elsewhere. So Verg. *Aen.* 8. 77; Ov. *Met.* 7. 701, 14. 602; *Am.* 3. 15. 17.
[278] 13. 121, 16. 114. Varro *RR* 1. 7. 7.; Columella 10. 403; Verg. *G.* 4. 119.
[279] 25. 94. Also Ov. *Met.* 14. 9.   [280] 16. 107, 19. 121.
[281] 10. 53.   [282] 31. 43.   [283] 3. 60; Sil. *Pun.* 4. 349.
[284] 29. 136.   [285] 16. 233.   [286] 9. 103.   [287] 20. 99 (very rare).
[288] 35. 74.   [289] 2. 118.   [290] 11. 255.   [291] 2. 66.
[292] 11. 261.   [293] 15. 121.   [294] 24. 120, 16. 21, 32. 22.
[295] 16. 116.   [296] 17. 189, 28. 183.   [297] *DRN* 3. 765.
[298] 33. 26.

to rhetoric. This is symptomatic of the age and of Roman education.

Pliny has been generally criticized for failing to be content with being informative. 'Frequently Pliny is carried away', wrote Goodyear, 'into bombast by enthusiasm for his theme, indignation, or a maudlin brand of moralizing... He drew no clear line between report and comment.' It has, furthermore, been suggested that Pliny should have tried to curb impulses which a scientist resists and that rhetoric has no place in a serious work of science. Goodyear continues: 'Instead of adopting the plain and sober style appropriate to his theme, Pliny succumbs to lust for embellishment.' A. Wallace-Hadrill[299] rightly observes, apropos this criticism, that Goodyear, like Pliny, is caught in the cultural categories of his day: 'Our cultural rules set up a sharp cleavage between scientific analysis and the passion of rhetoric, persuasion and moralization. But Pliny obviously did not.... The place of science in Roman culture was quite different from that of science in our contemporary society.... If Pliny's purpose is to persuade, and direct man towards a proper use of nature, then rhetoric is an essential tool, not a form of grotesque and tasteless ornamentation.'

The influence of rhetoric does, admittedly, sometimes give rise to affectation and artificiality, as can be observed in the apostrophe of Cicero,[300] or the eulogy of scientists (2.54 ff.):

O great men, way above mortal estate, who, by your discovery of the laws that govern such great divinities, have freed the miserable mind of men from fear—a mind which was afraid that eclipses of the planets signified some sort of crime, or death (those exalted poets Stesichorus and Pindar clearly experienced this fear because of an eclipse of the sun), or blamed poison when the moon 'died' and consequently went to her help by making discordant noises. Through such fear and ignorance of the underlying cause, the Athenian general Nicias was afraid to lead his fleet from harbour and so destroyed the Athenians' greatness. Praise be to your intellect, you interpreters of the heavens, you who comprehend the universe, discoverers of a theory by which you have bound gods and men!

---

[299] 'Pliny the Elder and Man's Unnatural History', *Greece and Rome*, NS 37 (1990), 80–96.   [300] Healy, 'Language and Style', 15.

viri ingentes, supraque mortalium naturam, tantorum numinum lege deprehensa et misera hominum mente metu soluta, in defectibus siderum scelera aut mortem aliquam pavente (quo in metu fuisse Stesichori et Pindari vatum sublimia ora palam est deliquio solis) aut in lunae veneficia arguente mortalitate et ob id crepitu dissono auxiliante (quo pavore ignarus causae Nicias Atheniensium imperator veritus classem portu educere opes eorum adflixit): macti ingenio, este, caeli interpretes rerumque naturae capaces, argumenti repertores quo deos hominesque vicistis!

Pliny often catalogues his facts in a dry style, using abrupt sentences which create a staccato effect. Equally, however, he can write in a vivid and entertaining manner. He is clearly aware of his own limitations when he describes his ability as extremely modest ('ingenium perquam mediocre'), adding (pref. 12) that his theme is somewhat lightweight ('levioris operae') and his subject barren ('sterilis materia'), for which an elevated style would be out of place. Moreover, he explains, 'they do not allow digressions, speeches, dialogues, marvellous accidents, or unusual occurrences—matters interesting to relate, or entertaining to read'. ('neque admittunt excessus, aut orationes, sermonesve aut casus mirabiles vel eventus varios, iucunda dictu aut legentibus blanda'.)

His style, as Campbell[301] observed, is an extraordinary mixture. Brevity and compression, which are hallmarks of Silver Latin, are, perhaps, the most noteworthy features of the *Natural History* and often add to our difficulty in interpreting Pliny's text and its nuances. The in-depth study of the wide range of stylistic figures[302] used in the *Natural History* provides a fascinating insight into Pliny's character and temperament, but has no primary role in this assessment of his contribution to science.

---

[301] *Natural History*, 2, p. 5.
[302] Healy, 'Language and Style', 13 ff. on Pliny's style and use of figures.

8

Science in Antiquity

8.1 INTRODUCTION

Lloyd,[1] in his work on early Greek science, somewhat illogically, points out that there is no exact equivalent to our modern definition of the term 'science', although Western science may still be said to originate with the Greeks. Greenaway[2] makes a similar point, although admitting that Pliny's world did recognize that the material substances which we must use are susceptible of consistent and rational examination for fitness of purpose and therefore for making everyday technical and economic decisions.

In a more recent, comprehensive treatment of the world of Nature, French[3] re-raises the question of whether we can legitimately, or meaningfully, refer to science in the context of the ancient world. He rightly concludes that although a 'science' (*scientia*),[4] for many centuries, could mean anything taught in schools (that is *all* forms of human knowledge), and that 'sciences' retains an older and more general usage, the terms 'Science' and 'Scientist', as defined from the mid-nineteenth century onwards, have connotations of purpose and methods that are quite out of place when describing the entirely different enterprises of the ancient world.[5]

---

[1] G. E. R. Lloyd, *Early Greek Science: Thales to Aristotle* (London, 1970).
[2] F. Greenaway, 'Chemical Tests in Pliny', in French–Greenaway, *Roman Science*, 159.
[3] French, *Ancient Natural History*, pp. x, ff.
[4] Other terms in use which embrace 'science' include: φιλοσοφία (philosophy, or love of learning); ἐπιστήμη (knowledge); ἱστορία (enquiry, research); θεωρία (speculation, contemplation); περὶ φύσεως ἱστορία (natural science).
[5] French (*Ancient Natural History*, p. xi, nn. 6 and 7) also decries the approach of certain historians of science and the use of modern categories within which to examine the achievements of the Greeks and Romans. See further Ch. 11 n. 1, for my apologia.

Such arguments may seem like an exercise in semantics, or inverted commas, but French's definition of 'modern science' does establish significant criteria.[6] Briefly, he writes: Science is (1) objective. The scientist puts his passions aside and relies on reason. (2) It is non-religious. No longer does an instinct veneration for a creator structure the search into nature. (3) It is experimental in its verification of its theories, (4) directed to the manipulation of nature, and (5), in its manipulative nature, strongly linked to technology. Finally (6), Science has universal law-like statements, often mathematical and with Boyle's Law as a paradigm.[7]

Very few of these criteria are found in the ancient *savant*[8] ('vocabulo cum honoris praefatione ponendo') but this in itself is no criterion for denigrating the extent of Pliny's contribution to our knowledge of the natural sciences and of early technology, including civil and mining engineering.

A typical lexicographical definition of science states: Science is the systematic study of the nature and behaviour of the material and physical universe, based on observation, experiment, and measurement. Alternatively, science is the knowledge so obtained or the practice of obtaining it. Finally, any particular branch of this knowledge is science.

---

[6] *Ancient Natural History*, p. xi. French follows D. C. Lindberg, *The Beginnings of Western Science: The European Scientific Tradition in Philosophical, Religious and Institutional Context, 600 BC to AD 1450* (Chicago, 1992), where, in ch. 1, he gives a similar list: each item is an alternative view of science, held by different groups. This allows Lindberg to find science in the past, based on one or more of these views. French adds: 'He is surely correct to see the advantages of also using the term "natural philosophy"?' Apropos W. H. Stahl, *Roman Science: Origins, Development and Influence to the Later Middle Ages* (Madison, 1962), 9 he comments (p. xxi, n. 10): 'Stahl opens his *Roman Science* refreshingly with doubts whether his subject is either Roman, or science; but nevertheless builds up a balance and useful picture of the Roman sources of medieval knowledge.'

[7] Assuming that the temperature remains constant, Boyle's Law tells us how to calculate the density of a gas at a given pressure.

[8] So Bailey, *Chemical Subjects*, i. 15, observes: 'We certainly cannot rank Pliny among the great thinkers of the Roman world. He recounts facts, but seldom attempts a generalization based on these facts. In reading his account of chemical substances we are struck by the contrast between the very considerable amount of accurate knowledge about their properties and the almost entire absence of theorising, however elementary.'

8.2 Natural Sciences—The Beginnings

'Science' did not burst upon the ancient world fully developed, like Athena from the head of Zeus,[9] but was the product of a long development.

Greek and Roman mythology defined five ages[10] of mankind and it is significant that four were named after metals. However, the absolute influence of Olympian religion and its morality, reflected in Homer's description of the Heroic Age, initially hampered the development of scientific thought until trade and colonization brought mainland Greece into ongoing contact with Ionia.

In the seventh century BC Ionia was the crossroads between East and West, and contact with new ideas, not least in the field of comparative religion and philosophy, caused the Greeks to re-examine their long-accepted beliefs.

What is, however, possibly the earliest account of natural phenomena and a form of natural history is to be found in two of the most notable stories of Hesiod's, *Theogony*.[11] Greene's detailed analysis of the text shows that, in the light of current geological, geographical, meteorological and historical knowledge, we can identify the specific locations and dates of these events.[12] Regardless of how the battles between Zeus and the Titans[13] and Zeus and Typhoeus[14] are interpreted—as origin stories or aetiological myths (whether of cosmic order[15] or local politics)—and whatever they may be taken to symbolize, their literal, superficial content is quite remarkable. Close attention to the sequences of events in the battles, to their appearance, their sound, and their effect in the physical world, leaves no doubt that the phenomena described are volcanic eruptions. The two battles represent the eruptions of the

---

[9] Hes. Theog. 924f. αὐτὸς δἰ ἐκ κεφαλῆς γλαυκώπιδα γείνατ᾽ Ἀθήνην δεινήν. See also Lloyd, *Early Greek Science*, 1ff. ('The Background and the Beginnings').

[10] So Hes. *WD*: Gold (109f.), Silver (127ff.), Bronze (143ff.), and Iron (176ff.).   [11] See further M. L. West (ed.) *Theogony* (Oxford, 1966).

[12] M. T. Greene, *Natural Knowledge in Preclassical Antiquity* (Baltimore, 1991), 46ff.   [13] Ibid. 47, and 49ff.   [14] Ibid. 63ff.

[15] G. S. Kirk, J. E. Raven, and M. Scholfield, *The Presocratic Philosophers*² (Cambridge, 1983), 31–41.

volcano at Thera (mod. Santorini)[16] in the fifteenth century BC, and that of Mt. Etna in Sicily (735 BC).[17]

In the course of succeeding centuries the role of the poet as educator and guardian of morals was gradually taken over by the natural scientist. The beginnings of rationalism saw the emergence of a number of philosophical explanations of the origin of the Universe.

The first attempts to explain the nature of the Universe (ὁ κόσμος) and Man's role in the natural order of things began with the research of Ionian 'natural scientists' (οἱ φυσιόλογοι).[18] They believed in the existence of four basic 'elements', fire, air, earth, and water, and each philosopher chose one—hence they were also known as 'Monists'—as the original material (ἡ ἀρχή), or underlying substance (τὸ ὑποκείμενον) of the Universe. Of these, Thales'[19] choice of water was the most logical in that it can readily be observed in the three states of matter, namely liquid, solid, and vapour (gas).

---

[16] Greene, *Natural Knowledge*, 63 ff. It is interesting, for example, to compare Hesiod's description of the battle with volcanologists' reconstructions of the Thera explosion which measured VEI (Volcanic Explosivity Index)6. In the following list, the events recorded by Hesiod are shown in quotation marks, and, in each case, the corresponding event at Thera follows after the colon. 1, 'a long war': premonitory seismicity; 2, 'both sides gather strength': increase of activity; 3, 'terrible echoes over sea': first-phase explosions; 4, 'ground rumbles loudly': tectonic earthquakes; 5, 'sky shakes and groans': air shock waves; 6, 'Mt. Olympus trembles': great earthquakes; 7, 'steady vibrations of ground': earthquakes; 8, 'weapons whistle through the air': pyroclastic ejecta; 9, 'loud battle cries': explosive reports; 10, 'Zeus arrives—lightning and thunder, fields and forests burn': volcanic lightning, heat of ignimbrites; 11, 'Earth and sea boil': magma chamber breach; 12, 'immense flame and heat': phreatomagmatic explosion; 13, 'sound of earth/sky collapse': sound of above explosion; 14, 'dust, lightning, thunder, wind': final ash eruptions; 15, 'Titans buried under missiles': collapsed debris. There is a one-to-one correspondence, and Zeus arrives at the climatic moment.

[17] Greene, *Natural Knowledge*, 48 and elsewhere.

[18] R. Halleux, 'Le Problème des métaux dans la science antique', *Bibliothèque de la Faculté de Philosophie et Lettres de l' Université de Liège*, fasc. 209 (1974), 65 ff., comments on another aspect of the natural scientists: 'Il reste toutefois que les premiers se sont intéressés aux pratique de chimie et du métallurgie. D'abord, ils n'étaient pas coupés du monde des artisans. Ensuite l'éventail des phenomènes observables n'était pas si grand que les chercheurs de i'époque pussent négliger des processus où apparaissait, de façon fascinante, la transformation de la matière.

[19] Thales, one of the Seven Sages, born in the second half of the 7th cent. BC.

Aristotle,[20] in fact, suggests that Thales, of Miletus, was the first to enquire into the causes of things, but there is no independent proof of such an interest. The speculation of the Milesian philosophers, however, makes a definite break with the past and justifies the claim that philosophy and science, as we interpret these disciplines, originated with them. The Milesians drew a distinction between the 'natural' and the 'supernatural': as Lloyd[21] explains, they stated that natural phenomena are not the products of random, or arbitrary, influences, but are regular and governed by determinable sequences of cause and effect. Homer, citing specific examples of earthquakes or lightning, attributed them to the anger of Zeus or Poseidon. The Milesians, by contrast, are concerned with these phenomena in general, in which, by their reasoning, the gods play no part. In other words they follow the principal of science that it investigates the universal and not the particular, although they had no conception of scientific method.

The myths in Hesiod, in Homer, and in their survivals in Ionian cosmology of the seventh century BC are the formative materials out of which the Greeks made, for the first time in human history, the transition from *mythic* to *rational*. As such, Greek myths become more than another mythology: they are the beginnings of our culture, our arts, our sciences, and our political forms of thought. From this point forward, the paths of natural science and religion diverge.

In the fifth century BC the Sophists took over higher education and claimed, like Hippias, to impart all subjects (αἱ τέχναι). Socrates, by contrast, apparently soon became disenchanted with 'Science', as Xenophon (*Mem.* 1. 1. 11) writes: 'He did not even discuss that topic favoured by other speakers, that is the "nature of the Universe" (περὶ τῆς τῶν πάντων φύσεως) and avoided speculation on how the so-called Kosmos of the professors works and on the laws that govern the phenomena of the heavens: indeed, he would argue that to trouble one's mind with such problems is sheer folly.'

In the fourth century BC, Aristotle refers to the 'branches of knowledge essential for a free man' (αἱ ἐλευθέριαι ἐπιστήμαι)—that

---

[20] Lloyd, *Early Greek Science*, 1–2.   [21] Ibid. 50 ff. and elsewhere.

is, astronomy, geometry, arithmetic, music, and grammar. The *Meteorologica* treats some of these subjects, namely, astronomy, geology seismology, and meteorology. The work is typical of a stage when the natural sciences had not become fully differentiated from philosophy, but none the less embraced a wide range of scientific knowledge.[22]

In the light of these more general criteria, even when he records his observations without understanding the principles involved in certain phenomena, or states of matter, Pliny may be described as a scientist—rather than a philosopher—however unsophisticated. Moreover, the *Natural History*'s contribution should be evaluated within the context of his aims and stated intentions, which is the purpose of the present reassessment.

To return to the concluding arguments advanced by French:[23] 'Fragments of world-views (like bricks) may certainly look scientific when represented in isolation. Fragments presented collectively, as in source books of ancient, or medieval 'science' and put (silently) into modern categories, take on an authority which none of the fragments had in its own context. More persuasive are examples of the 'scientific attitude' which are often used to show how the ancients, although getting their details wrong, were investigating nature in the right spirit. So much has been said about myth and magic, superstition and rationality, objectivity and science, largely by scientific historians, that the terms are largely debased currency. Some historians have recently recognized that to see science in antiquity we have to have so broad a definition of science as to be meaningless.'

Fortunately, this appears to be far from a universal view. The reader is, therefore, invited to make his, or her, own judgement.

---

[22] Greene, *Natural Knowledge*, 46.
[23] *Ancient Natural History*, pp. xii f.

9

The *Natural History* and Technical Literature

Schilling, assessing the place of Pliny's *Natural History* in technical literature,[1] states, 'La Naturalis Historia... n'est certainement pas la première œuvre latine qui traite de sujets techniques et "scientifiques", mais elle constitue sans doute l'ouvrage le plus représentatif.'

The work is certainly unique in the ancient world in that book 1 includes a list of contents and of authorities cited, both Roman and non-Roman.[2]

Pliny himself states, in his Preface (1) to the emperor Titus, that his work is a novel venture for the Muses who inspire your Roman citizens ('novicium Camenis Quiritium tuorum opus'), and adds that his 'path is not well worn by writers, nor the kind along which the mind aspires to wander. No Roman author has attempted the same project, nor has any Greek treated all these matters single-handedly' (pref. 14). ('praeterea iter non trita est auctoribus via nec qua peregrinari animus expetat; nemo apud nos qui idem temptaverit invenerit, nemo apud Graecos qui unus omnia ea tractaverit'.[3] Schilling continues, 'Dans la constellation littéraire de son temps, Pline occupe incontestablement un rang privilégié, si l'on place au point de vue de l'ampleur de l'œuvre.' Important authorities of

---

[1] R. Schilling, 'La Place de Pline l'Ancien dans la littérature technique', *Revue de Philologie*, 52 (1978), 272–83. [2] Ibid. 273.

[3] Cf. Lucretius, *DRN* 1. 926f., 'That is the spur which lends my spirit strength to pioneer through pathless tracks... where no foot has ever trod before' 'instinctius mente vigenti|avia Pieridum peragro loca nullius ante|trita solo'. He is here referring to his exposition of the Epicurean system as a novel one ('rerum novitatem'). Cf also *DRN* 4. 1–25.

the time who produced definitive works include, among others, Celsus (*De medicina*), Columella (*De re rustica*), Pomponius Mela (*De chorographia*), Seneca (*Quaestiones naturales*) and Papirius Fabianus (*Causae naturales*)—the latter, whose work unfortunately did not survive, Pliny describes as a very accomplished natural scientist ('rerum natura peritissimus').

Judged in the context of such authorities, Pliny exhibits weaknesses. For example, his style—said to possess 'l'allure heurtée et contournée'[4]—compares unfavourably with the graceful, classical style of Columella. Similarly, his scientific criteria fall short of those found in Seneca's *Questiones naturales*.

Pliny's literary output was immense:[5] but, although six of his works are lost, the surviving *Natural History* provides sufficient confirmation of his outstanding diligence. The Younger Pliny (*Letters* 3. 5) sums up his uncle's achievement, drawing attention to his penetrating intellect ('acre ingenium') and amazing powers of concentration ('incredibile studium'). All this, he observes, was achieved against a background of continuous public service. Pliny (pref. 18) is at pains to stress his fulfilment of his obligations. 'I do not doubt that many facts have eluded me. For I am only human and busy with official duties. I pursue my research in odd hours—that is, at night—just in case you think I pack up work then! The days I devote to you.' ('nec dubitamus multa esse quae et nos praeterierint; homines enim sumus et occupati officiis, subsicivisque temporibus ista curamus, id est nocturnis, ne quis vestrum putet his cessatum horis. dies vobis impendimus.')

Schilling,[6] in commenting on the title of Pliny's work, uses a striking image. 'L'Histoire Naturelle', he writes, 'est un pavillon qui ne saurait couvrir toute la marchandise'.

Pliny examines the idea of Nature (*natura*) in the light of terrestrial eruptions and exhalations and their origins. He concludes (2. 208): 'What other explanation could mortal man give except that these things are caused by the divine power of Nature, which is spread throughout the Universe and bursts out in different ways? ('quibus in rebus quid possit aliud causae adferre mortalium quispiam quam diffusae per omne

---

[4] See further, Healy, 'Language and Style', 13 ff.   [5] See Ch. 2. 2.
[6] 'La Place de Pline', 274.

subinde aliter atque aliter numen erumpens?') Nature shows a wonderful generosity (2. 25) towards man. This theme, developed elsewhere in the *Natural History*,[7] also reminds us that Nature can be responsible for bad things ('naturae improbitas and scelera naturae'). 'Man', states Pliny, 'is the highest species, and Nature appears to have created all other things for his benefit. Her very many gifts, however, are bestowed at a cruel price, so that we cannot confidently say whether she is a good parent to mankind, or a somewhat forbidding stepmother.' ('principium iure tribuetur homini, cuius causa videtur cuncta alia genuisse natura, saeva mercede contra tanta sua munera, ut non sit satis aestimare, parens melior homini an tristior noverca fuerit.') Pliny rightly observes that man experiences most ills at the hands of his fellow men (7. 5): 'at hercule homini plurima ex homine sunt mala.'

In his examination of the philosophical problem posed by the idea of Nature, Pliny explains the Universe in terms of four basic elements, fire, air, water, and earth (2. 10): 'I observe that there is no doubt about there being four elements. The uppermost is fire, source of those eyes of the great array of blazing stars. The next element is a vapour which the Greeks and ourselves call by the same name "air". This is the principle of life that permeates every part of the Universe and is intermingled with the whole. Suspended by its force, the earth is balanced in the middle of space, together with its fourth element—its waters.' ('nec de elementis video dubitari quattuor esse ea: igneum summum, inde tot stellarum illos conlucentium oculos; proximum spiritum quem Graeci nostrique eodem vocabulo aera appellant, vitalem hunc et per cuncta rerum meabilem totoque consertum; huius vi suspensam cum quarto aquarum elemento librari medio spatii tellurem'.) This explanation follows that of the Stoic Balbus (*Cic. ND* 2. 84.) But, as Schilling[8] points out, on closer examination, one is struck by two divergences in Pliny's account: 'La cohésion du monde se fondait sur la transmutation incessante de ces quatres éléments. Pline ne fait pas allusion à cette *vicissitudo corporum*.' The stability of the world derives from the

---

[7] 7. 2–5.   [8] 'La Place de Pline', 276 ff.

balance assured by contrary forces. Schilling also draws attention to a third divergence from orthodox Stoicism which states that fire in the upper regions or air is the dynamic principle ('vim vitalem per omnem mundum pertinentem', Cic. *ND* 2. 9. 24). It is πνεῦμα νοερὸν καὶ πυρῶδες οὐχ ἔχον μορφήν (*Placit.* 1. 7. 19 Diels). Pliny (2. 10) employs the term *spiritus* as a principal, endowed with life and penetrating all things ('vitalem hunc et per cuncta rerum meabilem'), but in this context it is simply 'air'. This has further consequences, as air is identified with the 'sky' (2. 102). Pliny also has recourse to the principle of the reaction of opposite forces, namely '*sympathia* and *antipathia*' (20. 1). He abandons the Stoic concept that the 'intelligible principle' gives life to the world, substituting a struggle between obscure forces. Then everything becomes possible. Pliny writes, 'when I have observed nature, she has always persuaded me to deem no statement about her incredibile' ('nam mihi contuenti semper suasit rerum natura nihil incredibile existimare de ea'). There is a certain obscurity in his definition of Nature and he does not clearly distinguish between her, God, and the world.

Schilling[9] concludes: 'Cette disposition d'esprit éclaire la démarche de Pline dans ses investigations. Il montre une étonnante disponibilité pour accueillir les faits: en ce sens, il pourrait être loué, même par un savant moderne, d'être libre de tout préjugé. En revanche, il ignore l'idée de loi générale, et préfère s'en tenir à des classifications de données analogues: ainsi il énumère les différentes variétés d'aimants.

Pliny's Critical Judgement

Although Pliny appears unduly credulous on numerous occasions, we may find in the *Natural History* not a few contributory factors to explain—but not to excuse—his apparent lack of critical judgement. There are, moreover, often specific reasons for his acceptance of the bizarre.[10]

While relying on other authorities, he tends to reproduce their errors, or misconceptions. Thus, because of his high

---

9 Ibid. 278.
10 See further Ch. 5 above.

regard for Papirius Fabianus, he quotes his reiteration of the widely held belief that marble regenerates itself (36. 125). He qualifies his acceptance of this assertion, however, by adding that, if it is true, there are grounds for hoping that there will always be sufficient marble to satisfy extravagant lifestyles.

Historical record (2. 140) asserts that thunderbolts were caused by, or occurred in answer to, certain rites and prayers. Lucius Piso, in his *Annals* 1, states that this practice was frequently followed by King Numa, although when Tullus Hostilius copied him, but used the wrong ritual, he was struck by lightning. Pliny concludes (2. 141) that, 'although such indications are certain in some cases but doubtful in others, and approved by some persons but, in the view of others, to be condemned, in accordance with Nature's will and pleasure, I for my part am not going to leave out the rest of the things worth recording in this department.' ('quamobrem sint ista ut rerum Natura libuit, alias certa alias dubia, aliis probata aliis damnanda, nos cetera quae sunt in his memorabilia non omittemus.') Here he is following an annalistic tradition and, similarly, he elsewhere (2. 147) accepts prodigies recorded in archives ('in acta eius anni relatum est'), as well (7. 86) as predictions attributed to gods ('visus et numinum fuere praesagia'). With regard to pronouncements by other authorities, he gives them credit according to his esteem for them, as in the case of Varro (29. 65): 'I should hesitate to put forward a remedy obtained from these creatures had not Marcus Varro, when aged 73, recorded that a very effective remedy for asp bites is for the victim to drink his own urine' ('ex his remedio, ni M. Varro LXXIII vitae anno prodidisset aspidum ictus efficacissime sanari hausta a percussis ipsorum urina').

In discussing remedies obtained from man, Pliny asks (28. 10), 'Have words and incantations any effect? If they have, it would be right and proper to give all the credit to mankind. As individuals, however, all our wisest men reject belief in them, although as a body the general public at all times believes in them subconsciously.' ('ex homine remediorum primum maximae quaestionis et semper incertae est, polleantne aliquid verba et incantamenta carminum, quod si verum est homini acceptum fieri oportere conveniat, sed viritim sapientissimi cuiusque respuit fides, in universum vero omnibus horis credit vita nec sentit.')

The *Natural History* and Technical Literature   111

In his account of remedies from the animal kingdom (30. 137), he writes: 'Certain details can scarcely be included as serious items, but I must not omit them, since they have been recorded' ('vix est serio conplecti quaedam, non omittenda tamen quia sunt prodita').

Tradition—which Pliny tends to follow—is the key word underlying many of his beliefs. There is, however, a partiality about his acceptance. On the one hand he writes (7. 8): 'Nevertheless, in most instances of these [apropos physiognomy] I shall not myself pledge my own faith and shall preferably ascribe the facts to the authorities who will be quoted for all doubtful points; only do not let us be too proud to follow the Greeks because of their far greater industry or older devotion to study'. ('nec tamen ego in plerisque eorum obstringam fidem meam, potiusque ad auctores relegabo qui dubiis reddentur omnibus, modo ne sit fastidio Graecos sequi tanto maiore eorum diligentia vel cura vetustiore.') Yet, more often than not, Pliny is extremely critical of the Greeks whom he describes (26. 15) as a race of very little consequence ('levissima gens'), untrustworthy ('vanitas'), boastful and very prone to praising themselves ('Grai, genus in gloriam sui effusissimum'). He also draws attention to their inventiveness in regard to tall stories (7. 174 and 36. 91) and is especially critical of the stories that appear in the Greek poets.

Pliny shares, with the rest of the ancient world, the view that knowledge derives from transmitted information (tradition is included in this category). This accounts, in part, for his desire not to omit anything that may be of benefit to his reader.

There have been many contrasting judgements relating to Pliny's contribution to the sum of human knowledge. Three chosen by Schilling,[11] namely those of Buffon, Cuvier, and Littré, give a representative picture of divergent assessments. Buffon praises the breadth of the *Natural History*, which is much greater than that of Aristotle's works: the *Natural History* embraces all natural sciences and arts: 'ce qu'il y a d'étonnant, c'est que dans chaque partie Pline est également

---

[11] 'La Place de Pline', 281.

grand.' Cuvier, by contrast, praises the erudition of a soldier and statesman, but criticizes the *mélange* of true and false found in his work which detracts from its value. Finally, Littré, who is more demanding of preciseness—the hallmark of the nineteenth century scientist—is the most critical. He asserts that there is no comparison between Pliny and earlier students of Nature and that Pliny and Aristotle have nothing in common!

Pliny does not know how to assemble his facts within the framework of a general law and admits as much (11. 8): 'Let each man form his own opinion, but my purpose is to point out the manifest properties of objects, not to search for doubtful causes' ('denique existimatio sua cuique sit, nobis propositum est naturas rerum manifestas indicare, non causas indagare incertas').

Secondly, Pliny provides for the student of folklore and history a unique treasure house of customs and rites, many of which, together with superstitions, are to be found in *NH* 28. Thirdly, the *Natural History* is a vast source of drug lore and is full of interest for the student of botany.

Finally, one of the ongoing interests Pliny has had, for the latter part of the twentieth century, is his concern for the environment and man's exploitation of natural resources. His preoccupation with ecological matters, however, has as much to do with man's greed and desire for extravagant lifestyles, as with the degradation of Nature herself.[12]

---

[12] See Ch. 19 below.

10

Science in the *Natural History*

A recently published diagram, illustrating 'Field of Research' classifications,[1] provides a useful framework within which to examine Pliny's contribution to the Natural Sciences (as well as to the Social Sciences and the Humanities).

In the Report, three major categories are listed:

1. Natural Sciences, Technologies, and Engineering (with ten subdivisions):
    Chemical sciences (organic and simple reactions)
    Physical sciences (acoustics and optical physics, magnetism)
    Earth sciences (geology, mineralogy, geochemistry, oceanography, hydrology, atmospheric sciences)
    General engineering (mining and ore-mineral processing, and civil engineering)
    Applied sciences (material sciences and technology)
    Biological sciences (botany, zoology, and 'ecology')
    Agricultural Sciences (soil and water sciences (hydrology), crop and pasture production, horticulture, and forestry).
    There is significantly no account of animal farming
2. Social Sciences:
    Sociology and anthropology[2]
3. Humanities:
    Language[3] and style,[4] literature

---

[1] See the *Report by the Department of Industry, Technology and Regional Development, Science and Technology Policy Branch* (Canberra, 1993), 2.

[2] Pliny provides a commentary on the mores and lifestyle of Romans under the early principate, often contrasting their materialistic attitude with the simpler pattern of behaviour under the Republic and with the *gravitas* of those then in authority.

[3] See Ch. 7 above.

[4] On style, see esp. Healy, 'Language and Style', 13 ff. Both language and style play an important part in the presentation of Pliny's material in the *Natural History*.

Archaeology
Philosophy and ethics
History of philosophy, science, and medicine[5]

Apart from the legendary material and *mirabilia* which Pliny includes in the *Natural History*, his work is a compendium of observed and reported fact.

Pliny's contribution to science and technology—however unsophisticated, and irrespective of any understanding of the principles underlying observed phenomena, chemical reactions, or the physical properties of matter—is discussed in the following chapters.

---

[5] In the following account of Pliny's contribution to science and technology, I have omitted any reference to medicine, pharmacology, and astronomy. Similarly there is no treatment of geography (currently defined as an 'Arts' subject). In these fields his originality is limited because of his reliance on secondary sources. On medicine see V. Nutton, 'Pliny and Roman Medicine', in French–Greenaway, *Roman Science*, 30 ff. and bibliography. Also R. French, ibid., 'Pliny and Renaissance Medicine', 252 ff.

11

Chemistry

11.1 Introduction

Chemistry,[1] as a closely defined area of study, dates from the eighteenth-century fashion for the ordering of knowledge.[2] It is the branch of physical science concerned with the composition, properties, and reactions of substances, including the accompanying energy changes. Analysis, which plays a major role in chemistry, involves the examination of substances of unknown composition in order to determine their composition. The modern chemist has a vast number of analytical techniques at his disposal but, even so, there are basically three

---

[1] French, *Ancient Natural History*, p. xxi, nn. 6–7, strongly opposing the use of modern categories in any description and assessment of ancient 'science', writes, 'Historians of the "physical", or "exact" sciences have been particularly prone to see breath-taking advances in Greek science that were nevertheless halted by things like their aversion to experiments and ignorance of statistics. Characteristically source-books omit the contexts, often including even the chronological, of their ancient extracts, and group them into modern categories: statics, dynamics, optics, acoustics, chemistry and chemical technology, biology (including natural selection), botany (including classification), physiological psychology.' The implication is, seemingly, that nothing worthwhile was discovered before the 19th cent.! Such a premise is, in my view, unsustainable. The observation and accurate recording of physical and other phenomena, even if their underlying principles were not fully understood, is surely part of scientific curiosity? The degree of sophistication is not a valid criterion. Examined dispassionately, much of Pliny's contribution, for example, in the field of 'Earth Sciences' was not improved upon until the late 18th cent.! Finally, it still seems to me—in spite of semantic, or other arguments—more meaningful to continue to examine the achievements of Pliny, and indeed of other natural scientists using modern, readily understood categories. ('cum honoris praefatione ponendis'!)

[2] F. Greenaway, 'Chemical Tests in Pliny', in French–Greenaway, *Roman Science*, 159 f., with n. 2. Cf. J. R. Partington, *A History of Chemistry*, i/1 (London, 1970), pp. xi–xviii.

ways in which a chemical species may be detected and estimated: (1) by isolation of the species itself, (2) by inference from a characteristic chemical reaction of the species, or (3) by inference from a characteristic physical property.[3]

'Pliny's treatment of subjects we would now distinguish as chemical', writes Greenaway,[4] 'ranges over a wide field: technical, pharmaceutical, metallurgical and much else.... It is impossible to pull them together by a common thread of theory.'

A positive and selective approach, however, may produce a more constructive picture of his contribution. There are two fundamental aspects of Pliny's 'chemistry': namely, (*a*) his observation of natural phenomena, among them the mechanical processes of evaporation and precipitation,[5] and (*b*) his use of discriminatory tests[6] which would, centuries later, form the basis of what may loosely be defined as 'industrial chemistry'. These tests involve reference to the senses, chemical change, and composition determined by process tests and the touchstone.[7]

11.2 SOLUBILITY AND SATURATION

Pliny (31. 67) is well aware of solubility and saturation: 'If more than a *sextarius* of salt is dropped into four *sextarii* of water, the water is saturated[8] and the salt does not dissolve. [Such a solution] gives the strength and properties of the saltest sea.' ('si plus quam sextarius in quattuor sextarios aquae mergatur, vinci aquam salemque non liquari.... salsissimi maris vim et naturam implet.')

---

[3] Greenaway, 'Chemical Tests', 148.   [4] Ibid. 147.
[5] As of the solute in a saturated solution.
[6] Greenaway, 'Chemical Tests', 150, lists eleven groups of tests with the number of examples of each: 1, smell (1); 2, taste (5); 3, feel (3); 4, colour (10); 5, melting point (2); 6, effect of heat (5); 7, density, estimated (1); 8, density, measured (2); 9, chemical change (6); 10, process test (1); 11, touchstone (1).   [7] See Ch. 17. 4. 9a and Ch. 14, s.vv. 'chert', 'lapis Lydius'.
[8] Bailey, *Chemical subjects*, 161 n. 73. This is surely one of the earliest estimates of solubility. The actual solubility of salt is 35.5 in 100 at 0 °C. *Sextarius:* see M. Darton and J. O. E. Clark, *The Dent Dictionary of Measurement* (London, 1996), 417. The *sextarius* is a basic measure of liquid capacity—especially in the preparation of wine and oil—and of dry capacity, very closely approximating to the modern pint (0.546 l.).

Chemistry

The salinity of the oceans varies between 4 and 6 per cent. The Dead Sea, however, the lowest point on earth, contains approximately 26 per cent of dissolved chemical salts, which accounts for its extreme buoyancy and the fact that no animals can survive in its waters. It is interesting to observe, therefore, that Pliny's saturated solution (25%), which he equates with the 'saltest sea', is surprisingly accurate, for whatever reason.

11.3 Evaporation and Precipitation

Some processes occur spontaneously in nature and are reproduced, or adapted, by man. In book 31, in which Pliny discusses aspects of 'water' and the qualities and properties of various waters and hot springs, he refers to evaporation, precipitation, and filtration.

In the context of the production of natural salt[9] (31. 73 ff.) Pliny uses a number of common words as interchangeable *technical terms*. Thus the verbs *cogi, densari, siccari, coqui,* and *(in)arescere* are synonymous and may be translated 'evaporate', since they refer simply to the concentration of brine by the heat of the sun until the salt (solute) is precipitated. Pliny explains (ibid.): 'Salt occurs naturally and is also produced artificially. Each type is formed in several ways, but there are two main processes involved, namely evaporation of the brine (sea water) to precipitate the salt. In summer, salt is so formed in the Tarentine Lake[10] by the sun: the whole expanse of water, always shallow and never more than knee-deep, becomes salt. The same thing happens in Sicily at Lake Cocanicus, and at another lake near Gela. Only the edges of these evaporate.' ('sal omnis aut fit aut gignitur, utrumque pluribus modis, sed causa gemina, coacto umore vel siccato. siccatur in lacu Tarentino aestivis solibus, totumque stagnum in salem abit, modicum alioqui, altitudine genua non excedens, item in Sicilia in lacu qui Cocanicus vocatur et alio iuxta Gelam.

---

[9] Varro, *RR* 1. 7. 8, and H. Blümner, *RE* 1 A, Suppl. 2075 ff. (s.v. 'Salz'). See generally R. König, J. Hopp, and W. Glockner, *Plinius: Naturkunde, Buch 31* (Zurich, 1994), esp. pp. 169 ff., Glockner on 'Kochsalz', a brief history of salt from its first use in 3000 BC; also L. A. Hutter, *Wasser und Wasseruntersuchungen*, (Frankfurt and Arau, 1994).
[10] On Tarentum cf. *NH* 3. 99 ff.

horum extremitates tantum inarescunt.)' Salt is still produced in this manner near Marsala[11] (western Sicily) and the heaps are kept dry by being covered with brightly decorated tiles—a very colourful sight.

Pliny (ibid.) continues: 'In Phrygia,[12] Cappadocia,[13] and at Aspendus,[14] the evaporation spreads as far as the centre. The amazing thing is that however much salt is removed during the day, that amount is replenished overnight.[15] All the salt from these natural "pans" is in the form of fine powder and not in lumps.' ('sicut in Phrygia, Cappadocia, Aspendi, ubi largius coquitur et usque ad medium. aliud etiam in eo mirabile quod tantundem nocte subvenit, quantum die auferas. omnis e stagnis minutus atque non glaeba[16] est.')

Pliny (31. 74) mentions 'another kind of salt [that] is spontaneously produced from sea water—namely the foam left on the shoreline and on the rocks by the sea'—as though this were different from that already described ('aliud genus ex aquis maris sponte gignitur spuma in extremis litoribus ac scopulis relicta'). This has a sharper taste ('acrior'). 'There are also three different kinds of native salt; for, in Bactra are two vast lakes, one facing the Scythians, the other, the Arii, which exude salt, while at Citium (Cyprus), and around Memphis [Egypt], precipitated salt is taken out of a lake and then dried in the sun' ('sunt etiamnum naturales differentiae tres. namque in Bactris duo lacus vasti, alter ad Scythas versus alter ad Arios, sale exaestuant, sicut ad Citium in Cypro et circa Memphim extrahunt e lacu, dein sole siccant').

Pliny (31. 75) adds: 'The surface of rivers may also evaporate and form salt, the rest of the streams flowing as if it were under ice: this happens near the Caspian Gates[17] and they are called "rivers of salt"' ('sed et summa fluminum densatur in salem amne reliquo veluti sub gelu fluente, ut apud Caspias portas quae salis flumina appellantur').

Elsewhere (31. 77), 'there are mountains of natural salt, such as at Oromenus (India), where it is cut out like blocks of stone from a quarry and ever replaces itself' ('sunt et montes

---

[11] *NH* 3. 89 and 91. [12] *NH* 5. 145. [13] *NH* 5. 146.
[14] Strabo 14. 4. 2. *RE* 2, Suppl. 1725.
[15] Cf. Ch. 14 below, p. 205, on the regeneration of minerals.
[16] See Ch. 14, s.v. 'sulphur' for this term applied to sulphur.
[17] *NH* 6. 40 and 43.

nativi salis, ut Indis Oromenus. in quo lapicidinarum modo caeditus renascens'). 'It is [also] dug out of the earth in Cappadocia, where it has evidently been formed by the evaporation of moisture. There it is split into sheets, like selenite'[18] ('effoditur e terra, ut palam est umore densato, in Cappadocia. ibi quidem caeditur specularium lapidum modo').

Pliny's account of evaporation and precipitation is, in its main detail, accurate and based on observation, but his use of such a variety of terms for these processes is puzzling. Furthermore, he asserts that salt is replaced at night,[19] implying that the moon has some part to play in this.

Pliny (31. 81) continues with a description of purpose-made salt pans. There are 'various kinds of artificially produced salt. The common one, and the most plentiful, is made in salt pools by running sea water into them, not without streams of fresh water, but rain helps very much and, above all, warm sunshine. In Africa, round about Utica, salt is formed in heaps like hills; when they have hardened under sun and moon, they are not melted [dissolved] by any moisture, and even iron cuts them with difficulty.' ('facticii varia genera, volgaris plurimusque in salinis mari adfuso non sine aqua dulcis riguis, sed imbre maxime iuvante ac super omnia sole *multo,* aliter non arescens. Africa, circa Uticam construit acervos salis ad collium speciem, qui ubi sole lunaque induruere, nullo umore liquescunt vixque etiam ferro caeduntur.')

Salt 'is also produced in Crete, without fresh water, by letting the sea flow into the pools, and around Egypt by the sea itself, which penetrates the soil soaked, as I believe, by the Nile. Salt is also made by pouring water from wells into salt pools.' ('fit tamen et in Creta sine riguis mare in salinas infundentibus et circa Aegyptum ipso mari influente in solum, ut credo, Nilo sucosum. fit et puteis in salinas ingestis.')

Salt pans are still a common feature of the Mediterranean region (although not all of them are still in use), as well as on

---

[18] See Ch. 14, s.v. 'lapis specularis'.
[19] The explanation is straightforward. The sun is the only direct agent, producing the temperature necessary for these processes: the moon, of course, plays no part in these. Salt continues to precipitate overnight because the vapour pressure of the brine does not vary much with temperature. In other words, the water continues to evaporate during the night and more salt is produced. (The solubility of salt does not vary much with differences in temperature and the decrease in temperature at night is not great.)

the west coast of Spain between Cadiz and Puerto de Santa Maria; they utilize the process of direct evaporation from shallow 'reservoirs' of sea water (*salinas*) within low dykes constructed immediately on shore.

'Near Pelusium[20] also, King Ptolemy found salt when he was making a camp [31. 78]. This led subsequently to the discovery of salt by digging away the sand[21] even in the rough tracts between Egypt and Arabia. It has also been found as far away as the Oracle of Ammon[22] through the parched deserts of Africa, where at night it increases, as the moon waxes.' ('invenit iuxta Pelusium Ptolemaeus rex, cum castra faceret. quo exemplo postea inter Aegyptum et Arabiam etiam squalentibus locis coeptus est inveniri detractis harenis, qualiter per Africae sitentia usque ad Hammonis oraculum, is quidem crescens cum luna noctibus.') The Ammoniac salt, so called because it is found under the sand in the region of Cyrenaica (31. 79), consists of chlorides of sodium, calcium, and magnesium.

The Amantes (5. 34), in Cyrenaica, built their houses of blocks of salt quarried out of the mountains, like stone ('domus sale montibus suis exciso ceu lapide construunt'). Similarly, at Gerra[23] (31. 78), a town in Arabia, the walls and houses are constructed of blocks of salt cemented with water ('Gerris Arabiae oppido muros domosque e massis salis faciunt aqua feruminantes').

Pliny clearly does not understand the role of fresh water in salt refining. It is, of course, used for redissolving the crude, rough sea salt, which occurs in different colours according to what impurities are present. These are filtered out from the brine which is then reprecipitated, by heat, to make the finer, pure white end-product.

There are yet other sources of salt (31. 82). 'In Chaonia[24] there is a spring the water from which they boil and, on cooling it, obtain a salt that is insipid and not white. In the

---

[20] *RE* 19, Suppl. 414. Cf. *NH* 4. 49.
[21] See Partington, *History of Chemistry*, i/1, 114 (31. 78). This is not ammonium chloride NH_4Cl_3, but a mixture of salts of sodium, calcium, and magnesium chloride.
[22] The Oracle of Ammon is in the oasis of Siwa, in Libya.
[23] *NH* 6. 147; Strabo 16. 3. 3.   [24] *NH* 3. 112. Cf. Arist. *Mete*. 2. 3. 42.

provinces of Gaul and Germany,[25] they pour salt water on burning logs. It is thought that the type of wood makes a difference to the quality of the salt; the best is oak.' He adds, rather curiously, that 'when [the brine] is poured on, even the burning wood [reading 'carbo'] turns into salt'. ('in Chaonia excocunt aquam ex fonte refrigerandoque salem faciunt inertem nec candidum. Galliae Germaniaeque ardentibus lignis aquam salsam infundunt. illi quidem et lignum referre arbitrantur. quercus optima. ita infuso liquore salso *carbo* etiam in salem vertitur.')

Pliny (31. 81) completes his account of salt production by referring to an interesting phenomenon allegedly found in Babylon where 'the first thickening [*densatio*] solidifies into a liquid bitumen[26] like oil, which is also used for lamps. When this is taken away, salt is found underneath' ('prima densatione Babylone in bitumen liquidum cogitur oleo simile, quo et in lucernis utuntur. hoc detracto subest sal'). The observation almost anticipates, in a very rudimentary way, the basis of modern fractional distillation.

11.4 'Osmosis' (?)—Filtration

Pliny (31. 70) refers to methods for the desalination of sea water, especially useful for those on ships when short of drinking water. 'Fleeces spread round a ship become damp by absorbing sea-spray: fresh water can then be squeezed out of their fleeces. Similarly, hollow balls of wax let down into the sea in nets, or empty containers with their openings sealed, collect fresh water inside.' ('expansa circa navem vellera madescunt accepto halitu maris, quibus dulcis umor exprimitur, item demissae reticulis in mare concavae ex cera pilae vel vasa inania opturata dulcem intra se colligunt umorem.') Such methods are, of course, not viable: the walls of the containers could not act as filters.

On land, sea water is made fresh by filtration through clay ('nam in terra marina aqua argilla percolata dulcescit').

The reference to salt from Agrigento (31. 85) which submits to fire and sputters in water—'Agrigentinus ignium patiens ex aqua exsilit'—is probably to *lime* not to salt.

---

[25] Cf. Varro, *RR* 1. 7. 8, and H. Blümner, *RE* 1 A, Suppl. 2075.
[26] *NH* 35. 178 ff.

11.5 Hydrometallurgy

Hydrometallurgy[27] is another method for concentrating minerals for smelting—making use of natural or artificial solutions. Metal may be recovered from these solutions by evaporation and crystallization, or by precipitation of a noble metal such as copper by its replacement in solution by a less noble one such as iron.

Natural weathering of mineral deposits occasionally results in caves and these may be traversed by streams and less constant flows of water. The water flowing out of such caverns, or disused mine-workings, will be coloured and may be seen to have deposited crystals of salts upon its banks. Rio Tinto, in Huelva province, in southern Spain, got its name in this manner.

Pliny (34. 123 ff.) alludes to hydrometallurgy in his description of chalcanthon ('flower of copper'), identified as ferrous ($FeSO_4$), or copper sulphate ($CuSO_4$), that is 'shoemaker's black' (*atramentum*):

> There is no substance that has an equally remarkable nature. It occurs in Spain in wells, or pools that contain that sort of water. This water is boiled with an equal quantity of pure water and poured into wooden tanks. Over these are firmly fixed cross-beams from which hang cords held taut by stones, and the mud clinging to the cords in a cluster of glassy drops has somewhat the appearance of a bunch of grapes. It is taken off and then left to dry for thirty days. Its colour is a brilliant blue and it is often thought to be glass. When dissolved it makes a black dye for colouring leather.

> appellant enim chalcanthon. nec ullius aeque mira natura est. fit in Hispaniae puteis stagnisve id genus aquae habentibus. decoquitur ea admixta dulci pari mensura et in piscinas ligneas funditur. immobilibus super has transtris dependent restes lapillis extentae, quibus adhaerescens limus vitreis acinis imaginem quandam uvae reddit. exemptum ita siccatur diebus XXX. color est caeruleus per quam spectabili nitore, vitrumque esse creditur; diluendo fit atramentum tinguendis coriis.

The process is as follows:[28] Chalcopyrite roasts to sulphate at about 500–600 °C. The heaps are broken up and the ore dumped into tanks where the liquid is concentrated by boiling,

---

[27] Tylecote, *Metallurgy*, 64 f.   [28] Ibid. 65.

Chemistry 123

and the sulphates are crystallized on ropes like grapes. The copper values could be recovered by subsequent smelting with a silicious flux giving rise to the usual iron silicate slag and copper.

Shoemaker's black is also made in several other ways—involving evaporation. 'Earth of the kind indicated is hollowed into trenches, droppings from the side of which form "icicles" in a winter frost which are called "drop-flower of copper" and this is the purest kind. ... It is also made in pans hollowed in the rocks, into which slime is carried by rain-water and freezes, and it also forms in the same way as salt when very hot sunshine evaporates the fresh water let in with it.' ('Fit et pluribus modis: genere terrrae eo in scrobes cavato quorum e lateribus destillantes hiberno gelu stirias stalagmian vocant, neque est purius aliud. ... fit et in saxorum catinis pluvia aqua conrivato limo gelante; fit et salis modo flagrantissimo sole admissas dulces aquas cogente.')

The first method described by Pliny is akin to the growth of a crystal by precipitation, while in the other processes, as in the case of salt, he suggests that fresh water somehow has a part to play in initiating the process.

11.6 PROPERTIES OF WATER

Some waters have remarkable properties according to the solutes[29] found in them. Pliny (31. 12) writes, 'The Tungri, a state of Gaul, has a remarkable spring that sparkles with innumerable bubbles, with a taste of iron rust, which yet cannot be detected until the water has been drunk'[30] ('Tungri civitas Galliae fontem habet insignem plurimis bullis stillantem,

---

[29] A. McLeish, *Geological Science* (London, 1991), 102, lists the salts in the sea as NaCl (sodium chloride, salt), 78.04%; $MgCl_2$ (magnesium chloride), 9.21%; $MgSO_4$ (magnesium sulphate), 6.53%; $CaSO_4$ (calcium sulphate), 3.48%; KCl (potassium chloride), 2.11%; $CaCO_3$ (calcium carbonate), 0.33%; $MgBr_2$ (magnesium bromide), 0.25%.

[30] As at Bath (Avon) and Treffiw (Conwy valley, North Wales). The water in the Pump Room at Bath, derived from local springs, was once described as having the taste of 'flat-irons'. The conduits into the Great Bath are coloured a yellowish brown because of the deposition of iron salts from the water over the years. Similarly, the Spa waters of Treffiw are rich in iron and other minerals.

ferruginei saporis, quod ipsum non nisi in fine potus intellegitur'). Some water corrodes bronze and iron (31. 28), 'aena etiam ac ferrum erodi illa aqua'. Yet other types are fatal to life (31. 26 f.). 'In Arcadia, for example, near the Pheneus, there flows from the rocks a stream called Styx, which, I have said, proves instantly fatal to life, but Theophrastus tells us that it contains small fish which are equally deadly' ('in Arcadia ad Pheneum aqua profluit e saxis Styx appellata, quae ilico necat, ut diximus, sed esse pisces parvos in ea tradit Theophrastus, letales et ipsos'). This is unlike any other kind of poisonous spring.

Other, health-giving, waters are found in spas, as in the hot springs of Mattiacum (31. 20), in Germany, across the Rhine ('sunt et Mattiaci in Germania fontes calidi trans Rhenum').

Perhaps the most interesting waters are those with petrifying properties,[31] a number of which are recorded by Pliny (31. 29 f.).

At Perperena [Mysia], a spring turns to stone any land it irrigates and the thermal waters of Adepsus, in Euboea, have the same effect. Whatever rocks reach the stream, (all) increase in size. At Eurymenae, garlands thrown into a spring turn to stone, bricks thrown into the river at Colossae are found to be of stone when retrieved. At the mine on Scyros all the trees washed by the river are petrified, branches and all. At Mieza, in Macedonia, drops of water form stalactites hanging from arched roofs, while, in a cave at Corinth, they become stalagmites after falling. In certain caves the water forms stalactites and stalagmites that join to make pillars,[32] as at Phausia,[33] on the peninsula facing Rhodes; these pillars are of different colours.

in Perperenis fons est quamcumque rigat lapideam faciens terram, item calidae aquae in Euboeae Adepso. nam quae adit rivus saxa in altitudinem crescunt. in Eurymenis deiectae coronae in fontem lapideae fiunt. in Colossis flumen est quo lateres coniecti lapidei

---

[31] McLeish, *Geological Science*, 189, describes the petrification of organic materials through impregnation by substances such as calcite, silica, and iron minerals. A well-known example of petrifying waters is found at Knaresborough (North Yorkshire).

[32] So at Gough's caves, Cheddar Gorge (Avon); at Nerja (Spain) and the Cuevas del Drach (Majorca); at Damlatas (southern Turkey), and many more examples.

[33] The location has not been identified; see *RE* 19. 1902 (s.v. 'Phausia').

extrahuntur. in Scyretico metallo arbores quaecumque flumine adluuntur saxeae fiunt cum ramis. destillantes quoque guttae lapide durescunt in antris, conchatis ideo, Miezae in Macedonia etiam pendentes, in ipsis camaris, at in Corinthio cum cecidere, in quibusdam speluncis utroque modo, columnasque faciunt, ut in Phausia Cherrhonesi adversae Rhodo in antro magno etiam discolori aspectu.

In this descriptive passage Pliny records a well-attested geological occurrence in limestone caves. Once caverns are isolated above the water table, percolating waters drip from the roof and form unusual deposits. Appropriately called dripstone, they consist of deposits of calcium carbonate. Two major kinds are recognized. Stalactites, are ice-like pendants that hang down from the cave roof. When dripping water is exposed to the air of the cave, some of the carbon dioxide contained in solution escapes and calcium carbonate is precipitated. In time a long pendant forms, customarily with a long narrow tube extending through its full length. Seldom though is such perfection achieved.... The stalagmites are deposits built upward from the cave floor. They usually form below stalactites, from saturated waters dripping from the latter, and, of the two structures, are normally the thicker and more diversified in shape. Stalagmites do not contain a central tube. Stalactite and stalagmite may eventually meet and fuse to form a column.

11.7 EFFLORESCENCE

Chemical change, or evaporation, may produce efflorescence[34] on the surface of a stone, that is in the form of powder, or crystals. Pliny (36. 133) is aware of this property and writes of the stone of Assos, also referred to earlier (ibid. 131) as the 'sarcophagus' stone, 'Belonging to the same stone is what is called efflorescence, which is soft enough to form powder[35]

---

[34] P. W. Birkeland and E. E. Larson (eds.), *Putnam's Geology*[5] (Oxford, 1989), 580.
[35] C. S. Smith, in Eichholz, *NH 36–37*, p. 107, note *e*. The powdery efflorescence may well be gypsum ($CaSO_4 2H_2O$) produced by the action of sulphuric acid which would, in turn, have resulted from the decomposition of pyrites upon the limestone.

and is just as effective as the stone for certain purposes' ('eiusdem lapidis flos appellatur, in farinam mollis ad quaedam perinde efficax').

The term 'efflorescence' clearly derives from the Latin *flos* ('flower') and this is just one of many examples where the objection, by some, to the use of the correct, 'modern' scientific term in referring to this readily observed chemical change is totally unjustified. Other compounds, among them sodium carbonate (Na_2CO_3), commonly exhibit such a surface change.

11.8 Chemical Reactions

Pliny (33. 84) records reactions with salt used in the cementation process to purify gold.[36] Gold 'is heated with twice its weight of salt and three times its weight of copper pyrites and again with two portions of salt and one of alum. Treated in this way it draws poison out, when the other substances have been burnt up with it in an earthenware crucible while it remains pure and uncorrupted itself.' ('torretur et cum salis gemino pondere, triplici misyis ac rursus cum II salis portionibus et una lapidis, quem schiston vocant. ita virus trahit rebus una crematis in fictili vase, ipsum purum et incorruptum.')

Salt is also used in the treatment of litharge (33. 109): '[when freed of impurities] they grind it in mortars for six days, three times daily washing it with cold water and, when these operations have ceased, with hot, and adding rock salt, an obol weight to a pound of "scum". Then on the last day they store it in a lead container.' ('postea sex diebus terunt in mortariis ter die abluentes aqua frigida et, cum desinant, calida, addito sale fossili in libram spumae obolo. novissimo die dein condunt in plumbeo vase'). Pliny's description includes the following operations: (1) the use of water to wash out the impurities; (2) the addition of salt, that is sodium chloride

---

[36] See further, J. F. Healy, 'Greek Refining Techniques and the Composition of Gold–Silver Alloys', *Revue belge de Numismatique*, 120 (1974), 25 ff. In laboratory experiments the gold content of a commercial alloy of 37.5% was upgraded to 93% gold, and the silver and copper formed a blue vitreous product with the brickdust, according to the following simplified equation: $Cu + Au + Ag + NaCl +$ brickdust at $800\,°C \rightarrow Au + AgCl_3 +$ vitreous slag.

(NaCl), and hot water (hot water is needed to initiate and sustain this reaction): $5PbO + H_2O + 2NaCl \ 2NaOH + PbCl_2, 7PbO$; (3) the residue is heated, and produces a pigment called Turner's yellow (lead oxychloride): $PbCl_2, 4PbO_2$. (See Partington, *Chemistry*, 525.) Pliny is here virtually describing a 'cementation' process designed to remove the silver present in what was probably argentiferous galena. This view is supported by the fact that Pliny has already mentioned the purification of gold (*NH* 33. 84). The use of lead in cupellation[37] is well attested in the ancient world[38] and certainly familiar to Pliny (33. 95) apropos the refining of silver.[39] 'It cannot be smelted except when combined with lead, or with the vein of lead called "galena" [lead ore], which is usually found running near veins of silver ore. Also, when submitted to the same process of firing, part of the ore precipitates as lead, while the silver floats on the surface, like oil on water.' ('excoqui non potest, nisi cum plumbo nigro aut cum vena plumbi—galenam vocant—iuxta argenti venas plerumque reperitur. et eodem opere ignium discedit pars in plumbum. argentum autem innatat superne, ut oleum aquis.')

Pliny (33. 131) also records the blackening effect of sulphur on silver and copper. 'With the silver is mixed one third of its amount of the very fine Cyprus copper called "chaplet copper" and the same amount of live sulphur as of silver and then they are melted together in an earthenware vessel smeared round with potter's clay; the heating continues until the lids of the vessels open of their own accord.' ('miscentur argento tertiae aeris Cyprii tenuissimi, quod coronarium vocant, et sulpuris vivi[40] quantum argenti; conflantur ita in fictili, circumlito argilla; modus coquendi, donec se ipsa opercula aperiant.') (The sulphur turns the mixture black.) The silver and copper become sulphides.

---

[37] Tylecote, *Metallurgy*, 88 ff.
[38] Forbes, *Ancient Technology*, viii. 172 ff.
[39] Conophagos, *Laurium*, 305 ff. Cupellation is a metallurgical operation by which silver is separated from argentiferous galena by oxidation of the lead, in contact with the air in a cupel in which the temperature is sufficient to melt the lead oxide and the silver.   [40] See Ch. 14, s.v. 'sulphur'.

Pliny continues (ibid.): 'Silver is also turned black by means of the yolk of a hard-boiled egg,[41] although the black can be rubbed off with vinegar and chalk' ('nigrescit et ovi indurati luteo, ut tamen aceto et creta deteratur'). The reaction is the same as in the previous example, and a superficial coating of silver sulphide causes the discolouration. This is cleaned by the abrasive action of the chalk. The role of the vinegar is possibly as a 'wetting agent'. Silver sulphide (the tarnish) could only be dissolved by cyanide solutions or hot dilute nitric acid—both unknown to the Greeks or Romans. The vinegar (acetic acid), however, would react weakly with the calcium carbonate.

11.9 Distillation

Theophrastus (Lap. 60) was the first to describe the refinement of mercury from cinnabar. 'Quicksilver, or mercury,' he writes, 'was made when cinnabar[42] [*minium*] mixed with vinegar was ground in a copper vessel with a pestle made of copper.'

This was not a mechanical method for the liberation of the metal from a natural mixture of mercury and cinnabar, but a true chemical process,[43] which depended on the displacement of the mercury from the cinnabar by the more active metal placed in contact with it. Experiment shows that this method is viable.[44] The reaction may be accelerated by warming the mixture.

The immediate products are copper sulphide and mercury. The mercury, however, soon forms an amalgam with the copper and an additional operation is needed to recover the metal in its pure state. The amalgam is heated and the metal is obtained by the condensation of the pure mercury volatilized by the heat—a simple form of distillation. Theophrastus, however, does not mention this final stage of the process.

The separation of mercury by distillation was known to Dioscorides (5. 95 Wellmann) and to Pliny (33. 123), who

---

[41] A common 'breakfast experience' when using a silver spoon.
[42] Lap. 58 f.
[43] E. R. Caley and J. F. Richards, *Theophrastus on Stones* (Columbus, Oh., 1956), 204.
[44] See further Bailey, *Chemical Subjects*, i. 223 (n. to 33. 12). Bailey adds that iron, being a more electropositive metal, would displace mercury from mercuric sulphide, thus assisting the preparation.

repeats Theophrastus' method but adds the second stage from Dioscorides:

Of secondary importance is the fact that experience has also discovered a way of getting hydrargyrus, or artificial quicksilver as a substitute for real quicksilver.... It is made in two ways, by pounding cinnabar [*minium*] in vinegar with a copper pestle, or it is put in an iron shell in flat earthenware pans and covered with a convex lid smeared on with clay, and then a fire is lit under the pans and kept constantly burning by means of bellows and so the surface moisture [mercury] (with the colour of silver and fluidity of water) which forms on the lid is wiped off. This moisture is also readily divided into drops and rains down freely with slippery fluidity.

ex secundario invenit vita et hydrargyrum in vicem argenti vivi. ... fit autem duobus modis: aereis mortariis pistillisque trito minio ex aceto aut patinis fictilibus impositum ferrea concha, calice coopertum, argilla superinlita, dein sub patinis accenso follibus continuis igni atque ita calici sudore deterso, qui fit argenti colore et aquae liquore. idem guttis dividi facilis et lubrico umore compluere.

In this passage, Pliny is clearly in error in considering hydrargyrus (mercury obtained from the treatment of cinnabar)[45] and *argentum vivum*[46] (mercury in its native form) as two *different* substances.

Mercury[47] boils at 357 °C, at atmospheric pressure, and is condensed. As the ore is relatively low grade, it is reheated in sealed shaft furnaces with grates, and the gases are led into a condensing system from which the mercury is recovered. Cinnabar, when heated, oxidizes to yield mercury vapour and sulphur dioxide as:[48] $O_2 + HgS \rightarrow Hg + SO_2$. Any oxide that is formed decomposes above 500 °C. On cooling the vapour (below 200 °C), the mercury is condensed and the sulphur dioxide escapes into the atmosphere.

In Ladik,[49] in Anatolia, remains of condensers have been found, but not the furnaces.[50] These clearly belong to the Graeco-Roman period and it is suggested that selected pieces of rich cinnabar were mixed with charcoal, fired, and covered

---

[45] M. Wellmann, *Dioscorides: De Materia Medica,* 3 Vols. (Berlin, 1906–14), iii. 71 f.   [46] *NH* 33. 123.   [47] See Ch. 14, s.v. 'cinnabar'.
[48] Tylecote, *Metallurgy,* 147 f. and fig. 4.23.
[49] See further J. W. Barnes *et al., Geology and Ore Deposits of the Sizma-Ladik District, Turkey* (Cento Rep., 1969).   [50] Tylecote, *Metallurgy,* 147.

by an inverted pot. But it seems likely that the pot was part of the condensing system.

The refining of cinnabar was carried out in factories known as *miniariae* and Pliny (33. 122) describes the safety precautions taken by the workers.

11.10 ACIDS

11.10.1 Solvents

With the exception of acetic acid (*acetum*),[51] that is vinegar, or sour wine, acid reagents were unknown in the ancient world. *Acetum* is mentioned by Pliny in three main contexts: (*a*) as an alleged solvent,[52] (*b*) in mining[53] and quarrying, in conjunction with fire-setting,[54] and (*c*) in medicine.[55]

Cleopatra had inherited from the kings of the East two of the largest specimen pearls (*uniones*)[56] of all time. In a wager, she boasted to Antony that she would spend 10,000,000 sesterces on one banquet. Pliny (9. 120 f.) records the story: 'Following her instructions, the servants set before her only a single vessel containing vinegar, the acidity of which [*asperitas*] can dissolve pearls. On her ears she was wearing that remarkable and truly unique work of Nature. Antony waited breathlessly to see what on earth she was going to do. Cleopatra took off one earring, dropped the pearl in the vinegar, and, when it had dissolved, swallowed it.' ('Ex praecepto ministri unum tantum vas ante eam posuere aceti. cuius asperitas visque in tabem margaritas resolvit. gerebat auribus cum maxime singulare

---

[51] Varro, *LL* 9. 66. [52] *NH* 9. 120.
[53] *NH* 33. 71, cf. Celsus 2. 18, 2.21.
[54] *NH* 23. 57. So Sil. *Pun.* 3. 642 refers to fire-setting without vinegar.
[55] *NH* 23. 56
[56] *NH* 9. 112: 'The value lies in their brilliance, size, roundness, smoothness, and weight—all such uncommon qualities that no two pearls are found exactly alike. This is why Roman luxury has given them the name *uniones* ('specimen pearls')—a term not in the vocabulary of the Greeks. Indeed foreigners who discovered this fact call pearls *margaritae*.' ('dos omnis in candore, magnitudine, orbe, levore, pondere, haut promptis rebus in tantum ut nulli duo reperiantur indiscreti: unde nomen unionum Romanae scilicet imposuere deliciae, nam id apud Graecos non est, nec apud barbaros quidem inventores rei eius, aliud quam margaritae.' Pliny uses the term in 9. 120, to avoid any play on words with *unicum* ('singulare illud et vere unicum naturae opus').

illud et vere unicum naturae opus. itaque expectante Antonio quidnam esset actura, detractatum alterum mersit ac liquefactum obsorbuit.')

No form of acetic acid can dissolve a pearl and this story, followed by Pliny, is probably just one of many surrounding Cleopatra and her notoriously extravagant lifestyle.[57] She may well have pretended to swallow the pearl, or performed some sleight of hand to make it disappear.

Pliny (23. 54), however, again implies that vinegar is a kind of solvent. 'Even when sour, wine still has uses as a remedy... it is equally efficacious as a solvent; earth in fact effervesces when vinegar is poured on it' ('vini etiam vitium transit in remedia... non tamen minor in discutiendo; ita fit ut infuso terra spumet').

11.10.2 Fire-setting

Pliny (23. 57), in his discussion of the power of vinegar, refers to its use in fire-setting. Poured on rocks it splits them when attempts to do so with fire have failed ('saxa rumpit infusum quae non ruperit ignis antecedens').

Pliny elsewhere (33. 71) describes the process in more detail. In both kinds of mining, masses of flint are encountered. These can be split by fire-setting, which involves the use of vinegar. Fire-setting in galleries, however, usually makes them suffocatingly hot and smoke-filled. Instead, therefore, the rocks are split by means of crushers weighted down with a hundred and fifty pounds of iron. ('occursant in utroque genere silices; hos igne et aceto rumpunt, saepius vero, quoniam id cuniculos vapore et fumo strangulat, caedunt fractariis CL libras ferri habentibus.') Vitruvius (*De arch.* 8. 3. 1) also writes that flints when heated in a fire and sprinkled with acid fly asunder and are dissolved.

The best known reference to this technique, however, occurs in Livy (21. 37. 2) in the passage in which he describes Hannibal's crossing of the Alps[58] and the employment of this

---

[57] One is reminded of the excesses of modern film stars of recent years.

[58] Cf. Juv. *Sat.* 10. 153: 'Nature puts the Alps and snow in his path; but he splits rocks and mountains with vinegar' ('opposuit natura Alpem nivemque | diducit scopulos et montem, rumpit aceto'). Casssius Dio, in the 2nd cent. AD, states that vinegar was also used for weakening a tower, Cass. Dio 36. 18: 'Those who had betrayed the city had, by night, repeatedly saturated with vinegar a large brick tower, most difficult to capture, so that it became brittle.'

method to clear a rock fall. 'They felled very large trees that grew nearby...and made an enormous pile of logs. They set this on fire as soon as the wind blew fresh enough to make it burn and, pouring vinegar over the glowing rocks, caused them to disintegrate.' ('arboribus circa immanibus deiectis...struem ingentem lignorum faciunt eamque, quum et vis venti apta faciendo igni coorta esset, succendunt ardentiaque saxa infuso aceto putrefaciunt.')

Recent laboratory experiments, by Shepherd,[59] subsequently confirmed in the field, have proved the viability of this method, which had been accepted for a long time without actual scientific verification. The essential constituent of vinegar is acetic acid, CH_3COOH. It is itself a weak electrolyte which turns blue litmus red, is colourless, and mixes freely with water in a warm room. In its concentrated, or pure form, it is often called glacial acetic acid because it has a moderately high freezing, or melting, point, 17 °C (62 °F). A very weak solution would freeze at, or a few degrees above, 0 °C, so that its use would be nearly equivalent to that of water. If Hannibal did actually heat the rock the action would have been very different. Vinegar boils at 118 °C (244 °F), but as the temperature of the rock, when heated would have been very much higher than this, the efficacy of the vinegar, compared to water, would depend on the time elapsing after the rock was heated before the liquid was applied. It is not known whether Hannibal's men threw the vinegar on to the rocks while the fire was still in progress, or whether it was thrown on the burning embers and rocks, or wood ash. It might be reasonable to assume in any case that the vinegar would have been more effective than water as the rocks cooled. At 118 °C the vinegar would cease to boil and start to increase in volume, whereas this state for water would not be reached before the attainment of 100 °C. The rocks most probably affected by the use of vinegar are limestones and their associated minerals containing calcium, such as limestone schists and marble.

The effect of acetic acid on the main constituent of limestone is: $CaCO_3 + 2CH_3COOH = Ca(COOCCH_3)_2 + CO_2 + H_2O$.

---

[59] R. Shepherd, 'Hannibal the Rockbreaker', *Minerals Industry International*, 1008 (Sept. 1992), 39–47.

Chemistry

The products resulting from a breakdown of the calcium carbonate are, therefore, calcium acetate, carbon dioxide, and water.

Four different types of rocks were used in a series of laboratory tests: (1) Jurassic limestone, (2) argillaceous lias; (3) carboniferous rock; (4) limestone schist. Briefly the results of these tests, confirmed by field experiments, showed that a 50 per cent acetic acid solution produced a greater number of cracks in the rock samples and that all broke easily on tapping. This dilution in fact proved to be almost as effective as greater concentrations of acid.

Each type of rock was also subjected to an application of red wine in the same quantity. The wine had been allowed to become sour when left uncorked in a warm atmosphere over a period of three months. In the tests carried out with sour wine (14% vol.), the reactions were very violent and the rocks fractured readily, 'with much sizzling'. It is interesting to note that Ovid (*Met.* 7. 107–8) vividly refers to the hissing of limestone burnt in a furnace, when water is poured on.

Shepherd did not specifically experiment with flint (*silex*), that is a form of chert, encountered by Roman gold miners, but quite clearly the technique of fire-setting combined with the use of vinegar would have been successful (cf. 33. 71).

11.10.3 *Medicine*

Vinegar is also extensively used in medicine[60] as a salve and as a draught and Pliny gives numerous examples. Its use as an antiseptic,[61] however, was apparently not known. Vitruvius (*De arch.* 8.3) claims that, when taken internally, vinegar can dissolve stones in the bladder. Were this true it would be a far easier treatment even than lithotripsy! The only result, however, of consuming quantities of vinegar would have been death from ulcers.

Pliny (14. 130) also mentions a reaction with lead.[62]

---

[60] See above, n. 55.
[61] Vinegar was used in the 19th cent. as an antiseptic—for example, at the island quarantine-station in the Mao estuary, Minorca. The Lazaretto remained in use until the late 19th cent. and, in recent years, it has been a holiday retreat for personnel working for the Spanish Health Authority.
[62] To form lead acetate.

11.11 Discriminatory Tests

The chemistry involved in technological progress—for example, in the firing of clay, the extraction of metals from their ores and subsequent purification, and the production of alloys—has a long history.[63]

Near Eastern civilizations had made great advances in technology in the fourth and third millennia BC: this knowledge was transmitted through Egypt and the Greek colonies of Asia Minor to reach the classical world.

Pliny examines material substances entering into commerce, or for everyday use, for fitness of purpose. Greenaway,[64] in his examination of the chemical part of the *Natural History*, considers decision-making and includes some examples of discriminatory tests.

Pliny (31. 90 f.) differentiates *flos salis* from common salt (sodium chloride), but does not identify it. The usual adulterant was red ochre, or powdered potsherds. The test, therefore, for adulteration was simple, namely washing the substance with water to remove the artificial colourant.

Aphronitrum (31. 113) may be a form of nitrum, that is soda, a mixture of soda and salt, or potassium nitrate. Pliny (ibid.) writes: 'The best is thought to be Lydian; the tests are that it should be the least heavy and the most friable and of an almost purple colour' ('optimum putatur Lydium; probatio, ut sit minime ponderosum et maxime friabile, colore paene purpureo'). 'The tests of *soda* are that it should be very fine and as spongy and full of holes as possible' ('nitri probatio, ut sit tenuissimum, et quam maxime spongeosum fistulosumque'). Bailey discusses nitrum because of the many references to it in classical literature. Pliny (31. 114) adds: 'It is burnt in an earthen jar with a lid, lest it should crackle out; otherwise soda does not crackle in fire.' Potassium nitrate would certainly crackle but so would other salts by decrepitation. The undoubted inconsistencies in this passage suggest that Pliny does not understand his source. On currently available evidence the problem cannot be satisfactorily resolved.

Sometimes Pliny treats what we recognize as two different substances as variants of the same. One such is *stimmi* (33. 101),

---

[63] See generally, Forbes, *Ancient Technology*.
[64] See Greenaway, 'Chemical Tests', 147 ff.

'a stone made of white and shiny, but not transparent, froth [but antimony sulphide is black or red]: several names are used for it, *stimmi, stibi, alabastrum* [this may account for the confusion in colour], and sometimes *larbasis*. It is of two kinds, male and female. The female variety is preferred, the male being more uneven and rougher to touch, as well as lighter in weight, not so brilliant and more gritty; the female on the contrary is bright and friable and splits in thin layers and not into globules.' ('spumae lapis candidae nitentisque, non tamen tralucentis; stimmi appellant, alii stibi, alii alabastrum, aliqui larbasim. duo eius genera, mas ac femina. magis probant feminam, horridior est mas scabriorque et minus pondersosus minusque radians et harenosior, femina contra nitet, friabilis fissurisque, non globis, dehiscens.') The two forms are antimony sulphide (SbS) and metallic antimony (Sb). Roasting with a paste of which cow-dung was a major constituent would convert the sulphide to an oxide and some free antimony would be deposited. The antimony oxide would subsequently be washed out. Antimony sulphide would possibly yield pure antimony following the method here described.

Pliny (33. 119 ff.), in describing *minium*—cinnabar and red lead—states that they were distinguished by the weight ('minium compendi ratio demonstrat') of a standard package (ibid. 120). This can hardly be rated as a density measurement,[65] although all the elements of this relationship are present. A more technical test for adulteration of cinnabar was to heat a specimen. Heating darkens the adulterated material. Pliny's chemistry is faulty here since both mercuric sulphide and red lead darken on heating and recover their colour on cooling, if neither has been heated to the point of decomposition. If, however, lime is present as an adulterant, the discolouration remains.[66]

Two tests for *atramentum* (ferrous sulphate)[67] are described by Pliny (34. 112): in test 1 verdigris (copper sulphate) can

---

[65] See below, pp. 182–3 Cf. Vitr. *De arch.* 7. 9. 5, and Zehnacker, *NH 33*, p. 208 n. 4.

[66] These tests do not allow a distinction to be made between mercury sulphide (HgS) and orthoplombate of lead (PbO). However, a mixture of cinnabar and chalk, heated violently, frees the mercury and darkens permanently. The equation is as follows (cf. Bailey, *Chemical Subjects*, i. 221 f.): $4CaO + 4HgS = 4Hg + 3CaS = CaSO_4$.

[67] *Atramentum* appears in three roles in the *NH*, as (1) ink—'atramentum librarium' (27. 52); (2) a black pigment (35. 41); and (3) a dye for colouring leather—'diluendo fit atramentum tinguendis coriis' (34. 123).

be tested on a hot shovel, since a specimen that is pure, keeps its colour (green), but what is mixed with *atramentum*, turns red. The first claim is incorrect, but the ferrous sulphate will decompose to yield red iron oxide. But, as Bailey explains, the copper acetate will also decompose to give firstly a red cupric oxide, then copper by reduction by the acetone given off from the acetate, and finally a black copper oxide. Pliny appears to have been quoting Dioscorides' account.

Test 2 is sound, and in this Pliny (34. 112) describes the first 'test paper' on record. Papyrus soaked in an infusion of gall nuts turns black in the presence of ferrous sulphate. The black colour was to be the basis of modern permanent inks.

11.11.1 Addendum

Other tests, included in Greenaway's survey, are based on physical characteristics, or properties, and have no place in a chemical context. The fire test for gold (assay) is unambiguous, and a high degree of accuracy could be obtained by means of the touchstone,[68] but Pliny's remarks apropos natural electrum are clearly born of an element of confusion. He writes (33. 81), 'natural electrum has the property of detecting poisons, for semicircles resembling rainbows run over the surface in poisoned goblets and emit a crack ling noise like fire, and so advertise the presence of poison in a twofold manner' ('quod est nativum, et venena deprehendit. namque discurrunt in calicibus arcus caelestibus similes cum igneo stridore et gemina ratione praedicunt'). Pliny may here be confusing white gold (*electrum nativum*) and amber (*electrum*), with the electrostatic properties[69] of which a crackling noise is associated!

Pliny's limited essay into the realms of 'chemistry' mainly relies on observation, whether his own, or that of other authorities. Although he was not a theorist, this is no reason to discount his achievement.

---

[68] See Ch. 17. 4. 9a; also Ch. 14, s.w. 'chert', 'lapis Lydius'.
[69] See Ch. 12. 2. 1b.

Appendix A: Niello

The Egyptians 'stained' portraits of their god Anubis and this probably refers to their practice of using sulphur to darken silver.[1]

Niello[2] is a matt black substance, composed of one or more sulphides, employed for the decoration of gold and, more commonly, silver. Although found in Hellenistic and Roman silver plate, niello occurs generally only in late Roman jewellery (of the fourth century AD), in gold rings, and silver phialai.[3] The production of niello involves a chemical process in which silver sulphide is produced. Pliny (33. 131) gives the basic formula for niello: 'With silver is mixed one third of its amount of the very fine Cyprus copper called "chaplet-copper" and the same amount of live sulphur as of silver and they are melted in an earthenware vessel smeared round with potter's clay; the heating goes on until the lids of the vessels open of their own accord' ('miscentur argento tertiae aeris Cyprii tenuissimi, quod coronarium vocant, et sulpuris vivi quantum argenti; conflantur ita in fictili circumlito argilla; modus coquendi, donec se ipsa opercula aperiant'). These are sulphides of silver and copper (AgS and CuS) in contrast to niellos of later date, which consist of sulphides of silver, copper, and lead.

The Roman type of niello cannot be applied in the molten condition because it decomposes before melting and so is applied hot as a soft paste to fill the incised design. 'Aes coronarium', as Pliny (34. 94) explains, was used, in imitation of gold, in crowns given to actors.

Interesting examples of niello inlay occur in horse-trappings from Fremington Hagg,[4] in North Yorkshire and from Xanten,[5] in Germany. Qualitative spectrographic analysis shows that the trappings from the former consist of a body of metal covered on one side with a bright metal overlay soldered on to the substrate, and, at many points, there are various types of inlay present in the overlay.[6] The comparable examples from Xanten are of a similar alloy except for the absence of tin in one. The decoration also consists of a complex

---

[1] See Zehnacker, *NH 33*, p. 215, comm. The idea of colouring statues of Anubis black relates to his role as god of the dead. In the Hellenistic period this role was assimilated by Hermes Psychpopompos. In the case of triumphal statues, Zehnacker suggests that the dark colour would have given them an air of antiquity.   [2] This contains silver sulphide and copper sulphide.

[3] R. Higgins, *Greek and Roman Jewellery* (London, 1961, and Berkeley and Los Angeles, 1980), 28.

[4] P. T. Craddock, Janet Lang, and K. S. Painter, 'Roman Horse-Trappings from Fremington Hagg, Reath, Yorkshire, NR', *British Museum Quarterly*, 37 (1973), 9–17.

[5] See Wolf-Dieter Heilmeyr, 'Titus vor Jerusalem', *Mitteilungen des deutschen Archäologischen Instituts, Römische Abteilung*, 82 (1975), 299–314.

[6] J. F. Healy, 'Problems in Mineralogy and Metallurgy in Pliny the Elder's Natural History', in *Atti del Convegno di Como—Technologia, economia, e società nel mondo romano* (Como, 1980), 179–80.

plant-motif inlaid with copper and niello set into a silver overlay. The exceptional trapping no. 5 was shown, by metallographic examination, to have been a very fine-worked structure with twinned grains, indicating that it was made from a sheet of brass hammered to shape. The hammer marks can be seen to the rear of the trapping, the front surface having been turned. The front of the trapping was smoothed and polished after it had been shaped on a lathe—there are deep circular scratches on the surface—while the decoration is chased.

The most interesting trapping (AD 57–9) is the one inscribed PLINIO PRAEF. EQ. and, seemingly, derives from Pliny's service in Germany. The dense black inlay was found to contain copper and silver, and matches the formula set out by Pliny, which is one of the earliest references to niello.

Appendix B: Dyes and Dyeing

Pliny writes (22. 4) 'Yet I should not have left out the craft of dyeing altogether, had it ever been included among the liberal arts' ('nec tinguendi tamen rationem omisissemus, si umquam ea liberalium artium fuisset').

Throughout the *Natural History* there are numerous references to natural dyes. 'The best quality of Asiatic [purple] is found at Tyre,[7] the best African, at Meninx and on the Gaetulian coast of the Ocean, and the best in Europe, in the region of Sparta' (9. 127, 'Tyri praecipuus hic Asiae, Meninge Africae, et Gaetulo litore oceani, in Laconica Europae').

Purples [9. 125] live seven years at most. They stay in hiding, like the murex, for thirty days at the rising of the Dog Star. They collect into shoals in springtime and their rubbing together causes them to discharge a sort of waxy slime. The murex also does this in a similar manner, but it has the famous flower of purple, sought after for dyeing robes, in the middle of its throat, where there is a white vein of very scanty fluid from which that precious dye, suffused with a dark rose colour, is drained; the rest of the body, however, produces nothing.

purpurae vivunt annis plurimum septenis. latent sicut murices circa canis ortum tricenis diebus. congregantur verno tempore, mutuoque attritu lentorem cuiusdam cerae salivant. simili modo et murices, sed purpurae florem illum tinguendis expetitum vestibus in mediis habent faucibus; liquoris hic minimi est candida vena unde pretiosus ille bibitur, nigrantis rosae colore sublucens; reliquum corpus sterile.

Pliny (9. 131 f.) continues with a description of the varieties of shellfish supplying purple, and their habits. 'The pebble purple, named after a pebble in the sea, is remarkably suitable for purple

---

[7] So Virgil, *Aen.* 4. 262, describes Aeneas' cloak: 'Tyrioque ardebat murice laena.'

dyes; and far the best of these is the melting purple that is, the one fed on a varying kind of mud' ('calculense appellatur a calculo in mari mire aptum conchyliis; et longe optimum purpuris dialutense, id est vario soli genere pastum').

The preparation and blending of the raw dye is a mechanical process and no chemical reaction is involved (133 f.).

The vein is removed and to this salt has to be added in the proportion of about one pint for every hundred pounds. It should be left to dissolve for three days, since the fresher the salt the stronger. The mixture is then heated in a lead pot, with about seven gallons of water to every fifty pounds and kept at a moderate temperature by a pipe connected to a furnace some distance away. This separates the flesh which will have adhered to the veins, and, after about nine days, the cauldron is filtered and a washed fleece is dipped by way of trial. Then the dyers heat the liquid until they feel confident of the result. A red colour is inferior to black. The fleece is soaked for five hours, carded, and again dipped until it absorbs all the dye.

eximitur postea vena... cui addi salem necessarium, sextarios ferme centenas in libras; macerari triduo iustum, quippe tanto maior vis quanto recentior, fervere in plumbo, singulasque amphoras aquae, quinquagenas medicaminis libras aequali ac modico vapore torreri adducto longinquae fornacis cuniculo. ita despumatis subinde carnibus quas adhaesisse venis necesse est, decimo ferme die liquata cortina vellus elutriatum mergitur in experimentum et, donec spei satis fiat, uritur liquor. rubens color nigrante deterior. quinis lana potat horis rursusque mergitur carminata, donec omnem ebibat saniem.

There is also another kind of indigo (35. 46), a by-product of purple, which floats on the pans in the purple dye factories, and this is called 'scum of purple': 'alterum genus eius est in purpurariis officinis innatans cortinis, et est purpurae spuma.'

Pliny also describes (9. 140), in passing, the use of the 'kermes' (*coccum*), the nature of which he misunderstands.[8] There is also a method to blend minerals and dye with Tyrian a fabric already dyed with scarlet, to produce the colour called 'hysgine': 'quin et terrena miscere coccoque tinctum Tyrio tinguere et fieret hysginum.' Woad[9] (*vitrum*) is also used to stain minerals (35. 46).

Pliny (35. 139) clearly regards the changing of colour, elsewhere, as the equivalent of 'dyeing'. Thus he includes the staining of tortoiseshells and the alloying of gold with silver to produce electrum[10] and, with the addition also of copper, to produce Corinthian metal,[11] as

---

[8] The *coccum* is really a scale-insect which lives on the so-called 'scarlet' oak, Pliny, and most ancient authorities, confused it with seed. In modern botanical terminology, *Kermes* is the insect (cochineal).
[9] See below, n. 14.   [10] 16. 232.   [11] 34. 6.

'dyeing': 'sicut testudines tinguere, argentum auro confundere ut electra fiant, addere his aera ut Corinthia.'

Minerals like azurite are enhanced by Tyrian purple. Similarly, gemstones (37. 79) can be produced by dyeing rock-crystal[12] or glass. 'The Indians have found a way of counterfeiting various precious stones, and beryls in particular, by staining rock-crystal' ('Indi et alias quidem gemmas crystallum tinguendo adulterare invenerunt, sed praecipue berullos'). Fake opals (ibid. 83) and 'carbunculi' (ibid. 98) can be manufactured from tinted glass.

Glass (36. 198) is also tinted. 'There is also the artificial 'obsian' which is used as a material for tableware, this being produced by a colouring process, as is also the case with a completely red, opaque glass called 'blood-red ware' ('fit et tincturae genere obsianum ad escaria vasa et totum rubens vitrum atque non tralucens, haematinum appellatum').

Dyes are also obtained from plants, as in the case of a variety of laserwort, called *magydaris*, which is used for dyeing (19. 46 f.). 'Madder[13] is indispensable for dyeing woollens and leather; the most highly esteemed is the Italian and especially that grown in the vicinity of Rome. Almost all the provinces grow this in abundance.' ('in primis rubia tinguendis lanis et coriis necessaria: laudatissima Italica et maxime suburbana, et omnes paene provinciae scatent ea.') Similarly, the *erythrodanum* (24. 94) is used to dye wool and tan leather: 'erythrodanum qua tinguntur lanae pellesque perficiuntur'.

Ancient peoples, among them the Britons, used woad as body paint[14] (22. 2 ff.)

I observe that some foreign races use certain plants as body paint, both to make themselves more attractive and also to keep up tradition. At any rate barbarian women stain their faces, using a variety of plants. Dacian men, and the Saramatae as well, tattoo their bodies. In Gaul there is a plant like the plantain, called 'woad' and the wives and daughters-in-law of the Britons stain all their body with it and, at certain religious ceremonies, march along naked, with a colour resembling that of Ethiopians.

equidem et formae gratia ritusque perpetui in corporibus suis aliquas exterarum gentium uti herbis quibusdam adverto animo. inlinunt certe

---

[12] 37. 79. E. H. Warmington, *The Commerce Between the Roman Empire and India*² (Cambridge, 1974), 251, refers to good forgeries of beryls manufactured in this way.

[13] Cf. 24. 94. Pliny also mentions a plant called 'little root' (*radicula*) which is used as a fabric softener in washing.

[14] Caes. *BG* 5.14 (for *vitrum,* cf. Vitr. *De arch.* 7. 14): 'All Britons dye their bodies with woad because it produces a colour like indigo. As a result of this, they have a more savage appearance in battle.' ('omnes vero se Britanni vitro inficiunt, quod caeruleum efficit colorem, atque hoc horridiores sunt in pugna aspectu.') The botanical name for woad is *Isatis tinctoria.*

aliis aliae faciem in populis barbarorum feminae, maresque etiam apud Dacos et Sarmatas corpora sua inscribunt, similis plantagini glastum in Gallia vocatur, Britannorum coniuges nurusque toto corpore oblitae quibusdam in sacris nudae incedunt Aethiopum colorem imitantes.

Pliny continues (22. 3 f.) with other vegetable dyes.

Moreover we know that clothes are dyed with a wonderful dye from a plant and, to say nothing of the fact that of the berries of Galatia, Africa, and Lusitania the 'kermes'[15] is specially reserved to colour the military cloaks of our generals, Transalpine Gaul can produce with vegetable dyes Tyrian purple, oyster purple,[16] and all other colours. To get these, nobody seeks the murex oyster in the depths, offering his body as bait to sea monsters, while he hastens to snatch his booty, and exploring a sea-bed that no anchor has yet touched.

iam vero infici vestes scimus admirabili fuco, atque ut sileamus Galatiae, Africae, Lusitaniae e granis coccum imperatoriis dicatum paludamentis, transalpina Gallia herbis Tyria atque conchylia tinguit et omnes alios colores. nec quaerit in profundis murices, seque obiciendo escam, dum praeripit beluis marinis, intacta ancoris scrutatur vada.

Dyes, like crops, are harvested on land, but, in the case of vegetable dyes, the colours are not fast. Mordants, however—for example alum[17]—were used to 'fix' colours, as Pliny (33. 88) explains, in the case of flax or wool: 'Then it is dyed by means of alum and the plant above mentioned (yellow-weed) and so given a colour before it serves as a colour itself. It is important how absorbent the material is and ready to take the dye; for if it does not at once catch the colour, *scytanum* and *turbistum* must be added as well, those being the names of two drugs producing absorption [in other words, mordants].' ('tum tinguitur alumine schisto et herba supra dicta pinguiturque antequam pingat. refert quam bibula docilisque sit. nam nisi rapuit colorem adduntur et scytanum atque turbistum; ita vocant medicamenta sorbere cogentia.')

Elsewhere, (31. 110) Pliny appears to confuse dyes and mordants: 'But for some purposes the impure [soda] is good, for example, for colouring purple materials and all kinds of dyeing' ('ad aliqua tamen sordidum [nitrum], tamquam ad inficiendas purpureas tincturasque omnes').

Dyes, especially purple, were an important part in the symbolism of Roman political life[18] and society, while, in the world of precious stones, they made possible a lucrative trade in counterfeit gems.[19]

---

[15] See above, App. n. 8. [16] There is little, if any distinction between the kinds of purple. [17] See Ch. 14, s.v. 'alumen'.
[18] 9. 127 and 136. Magistrates, and boys of free birth who had not yet assumed the toga virilis, wore a purple-bordered toga. [19] 37. 197 ff.

12

Physics

Physics is the branch of science concerned with the properties of matter and energy and the relationships between them. It is based on mathematics and, traditionally, includes optics, electricity, magnetism, acoustics, heat, and mechanics.

Although Pliny does not define such as an independent category of science, *physica* (τὰ φυσικά) applies to all aspects of natural sciences, and there are numerous references to physical properties that are part of physics, in however elementary a form. His debt to earlier authorities is very evident.

12.1 Optics

12.1.1 Reflection

The phenomenon of reflection[1] was well known to ancient peoples, and Lucretius (*DRN* 4. 269 ff.)[2] provides the definitive

---

[1] Light undergoes reflection, refraction, interference, and diffraction. Light waves reflect from a barrier which must be flat, as a mirror. The laws of reflection are: (1) the incident ray, the normal, and reflected ray all lie in the same plane—that is they are all level; (2) the angle of incidence equals the angle of reflection (i = r). In his introduction to the topic of reflection, Lucretius (*DRN* 4. 54 ff.) states that many objects give off particles—he likens this to cicadas shedding their jackets in summer—and these add up to an image of the object (or film); ... he concludes (4. 98 ff.): 'The reflections that we see in mirrors, or in water, or any polished surface, have the same appearance as the actual objects. They must necessarily be composed of films given off by these objects. There exist, therefore, flimsy but accurate replicas of objects, individually invisible, but such that, when flung back in a rapid succession of recoils from the flat surface of mirrors, they produce a visible image. That is the only conceivable way in which these films can be preserved so as to reproduce such a perfect likeness of each object.'

[2] Translations of *DRN* in this chapter are from R. E. Latham, *Lucretius: On the Nature of the Universe* (London, 1997).

account of his time:

> Let us now consider *why the image is seen beyond the mirror*—for it certainly does appear to be some distance behind the surface. It is just as though we were readily looking out through a doorway, when the door offers a free prospect through it and affords a glimpse of many objects outside the house. In this case also the vision is accompanied by a double dose of air. First we perceive the air within the doorposts; then follow the posts themselves to right and left; then the light outside and a second stretch of air brushes through the eyes, followed by the objects that are really seen out of doors. A similar thing happens when a mirrored image projects itself upon our sight. On its way to us the film pushes and drives before it all the air that intervenes between itself and the eyes, so that we feel all this before perceiving the mirror. When we have perceived the mirror itself, then the film that travels from us to it, and is reflected, comes back to our eyes, pushing another lot of air in front of it, so that we perceive this before the image which thus appears to lie at some distance from the mirror.[3] Here then is ample reason why we should not be surprised at this appearance of objects reflected in the surface of a mirror, since they involve a double journey with two lots of air.
>
> Now for the question *why our right side appears in mirrors on the left*. The reason is that, when the film on its outward journey strikes the flat surface of the mirror, it is not slewed round intact, but flung straight back in reverse. It is just as if someone were to take a plaster mask before it had set and hurl it against a pillar or beam, so that it bounced straight back, preserving the features imprinted on its front but displaying them now in reverse. In this case what had been the right eye, would now be the left and the left, correspondingly, would have become the right.

> > nunc age, cur ultra speculum videatur imago
> > percipe; nam certe penitus semota videtur. 270
> > quod genus illa foris quae vere transpiciuntur,
> > ianua cum per se transpectum praebet apertum,
> > multa facitque foris ex aedibus ut videantur.
> > is quoque enim duplici geminoque fit aere visus.

---

[3] Lucretius, here, and later, accurately describes what he observes. The image formed in a plane mirror has the following properties: (1) it is the same size as the object, (2) is the right way up, (3) is as far behind the mirror as the object is in front, and (4) is laterally inverted—so that, as one looks in a mirror one's right side is on the left and vice versa. Optometrists make use of phenomenon (3) to extend the apparent distance of their sight-testing charts from the patient.

> primus enim citra postis tum cernitur aer, 275
> inde fores ipsae dextra laevaque sequuntur,
> post extraria lux oculos perterget et aer
> alter et illa foris quae vere transpiciuntur.
> sic ubi se primum speculi proiecit imago,
> dum venit ad nostras acies, protrudit agitque 280
> aera qui inter se cumquest oculosque locatus,
> et facit, ut prius hunc omnem sentire queamus
> quam speculum. sed ubi speculum quoque
> sensimus ipsum,
> continuo a nobis in idem quae fertur imago
> pervenit et nostros oculos reiecta revisit 285
> atque alium prae se propellens aera volvit
> et facit ut prius hunc quam se videamus, eoque
> distare ab speculo tantum semota videtur.
> quare etiam atque etiam minime mirarier est par,
> illis quae reddunt speculorum ex aequore visum, 290
> aeribus binis quoniam res confit utraque.
> nunc ea quae nobis membrorum dextera pars est,
> in speculis fit ut in laeva videatur eo quod
> planitiem ad speculi veniens cum offendit imago,
> non convertitur incolumis, sed recta retrorsum 295
> sic eliditur, ut siquis, prius arida quam sit
> cretea persona, allidit pilaeve trabive,
> atque ea continuo rectam si fronte figuram
> servet et elisam retro sese exprimat ipsa.
> fiet ita, ante oculus fuerit qui dexter, ut idem 300
> nunc sit laevus et e laevo sit mutua dexter.

Pliny's account of the phenomenon of reflection follows Lucretius' explanation, but is not as comprehensive. He writes (33. 128 f.).

It used to be believed that only the best silver could be beaten into plates to produce a reflected image. This was once true, but even now this test is open to fraud. However, the power to reflect images is marvellous. The general consensus of opinion is that this phenomenon is due to air bouncing back and making contact with the eyes. Similarly, by using a mirror in which the thickness of the metal has been polished and beaten into a slightly concave shape, objects are greatly magnified.[4] Such a big difference does it make whether the surface absorbs the air, or reflects it.

---

[4] See below, n. 13.

lamnas duci, speciem fieri non nisi ex optimo [argento] posse creditum. fuerat id integrum, sed id quoque iam fraude corrumpitur. est natura mira imagines reddendi, quod repercusso aëra atque in oculos regesto fieri convenit. eadem vi sic in speculi usu polita crassitudine paulumque propulsa dilatatur in inmensum magnitudo imaginum. tantum interest, repercussum illum excipiat an respuat.

Without reference to Lucretius' explanation of the reason for the reversal of images in a mirror (*DRN* 4. 292 ff.), Pliny (33. 129) continues:

It is possible to make goblets with a number of mirrors beaten, as it were, outwards from the inside in such a shape that a single face sees itself reflected as a multiplicity of images equal in number to the reflecting surfaces. Vessels like those dedicated in the temple of Smyrna, are devised so as to achieve strange effects. These are brought about by the shape of the material and it makes a significant difference whether the vessels are concave and shaped like a bowl, or convex like a Thracian shield, whether their centre is recessed, or projecting, whether the oval[5] is horizontal, oblique, laid flat, or placed upright, as the quality of the shape receiving the shadows twists them as they come: for, in fact, the image in a mirror is merely the shadow arranged by the brilliance of the material receiving it.

quin etiam pocula ita figurantur expulsis intus crebris ceu speculis, ut vel uno intuente totidem populus imaginum fiat. excogitantur et monstrifica, ut in templo Zmyrnae dicata. id evenit figura materiae. plurimum refert concava sint et poculi modo an parmae Threcidicae, media depressa an elata, transversa an obliqua, supina an infesta, qualitate excipientis figurae torquente venientes umbras; neque enim est aliud illa imago quam digesta claritate materiae accipientis umbra.

The idea of the creation of multiple images by facets, or mirrors, is also found in Seneca (*QN* 1. 5. 5) and, similarly, he describes (*QN* 1. 5. 14) mirrors which reflect distorted images.[6] Pausanias (8. 37. 4) makes reference to mirrors in temples.

---

[5] Rackham, *NH 33–35*, p. 96 note *c*, ad loc.
[6] 'neque enim omnia ad verum specula respondent, sunt quae videre extimescas, tantam deformitatem corrupta facie, visenti reddunt, servata similtudine in peius.' Also *QN* 1. 15. 8. These are large mirrors (*QN* 1. 17. 8 'specula totis paria corporibus'), but small distorting hand-mirrors are also known.

Zehnacker[7] provides a comprehensive commentary on these passages, noting that spherical concave mirrors are not affected by their position. Those like Thracian shields are cylindrical, or parabolic, square, or rectangular, and divided as follows: (*a*) concave (*media depressa*); (*b*) convex (*elata*).[8] Their axes are (1) vertical (not described here); (2) horizontal (*transversa*); and (3) oblique (*obliqua*). Parabolic mirrors have analogous axes.

Pliny's account does not include the use of mirrors in series, which can, by reflection, bring to light objects hidden in the recesses of a house, or mirrors that give back unreversed images, as Lucretius (4. 302 ff.) explains:

It may also happen that a film is passed on from one mirror to another, so that as many as five or six images are produced. Objects tucked away in the inner part of a house, however long and winding the approach to their hiding place, can thus be brought into sight along devious routes by a series of mirrors.[9] So the image is flashed from mirror to mirror. And on each occasion what is transmitted as the left becomes the right and is then again reversed and returns to its original relative position.

>   fit quoque de speculo in speculum tradatur imago,
>   quinque etiam sex ⟨ve⟩ ut fieri simulacra suerint.
>   nam quaecumque retro parte interiore latebunt,
>   inde tamen, quamvis torte penitusque remota,   305
>   omnia per flexos aditus educta licebit
>   pluribus haec speculis videantur in aedibus esse.
>   usque adeo ⟨e⟩ speculo in speculum translucet imago,
>   et cum laeva data est, fit rursum ut dextera fiat,
>   inde retro rursum redit et convertit eodem.   310

Again (4. 311 ff.), 'mirrors with projecting sides, whose curvature matches our own, give back to us unreversed images. This may be because the film is thrown from one surface of the mirror to the other and reaches us only after a double

---

[7] *NH 33*, p. 214, apropos 33. 129 (Thracian shields).

[8] In the concave mirror, the reflecting surface curves inwards and in the convex mirror, the reflecting surface curves outwards. Both types reflect light according to the laws of reflection. Concave mirrors concentrate light to a focus, while convex mirrors increase the field of view.

[9] This use of mirrors anticipates the periscope used to look over obstacles and to help us to read the scales of instruments with a greater degree of accuracy.

rebound. Alternatively, it may be that on reaching the mirror the film is slewed round, because the curved surface gives it a twist towards us.'

Pliny (37. 64), describing emeralds as generally concave in shape, adds that they concentrate the vision: 'iidem plerumque concavi, ut visum conligant'. It is uncertain what Pliny means but, on generally available evidence, it is unlikely that he is referring to the use of emeralds as lenses, since a green gemstone would hardly serve such a purpose.

He continues (ibid.), 'when emeralds that are tabular in shape are laid flat, they reflect objects, just as mirrors do. The emperor Nero used to watch fights between gladiators in a reflecting emerald.'('quorum vero corpus extentum est, eadem qua specula ratione supini rerum imagines reddunt. Nero princeps gladiatorum pugnas spectabat in smaragdo.') An emerald, however, even one as large as those in the well-known Topkapi dagger,[10] can hardly have been used as an effective 'mirror'. Nor is the suggestion that Nero rested his eyes from the glare of the sun by gazing at the green stone any more plausible. Further speculation is pointless without additional evidence.

12.1.2 Refraction

The change in direction of a propagating wave, such as light, or sound, passing from one medium to another, in which it has a different velocity, gives rise to refraction. This phenomenon was universally known from observation. Lucretius, for example, writes (4. 436):

To landsmen ignorant of the seas, ships in harbour seem to be riding crippled on the waves, their curved sterns broken. So much of the oars as projects above the water-line is straight and, similarly, the upper part of the rudder. But all the submerged parts appear refracted, wrenched round in an upward direction and almost as though bent back so as to float on the surface.

> at maris ignaris in portu clauda videntur
> navigia aplustris fractis obnitier undae.

---

[10] In Topkapi Sarayi, Istanbul.

> nam quaecumque supra rorem, salis edita pars est
> remorum recta est, et recta superne guberna.
> quae demersa liquorem obeunt, refracta videntur
> omnia converti sursumque supina reverti.

'The 'rainbow' stone [*NH* 37. 136] is, in all but one respect, rock-crystal;[11] and some authorities have referred to it as 'root-crystal'. It is called 'rainbow' stone from the fact that, when it catches the sunlight in a room it refracts the light and throws the colours of a rainbow on the nearby walls; it continually changes its colours and this kaleidoscopic effect arouses ever-increasing astonishment. The stone has hexagonal faces like rock-crystal.' ('cetera sui parte crystallus. itaque quidam eam radicem crystalli esse dixerunt. ex argumento vocatur iris, nam sub tecto percussa sole species et colores arcus caelestis in proximos parietes eiaculatur, subinde mutans magna varietate admirationem sui augens. sexangulam esse ut crystallum constat.') In other words, it is a quartz prism that resolves light into the colours of the spectrum.

Pliny continues (ibid.), 'there are some who claim that the faces are rough and that the angles are unequal. They say that, in full sunlight, this stone refracts the beams that shine on it, while simultaneously lighting adjacent objects by throwing out a kind of bright light in front of it. As I have said, however, it only produces colours in a darkened place. The best stone is that which produces the largest rainbow effect, most like the natural phenomenon.' ('sed aliqui scabris lateribus et angulis inaequalibus dicunt, in sole aperto radios in se candentes discutere, aliquo vere ante se proiecto nitore adiacentia inlustrare, colores autem non nisi ex opaco, ut diximus, reddunt, nec ut ipsae habeant, sed ut repercussu parietum elidant; optima quae maximos arcus facit simillimosque caelestibus.') Pliny's claim here, that a darkened room is necessary, is not true; nor has he made any prior reference to this.

To sum up, a prism is a transparent, polygonal solid, often having triangular ends and rectangular sides for dispersing light into a spectrum, or, indeed, for refracting light.

---

[11] Ch. 14, s.v. 'crystallum'.

12.1.3 Magnification

A simple means of magnification must have been available for coin and die-engravers[12] and craftsmen who made miniature models in the ancient world, but there are few references to lenses, or their equivalent, in ancient literature. Pliny (7. 85) states that

> keenness of sight achieved things that transcend belief, but clearly such miniaturization would not have been possible without some mechanical means of magnification. Cicero records that a parchment copy of Homer's *Iliad*, was kept in a nutshell. ... Callicrates used to make such small ivory models of ants and other creatures that to anyone else their parts were invisible. A certain Myrmecides was renowned as a model-maker for his four-horsed chariot, carved from ivory, so small that a fly's wings could cover it and for a ship that a small bee could hide with its wings.
>
> oculorum acies vel maxime fidem excedentia invenit exempla. in nuce inclusam *Iliadem* Homeri carmen in membrana scriptum tradit Cicero. ... Callicrates ex ebore formicas et alia tam parva fecit animalia ut partes eorum a ceteris cerni non possent. Myrmecides quidam in eodem genere inclaruit. quadriga ex eadem materia quam musca integeret alis fabricata et nave quam apicula pinnis absconderet.

Miniaturized exhibits are still popular: these can be seen, for example, at Micro World, St Ouen (Jersey). They include a horse standing on the head of a real ant, and Adam and Eve carved on the tip of a pencil.

In a rare reference (36. 199) Pliny states that 'glass globes containing water become so hot, when they face the sun, that they can burn clothes' ('cum addita aqua vitreae pilae, sole adverso, in tantum candescant ut vestes exurant'). This is not, however, because of the small increase in the heat of the glass and of the liquid, but because of the concentration of the sun's rays. The water-filled globes act like a primitive magnifying glass.

---

[12] C. T. Seltman, *Masterpieces of Greek Coinage* (Oxford, 1949), 20. One cannot positively state that no artificial aid to eyesight existed. A drop of water caught in a loop of wire is found to magnify, and a Greek would surely be quick to imitate his water-drop on a larger scale by carefully grinding a piece of rock crystal.

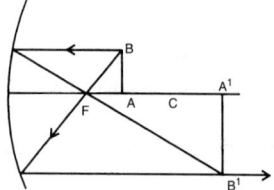

Key F = focus (that is the point through which all light is reflected)
C = the centre of curvature of the concave mirror (the centre of a sphere of which the mirror would form a part)
A/A¹ and B/B¹ illustrate how a concave mirror inverts an image and enlarges it:

The image A^1B^1 is inverted relative to AB
Diagram reproduced courtesy of Professor H. Zehnacker

Fig. 12.1. Concave mirror

In discussing mirrors, Pliny (33. 128) records the power of magnification of a concave mirror: 'dilatatur in immensum magnitudo imaginum'[13] (see Fig. 12. 1 above). He also writes (2. 239): 'doubtless all marvels will be surpassed by the fact that there has ever been a single day on which there has not been a universal conflagration, when also hollow mirrors, facing the sun's rays, set things alight more easily than any other fire' ('...cum specula quoque concava adversa solis radiis facilius etiam accendant quam ullus alius ignis'). This may refer to hollow mirrors, or, possibly, concave lenses, acting as magnifying glasses.

12.1.4 *The Earth's Curvature*

Pliny (2. 164) is aware of the curvature of the earth which gives rise to the horizon. 'The same reason [convexity] explains why land that is not visible from the deck of a ship can nevertheless be seen from the mast, and why, as a ship

---

[13] Cf. Zehnacker, *NH 33*, p. 213, on 33. 128.

recedes into the distance, if some shining object is tied to the top of the mast, it appears to sink slowly and, finally, disappears from view.' ('eadem est causa propter quam e navibus terra non cernatur e navium malis conspicua, ac procul recendente navigio, si quid quod fulgeat religetur in mali cacumine, paulatim descendere videatur et postremo occultetur.')

In a further fascinating passage, among those in which he discusses unusual physical attributes and examples of outstanding strength, sight, voice, and memory, Pliny (7. 85) states that Cicero records the story of a man 'who could see a distance of 123 miles! Marcus Varro gives his name as Strabo and asserts that during the Punic Wars, he used to stand on the headland of Lilybaeum, in Sicily, and would tell even the number of ships leaving harbour at Carthage.' ('idem fuisse qui pervideret $\overline{\text{CXXXV}}$ passuum. huic et nomen M. Varro reddit, Strabonem vocatum; solitum autem Punico bello a Lilybaeo Siciliae promunturio exeunte classe e Carthaginis portu etiam numerum navium dicere.') Apart from the fact that his name, 'cross-eyed', seems strangely inappropriate in such a context, an important problem is raised by this claim, even assuming that anyone, without a telescope or binoculars, enjoyed such long-sightedness. The problem involves the restricting factor of the earth's curvature, or to put it in a simpler form, the distance of the horizon from a person standing at whatever height from sea-level upwards. For example, from the bridge of a ship about 50 feet above sea-level, the horizon is approximately sixteen miles away!

We can, of course, easily put Pliny's story to the test by simple trigonometrical calculations.[14] The actual distance between Lilybaeum (Cape Boeo), on the south coast of Sicily, and Carthage—some eight miles from Tunis—is about 130 miles confirming, within a few miles at least, the figure quoted by Pliny. However, the height necessary to overcome the earth's curvature at that distance, is approximately 11,000 feet. Strabo might just have been able to see the tops of the masts of the Carthaginian ships setting out from Carthage—from the *summit of Mt. Etna!*

---

[14] I am indebted to Emma Town for the calculations shown in the appendix to this chapter: see pp. 171–2 below.

Strabo's alleged achievement should be deleted from the 'Record Books'!

12.2 Electricity

Electricity, as we define the term, namely a form of energy associated with stationary or moving electrons, protons, or other charged particles, was unknown in the ancient world: its discovery and practical use stems from early nineteenth century experiments when the term 'current electricity' was used to distinguish it from 'frictional'.

12.2.1 Static electricity

The Greeks and Romans experienced two forms of static electricity,[15] namely (1) St Elmo's fire and (2) that produced by friction on fossil resin.[16] This phenomenon, however, was not properly investigated until the sixteenth century, when William Gilbert, while exploring the nature of terrestrial magnetism, experimented with amber and showed that resins, crystals, sulphur, and glass, when rubbed, acquire the power to attract substances other than iron.

12.2.1a St Elmo's Fire

St Elmo's[17] fire is the glow accompanying atmospheric static-electric brush discharges which appear on the tips of wooden masts and spars in an electric storm. Pliny (2. 101) describes two such occurrences. 'Stars become visible on land and at sea; I have seen them with the appearance of lightning clinging to the spears of soldiers on guard duty at night in front of the

---

[15] All substances are composed of atoms. Each atom consists of a tiny central core called the nucleus surrounded by electrons which are in orbit around it. The nucleus carries a positive charge and the electrons, a negative. The positive charge of the nucleus is exactly equal to the negative of the electrons. In some substances, some of the orbiting electrons can be freed from the atoms by rubbing. The electrons cannot move around from where the rubbing occurs—and are known as static. However, they can move on to the rubbing material. The substance loses electrons, while the material which does the rubbing gains them.

[16] See Ch. 14, App. A.

[17] St Elmo is the patron saint of sailors. See also above, n. 15.

rampart. At sea, I have observed St Elmo's Fire on the yard-arms and other parts of a ship jumping about with a sound like a voice, just as birds hop from perch to perch.' ('existunt stellae et in mari terrisque. vidi nocturnis militum vigiliis inhaerere pilis pro vallo fulgurum effigie eas, et antemnis navigantium aliisque navium partibus cum vocali quodam sono insistunt ut volucres sedem ex sede mutantes.')

12.2.1b *Pyroelectricity*

Pliny (37. 48) describes the pyroelectric properties[18] of amber: 'When the hot exhalation is released by rubbing amber with the fingers, it attracts straw, dry leaves, and bark from the linden tree, like a magnet attracts iron' ('ceterum attritu digitorum accita caloris anima trahunt in se paleas et folia arida et philyras, ut magnes lapis ferrum'). He does not understand the difference between this kind of attraction, produced by temporary electrical charges, and magnetism, but his analogy deserves credit. Plutarch (*Quaest. Plat.* 7.7) is equally wide of the mark when he suggests that the hot exhalations released by rubbing amber, act in the same way as emanations from a magnet, namely that they displace air, forming a vacuum in front of the attracted object and driving the air to its rear. Elsewhere, Pliny (37. 37) again refers to the powers of amber which, Nicias states, the Egyptians and Indians call *sacal*. 'In Syria, the women make whorls of this resin and call it *harpax*, or the "snatcher", because it picks up leaves, straw, and the fringes of garments' ('in Syria quoque feminas vericillos inde facere et vocare harpaga, quia folia, paleasque et vestium fimbrias rapiat'). Friction produces static electricity, but a change of temperature also allows some minerals to be charged more readily. Certain di-electric (that is electrically non-conductive) crystals develop an electric polarization when they are subjected to a uniform temperature change. In other words, when one end of such a crystal is heated, the other end develops an electric charge and is said to be pyroelectric. This effect only occurs on crystals which lack a centre of symmetry and also

---

[18] Whitten–Brooks, *Geology*, 371, and above, n. 15.

have polar directions—that is a polar axis: it is intimately related to the hemimorphic, or polar development of the crystals. These conditions are fulfilled in ten of the thirty-two crystal classes. Typical examples of pyroelectric crystals[19] include tourmaline,[20] lithium sulphate monohydrate, and ferroelectric barium titinate. Of these only tourmaline was known in the ancient world. Pliny (37. 103) discusses the mineral *lychnis*,[21] which is clearly identifiable with tourmaline, a borosilicate of aluminium containing an alkali metal, iron, or magnesium: 'It is found around Orthosia and throughout Caria and the neighbouring regions, but the best quality occurs in India. ... I find that there are other varieties such as violet-red and rose-red tourmaline. When these are heated in the sun, or are rubbed between the fingers, they are said to attract straws and papyrus fibres.' ('nascitur circa Orthosiam totaque Caria ac vicinis locis, sed probatissima in Indis. ... et alias invenio differentias: unam quae purpura radiet, alteram quae cocco; has sole excalfactus aut attritu digitorum paleas et chartarum fila ad se rapere.') There is also a 'Carthaginian' stone (37. 104) with the same power, although it is far less valuable. This is named after its country of origin and is often identified with the garnet, although this attribution is uncertain. Topaz, like tourmaline, is remarkable for the strength of the pyroelectrical charge it may acquire and for the length of time it can hold this—about thirty hours. Other precious stones are weakly pyroelectric and their charge soon disappears.

12.2.1c Piezoelectricity

When a plate, cut perpendicularly to the principal axis, is subjected to variations in pressure, it develops positive and negative charges (piezoelectricity[22]) on the two surfaces. Although tourmaline and topaz also have such properties, neither Pliny, nor other ancient authorities, appear to have been aware of this.

---

[19] See above, n. 15.   [20] See Ch. 14, s.v. 'Lychnis'.
[21] See Ch. 14, s.v. See further Partington, *Chemistry*, 764 (aluminium borosilicate).
[22] Whitten–Brooks, *Geology*, 350.

12.3 Magnetism

In the ancient world there was ongoing interest in magnetism[23] and the powers of attraction of some iron ores. Pliny (36. 126), following earlier authorities such as Lucretius, writes: 'What is more strange than magnetite' ('quid enim mirabilius')? 'In what field has Nature displayed a more perverse wilfulness, for she has endowed the magnet with senses and hands' ('ecce sensus manusque tribuit illi').

Although he describes magnetic properties, Pliny is unaware of the basic principle of polarity which all bodies that exhibit magnetic attraction possess—namely that *like poles repel* and *unlike attract* and that *polarity can be reversed*. He states, for example, that an Ethiopian magnet can be distinguished by its ability to attract another magnet: 'Aethiopici argumentum est, quod magneta quoque alium ad se trahit.' This, of course, is true of all magnets if unlike poles are brought together.

In the Western world, Thales,[24] *c*.600 BC, is reputed to have known that iron ores, such as those found near Magnesia in Asia Minor, attract particles of iron. Plato (*Ion* 533d–e) uses the imagery of magnetism in describing Ion's gift of speaking well on Homer: 'it is a power divine, impelling you, like the power in the stone Euripides[25] called the magnet, which most men call the "stone of Heraclea". This stone does not simply attract the iron rings just by themselves; it also imparts to the rings a force enabling them to do the same thing as the stone itself. That is, to attract another ring, so that sometimes a chain is formed, quite a long one, of iron rings, suspended from one another. For all of them, however, their power depends upon that lodestone.'

Lucretius (*DRN* 6. 910–16) repeats Plato's account of the rings:[26]

Men are amazed to see it [the magnet] form a chain of little rings hanging from itself. Sometimes you may see as many as five in

---

[23] See generally Albert Radl, *Der Magnetstein in der Antike: Quellen und Zusammenhänge* (Boethius: Texte und Abhandlungen zur Geschichte der exacten Wissenschaften, 19; Stuttgart, 1988). [24] Ibid. 48 ff.
[25] Fr. 567 (A. Nauck (ed.), *Tragicorum Graecorum Fragmenta*² (Leipzig, 1889; repr. with suppl. by B. Snell, Hildesheim, 1964).
[26] Plato is here describing induced magnetism. When a piece of ferromagnetic substance is made to touch, or brought near to the pole of a magnet, it becomes a magnet itself and has the power of attraction.

pendent succession swaying in the light puffs of air; one hangs from another, clinging to it underneath, and one derives from another the cohesive force of the stone. Such is power of this force to permeate.

> hunc [Magnetum] homines lapidem mirantur; quippe catenam
> saepe ex anellis reddit pendentibus ex se.
> quinque etenim licet interdum plurisque videre
> ordine demisso levibus iactarier auris,
> unus ubi ex uno dependet subter adhaerens
> ex alioque alius lapidis vim vinclaque noscit:
> usque adeo permananter vis pervalet eius.

Lucretius (6. 1002–8) continues with his attempt to explain magnetic attraction.

Firstly this stone must emit a dense stream, or emanation of atoms, which dispels by a process of bombardment all the air that lies between the stone and the iron. When this space is emptied and a large tract in the middle is left void, then atoms of the iron all tangled together immediately slide and tumble into the vacuum. The consequence is that the ring itself follows and so moves in with its whole mass.

> principio fluere e lapide hoc permulta necessest
> semina sive aestum qui discutit aera plagis,
> inter qui lapidem ferrumque est cumque locatus.
> hoc uni inanitur spatium multusque vacefit
> in medio locus, extemplo primordia ferri
> In vacuum prolapsa cadunt coniuncta; fit utqui
> anulus ipse sequatur eatque ita corpore toto.

Pliny (36. 127) briefly alludes to the idea of the vacuum:

iron is attracted by the magnet. The substance that vanquishes all other things rushes into a kind of vacuum [inane nescioquid], and, as it approaches the magnet, leaps towards it and is held fast and embraced by it. Some Greeks call the magnet 'ironstone', some the stone of Heracles. According to Nicander, it was known as magnetite, after its discoverer Magnes. It is said to have been discovered when the nails of Magnes' sandals and the ferule of his staff stuck to the stone, as he was grazing his herds on Mt. Ida. It is, incidentally, found in many places, including Spain.

trahitur namque magnete lapide, domitrixque illa rerum omnium materia ad inane nescioquid currit atque, ut propius venit, adsilit, tenetur amplexuque haeret. sideritim ob id alio nomine vocant, quidam Heraclian. magnes appellatus est ab inventore, ut auctor est Nicander—in Ida repertus, namque et passim inveniuntur in

Hispania quoque; invenisse autem fertur clavis crepidarum baculi cuspide haerentibus, cum armenta pasceret.')

According to Pliny (36. 128), Sotacus, the author of a work *On Stones*, describes five kinds of magnetite, the highest grade of which is worth its weight in silver. Although it is possible that the one from Magnesia in Asia Minor is talc, previously described by Theophrastus (*Lap.* 41), the mineral known as magnesian stone occurs elsewhere (36. 129): it is used in the manufacture of glass[27] and may, therefore, be limestone, or dolomite, a source of lime.

Pliny does not make any reference to the interesting fact recorded by Lucretius (*DRN* 6. 1042–55) that lodestone will

excite Samothracian rings of gilded iron and iron filings in a copper vessel when the lodestone is placed beneath.[28] So eager, it seemed, is the iron to run from the stone. The reason why the interposition of copper causes such a turmoil is doubtless this. After the effluence of the copper has first taken possession of the open passage-ways in the iron and occupied them, along comes the effluence of the magnet and finds everything full in the iron and so has no way of passing through as before. It is therefore compelled to pelt and batter the texture of the iron with its stream. In this way it repels the iron from itself and through the copper drives away what otherwise it normally attracts.

> fit quoque ut a lapide hoc ferri natura recedat
> interdum, fugere atque sequi consueta vicissim.
> exsultare etiam Samothracia ferrea vidi
> et ramenta simul ferri furere intus aenis   1045
> in scaphiis, lapis hic Magnes cum subditus esset:
> usque adeo fugere a saxo gestire videtur.
> aere interposito discordia tanta creatur
> propterea quia nimirum prius aestus ubi aeris
> praecipit ferrique vias possedit apertas,   1050
> posterior lapidis venit aestus et omnia plena
> invenit in ferro neque habet qua tranet ut ante.
> cogitur offensare igitur pulsareque fluctu
> ferrea texta suo; quo pacto respuit ab se
> atque per aes agitat, sine eo quod saepe resorbet. 1055

---

[27] See Ch. 18.2.
[28] The filings would form a pattern round the pole(s) of the magnet, indicating the magnetic field.

Lucretius is led to the erroneous conclusion that emanations from the copper cause the filings to be repelled, rather than attracted by the magnet.

Pliny, likewise, is mystified by 'polarity'. He states (36. 130) that, in Ethiopia, there is a mountain which repels and rejects all iron.[29] There is no obvious explanation for such a phenomenon. Possibly iron-studded boots might acquire a temporary *static* charge, with the same polarity—which would result in rejection—but this is only speculation.

In a brief but fascinating reference to an imaginative practical application of magnetic attraction, Pliny (34. 148) describes a project of 'the architect Timochares [who] had begun to use lodestone in the construction of the vaulting in the temple of Arsinoe[30] at Alexandria, so that the iron statue contained in it might have the appearance of being suspended in mid-air. The project was interrupted by the deaths of Timochares and of King Ptolemy who had commissioned the work in honour of his sister.' ('magnete lapide architectus Timochares Alexandriae Arsinoes templum concamarare inchoaverat, ut in eo simulacrum e ferro pendere in aëre videretur. intercessit ipsius mors et Ptolemaei regis qui id sorori suae iusserat fieri.') Pliny does not state whether any attempt to complete the work was undertaken. It would seem unlikely that such a project would have been practicable. However, the principle of magnetic suspension has, in recent years, been applied in the Maglev[31] project in which a mass people-transporter in the form of a train runs above the track, having no physical contact with it.

Magnetism continued as an ongoing, inexplicable *mirabile* until the experiments of William Gilbert in the sixteenth century.

---

[29] A bar of iron rubbed in a single direction across a magnetic ore will acquire a *static* charge.
[30] The statement is the result of a misunderstanding on the part of Pliny. See further Radl, *Der Magnetstein*, 51. Arsinoe was the wife of Ptolemy II Philadelphus, king of Egypt, 286–247 BC.
[31] One of the first operational trains to use the principle of magnetic levitation was installed to provide rapid transportation between the north and south terminals at Gatwick airport.

12.4 ACOUSTICS

By definition acoustics is the scientific study of sound and sound waves, including the characteristics of an auditorium, or room, which may affect the quality of the sound heard.

Examples of the relay of sound over long distances,[32] presumably under certain favourable atmospheric conditions, are recorded by Pliny. (7. 86) The battle in which Sybaris was destroyed was heard at Olympia on the day that it took place ('auditus unum exemplum habet mirabile, proelium quo Sybaris deleta est eo die quo gestum erat auditum Olympiae').

In the case of the two other victories he mentions (ibid.), namely that of Marius over the Cimbri at Campus Raudius, and of Aemilius Paulus over Perseus, the news was learned on the same day as result of visions and warnings sent by divine power—in the latter example, brought by Castor and Pollux. Elsewhere (2. 148), Pliny refers to a noise of clanging armour and the sound of a trumpet heard from the sky, during the wars against the Cimbri. He implies that this was not an isolated incident 'armorum crepitus et tubae sonitus auditos e caelo Cimbricis bellis accepimus, crebroque et prius et postea.'

Pliny does not offer any explanation why sound could, allegedly, travel such great distances, nor, somewhat surprisingly, is there any reference in the *Natural History* to the outstanding acoustical properties of the Greek and Roman theatre,[33] odeon, or other buildings designed for public assembly.

Echoes, Pliny explains, may be produced naturally, or artificially (36. 99 f.): 'In the city [Cyzicus], close to the so-called Thracian Gate, there are seven towers that repeat, with numerous reverberations, any sound that strikes upon them. The Greek term for this phenomenon is "Echo". It is caused, in nature, by the configuration of the landscape and especially by deep valleys. But, at Cyzicus, it occurs by pure chance, while at Olympia, it is reproduced artificially in a remarkable manner

---

[32] There is no evidence to suggest that there is any basis of truth in these stories and the suggestion of divine messengers having brought the news is, I believe, an attempted rationalization of this alleged 'phenomenon' by Pliny! In nature, thunder is rarely heard beyond a localized area.

[33] A. J. Brothers, 'Buildings for Entertainment', in Barton, *Roman Buildings*, 97 ff. Cf. P. D. Arnott, 'Drama', in Grant–Kitzinger, *Mediterranean Civilization*, i. 480.

within the portico known as the "Seven Voices", so called because the same sound re-echoes seven times.' ('eadem in urbe iuxta portam quae Thracia vocatur turres septem acceptas voces numeroso repercussu multiplicant. nomen huic miraculo Echo est a Graecis datum, et hoc quidem locorum natura evenit ac plerumque convallium; ibi casu accidit, Olympiae autem arte mirabili modo, in porticu quam ob id Heptaphonon appellant, quoniam septiens eadem vox redditur.')

Pliny bases his description on observation without actually explaining that the sound is 'reflected' in a manner akin to the reflection of light.

12.5 HEAT

The concepts of conduction, convection, and latent and specific heat are not found in the *Natural History*, although the practical application of convection in heating space had long been incorporated in the hypocaust,[34] and steam had been used in Hero's[35] well-known device at Alexandria. In the latter, pressure built up in a sealed cauldron passed through a pipe into a sphere, from which it escaped at various points, but mainly through bent tubes. As the steam is forced out in one direction (from the outlets), it causes a reaction thrust in the opposite direction and makes the sphere revolve. The principle is, of course, the same as that on which jet propulsion is based. It is a constant source of surprise that neither the Greeks nor the Romans ever exploited steam as a power source and that this possibility remained latent for so many centuries!

The basic functions of fire (or heat), in raising the temperature of liquids, in the evaporation of solutions,[36] in crystallization,[37] in metallurgical operations[38]—melting, or smelting, and in manufacturing processes (such as the production of glass[39]), are well attested in the *Natural History* as elsewhere in ancient literature.

---

[34] Pliny (26. 16) refers to the first time (*c.* mid-1st cent. BC) that hot-air baths, heated from below, were used: 'tum primum pensili balnearum usu.' Elsewhere (9. 168) he attributes their invention to Sergius Orata (91–87 BC, the time of the Marsian War). See also Vitruvius, *De arch.* 5. 10. 1 ff. and the Younger Pliny, *Letters* 2. 17. 51, cf. 3. 14. 8. Under-floor heating relied on the circulation of hot air (convection).
[35] Landels, *Ancient Engineering*, 28 ff.   [36] See Ch. 11.3.
[37] See Ch. 11.3.   [38] Cf., generally, Ch. 17.   [39] See Ch. 18.2.

12.5.1 Friction

Apart from widespread references to the functions of heat, Pliny (33. 66) is aware of friction (*attritus*): 'There is no gold that is in a more perfect state, as it is thoroughly polished by friction' ('nec ullum absolutius aurum est, ut cursu ipso attrituque perpolitum'). Friction also produces heat (*calor* and *fervor*): 'Rubbing with the fingers draws forth the hot exhalation...' ('ceterum attritu digitorum accita caloris anima'). Similarly, heat is produced in the engraving of gemstones (37. 200)—what is most effective in working gemstones is the heat generated by the drill,[40] 'plurimum vero in iis terebrarum proficat fervor.' Friction also underlies the technique of 'cutting' marble into veneers (36. 51). The cutting is effected, seemingly by iron, but actually by sand, for the 'saw' merely presses the sand on a thinly traced line and then the passage of the instrument, owing to the rapid movement to and fro is, in itself, enough to cut the stone. ('harena hoc fit et ferro videtur fieri, serra in praetenui linea premente harenas versandoque tractu ipso secante.') This technique is still employed today by stonemasons.

12.6 Mechanics

Mechanics[41] includes statics, dynamics, and kinetics, to which there is no specific reference in the *Natural History*. (Pliny often refers to melting and evaporation[42] but has no understanding of the kinetic energy produced in these changes of state.)

---

[40] Modern grinding methods produce enough heat to modify the surface structure of some stones and cause the formation of a 'Beilby layer', a microcrystalline or amorphous layer formed on the surface of metals by polishing.

[41] Arist. *Metaph.* 1078^{a}16. Landels, *Ancient Engineering*, 205 f., describes Hero's treatise on mechanics (τ ὰ μηχανικά) which survives in an Arabic version of the mid-9th cent. AD. See further A. G. Drachmann, *The Mechanical Technology of Greek and Roman Antiquity* (Copenhagen and Madison, 1963) for a more detailed discussion of Hero's lost works. For a chronology of ancient mechanics, see P. Fleury, *La Mécanique de Vitruve* (Caen, 1993), 17, table 1.

[42] Not all the molecules of a liquid have the same speed. As a result, they will not all have the same kinetic energy. There will be some molecules near the surface of a liquid which will have enough kinetic energy to overcome the attractive forces of the neighbouring molecules. They will escape from the surface of the liquid (evaporation) and the space above the liquid becomes filled with vapour.

The Greeks had been concerned with the practical application of 'mechanics' especially in building and construction work, from as early as the time of Chersiphron (c. 560–550 BC), the architect of the first temple of Artemis, built at Ephesus. A work on mechanics (τὰ μηχανικά) is attributed to Aristotle.

Unlike Vitruvius, Pliny appears only to have had a superficial interest in mechanics. He uses the term 'mechanics', *machinalis scientia* (7. 125), in connection with Archimedes, and, in the same section, merely records Ctesibius' discovery of the theory of the pneumatic pump and hydraulic engines, without comment: 'Ctesibius pneumatica ratione et hydraulicis organibus repertis.' The works of earlier Greek authorities on mechanics must have been available in the early imperial period and this is a possible further reason why Pliny omitted to include this branch of science in his work.

Machina[43] and comparable terms, derived from the Greek μηχανή[44] and μηχάνημα, widely used by Vitruvius, are somewhat loosely employed by Pliny (19. 30) in the sense of mechanical appliances[45] ('machinis aedificationum')—in this context appearing to signify little more than 'scaffolding'—and, elsewhere (7. 202), 'siege engines' ('murales machinae'), which are merely tower-like structures placed against enemy walls.

---

[43] For a discussion of these terms (machines and types of machines) see Fleury, *Mécanique*, 35 ff. Vitruvius (*De arch.* 10. 1. 1) defines *machina* as follows: 'A "machine" is a device made of solid pieces of wood, with a very great power to accomplish tasks following the principle of circular movement as the Greeks define this.' ('Machina est continens e materia coniunctio maximas ad onerum motus habens virtutes, ea movetur ex arte circulorum rotundationibus quam Graeci κυκλικὴν κίνησιν appellant').

[44] As in Hdt. 3. 83, 8. 57, and elsewhere, in the sense of a mechanical device that transmits, or modifies, force in order to perform useful work. It is often used of a crane, or hoist (Hdt. 2. 125). In addition, it can refer to theatrical machinery (*deus e machina*), Pl. *Cra.* 425d and Arist. *Poet.* 76^a 24; also to irrigation devices, as *P Oxy.* 985 (1st cent. AD; ed. B. P. Grenfell and A. S. Hunt, *The Oxyrhynchus Papyri* (London, 1898–); and to siege engines: Dem. 18.87; Polyb. 1. 48. 2; Plut. *Marcell.* 14. 2.

[45] See generally, Fleury, *Mécanique*, 28, table 3, lists mechanical devices discussed in Vitruvius, *De arch.* 10.

12.6.1 Lifting Devices

In the ancient world, lifting devices,[46] such as cranes,[47] pulleys, blocks and tackle, were all important and their use is well attested by the nature of surviving buildings, as well as in literary sources.

Pliny (36. 66) describes the erection of an obelisk.[48] 'When it was about to be erected, the king [Rameses II] feared that the hoists [this is a more appropriate translation than 'scaffolding' here] would not be strong enough for the weight and, in order to force an even greater danger upon the attention of the workmen, tied his son to the pinnacle, intending that the stone should share the benefit of his deliverance at the hands of the labourers.' ('ipse rex, cum surrecturus esset verereturque ne machinae ponderi non sufficerent, quo maius periculum curae artificum denuntiaret, filium suum adalligavit cacumini, ut salus eius apud molientes prodesset et lapidi.')

Pliny (36. 96 f.) was unable to explain how pyramids were built—he was not alone in this—and gives an apparently fanciful account of the method used in putting in place the architraves of the temple of Diana at Ephesus, without reference to the use of mechanical lifting devices.[49]

The crowning marvel was [Chersiphron's] success in lifting the architrave of this massive building into place. This he achieved by filling bags of plaited reed with sand and constructing a gently graded ramp which reached the upper surfaces of the capitals of the columns. Then, little by little, he emptied the lowest layer of bags so

---

[46] See Fleury, *Mécanique*, 95 ff. Cf. also Hero, *Mechanics* 2: a large part of Drachmann, *Mechanical Technology* (pp. 19–140) is devoted to Hero's *Mechanics*, with translations of the crucial passages, and commentary. See further B. Gille, *Les Mécaniciens grecs* (Paris, 1980).

[47] See Landels, *Ancient Engineering* 84 ff., on cranes and hoists. Cf. also Hero, *Mechanics* 2.

[48] Modern experiments have been carried out in an effort to reproduce the technology involved in erecting trilitha of the type found at Stonehenge. Cf. also Fleury, *Mécanique*, ch.1, 'Machines de soulèvement et systèmes de traction', 95 ff., esp. 138 ff. ('Le Transport des mégalithes').

[49] *NH* 36. 96. Landels, *Ancient Engineering* 92, comments, 'It is much more likely that the architrave was hoisted by crane and placed roughly in position, when it could then be adjusted very finely by letting sand out of the lower bags thus causing the pile to settle very gradually in the required direction.' See also J. J Coulton, 'Lifting in Early Greek Architecture', *Journal of Hellenic Studies*, 94 (1974), 11.

that the fabric gradually settled into its right position. But the greatest difficulty was encountered with the lintel itself when he was trying to place it over the door; for this was the largest block and it would not settle on its bed.

summa miraculi epistylia tantae molis attolli potuisse; id consecutus ille est aeronibus harenae plenis, molli clivo super capita columnarum exaggerato, paulatim exinaniens imos ut sensim opus in loco sederet, difficillime hoc contigit in limine ipso quod foribus imponebat; etenim ea maxima moles fuit nec sedit in cubili.

Pliny repeats the legend that the problem was resolved by the intervention of Diana herself. Whatever happened during the night, the next day the lintel was apparently found to be in place.

12.6.2 Levers

The practical application of leverage[50] was known from a very early period in the Near East and Egypt. In Roman times, Caesar (BC 2. 11 and 3. 40) refers to the movement of the largest possible stones by levers: 'saxa quam maxima possunt vectibus promovent.' In the *Natural History*, levers are employed, among other uses, as part of the machinery of winepresses[51] (18. 317).

Some press the grape with a single press-beam, but it pays better to use a pair, however large the single beams may be. It is the length that matters in the case of the beams, not the thickness [i.e. the work

---

[50] Cf. Landels, *Ancient Engineering* 189 ff., and Healy, *Pliny, NH: A selection*, 95 n. 7. Archimedes understood the principle of leverage. His assertion, 'Give me but one firm spot on which to stand, and I will move the earth', is well known (see Pappus, *Synagoge* 8, proposition 10, section 11).

[51] K. D. White, *Roman Farming* (London, 1970), 425, quotes Vitruvius (*De arch.* 6.6.3) as stating that the single-lever press requires a beam of some forty feet in length. On the types of press here described by Pliny see further, K. D. White, 'Farming and Animal Husbandry', in Grant–Kitzinger, *Mediterranean Civilization*, i. 230. The lever-press was scrapped, according to Pliny, in the mid-1st cent. AD and replaced by a direct-screw press. White (loc. cit.) writes, 'The earliest form resembled the Victorian letterpress, with a single screw bearing down on the mush within a strong frame capable of taking the counter-thrust. The subsequent development of a twin-screw model and a more heavily loaded frame follows the pattern of other well-documented inventions. The established lever-press will not have been displaced until screw-presses of similar capacity had been developed.

is done by leverage, not by the mere weight of the beam]; but those of ample width press better. People in olden times used to drag down the press-beams with ropes and leather straps, and by means of levers: but within the last hundred years, the Greek pattern has been invented, with the grooves of the upright beams running spirally. ... within the last twenty-two years, a plan has been devised to use smaller presses and a smaller pressing-shed with a shorter upright beam running straight down into the middle, and to press down the drums placed on top of the grape-skins with the whole weight and to pile a heap of stones above the presses.

premunt aliqui singulis, utilius binis, licet magna sit vastitas singulis. longitudo in his refert, non crassitudo; spatiosa melius premunt. antiqui funibus vittisque loreis ea detrahebant et vectibus; intra C annos inventa Graecianica, mali rugis per cocleam ambulantibus. ... intra XXII hos annos inventum parvis prelis et minore torculario aedificio, breviore malo in media derecto, tympana inposita vinaceis superne tot pondere urguere et super prela construere congeriem.

12.6.3 Curio's Revolving Theatre

Curio (36. 117 f.) built, close to each other, two very large wooden theatres,[52] each poised and balanced on a revolving pivot.[53] During the forenoon, a performance of a play was given in both theatres; they faced in opposite directions so that the two casts should not drown each other's words [see Fig. 12. 2 below]. Then, suddenly, the theatres revolved and their outermost points met. (It is agreed that, after the first few days, they revolved with some of the audience actually remaining in their seats.). Thus Curio provided an amphitheatre in which he produced fights between gladiators and the Roman people itself was brought into greater danger than the gladiators, as the theatre whirled round.

Large screw-presses were in use until very recent times for wine in Sicily, and for oil, in North Africa. To complete the list of presses, there is the wedge-and-beam press, not mentioned by any of the authorities, but displayed on two well-known paintings from Pompeii and Herculaneum. Here a different power source, the wedge, was used. A strong wooden frame like that of the single-screw press is filled with a series of horizontal floors, separated from each other by cone-shaped members, which are driven in by mallets so as to force the floors apart, causing the juice to be expressed into a receptacle.'

[52] Curio was consul in 76 BC. The theatres were allegedly built in 52 BC.
[53] See W. Beare, *The Roman Stage* (London, 1968), 172; Brothers, 'Buildings for Entertainment', 114; and Healy, *Pliny, NH: A Selection*, 358 n. 1.

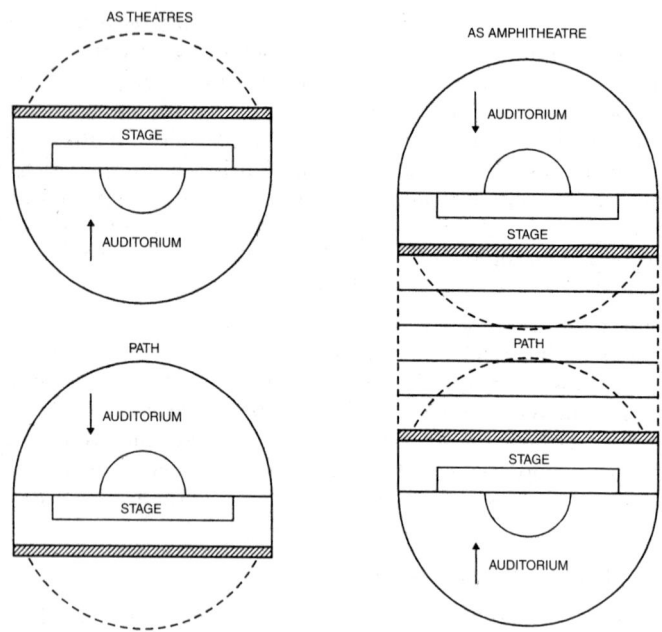

Fig. 12.2. Simplified plan of Curio's revolving theatres (*NH* 36. 116 f.)

theatra iuxta duo fecit amplissima ligno cardinum singulorum versatili suspensa libramento, in quibus utrisque antemeridiano ludorum spectaculo edito inter sese aversis ne invicem obstreperent scaenae, repente circumactis—ut constat, post primos dies etiam sedentibus aliquis—cornibus in se coeuntibus faciebat amphitheatrum gladiatorumque proelia edebat, ipsum magis auctoratum populum Romanum circumferens.

12.6.4 Hydrostatics

Archimedes[54] was the first to explore, in detail, problems relating to hydrostatics,[55] that is the science of the mechanical properties and behaviour of fluids that are not in motion.

---

[54] Archimedes (287–212 BC), born at Syracuse, mathematician, natural philosopher, and inventor. A close friend of Hiero II (tyrant of Syracuse) for whom he constructed a number of engines of war. Many of his discoveries,

The story of the displacement of water as a means of determining the volume of an irregularly shaped object—Hiero's crown—is well known. Archimedes' treatise *On Floating Bodies* is a remarkable achievement.

Pliny gives no account of hydrostatics in the strict definition of the term, but includes (36. 67) an interesting description of the transportation of a huge obelisk, some 120 feet tall, which depended on buoyancy.

According to some authorities, it was carried downstream by the engineer Satyrus, on a raft, but Callixenus states that it was conveyed by Phoenix. The latter, by digging a canal, brought the Nile right up to where the obelisk lay. Two wide barges were loaded with cubes of the same granite as the obelisk—each one a cubic foot in volume—until the weight of the blocks was double that of the obelisk and the total volume the same. The ships were thus able to float beneath the obelisk which was suspended by its ends on both banks of the canal. Then the blocks of granite were unloaded and the lightened barges took the weight of the obelisk.

a Satyro architecto aliqui devectum tradunt rate, Callixenus a Phoenice, fossa perducto usque ad iacentem obeliscum Nilo, navesque duas in latitudinem patulas pedalibus ex eodem lapide ad rationem geminati per duplicem mensuram ponderis oneratas ita, ut subirent obeliscum pendentem extremitatibus suis in ripis utrimque; postea egestis laterculis adlevatas naves excepisse onus.

12.7 HYDROLOGY

Although Pliny does not refer to hydrostatics, he is certainly much concerned with hydrology, that is the study of water, its distribution, conservation, and use.

The whole of book 31 is devoted to sources and properties of water which was obtained from springs, rivers,[56] and wells. Vitruvius (*De arch.* 8. 1 and 4–5) discusses the types of soil in which water can be found and Pliny (31. 44) also mentions

---

including the helical Archimedian screw, were highly important. In addition to his knowledge of levers and hydrostatics, he worked out the value of Π, and made many other discoveries relating to conoids, spirals, the centres of gravity of planes, and the quadrature of the parabola.

[55] Landels, *Ancient Engineering*, 189 ff.
[56] Ibid., 35 ff.

indicators which suggest the presence of water sources. He outlines a test for discovering water (ibid. 46): 'They dig a hole to the depth of five feet, covering it with jars of unbaked potter's clay, or else with a well-oiled bronze basin, and also a burning lamp arched over with foliage and earth on top; if the clay is found to be wet, or broken, or if moisture covers the bronze, or the lamp goes out without any failure of oil, or a fleece of wool is wet, then the finding of water is assured. Some light a fire first and dry the hole, making yet more conclusive the evidence of the vessels.' ('loco in altitudinum pedum quinque defosso ollisque e figlino opere crudis aut peruncta pelvi aerea, cooperto, lucernaque ardente concamarata frondibus, dein terra, si figlinum umidum ruptumve, aut in aere sudor vel lucerna sine defectu olei restincta aut etiam vellus lanae madidum repperiatur, non dubie promittunt aquas. quidam et igni prius excocunt locum tanto efficaciore vasorum argumento.')

Landels[57] writes: 'The water supply represented one of the most serious problems for Greek and Roman urban communities. Its management was an important part of Roman civil engineering.' Two forms of conduits were employed to convey water over distances by gravity flow. The open conduit—much the more common in Roman systems—consists of a channel, usually built into a stone structure and waterproofed with plaster or cement. This was the aqueduct,[58] many examples of which survive, in part at least—one of the most famous being the Pont du Gard,[59] at Nîmes. To keep the level of the water even, the aqueduct has to slope at a more or less consistent angle along its entire length. The gradient was normally between 1 in 150 and 1 in 500. (Vitruvius recommends 'not less than 1 in 200').

Pliny (36. 123) discussing the aqueducts which supply Rome with water, states, 'If we carefully consider... the distances traversed by the water before it arrives, the construction of the arches, the tunnelling of mountains, and the building of level routes across deep valleys, we shall readily admit that there has

---

[57] Ibid. 37.
[58] See C. E. Bennett (trans.), *Frontinus: Aqueducts* (Loeb: London, 1925); A. T. Hodge, 'Aqueducts', in Barton, *Roman Buildings*, 127 ff.; and Landels, 'Engineering' in Grant–Kitzinger, *Mediterranean Civilization*, i. 338 ff.
[59] See Landels, *Ancient Engineering*, 43 f. Cf. also the aqueduct at Aspendus, in southern Turkey.

never been anything more remarkable in the whole world.' ('quod si quis diligentius aestumaverit... spatia aquae venientis, exstructos arcus, montes perfossos, convalles aequatas, fatebitur nil magis mirandum fuisse in toto orbe terrarum.')

The second method of supplying water, a closed conduit, usually took the form of a round pipe made of earthenware,[60] or a pipe with a triangular cross-section made of lead and sealed by an overlap at its top. This could slope up, or down, at any angle provided that it did not rise at any point above the level of the intake. The main problem with an open conduit was maintaining a consistency of gradient over the rises and falls of ground level between the source and the delivery point. Pliny (33. 74) describes the problems encountered in providing water for *hushing*,[61] or *hydraulicking*,[62] in fold-mining operations in north-west Spain.[63]

Although the Romans used three types of drainage[64] in their mines—cross-cuts, baling, and mechanical devices including the Archimedian screw[65] and water-drainage wheels[66]—Pliny (33. 97), unaccountably, only mentions baling: 'Along the whole of this distance [in the mine at Baebalo] watermen are posted who, night and day in spells measured by lanterns, bale out the water and make a stream' ('per quod spatium aquatini stantes noctibus diebusque egerunt aquas lucernarum mensura amnemque faciunt').

Finally, the increase in water pressure obtained by using inverted funnels (*tuyères*) as outlets from cisterns greatly improved the gravity separation methods employed in concentrating ground argentiferous ore from the Athenian mines at Laurium[67] and elsewhere. The water supplied washing tables

---

[60] See Landels, *Ancient Engineering*, 43 ff.
[61] Healy, *Mining*, 87. See also P. R. Lewis and G. D. B. Jones, 'Roman Gold-mining in North-west Spain', *Journal of Roman Studies*, 60 (1970), 177 f.
[62] S. V. Griffith, *Alluvial Prospecting and Mining*² (Oxford, New York, and Paris, 1961), 150 ff., and Healy, *Mining*, 88.
[63] J. F. Healy, 'Greek and Roman Gold Sources: The Literary and Scientific Evidence, in Domergue, *Mineria y metalurgia*, ii. 120.
[64] Healy, *Mining*, 94 ff. It would appear, from his description of bailing, that Pliny had not encountered mechanical forms of drainage in the gold-mines of Hispania Tarraconensis.
[65] Healy, *Mining*, 94 ff.   [66] Ibid.
[67] Conophagos, *Laurium*, 241 ff. Both flat-table (at Agrileza) and helicoidal washeries (at Demoliaki) were used in Greece.

and helicoidal washeries and was recycled. The pressure in a fluid increases with depth and this is a further reason, other than the obvious one, for the low position of outlets in the cisterns.[68]

12.8 Gravity

Finally, although gravity is not mentioned in *Natural History*, Hesiod[69] (Theog. 722 ff.) throws an interesting sidelight on the sort of scale on which an early Greek thinker conceived of the Universe. He also indirectly describes the effect of gravitational forces.

> An anvil made of bronze, falling from heaven,
> would fall nine nights and days, and on the
> tenth would reach the earth; and if the anvil
> fell from earth, would fall again nine nights and days
> and come to Tartarus upon the tenth.[70]

12.9 Conclusions

Pliny describes many observed facts which would, centuries later, become incorporated in the *Laws of Physics*. He is herein particularly indebted to Lucretius as a source. But, although his degree of originality in this field is limited, Pliny's collation of what was known, in the first century AD, apropos the properties of matter and energy and the relationships between them, is an invaluable contribution to our understanding of the 'science' of that period.

---

[68] Conophagos, *Laurium*, 238 and fig. 10.21, and pp. 246 f. See especially 'Le Débit d'eau à la tuyère', in which Conophagos calculates the flow rates.
[69] M. L. West (ed.), *Hesiod: Theogony* (Oxford, 1966).
[70] Translation from Dorothea S. Wender, *Hesiod: Theogony, Works and Days;* and *Theognis: Elegies* (London, 1979), p. 46. Newton formulated the Law of Gravity, and Galileo, by experiments from the top of the Leaning Tower of Pisa, showed that objects fall at the same speed, independent of their weight.

Appendix

APPENDIX: FROM OBSERVER TO HORIZON

Earth is an oblate spheroid with equatorial radius of 3,963 miles (6,390 km.) and polar radius of 3,950 miles (6,320 km.).

ASSUMPTIONS FOR THE CALCULATIONS:

the radius of the earth is 3,957 miles (6,331 km.). All parts at sea-level are that distance from the centre of the earth: there is accordingly no allowance for tidal changes.

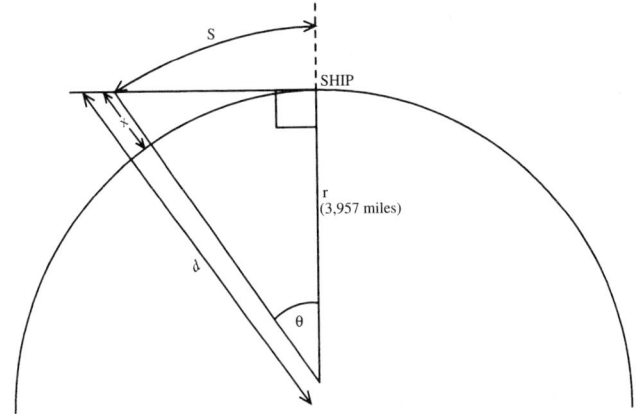

$s = r\theta$ where s = arc length, r = radius, θ = the angle subtended at the centre in radians, and x = the height of the observer's eye—that is Strabo's—above sea-level. It is assumed that he saw the whole ship and not just part of the mast.

$$s = r\theta \Rightarrow \theta = \frac{s}{r}, \cos\theta = \frac{3957}{d} \Rightarrow d = \frac{3957}{\cos\theta}, d = x + 3957.$$

$$d = \frac{3957}{\cos\theta} \Rightarrow x + 3957 = \frac{3957}{\cos(s/r)} \left[\text{or } x + r = \frac{r}{\cos(s/r)} \right]$$

$$\Rightarrow x = \frac{3957}{\cos(s/r)} - 3957.$$

(i) If $s = 135$ miles, $x = \dfrac{3957}{\cos(135/3957)} - 3957 = 2.30$ miles;

(ii) If $s = 100$ miles, $x = \dfrac{3957}{\cos(100/3957)} - 3957 = 1.26$ miles;

(iii) If $s = 85$ miles, $x = \dfrac{3957}{\cos(85/3957)} - 3957 = 0.91$ miles.

(iv) The top of Mt. Etna is 11,000 feet above sea-level, and the number of feet in one mile is 5,280 $\Rightarrow x = \dfrac{11000}{5280} = 2.08$ miles;

$$x + r = \frac{r}{\cos(s/r)} \Rightarrow \cos\left(\frac{s}{r}\right) = \frac{r}{x+r} \Rightarrow \frac{s}{r}$$

$$= \cos^{-1}\left(\frac{r}{x+r}\right) \Rightarrow s = r \cdot \cos^{-1}\left(\frac{r}{x+r}\right);$$

here $r = 3957$ miles, $x = 2.08$ miles $\Rightarrow s = 3957 \cdot \cos^{-1}\left(\dfrac{3957}{3959.08}\right) = 128$ miles.

(v) At sea-level, assuming an observer on a boat at a height above water of 10 feet (if at exactly sea-level, i.e. $x = 0$, then $s = 0$) $\Rightarrow s = 3957 \cdot \cos^{-1}\left(\dfrac{3957}{3957+x}\right);$

$$x = \frac{10}{5280} = 0.00189 \text{ miles} \Rightarrow s = 3.87 \text{ miles.}$$

13

Earth Sciences

13.1 Introduction

Pliny devotes the major part of books 33–7 to minerals (including ore-minerals), mineral earths, types of common stone, and gemstones. In addition there are many random references to these throughout the *Natural History*.

His descriptions and record of their properties embrace a whole range of physical characteristics, although there is no systematic study and classification of minerals. Pliny's observations, however provide detailed information about the crystal systems of quartz, diamonds, emeralds, and the iris stone which were to form the basis of the study of crystallography[1] in the eighteenth and succeeding centuries, but, at the same time, he includes *mirabilia*[2] (for example, the sarcophagus stone) which are derived from legendary accounts rather than observed facts.

Although he repeats, litcratim, information from Theophrastus, *De lapidibus*, and from other authorities, whose works are no longer extant, Pliny adds new substances to those previously described, for example, sulphur.

Minerals, by modern definition, are naturally formed, structurally homogeneous,[3] almost exclusively solid[4] constituents of

---

[1] See below, n. 52.   [2] See Ch. 5 above.
[3] This implies that the fundamental atomic structure is continuous and constant throughout the mineral unit; in silicates, for example, the silicon–oxygen lattice will be constant in character.
[4] Mercury, which is a liquid, is the only exception.

the Earth and of extraterrestrial bodies:[5] they possess definite, but not necessarily fixed, or constant, chemical composition[6] and are either uncombined elements in a native state,[7] or compounds of elements formed in accordance with chemical laws. Minerals resulted from a sequence of complex processes which began with crystallization in rocks, or ore bodies controlled by trivial local factors. They have a characteristic atomic structure which is expressed in their external crystalline form, together with other distinctive physical qualities such as colour, hardness, lustre, and fracture habit.[8]

At the present time approximately 2,500 minerals are known.[9]

13.2 EARLY THEORIES OF THE ORIGIN OF MINERALS

The Presocratics[10] expressed no interest in the nature and composition of physical substances and made virtually no contribution to the Greeks' understanding of minerals and their genesis. Anaxagoras' comparison of the sun to $\mu\acute{v}\delta\rho o\varsigma$ $\delta\iota\acute{a}\pi \upsilon \rho o\varsigma$, a red-hot mass of stone, or metal, is very much an exception.[11]

---

[5] As e.g. iron in meteorites. Healy, *Mining*, 35. Meteoric iron forms the entire mass of the meteorite, or is a spongy cellular matrix in which are embedded grains of chrysolite, or other silicates: it also occurs, more or less freely, throughout a stony matrix.

[6] Many minerals have compositions which are variable between certain limits, that are defined in terms of end members: the composition of common olivines is expressible in terms of the two compounds Mg_2SiO_4 (fosterite) and Fe_2SiO_4 (fayalite).

[7] Hurlbut, *Mineralogy*, 222. The more common native metals constitute three isostructural groups: gold, silver, copper, and lead; platinum, palladium, iridium, osmium, and rhodium (PGE); iron and nickel-iron. In addition, tantalum, tin, and zinc have been found in their native form. The native semi-metals (metalloids) include arsenic, antimony, and bismuth. Finally mercury—not a metal—occurs in its native state in specific regions.

[8] Cf. Healy, *Mining*, 20.

[9] P. W. Birkeland and E. E. Larson (eds.) *Putnam's Geology*[5] (Oxford, 1989), 57.

[10] Healy, *Mining*. Cf. also Socrates' disenchantment with science, as Xenophon (Mem. 1.1. 11) informs us: 'He did not even discuss that topic favoured by other speakers, that is, the nature of the Universe, and avoided speculation on how the so-called cosmos of the professors works and on the laws that govern the phenomena of the heavens: indeed he would argue that to trouble one's mind with such problems is sheer folly.'

[11] *Apud* Diog. Laert. 2. 8. The origin of Anaxagoras' imagery is not certain. It may have been the sight of a mass of red-hot iron on an anvil

The first tentative mention of the formation of rocks is found in Plato (*Ti.* 60b–c6): 'Of the different kinds of earth, that which is strained through water becomes a stony mass in the following way. When the commingled water is broken up in the mixing, it changes into the form of air; and, having become air, it darts up to its own region. Now there is no void surrounding it; accordingly, it gives a thrust to the neighbouring air. And the air, being weighty, when it is thrust and poured around the mass of earth, presses it hard and squeezes it into the space which the new-made air quitted. Thus the earth, when compressed by air into a mass that will not dissolve in water, forms stone.' Other formations, which Plato[12] defines, are mixtures either of one and the same element in its different grades and sizes, or of different elements.

Aristotle's conclusions are typical of a stage when the natural sciences had not become fully differentiated from natural philosophy, but embraced a wide range of scientific knowledge. He partly agrees (*Mete.* 3, 378^{a-b}) with Plato's theory about the formation of metals: briefly, that which is 'soluble' by heat is water, or, at any rate, is mainly composed of water. There are two exhalations, one vaporous, the other smoky; and there are two corresponding kinds of body produced within the Earth—fossils (quarried material) and metals. the dry exhalation produced all the fossils by the action of its heat: for example, realgar (a red disulphide of arsenic), ochre and ruddle (a red variety of ochre), and sulphur. Metals are the product of the vaporous exhalation. In Aristotle's account, however, the part played by the dry exhalation in producing the fossils is obscure in every respect.

Eichholz[13] concludes that the fire and heat from the dry exhalation form the fossils by reducing earth to the consistency of fire ash and perhaps causing it to take on bright colours—although this is less certain. Those of the fossils that are stones must furthermore have been hardened by this heat (*Mete.* 3, $383^{b}10$ ff.) The conclusion that the dry exhalation is the efficient and not the material cause of Aristotle's fossils

---

(cf. Callimachus, *Hymn to Artemis*, 49). Although it is interesting to compare the description of stones thrown out by Mt. Etna ($\mu\acute{\upsilon}\delta\rho o\iota\ \delta\iota\acute{\alpha}\pi\upsilon\rho o\iota$, [Arist.] *Mund.* $359^{b}23$, and Strabo 6. 2. 8), there are no grounds for inferring that Anaxagoras had in mind any product of volcanic activity.

[12] *Ti.* 58d–59d. See Healy, *Mining*, 16.   [13] Eichholz, *Lap.* 42.

seems to be corroborated by Theophrastus, who discusses the origin of a group of mineral earths, of which three, namely ochre, ruddle, and realgar, are cited by Aristotle as fossils. Some of them have been exposed to fire ($\pi\epsilon\pi\upsilon\rho\omega\mu\acute{\epsilon}\nu\alpha$) and burnt ($\kappa\alpha\tau\alpha\kappa\epsilon\kappa\alpha\upsilon\mu\acute{\epsilon}\nu\alpha$) such as realgar, orpiment (the yellow trisulphide of arsenic), and others of the same kind. To put it generally, all of these result from a dry and smoky exhalation. This passage makes sense only if the dry exhalation is the efficient cause of the formation of these earths.

Theophrastus' *De lapidibus* is the oldest scientific treatise dealing expressly with minerals. It is important not only in the history of mineralogy but also of chemical technology.[14] Although he owes a debt to his predecessors, his views are often more advanced. He writes (*Lap.* 1): 'Some of the substances found in the Earth are composed of water, and some of earth. Metals, such as silver, gold, and the rest, come from water; from earth come ordinary stones and the more unusual kinds, and those varieties of minerals that are distinctive in colour, smoothness, density, or some other quality.' Theophrastus, like Strato,[15] reverted to the theories of Plato on the origin of non-metallic mineral substances, but followed Aristotle for the explanation of their solidification. He lists and describes sixteen mineral species grouping them, as 'metals, stones, and earths'. It is noteworthy that the treatment is essentially from the economic aspect and deals with minerals in a practical way, giving an account of their uses. He attempts to classify minerals on the basis of Aristotelian principles: those he records are by way of illustration and do not represent the total known in his day. The main interest, for us, lies in the fact that Theophrastus was the first to try to study minerals in a systematic manner. His method of treatment was logical enough, but his classification, based on superficial characteristics and behaviour, rather than any concept of chemical composition, had marked limitations. One noteworthy feature of this work is its comparative freedom from fable and magic. It remained, for

---

[14] Eichholz. 42.
[15] R. Halleux, 'Le Problème des métaux dans la science antique', *Bibliothèque de la Faculté de Philosophie et Lettres de Université de Liège*, fasc. 209 (Liège, 1974), 115 ff.

nearly two thousand years, the most rational and systematic treatment of the subject!

Lucretius (*DRN* 5. 460–5) describes the formation of earth, sky, and heavenly bodies from exhalations, but his poetic imagery adds little of consequence.

Pliny[16] follows Aristotle, in the main, but also inherits data from a number of other Greek philosophers and 'natural scientists' whose works have either not survived, or exist only in tantalizing fragments. There are random allusions to a theory regarding the formation of transparent and semi-transparent stones, which seems to have been developed by Posidonius. According to this, the raw material of such stones was water, possibly impregnated with earthy particles; this liquid was compacted either by cold in the atmosphere, or by one or other of the two exhalations, colours being imparted to these stones by the dry exhalations, which also hardened them. Posidonius' interest in natural phenomena left its mark on Pliny's sources, although the theory as a whole was unknown to Pliny himself.

Seneca (QN 3. 25. 10) describes rock crystal as 'rain water containing a very little earth matter' ('aqua caelestis minimum in se terreni habens');[17] Diodorus Siculus (2. 52. 1–4) says 'pure water' ($ὕδατος\ καθαροῦ$). Thus Pliny (37. 21) claims that fluorspar is a liquid solidified underground by heat and Aristotle's dry exhalation ('caloris anima'); this is specifically referred to as being responsible for the hardening process and colouration of the mineral. Elsewhere (37. 23), rock-crystal ($κρύσταλλος$) is described as being hardened by excessively intense freezing, although Pliny appears to express scepticism with regard to this theory. Pliny (36. 161) also states that selenite (*lapis specularis*) is formed when a liquid is frozen and petrified by an exhalation in the Earth ('terrae quadam anima'). Seneca confirms that the two exhalations played a part in Posidonius' physical theories.

Pliny (36. 125) repeats the commonly held belief in the ancient world that minerals grow. Among the many other marvels of Italy itself, is that marble actually grows in the quarries. ... The quarrymen, moreover, assert that the scars on

---

[16] *NH* 36. 161, 37. 21, 23. 26–8, and 48.
[17] Eichholz, *NH 36–37*, pp. xiv f. Cf. Forbes, *Ancient Technology*, vii 86, and Theophr. *Lap.* generally.

the mountain sides fill up of their own accord. If this is true, there is reason to hope that there will always be marble sufficient to satisfy luxury's demands. ('inter plurima alia Italiae ipsius miracula marmora in lapidicinis crescere.... exemptores quoque adfirmant compleri sponte illa montium ulcera, quae si vera sunt, spes est numquam defutura luxuriae.')

Such instances of the regeneration of stone, or metalliferous ores, occur throughout Greek and Roman literature. Thus Strabo (5. 2. 6) comments when writing about Elba: 'This is not the only remarkable thing about this island: there is also the fact that the diggings, which have been mined, are in time filled up again, as is the case with the ledges of rock on Rhodes...and elsewhere.' Again, Pliny (34. 164), while discussing lead, observes: 'It is remarkable that, in the case of these mines only, when they have been abandoned, they replenish themselves and become more productive' ('mirum in his solis metallis, quod derelicta fertilius revivescunt').

Similar beliefs have endured in India to the present day, where diamonds are thought to regenerate themselves. Diamantiferous[18] sandstone, which has been removed from its natural bed, and from which the diamonds have been extracted, is often allowed to be exposed to the various atmospheric weathering agencies for some time, and is then worked over, when a further yield of diamonds may be found: this is sometimes repeated. Such occurrences have given rise to a belief among natives that this second crop of diamonds has originated in the waste rock, or that it is the result of a fusion together of smaller diamonds originally left behind; similar beliefs are encountered in South Africa. The diamonds have, of course, simply been released by the weathering of larger fragments of rock in which they are embedded.

13.3 PHYSICAL CHARACTERISTICS OF MINERALS

13.3.1 Colour

Minerals have a number of properties that are derived from their chemical composition or crystalline structure, but for the

---

[18] J. F. Healy, 'Pliny on Mineralogy and Metals', in French–Greenaway, *Roman Science*, 116 f.

Greeks and Romans, one of the most important was colour.[19] Theophrastus and other writers before Pliny refer especially to transparency (or opacity), lustre, and colour. In the search for ore-minerals, colour was vital for the purpose of identification. Prospecting for silver,[20] however, raised problems, as Pliny (33. 95) observes: 'Silver is only found in deep shafts and presents no evidence of its existence: it does not give off shining sparkles such as are seen in the case of gold. Its ore is sometimes red, sometimes the colour of ash.' ('non nisi in puteis reperitur [argentum] nullaque spe sui nascitur, nullis, ut in auro, lucentibus scintillis. ... terra est alia rubra, alia cineracea.')

In Attica, the First Contact outcropped at a number of points on the slopes from Dipileza to Cape Sunium[21] and these outcrops were indicated by the red colour of iron oxides (haematite) found on the surface.

Colour may be due to one of the mineral's major constituent elements and therefore constant and characteristic. Such minerals are known as idiochromatic[22] and colour serves as a primary means of identification. Malachite, for example, is green, and azurite, blue (both forms of copper carbonate), while rhodonite and rhodochrosite are red and pink. The Median stones described by Pliny (37. 71) are malachite intergrown with azurite—a common phenomenon.

In allochromatic[23] minerals, the colour is derived from appreciable amounts of a single element such as iron, which has a strong pigmenting power. So, in sphalerite (zinc sulphide), the progressive substitution of iron for zinc changes the colour from white, through yellow and brown, to black. Ore-minerals, although their colour is constant—as the brass-yellow of chalcopyrite, the brownish bronze of bornite, and the copper-red of niccolite—may, on exposure to air, produce a coloured tarnish different from their original true colour and this is especially noticeable in the case of copper minerals. Bornite, found in association with other forms of copper, in primary deposits, is called 'peacock' ore because of this phenomenon: it acquires a blue-violet surface film when exposed to the air. Copper pyrites is similarly affected.

---

[19] Healy, *Mining* 28 f. So colour continued to be important in the speculations of Cheng Ssu-hsiao in the 14th cent. [20] Ibid. 71 f.
[21] Conophagos, *Laurium*, 155 ff. [22] Healy, *Mining* 28 f. [23] Ibid. 29.

In the *Natural History*, Pliny shows evidence of a much wider understanding of the whole range of physical properties of minerals, among them, cleavage, hardness (and softness), tenacity, density (specific gravity), streak, and magnetic and electrical properties.

13.3.2 Cleavage

A mineral is said to have cleavage if it breaks along definite plane surfaces.[24] Pliny (36. 131) describes 'sarcophagus stone' which has this characteristic: it is found at Assos in the Troad. The best known example of perfect cleavage, however, occurs in selenite[25] (*lapis specularis*): this has a far more amenable character which allows it to be split into plates as thin as may be wished (36. 160–2). The Younger Pliny (*Letters* 2. 17. 4 and 21) and other authorities refer to its use for windows (*specularia*).[26]

13.3.3 Hardness (duritia)

The resistance that a smooth surface of a mineral offers to scratching constitutes its hardness.[27] The degree is determined, therefore, by observing the comparative ease, or difficulty, with which one mineral is scratched by another, or by a file, or knife. Mineral hardness is sclerometric, as opposed to indentation[28] hardness measured in metals. Hardness was an important property in the ancient world and well attested. Pliny (37. 23) mentions it in connection with fluorspar, onyx marble—*chernites*—(36. 132), and other minerals. In his account of the diamond (37. 57) Pliny discusses hardness and tenacity, that is cohesiveness: the diamond is described as defying

---

[24] M. H. Battey, *Mineralogy for Students*² (London, 1981), 84. The tendency of many minerals to break smoothly along particular planes of atoms in the structure is a manifestation of weaker bond strength across those planes than in other directions in the structure. Cleavage planes are always planes with a high density of atoms, and so are always parallel to a possible crystal face.

[25] See Ch. 14, s.v. 'lapis specularis'.

[26] Cf. Sen. *EP.* 90. 25, 86. 11; Mart. *Epigr.* 8. 14; Juv. *Sat.* 4. 21; Columella 1. 3. 52. It was occasionally used in windows, as a substitute for glass, until the 18th cent.

[27] Battey, *Mineralogy*, 83; Whitten–Brooks, *Geology*, 221 f.

[28] Battey, *Mineralogy*, 138.

description ('inenarrabilis'). It has the highest hardness value (10 on the Mohs[29] scale), being at least ninety times as hard as corundum (9 on the scale). Yet, he continues (37. 60), 'when a diamond is successfully broken, it disintegrates into splinters so small as to be scarcely visible' ('cum feliciter contigiit rumpere in tam parvas friatur crustas ut cerni vix possint'. Pliny later (37. 189) describes *chalazias*[30] (an unidentifiable stone) as being as hard as *adamas*.

Other hardness values[31] are topaz (8 on the Mohs scale), quartz (7), orthoclase (6), apatite (5), fluorite (4), calcite (3), gypsum (2), and talc (1). At the other extreme to the diamond is the softness (*mollitia*) of Eretrian earth (35. 192). Stibnite (33. 101), galena (34. 1723), and haematite (36. 144) are all easily crumbled (*friabilis*).

13.3.4 Tenacity

The resistance that a mineral offers to breaking, crushing, and bending is known as its tenacity.[32] This quality is further defined by the following terms: brittle, malleable, sectile, ductile, flexible, and elastic. Some of these properties are known to Pliny. Jet (*gagates*, 36. 141 f.) is brittle (*fragilis*). Likewise, emeralds from Chalcedon (37. 72) are noted for their brittleness.

[Pliny (34. 146) also alludes to brittleness (*fragilitas*) in the case of iron. Oil is used to quench small iron-forgings, for fear that water might harden them and make them brittle ('tenuiora ferramenta oleo restingui mos est, ne aqua in fragilitatem durentur'). He also (33. 59 and 61) recognizes malleability (*facilitas*), although incorrectly asserting that lead is more malleable than gold ('laxius dilatatur'). Orpiment (33. 79), yellow arsenic sulphide As_2S_2 is classed as brittle (*fragilis*).]

---

[29] The Austrian mineralogist who, in 1824, used the criterion of scratch hardness to determine values for a selection of minerals (from 1 to 10). See also n. 31.

[30] '*Chalazias* has the whiteness and shape of hailstones, and is as hard as *adamas*', ('Chalazias grandinum et candorem et figuram habet, adamantinae duritiae'). This statement is misleading. *Chalazias* cannot be as hard as *adamas*, if the latter is a diamond—10 on the Mohs scale.

[31] Battey, *Mineralogy*, 83.   [32] Ibid. 86.

13.3.5 Density—Specific Gravity

Density, or specific gravity[33]—that is the number adopted to express the ratio of the weight of a substance to the weight of an equal volume of water at 4 °C—in the case of a crystalline substance, depends on (1) the kind of atoms of which it is composed, and (2) the manner in which the atoms are packed together. In isostructural compounds, in which the packing is constant, those with elements of higher atomic weight usually have higher specific gravities. From a determination of the specific gravity it is, therefore, possible to obtain a close approximation of the chemical composition.

Although the equation $D = M/V$ was unknown in the ancient world, the practical application of this principle played an important part in metallurgical operations. Diodorus Siculus (5. 27. 2) states that miners washed ore-minerals to free them from adhering earth and Strabo (3. 2. 10) refers to silver-bearing ore being crushed and disengaged in water by means of sieves. However, the separation, on a large scale, of the metalliferous fraction of ore-minerals from accompanying gangue minerals required the development of a more sophisticated technique for washing. The Greeks devised flat 'tables'[34] and helicoidal[35] washeries for the concentration of crushed argentiferous galena, after it had been ground and sieved. The process relied on gravity separation and modern experiments[36] confirm that it was very effective. There is no reason, however, to assume that the technique was other than empirical.

Pliny, like other ancient authorities, makes no reference to the relative densities, or specific gravity, of minerals, or metals,

---

[33] Battey, *Mineralogy*, 75 ff.   [34] Conophagos, *Laurium*, 224 ff. (fig. 10.1).
[35] Ibid. 248 ff. (figs. 10.32 ff.)
[36] H. F. Mussche and C. E. Conophagos, 'Ore-washing Establishments and Furnaces at Megala Pevka and Demoliaki', in *Thorikos VI—1969* (Ghent, 1973), 60–72, esp. 65. During the test 60 kg. of galena with a lead content of 16% were crushed to grains of less than 1 mm. which were washed in the reconstructed sluice; 13.5 kg. of concentrated ore, with a lead content of 45% and 46.5 kg. of low-grade ore with a lead content of 7.5% were obtained. The output in weight is: $R = 13.5/60 = 22.5\%$. The output of metal is $r = (22.5 \times 0.45)/(100 \times 0.16) = 63\%$. This is a very positive result, taking into account the fact that no previous concentration by sieving had been done.

but observes (34. 157). Tin[37] 'is detected only by its weight. ... it is found in gold mines called "alutiae", through which a current of water is sent to wash out the black pebbles with small white spots [cassiterite, SnO_2]. The pebbles weigh the same as the gold and so remain with it in the bowls in which they are collected. Afterwards, the pebbles are separated in furnaces and fused into tin.' ('pondere tantum ea deprehenditur. ... invenitur et in aurariis metallis, quae alutias vocant, aqua immissa eluente calculos nigros paullum candore variatos, quibus eadem gravitas quae auro; et ideo in catillis quibus aurum colligitur, cum eo remanent; postea caminis separantur conflatique in plumbum album resolvuntur.')

There is a basic misconception in this account of tin-streaming. The specific gravity of tin and gold differ widely, being 7.3 and 19.3[38] respectively, and Pliny fails to realize that, although their weight, or density (*gravitas*), is not the same, both are relatively heavier than the accompanying gangue: quartz, for example, has a specific gravity of only 65.[39]

Elsewhere (33. 59), Pliny draws the wrong conclusion in his comparison of gold and lead: gold is not preferred to other metals for its weight (and malleability).[40]

His apparent lack of understanding that there is a clear relationship between Mass (weight) and Volume is all the more surprising in the light of the fact that the Greeks must surely have been aware of this relationship, which is the basis of density, or specific gravity measurements. As early as the fifth century BC, Herodotus (1. 50. 2)[41] had recorded the difference in weight (two and a half and two talents respectively) of ingots of refined gold (ἄπεφθος χρυσός) and of 'white gold' (λευκὸς χρυσός), all of the same volume. In the case of ingots the volume could easily have been determined by linear measurement, although it remained for Archimedes to discover a method of calculating the volume of an irregularly shaped object (Hiero's

---

[37] Cf. Healy, *Mining*, 89 ff.   [38] Ibid. 226.
[39] Battey, *Mineralogy*, 279 f. (quartz).
[40] See further above, section 13.3.4.
[41] ἀριθμὸν δὲ ἑπτακαίδεκα καὶ ἑκατόν, καὶ τούτων ἀπέφθου χρυσοῦ τέσσερα, τρίτον ἡμιτάλαντον ἕκαστον ἕλκοντα, τὰ δὲ ἄλλα ἡμιπλίνθια λευκοῦ χρυσοῦ, σταθμὸν διτάλαντα. See J. F. Healy, 'Greek and Roman Gold Sources: The Literary and Scientific Evidence', in Domergue, *Mineria y Metalurgia*, ii . 10.

crown) by its displacement of water.[42] If we assume that the 'white gold' ingots dedicated at Delphi were of a consistent mixture of gold and silver, without significant impurities (such as copper), the theoretical specific gravity gives a gold content of approximately 70 per cent.[43] The fact that the Greeks exercised a quality control in the production of the artificial alloy electrum—a basically binary alloy—by the end of the fifth, or the beginning of the fourth, century BC[44] presupposes that there must have been some means of checking ingots against a standard[45] before blanks were prepared for striking electrum coin series of Mytilene, Phokaia, and Carthage.

Furthermore, Plato (Tim. 59b–c), in his discussion of gold, which he describes as the 'most compact' of the fusile materials ($\pi\upsilon\kappa\nu\acute{o}\tau\alpha\tau\sigma\varsigma$), and in his observations about copper, appears to have a rudimentary inkling of density.

13.3.6 Streak

The colour of a mineral in a finely divided state is its streak[46] (*sucus*, or, simply, *color*). Although the colour may vary superficially, because of tarnish, the streak is usually constant and thus a useful factor in establishing a positive identification.[47]

Minerals may be divided into five groups, according to their streak characteristic: (1) minerals which leave no streak; (2) White, or pale coloured minerals, with a white streak (gypsum and calcite); (3) black, or strongly coloured minerals, with a white streak (hornblende); (4) black, or strongly coloured minerals which leave a streak of the same colour (malachite,

---

[42] Landels, *Ancient Engineering*, 190 f.   [43] Healy, *Mining*, 207.

[44] J. F. Healy, 'Notes on the Monetary Union between Mytilene and Phokaia', *Journal of Hellenic Studies*, 77/2 (1957), 267 f.

[45] That is, ingots of standardized weight. Cf. the imperial standards preserved at Greenwich.

[46] Battey, *Mineralogy*, 88. The colour of the powdered mineral produced by rubbing it on a piece of ungazed porcelain, or by scratching some powder from the mineral with a knife (or file), or by crushing a small portion is more consistent and reliable than the body colour (cf. Eretrian earth, 33. 163).

[47] Birkeland and Larson, *Putnam's Geology*, 61. Haematite and magnetite superficially resemble each other in physical properties. However, the streak may be the most definitive aspect of the minerals, since the former produces a red-brown streak, whereas the latter leaves a black-brown streak.

graphite, and pyrolusite); (5) black, or strongly coloured minerals which give a streak of a different colour: haematite (a black-brown colour) leaves a red-brown streak, limonite (a black-brown colour) gives a yellow streak, and chalcopyrite (a brass-yellow colour) gives a greenish-black streak.

Pliny writes of Eretrian earth (35. 192) that it is tested by its softness and by the fact that it leaves a violet tint (*violacium*) when rubbed on copper (here acting as a streak plate): 'probatur mollitia et quod si aere perducatur, violacium reddit colorem.'

One of the tests for specular iron ore (*andromas*, 36. 147) is to rub it on a whetstone of slate (*lapis basanites*) when, if genuine, it produces a blood-red streak: 'experimentum eius esse in cote ex lapide basanite—reddere enim sucum sanguineum.' Sotacus (quoted by Pliny, ibid.) records a variety of iron allegedly from Arabia. This is similarly hard and produces scarcely any trace on a hone used with water although, on occasion, it leaves a saffron coloured streak: 'tertium genus Arabici facit, simili duritia, vix reddentis sucum ad cotem aquariam, aliquando croco similem.' A further example described by Sotacus (36. 148) is a black stone found in Africa (*anthracitis*), which, when rubbed on a water hone, produces, on what was orginally the lower end, a black mark and, at the other end, a saffron coloured streak: 'nasci in Africa nigrum, attritum aquariis cotibus reddere ab ea parte quae fuerit ab radice nigrum colorem, ab altera parte croci.'

Stones which produce a streak (36. 157) are considered to be useful for making eye-salves: 'ii lapides qui sucum reddunt oculorum medicamentis utiles existimantur.'

13.3.7 *Magnetism and Pyroelectric Properties*

Pliny records these further important properties of minerals which are discussed in detail in Chapter 12 (in sections 3 and 2, respectively).

(1) The ore-mineral magnetite[48] is a natural ferrimagnet (36. 127 ff.). (2) Pyroelectricity[49] and piezoelectricity,[50] whereby certain minerals have the power to attract other non-metallic

---

[48] See Ch. 14, s.v. [49] Whitten–Brooks, *Geology*, 371.
[50] Battey, *Mineralogy*, 67 and 89 f.

substances, are related phenomena in which stress and heat, respectively, cause the separation of electric charges on the surface of the crystal. The resin, amber[51] also has pyroelectric properties.

In the *Natural History* Pliny describes and records many aspects of minerals—some, like sulphur, for the first time. Individual minerals are, for convenience of access, listed and discussed alphabetically in Chapter 14 below.

Pliny adds much to our understanding of the state of mineralogy under the early empire, but, arguably, his major contribution to Earth Sciences stems from his careful observation of the nature and form of crystals.

13.4 Crystal Systems

Crystallography[52] is the branch of science that deals with the geometric description of crystals, their internal arrangement, and their properties. Symmetry is the basic property of crystals.

The seven groups[53] obtained when only the dominant symmetry elements of the classes are taken into account, are called crystal systems. In order of decreasing symmetry, they are called cubic, hexagonal, tetragonal, trigonal orthorhombic, monoclinic, and triclinic.

Quartz, or silicon dioxide (SiO_2), is one of the earth's most abundant minerals.[54] As a result of careful observation, Pliny accurately describes its crystal system and that of the 'rainbow' stone, diamond, and emerald (beryl).[55]

---

[51] See Ch. 14, App. A.
[52] René-Just Haüy, the French physicist and mineralogist, was the first scientist to make a detailed study of the crystallographic forms of minerals, on the basis of geometric laws, in his *Traité de minéralogie* (Paris, 1801). See further *Larousse Dictionary of Scientists* (New York, 1994), 237. For a general introduction to Crystallography, see W. L. Bragg and G. F. Claringbull, *Crystal Structures of Minerals* (London, 1965). The study of the properties of crystals has become a new science of solid state physics with which mineralogy has close ties. Cf. Battey, *Mineralogy*, p. xi.
[53] Battey Mineralogy, 34 ff.
[54] Whitten–Brooks, *Geology* 413. Silicates make up 95% of the earth's crust.
[55] See Ch. 14, s.vv. 'iris', 'adamas', 'smaragdus' and 'beryllus'.

Quartz presents a number of problems to the superficial observer. Pliny (37. 23 ff.) writes: 'An opposite cause produces rock-crystal: for it is hardened by intense cold. At any rate, this mineral is found only where the winter snows freeze solid and it is undoubtedly a kind of ice, which is the reason for the Greek name [κρύσταλλος]. ... Why rock-crystal is formed with hexagonal faces is hard to explain, and the difficulty is complicated by the fact that its terminal points are not symmetrical, while its six faces are so perfectly smooth that not even the most skilful lapidary could achieve such a finish.'[56] ('contraria huic causa crystallinum facit, gelu vehementiore concretum. non aliubi certe reperitur quam ubi maxime hibernae nives rigent. glaciemque esse certum est, unde nomen Graeci dedere. ... quare nascatur sexangulis lateribus, non facile ratio iniri potest, eo magis quod neque in mucronibus eadem species est et ita absolutus laterum levor est ut nulla id arte possit aequari.')

Pliny appears to imply in his description that there is some dependence of external form on the internal structure, but obviously does not understand exactly how minerals are formed. The presence of asymmetrical points leads him to suspect, quite correctly, that there is an underlying natural cause. He compares with quartz the 'rainbow' stone and is already on the threshold of 'discovery crystallography'.

Emeralds and beryls also belong to the hexagonal[57] system and Pliny (37. 79) repeats the view, correctly held by some, that they are formed, from the very outset, as prisms: 'quidam et angulosos statim putant nasci.' They are, of course, the result of natural crystalline growth, the prisms representing the orderly arrangement of atoms within the structure.

Pliny (37. 56) recognizes the octahedral[58] system of the Indian diamond.[59] Briefly, the extraordinary difference in appearance of the diamond and that of other forms of carbon, depends solely on the crystallization of the material and the

---

[56] See Ch. 14, s.v. 'crystallum'. [57] Battey, *Mineralogy*, p. 279.
[58] Ibid. 34 and 37.
[59] 62, fig. 66. The diamond has eight faces on the orthorhombic system (sixteen on the tetragonal). See also J. F. Healy, 'Pliny the Elder and Ancient Mineralogy', *Interdisciplinary Science Reviews*, 6/2 (June, 1981), 173.

physical characteristics that result therefrom, and this is the reason why it has always been so sought after. Almost every single stone is bounded by more or less regularly developed faces. These differ from those of most other crystallized minerals in that they are, as a rule, much curved and rounded instead of being perfectly plane, as is usually the case. The curvature is due to the mode of growth of the crystal and not to subsequent attrition, as might be thought.

13.5 Nomenclature

The identification of minerals by name is an area in which Pliny, like most ancient authorities, is especially prone to inconsistency.

The common terminations in -itis, -ites, which had been used by the Greeks, were in general, adopted by the Romans without problems.[60]

Names often indicate some characteristic feature of the minerals.[61] So haematite (37. 169 and 36. 144) is named after its blood-red colour (αἷμα) hepatite (37. 186) reflects its liver-like appearance (ἧπαρ), and steatite (ibid.), soapstone, is named after fat (στέαρ).

Some minerals, like *carcinias* (37. 187), are named after an animal (καρκινίας, 'crab'). Others, after inanimate objects, as *cenchros* (ibid. 188), which looks as if it has been sprinkled with grains of millet (κέγχρος) and *ammochrysos* (ibid.), like gold mixed with sand (ἀμμόχρυσος).

Yet other terms, however, appear to be generic; *schistos*[62] (29. 124, 31. 79, and elsewhere) covers all lamellar minerals;[63] *galena*, or *molybdaena*, is lead (as 33. 95 and 34. 159). *Anthracites*[64] (37. 148) is a mixture of limonite and magnetite. *Adamas*[65] describes any extremely hard substance, including diamonds, possibly rock-crystal (37. 57), and iron pyrites (ibid. 58).

Occasionally terms are confused. Pliny (36. 14) interchanges magnetite (36. 127) and magnesian limestone (dolomite)—both *magnes lapis*. Occasionally, too, it is impossible to identify

---

[60] Ibid. 172.   [61] Ibid.   [62] See Ch. 14, s.v.
[63] Healy, 'Pliny on Mineralogy', in French–Greenaway, *Roman Science*, 118.
[64] See Ch.14, s.v. 'magnetite', and Ch. 17.9.1.   [65] See Ch. 14. s.v.

a mineral until its second, or third appearance in the text; so *andromas* (36. 146)—iron pyrites. A more serious problem, however, arises from the fact that modern nomenclature sometimes differs significantly from ancient. *Chrysocolla*,[66] for example, was used by the Greeks and the Romans as a generic term referring to any bright-green copper mineral. Theophrastus[67] and other authorities apply this name to any material used in soldering gold. The mineral currently known as chrysocolla, however, is a hydrous copper silicate which occurs as a decomposition product of copper ores; it is found as encrusting and botryoidal masses.[68]

Elsewhere (in book 37) Pliny confronts the equally complicated problems he encounters with the identification of gems or precious stones.[69]

---

[66] *NH* 33. 4 and 86. [67] Eichholz, *Lap.* 26(cf. 39 and 51).
[68] Hurlbut, *Mineralogy*, 310 f. [69] See Ch. 16 below.

14

Minerals

In this chapter the minerals treated, or referred to, by Pliny are arranged in alphabetical order.

adamas The term *adamas* (ἄδαμας) has a long pedigree, being indiscriminately used in the ancient world to describe any unusually hard substance, whether mineral, or metal—for example, iron, corundum, lodestone, and diamonds.[1]

Diamonds were known to the Romans certainly from the principate of Augustus. The first literary reference to diamonds occurs in Manilius (*Astronomicon* 4. 926): 'The diamond, a stone no bigger than a dot, is more precious than gold.'

Roman trade with the East flourished under the empire and the earliest diamonds were brought to Rome from India, where they were found in the diamantiferous sandstone which belongs to the oldest division of the sedimentary formations of the country that usually rest directly on the still older crystalline rocks such as granite, gneiss, mica and other schists (hornblende, chlorite and talc). The diamond river mentioned by Ptolemy may have been the Mahandi (south-west of Calcutta) in the Hyderabad region. That river also yields an appreciable amount of alluvial gold. See generally E. H. Warmington, *The Commerce between the Roman Empire and India*[2] (Cambridge, 1974).

Pliny (37. 56 ff.) writes:

Now, for the first time, as many as six varieties of diamond are known. There is the Indian octahedral diamond which is not formed in gold and has a certain affinity with rock crystal, which it resembles

---

[1] J. F. Healy, 'Pliny the Elder and Ancient Mineralogy', *Interdisciplinary Science Reviews*, 6/2 (1981), 172 ff.

in its transparency and in its smooth faces meeting at six corners. It tapers to a point in two opposite directions and is all the more remarkable because it is like two spinning-tops joined at their bases. It can even be as large as a hazel nut. Similar to the Indian, but smaller, is the Arabian diamond which is formed under similar conditions. The other kinds have the pale appearance of silver and are formed only in gold of the finest quality.

nunc primum genera eius sex noscuntur: Indici non in auro nascentis et quadam crystalli cognatione, siquidem et colore tralucido non differt et laterum sexangulo levore, turbinati in mucronum e duabus contrariis partibus, quo magis miremur, ut si duo turbines latissimis partibus iungantur, magnitudine vero etiam abellani nuclei. similis est huic Arabius, minor tantum, similiter et nascens. ceteris pallor argenti et in auro non nisi excellentissimo natalis.

Pliny accurately describes the crystal system of the diamond[2] and observes its similarity to that of rock-crystal.[3] The term Arabian may be a trade name, or refer to diamonds of inferior quality, originating in India, but re-exported by way of Arabia.

The identification of the other varieties of diamonds presents a number of challenging problems. It has been suggested that *cenchros* may have been the name for a small diamond of the type mentioned by Manilius.

In his account of the physical properties of the diamond, Pliny (37. 57) confuses hardness and brittleness: 'All these stones can be tested upon the anvil and they repel blows so that an iron hammer head may be split in two and even the anvil unseated. Indeed the hardness of the diamond is not able to be described.' ('incudibus hi deprehenduntur ita respuentes ictus ut ferrum utrimque dissultet. incudes ipsae etiam exiliant. quippe duritia est inenarrabilis.')

The term *adamas* is singularly inappropriate when a diamond's extreme brittleness is considered. A diamond is easily fractured—a very moderate blow from a hammer suffices to break it because of its perfect cleavage. Pliny continues: 'Also indescribable is that property whereby it conquers fire and never becomes heated. Hence it derives its name, because, according to the meaning of the term in Greek, it is the

---

[2] Ch. 13.4 above.   [3] See, generally, *NH* 37. 23–9.

"unconquerable force"' ('simulque ignium victrix natura et numquam incalescens, unde et nomen interpretatione Graeca— indomita vis—accepit'). Diamonds are unaffected by heat unless the temperature reaches $c.700\,°C$.

Pliny (37. 60) contradicts himself by stating that a diamond, when broken, disintegrates into splinters so small as to be scarcely visible. These are much sought after by engravers of gems and are inserted by them into iron tools because they make hollows in the hardest materials without difficulty. This is the first reference to what we would call 'industrial' diamonds.

A further variety [37. 58], known as Macedonian, is found in the gold-mines of Philippi. This is equal in size to a cucumber seed. Next there is the Cyprian kind, found in Cyprus and tending towards the colour of copper. ... after this, ironstone [*siderites*], which shines like iron and exceeds the rest in weight, but has different properties. This can not only be broken by hammering but can also be pierced by another diamond. This can happen to the Cyprian kind, and in a word, these stones, being untrue to their kind, possess only the prestige of the name they bear.

alterum Macedonicum in Philippico auro repertum; hic est cucumis semini par. post hos Cyprius vocatur in Cypro repertus, vergens ad aereum colorem. ... post hunc siderites ferrei splendoris, pondere ante ceteros, sed natura dissimilis. namque et ictibus frangi et alio adamante perforari potest, quod et Cyprio evenit, breviterque ut degeneres nominis tantum auctoritatem habent.

It is likely that Pliny is referring to other substances. The Macedonian 'diamond' may simply be quartz (*crystallum*), the Cyprian, *analcime*—that is, hydrous sodium aluminium silicate—and finally, ironstone, iron pyrites.

Pliny (37. 55) also states that *adamas* is used as a term for the substance otherwise known as 'auri nodus', very occasionally found in mines, in association with gold, and, he adds, seemingly only formed in gold: 'ita [adamas] appellabatur auri nodus in metallis repertus perquam raro comes auri nec nisi in auro nasci videbatur.' A tentative identification of this with platinum has been considered.[4]

---

[4] See Ch. 17. 13 below.

Minerals

alum See below, s.v. 'alumen'.

alumen Alum[5] is a colourless, soluble hydrated double sulphate of aluminium and potassium (K_2SO_4, $Al_2(SO_4)_3$).

Pliny (35. 183 ff.) gives a lengthy description of this mineral and the principal locations in which it is found.

> No less important are the uses of alum which is a salt exudation from the earth. There are several kinds. In Cyprus there occurs a white alum and a variety of a somewhat darker colour. Although the difference in colour is only marginal, the use to which each is put is very different: the white, liquid kind of alum is useful for dyeing woollens a bright colour, whereas the black is best for dark, or dull colours.[6] Black alum ($AlSO_4$) is employed in the purification of gold.

> nec minor est aut adeo dissimilis aluminis opera, quod intellegitur salsugo terrae, plura et eius genera. in Cypro candidum et nigrius, exigua coloris differentia, cum sit usus magna, quoniam inficiendis claro colore lanis candidum liquidumque utilissimum est contraque fuscis aut obscuris nigrum. et aurum nigro purgatur.

He continues (35. 184):

> All [alum] is produced from water and slime, which is a substance exuded by the earth. This collects naturally in hollows during the winter, and then the summer sunshine causes it to crystallize.[7] The first to precipitate is white in colour. The mineral occurs in Spain, Egypt, Armenia, Macedonia, Pontus, Africa, and the islands of Sardinia, Melos, Lipara, and Strongyle. The most highly prized alum is Egyptian:[8] the next best is Melian, of which there are two types—namely liquid and solid.

> fit autem omne ex aqua limoque, hoc est terrae exudantis natura. conrivatum hieme aestivis solibus maturantur quod fuit ex eo praecox, candidius fit. gignitur autem in Hispania, Aegypto, Armenia, Macedonia, Ponto, Africa, insulis Sardinia, Melo, Lipara, Strongyle,

---

[5] See Bailey, *Chemical Subjects*, ii. 233 f. Several astringent substances appear to have been included in the term *alumen*, especially aluminium sulphates, sulphate of iron, and common potash alum; also kaolinite and, perhaps, certain halotrichites ($FeSO_4 \cdot Al_2(SO_4)_3 \cdot 24H_2O$).

[6] Sulphate of aluminium would be useful for dyeing; potash-alum and alunogen could provide the bright colours and alums containing the metals, the sombre.   [7] See Ch. 11.3 above.

[8] Pliny is possibly confusing this substance with στυπτηριώδης γῆ.

laudatissimum in Aegypto, proximum in Melo. huius quoque duae species, liquidum spissumque.

Liquid alum has astringent, hardening and corrosive properties.[9] It is also an anti-perspirant.

One kind of solid alum [35. 186], which Greeks call "schistos',[10] breaks up into a kind of whitish filament, and because of this, some people call it "hair-like" alum [*trichitis*]. This is produced from the same ore-mineral as the copper known as *chalcitis*:[11] it is like a sort of sweat coagulated into foam.' ('concreti aluminis unum genus σχιστόν appellant Graeci, in capillimenta quaedam canescentia dehiscens unde quidam trichitim potius appellavere. hoc fit e lapide ex quo et aes—chalcitim vocant—ut sudor quidam eius lapidis in spumam coagulatus.')

Alum is important as a mordant in dyeing: it is a substance used to stabilize, or fix colour. Pliny (35. 150) describes

a remarkable Egyptian process for dyeing cloth. They first thoroughly rub white fabrics and then smear them not with colours, but with chemicals that absorb colour [*medicamenta*]. When this has been done, the fabrics show no sign of the treatment, but, after being plunged into a cauldron of boiling dye, they are drawn out a moment later dyed. And the remarkable thing is that, although the cauldron contains only one colour, it produces a series of different colours in the fabric, the hue changing with the quality of the chemical employed, and it cannot afterwards be washed out. Thus the cauldron, which would undoubtedly produce a blend of colours if coloured fabrics were put into it, produces several colours from one, and dyes the material when heated. The fabrics subsequently become more resistant to wear than they would have been had they not been subjected to heat.

pingunt et vestes in Aegypto inter pauca mirabili genere, candida vela, postquam attrivere, inlinentes non coloribus, sed colorem, sorbentibus medicamentis. hoc cum fecere, non apparet in velis, sed in cortinam pigmenti ferventis mersa post momentum extrahuntur picta. mirumque, cum sit unus in cortina colos, ex illo alius atque alius fit veste accipientis medicamenti qualitate mutatus, nec postea

---

[9] NH 35. 185, 'liquidi aluminis vis adstringere, indurare, rodere'. The medical applications of alum are like those in use today. [10] See below, s.v.
[11] Dioscorides 5. 127, and cf. 106: σχιστὸς λίθος. He states that this is the best variety of alum.

ablui potest. ita cortina, non dubie confusura colores, si pictos acciperet, digerit ex uno pingitque, dum coquit, et adustae eae vestes firmiores usibus fiunt quam si non urerentur.

In this description, Pliny appears to refer to the action of a mordant (*medicamentum*) in preparing the cloth for dyeing, without it visibly affecting the white colour of the fabric. The mordant ensures that the colour remains fast. So far the chemistry is sound, but it is not possible to explain the alleged multicoloured result from a single dye substance unless Pliny intended to say that the dipping-time produced different *shades* of the *same* colour. There is no reason either why dress fabrics should become stronger after heating! He may well have misunderstood, or misquoted, some obscure text.

Laboratory experiments show that Pliny's implied formula (33. 89) would have resulted in a green colour, reproducing, as closely as possible, that of a bright green blade of corn, 'ut in herba segetis laete virentis quam simillime reddat'.

The metallurgical uses of alum are equally interesting. Alum is used in the quenching of copper in preparation for gilding.[12]

The copper [33. 65] is first subjected to a hot fire; then, when it is red hot, it is quenched with a mixutre of brine, vinegar, and alum and afterwards put to a test, its brilliance of colour showing whether it has been sufficiently heated. Then it is again dried in the fire, so that, after a thorough polishing with a mixture of pumice[13] and alum, it is able to take the gold leaf laid on with quicksilver. Alum has the same cleansing property here that we said is found in lead.

namque aes cruciatur in primis accensumque restinguitur sale, aceto, alumine, postea examinatur an satis recoctum sit, splendore deprehendente, iterumque exhalatur igni, ut possit, edomitum mixtis pumice et alumine, argento vivo inductas accipere bratteas. alumen et in purgando vim habet qualem esse diximus plumbo.

At the present time the quenching liquid is hot, whereas, in the ancient process it was the metal that was heated. Together with pumice, alum is also used to polish copper to a bright finish.

---

[12] See further below, Ch. 17.4.8.   [13] See below, s.v. 'pumex'.

The second usage noted by Pliny is in the purification of gold, where it has a similar role to lead in the cupellation of gold (33. 84). Gold 'is also heated with twice its weight of salt and three times its weight of copper pyrites[14] and again with two portions of salt and one of the mineral called alum' ('torretur et cum salis gemino pondere, triplici misyis ac rursus cum II salis portionibus et una lapidis, quem schiston vocant').

amiantus The original meaning of ἀμίαντος referring to water (Theognis 447), is 'pure', or 'undefiled'. In this sense it is also applied to the sea, ἡ ἀμίαντος θάλασσα (Aesch. *Pers.* 578.)

It is first used of the mineral substance known as asbestos (ὁ ἀμίαντος λίθος) by Aristotle (fr. 495 Rose (Leipzig, 1886)) and subsequently, Dioscorides (5. 138). Surprisingly, it is not mentioned by Theophrastus in his *De lapidibus*.

Pliny (36. 139) describes asbestos.[15] It 'looks like alum and is completely fire-resistant. It affords protection against spells, especially those of the Magi.' ('amiantus alumini similis nihil; igni deperdit. hic veneficiis resistit omnibus privatim Magorum.')

In a further passage (37. 146) he writes, 'asbestos is found in the mountains of Arcadia and has the colour of iron' ('asbestos in Arcadiae montibus nascitur coloris ferrei'). This is one of the only two references in which he uses the term *asbestos*, rather than *amiantus*, and it has been suggested that the mineral here described is chrysotile, a fibrous form of serpentine.[16]

Asbestos is the modern name given to fibrous varieties of several distinct mineral species: all are silicates. Their importance

---

[14] See below, s.v. 'chalcitis'.

[15] Asbestos (ἄσβεστος) means 'unquenchable', 'inextinguishable', or, in the case of fire, 'not quenched'. Also, ἄσβεστος (sc. τίτανος) refers to unslaked lime (cf. Dioscorides 5. 115), a form of gypsum (see below, s.v.). The colour, however, suggests a variety of asbestos.

[16] Whitten–Brooks, *Geology*, 38 f. Asbestos in a fibrous amphibole and the name given to fibrous varieties of several distinct mineral species. All are silicates and common types are tremolite and chrysotile (fibrous serpentine). Serpentine ($Mg_6Si_4O_{10}(OH)_8$), the main alteration product of olivines and pyroxenes, is a layer-lattice of which two forms occur, (*a*) a fibrous one known as chrysotile, and (*b*) a lamellar one, antigorite. Chrysotile has been the most important of the asbestiform minerals in commerce.

lies in the fact that they can be felted and woven in the same way as any other fibre to produce a fireproof material.

Elsewhere (19. 19 f.) in his discussion of flax and linen produced in Spain and Italy, Pliny erroneously believes that asbestos is derived from a plant: 'Also a linen has now been invented that is incombustible. It is called "live" linen and I have seen napkins made from it, glowing on the hearth at banquets and burnt more brightly clean than they could be by being washed in water.' ('inventum est etiam quod ignibus non absumeretur. vivum[17] id vocant, ardentesque in focis conviviorum ex eo vidimus mappas sordibus exustis splendescentes igni magius quam possent aquis.') This linen, he continues, is used for making shrouds for royalty which keep the ashes of the corpse separate from the rest of the pyre: 'regum inde funebres tunicae corporis favillam ab reliquo separant cinere.' The plant 'grows in the deserts and sun-scorched regions of India.... it is rarely found and difficult to weave into cloth because of its shortness. Its colour is normally red but turns white by the action of fire. When any of it is found, it rivals the prices of exceptionally fine pearls. The Greek name for it is "*asbestinon*" derived from its peculiar property.' ('nascitur in desertis adustisque Indiae locis. ... rarum inventu, difficile textu propter brevitatem; rufus de cetero colos splendiscit igni. cum inventum est, aequat pretia excellentium margaritarum. vocatur autem a Graecis ἀσβέστινον ex argumento naturae suae.') There is, of course, no plant that produces such fibres: the substance can only be the actual mineral asbestos. Nor is there any evidence to confirm that the Greeks ever used this term.

Although Pliny mentions other ore-minerals, especially of copper, mined on Cyprus, he nowhere refers to the deposits of asbestos found on that island. It has only been produced (in open-cast mines) at Pano Amiandos[18] in the interior, not far from Troödos, since 1907.

anthracitis See Appendix B below, s.v. 'coal'.

---

[17] The epithet 'vivum' is also applied to sulphur (35. 175) and to mercury (33. 119). [18] I. Robertson, *Blue Guide to Cyprus* (London, 1981), 111.

aphronitrum There are some names which we cannot associate with any degree of certainty with any chemical, or geological, species.[19] *Aphronitrum*[20] falls into this category.

Greenaway[21] observes that this could be any one of a variety of substances including soda (Na_2CO_3), soda with salt, (NaCl), or potassium nitrate (KNO_3). It is commonly translated as the efflorescence of potassium nitrate, although sodium carbonate[22] also readily effloresces.

Following his discussion of 'foam of soda' ('spuma[23] nitri') Pliny (31. 113) writes:

The last generation of physicians said that *aphronitrum* was gathered in Asia oozing out in soft caves—the latter are called 'colligae'[24]—and then dried in the sun. The best is thought to be Lydian. The tests are that it should be least heavy and the most friable and of an almost purple colour. The last kind is imported in lozenges, but the Egyptian in vessels lined with pitch, lest it melt. These vessels too are finished off by being dried in the sun. The tests of soda [nitrum] are that it should be very fine and as spongy and full of holes as possible.

proxima aetas medicorum aphronitrum tradidit in Asia colligi in speluncis mollibus destillans—specus eos colligas vocant—dein siccant sole. optimum putatur Lydium; probatio, ut sit minime ponderosum et maxime friabile, colore paene purpureo. hoc in pastillis adfertur. Aegyptium in vasis picatis ne liquescat. vasa quoque ea sole inarescentia perficiuntur, nitri probatio, ut sit tenuissimum et quam maxime spongeosum fistulosumque.

Certain identification is difficult, if not impossible, because of a number of conflicting factors. Pliny interchanges the

---

[19] See generally M. P. Crossland, *Historical Studies in the Language of Chemistry* (London, 1962).

[20] *NH* 31. 113; cf. Mart. *Epigr.* 14. 58: 'Rusticus es? nescis quid Graeco nomine dicar: | spuma vocor nitri. Graecus es? aphronitrum.'

[21] F. Greenaway, 'Chemical Tests in Pliny', in French – Greenaway, *Roman Science*, 147.   [22] See Partington, *Chemistry*, 306.

[23] Cf. R. König, J. Hopp, and W. Glockner, *Plinius: Naturkunde, Buch 31* (Zurich 1994), p. 122 ad loc.(112).

[24] Clearly a manuscript error, perhaps influenced by the previous verb 'colligi'?

terms *aphronitrum* and *nitrum*[25] which would support the view that they both refer to the same substance, in which case the prefix *aphro-*[26] would well describe the appearance of efflorescence.[27] One further observation of Pliny, however, goes against the identification of *aphronitrum* as potassium nitrate. He writes: 'It is burnt in an earthen jar with a lid, lest it should crackle out; otherwise soda does not crackle in fire' ('uritur in testa opertum ne exultet, alias igni non exilit nitrum').

Greenaway comments that potassium nitrate—and some other salts—would crackle in the fire, which Pliny states does not happen in the case of this substance. He concludes, therefore, by assuming that Pliny is badly reporting someone's distinction between a 'nitrate nitrum' and some other kind, but also adds what he tentatively puts forward as a 'speculative explanation' that *aphronitrum* was indeed an efflorescence of potassium nitrate[28] brought about by decomposition.[29]

Finally, in the light of Pliny's statement that *aphronitrum* was purple in colour, it has been suggested that this was due to the presence of an unrefined impurity[30] such as ferric oxide (Fe_2O_3). However, *purpureus* is open to a number of translations and, in the ancient world, fine shades of colour were not able to be meaningfully described.[31]

---

[25] Pliny (31. 106) states that Theophrastus gives the most careful description of *nitrum*: 'nec quemquam Theophrasto diligentius tradidisse [nitri naturam].' See H. Schramm, *RE* 17. 775–80, s.v. 'nitrum'. It is described as sodium bicarbonate ($NaHCO_3$), mixed with sodium sulphate ($NaSO_4$), sodium carbonate (Na_2CO_3), and sodium chloride NaCl.

[26] From the Greek ἀ φρός, 'foam'.   [27] See also Ch. 11.7 above.

[28] Greenaway, 'Chemical Tests', 151: 'There is a tempting speculative explanation: if we suppose that the cave referred to [by Pliny at 31. 113, see above] was, in fact, some excavation, or structure used as a human habitation or place for livestock. It has been established that in middle ages, and later, potassium nitrate was found in and collected from country dwellings where walls and mounds were impregnated with animal and human urine, which, after decomposition, produced an efflorescence of potassium nitrate.' Cf. A. R. Williams, 'The Production of Saltpetre in the Middle Ages', *Ambix* (*Journal of the Society for the History of Alchemy and Chemistry*), (1975) 22, 125–33.

[29] König, Hopp, and Glockner, *Naturkunde, 31,* p. 123 ad loc. (31. 113).

[30] As in the case of salt. See F. Sherwood Taylor, 'A Survey of Greek Alchemy', *Journal of Hellenic Studies*, 50 (1930), 109–39.

[31] Cf. Greenaway, 'Chemical Tests', 151.

armenium See Ch. 15 s.v.

arsenopyrite Arsenic (As) is a silver-grey, allotropic, semi-metal,[32] akin to phosphorus. It occurs in a native state and in three ore minerals.[33] The pure metal, however, was unknown to the Greeks and Romans.

Arsenopyrite[34] (FeAsS) is the most common mineral containing arsenic. It is found in large amounts and occurs together with tin and tungsten in high temperature hydrothermal deposits.[35] It is associated with silver, lead, and copper and, frequently, with gold. Realgar[36] (AsS) is found in veins of lead, silver, and gold together with orpiment, other arsenical minerals, and stibnite. It also occurs as a volcanic sublimation product as a deposit from hot springs. Orpiment[37] (As_2S_2) is rare; it is usually associated with realgar and formed under similar conditions.

The mining of soft minerals was very hazardous,[38] especially because of the danger of the roof of the gallery collapsing. Theophrastus (Lap. 52), discussing the ochre mines of Cappadocia, writes: 'The risk of suffocation, they say, is a serious matter for the miners since this can happen to them quickly and takes a very short time.'

Strabo (12. 3. 40) likewise describes, in vivid detail, the terrible conditions of work in the mines at Mt. Sandaracurgium in Pontus: 'In this city [Pompeiopolis] is Mt. Sandaracurgium, which is hollowed out in consequence of the mining done there, since the workmen have excavated great cavities beneath it. The mine was worked by *publicani* [state contractors] who used, as their workforce, the slaves sold in the market because of their crimes. In addition to the painfulness of the work, they say that the air in the mines is both deadly and hard to endure on account of the grievous odour of the ore [realgar], so that the miners are doomed to a quick death. What is more, the mine is often left idle because of its unprofitableness, since

---

[32] Healy, *Mining*, 30.
[33] Tylecote, *Metallurgy*, 42 f. Arsenic is comparatively rare, being found in veins in crystalline rocks associated with silver, cobalt, or nickel ores.
[34] Ibid. 43. See also Healy, *Mining*, 42.   [35] Ibid.
[36] Hurlbut, *Mineralogy*, 258 f.   [37] Ibid.   [38] Healy, *Mining*, 137.

the miners are not more than two hundred in number and are continually spent by disease and death.'

Realgar was also found in Paphlagonia,[39] Mysia, Cappadocia, and Carmania.[40] Minium and cinnabar were both sometimes confused with realgar. Compounds of arsenic were also used as pigments.[41]

Pure arsenic was not refined by the Greeks or Romans, although this could have been achieved by the roasting and reduction of realgar, orpiment, or the decomposition product of these two minerals, taking great care to avoid volatilization.[42] White arsenic, produced when the sulphides are roasted, is of course, a very poisonous compound—which, no doubt, was the reason why the ancients did not attempt such a process.

Arsenic also occurs in copper from the Early Bronze Age,[43] in Crete and in the Cyclades, in addition to small quantities of lead and antimony. On the mainland, arsenic was plentiful. So arsenical coppers continued to be used over most of eastern and south-eastern Europe until the hierarchical societies could encourage, if not organize, trade routes and exchange mechanisms.[44]

The presence of arsenic and the actual amount in copper–arsenic bronzes is often significant of their origin. The naturally occurring impurity may be distinguished from the deliberately added arsenic (defined as amounts above 2 per cent).[45] An interesting example of arsenical copper is found at Alderley Edge,[46] in Cheshire, where streaks of bright yellow sandstone owe their colour to a mixture of cerussite and olivenite, that is copper arsenate ($Cu_3As_2O_8.Cu(OH_2)$), or a mixture of arsenates.

---

[39] Strabo 12. 3. 40.   [40] Forbes, *Ancient Technology*, ix. 167.
[41] See Ch. 15 below, s.vv. 'orpiment', 'sandaraca'.
[42] Forbes, *Ancient Technology*, ix. 168.
[43] N. H. Gale and Z. A. Stos-Gale, 'Some Aspects of Early Cycladic Metallurgy, in Domergue, *Mineria y metalurgia*, i. 21–36. Also K. Brannigan, 'Metal Objects and Metal Technology of the Cycladic Cultures', in J. Thimme (ed.), *Art and Culture of the Cyclades* (Karlsruhe, 1977), 117–22: J. A. Charles, 'Early Arsenical Bronzes—a Metallurgical View', *American Journal of Archaeology*, 71 (1967), 21–6.   [44] Tylecote, *Metallurgy*, 7.
[45] Healy, *Mining*, 219.   [46] Tylecote, *Metallurgy*, 29

asbestos See above, s.v. 'amiantus'.

atramentum See Ch. 15, s.v. p. 260.

azurite (armenium) See Ch. 15, s.v. p. 259.

basanites See below, s.v. 'chert'.

beryl See below, s.v. 'beryllus'.

beryllus Pliny continues with a description of the beryl, a cyclosilicate,[47] hexagonal holohedral, the green variety of which is the emerald. 'Many consider that beryls are similar to, if not the same as, emeralds. They occur in India, but are rarely found elsewhere. All are cut to a hexagonal shape by skilled craftsmen since their colour, which is dull when the surface is unbroken, is enlivened by their reflection from the facets. If they are cut in any other way, they lack brilliance. The most esteemed beryls are those that imitate the pure green of the sea.' ('eandem multis naturam aut certe similem habere berulli videntur. India eos gignit, raro alibi repertos. poliuntur omnes sexangula figura artificum ingeniis, quoniam hebes unitate surda color repercussu angulorum excitetur. aliter politi non habent fulgorem. probatissimi ex iis sunt qui viriditatem maris puri imitantur.')

In addition to India, a further source of beryls is said to have been the Black Sea: the latter gems came from workings in the Urals, about fifty miles east of Sverdlovsk. The beryls were exported by way of the river Don and the Black Sea.

Pliny continues (37. 78): 'The Indians are passionately fond of elongated beryls. ... They prefer to make beryls into long prisms rather than gemstones, because they equate excellence with length. Some people think that beryls occur as natural prism. ... The Indians have discovered a means of counterfeiting gemstones, especially beryls, by colouring rock-crystal.' ('Indi mire gaudent longitudine eorum. ... ideo cylindros ex iis malunt facere quam gemmas quoniam est summa

---

[47] Whitten–Brooks, *Geology*, 115 and appendix.

commendatio in longitudine. quidam et angulosos statim putant nasci.... Indi et alias quidem gemmas crystallum tinguendo adulterare invenerunt, sed praecipue berullos.')

Pliny is correct in stating that emeralds are deep-green beryls. Although wrong in his assertion that their hexagonal shape is achieved by lapidaries, he is aware that some believe that the beryl is, from its origin, a naturally formed crystal. It is interesting to compare this 'afterthought' with his conclusion apropos the nature of rock-crystal. Beryl crystallizes in the hexagonal system,[48] usually taking the form of long six-sided prisms, striated vertically and terminated by the basal plane, sometimes associated with pyramidal faces.

Quartz was widely used in the manufacture of forged beryls which were in circulation in the first century AD.

The mineral beryl affords a good example of the difference in attitude to classification on the part of the mineralogist and the gemmologist. The mineralogist includes the deep-green, bluish-green, greenish-blue, and yellow specimens all in the same species, to which he gives the name beryl, since all agree in chemical composition ($Be_3Al_2Si_6O_{18}$), being a silicate of the metals aluminium and beryllium, crystalline in form and differing only in colour. The gemmologist, however, refers to the deep-green variety as emeralds, to the greenish-blues as aquamarine, and only to the yellows as beryl.

cadmea Copper 'is produced from an ore that the Greeks call' "cadmea" [34. 2]: a highly reputed variety comes from overseas. Formerly found in Campania it now occurs in the region of Bergamum on the very border of Italy. It has also recently been reported in the province of Germany [that is, the left bank of the Rhine].' ('fit e lapide aeroso, quem vocant Graeci cadmean, celebri trans maria et quondam in Campania, nunc et in Bergomatium agro extrema parte Italia. ferunt nuper etiam in Germania provincia repertum.')

Cadmea is a mixture of carbonate of zinc (smithsonite) and hydrosilicate of zinc;[49] the latter was formerly known as

---

[48] Healy, 'Pliny and Ancient Mineralogy', 177–8.
[49] Rackham *NH 33–35*, p. 126, note *c*.

calamine and is basically a whitish colour (tinged with green, brown, and grey. It is found in the oxidized zone of ore deposits, or in hydrothermal veins and is characterized by a white streak. Hydrosilicate of zinc, or hemimorphite[50] ($Zn_4Si_2O_7(OH)2H_2O$) is associated with it.

Pliny (34. 100) describes another kind of *cadmea*, that is oxide of zinc. This is 'found in furnaces,[51] which is given a name indicating its origin. ... It is called "smoky" *cadmea* from its burnt appearance.' ('sic rursus in fornacibus existit... alia, quae originis suae nomen recipit....') Substances processed from this residue have a variety of medicinal uses (34. 108 ff.) There are two other kinds of *cadmea,* namely 'grape cluster', *botryitis* (34. 101), and 'caked residue', *placitis*.

calamine See above, s.v. 'cadmea'.

calcite Marble is a metamorphosed form of calcite ($CaCO_3$) produced by recrystallization under conditions of thermal metamorphism. The presence of impurities in the form of dolomite, silica, and iron compounds give certain marbles their characteristic appearance.[52]

The Greek term μάρμαρος[53] (Latin, *marmor*) was used to include not only types of marble, but also of granite, porphyry, and all stones, in fact, capable of taking a high degree of polish.

Marble[54] was first used, on a limited scale, for Cycladic sculptures in the third millenium BC. By the fifth century BC it had virtually replaced limestone. The main sources of marble were on Paros (Mt. Marpessa) and Naxos, and, in Attica, Mt. Pentelicus—with twenty-five quarries at Spilia—and Mt. Hymettus. Marble was also quarried in the Peloponnese

---

[50] Whitten–Brooks, *Geology*, 225.
[51] On furnace calamine, see further Bailey, *Chemical Subjects*, ii. 166–7.
[52] Whitten–Brooks, *Geology*, 412.
[53] In Homer, *Il*. 12. 380 and *Od*. 9. 499, it is simply used to refer to a shiny stone. Later, however, in Theophrastus, *Lap*. 9, and Strabo 9. 1. 23 and 14. 1. 35, it is marble.
[54] J. F. Healy, 'Mines and Quarries', in Grant–Kitzinger, *Mediterranean Civilization*, ii. 789. On quarries generally see also Krystyna Kozlowska (trans.), *Quarries in Ancient Greece* (Warsaw, 1975).

at Doliana, near Tegea. Green marble (porphyry) is found at Croceae, near Gythion. Theophrastus (*Lap.* 6) adds Chios and Thebes (Egypt), as sources, although the latter probably refers to mottled-red granite (syenite) at Aswan (the ancient name for Syene). To a very limited extent the different types of marble are distinguishable in appearance and petrographically. Parian marble is formed of large transparent crystals and has a glistening white surface; Naxian is grey. Pentelic is the only marble containing traces of iron, so that, when its surface is exposed to the weather, it gradually acquires a golden patina, as in the case of the Parthenon. Hymettan marble is grey, or bluish-grey, and is a more recent geological formation.

Pliny (36. 125) quotes Papirius Fabianus as his authority for asserting the commonly held belief that marble actually grows in quarries: 'marmora in lapicidinis crescere auctor est Papirius Fabianus.'

Under the Romans, although Parian and Pentelic marble continued to be in demand, the taste for coloured marble from Euboea and Thasos, and for serpentine from Tenos and Egypt, increased. Tibullus (2. 3. 43–5) observes that the streets of Rome are loaded with marble, and Pliny (loc. cit.) adds, '[If marble regenerates] there is reason to hope that there will always be marble sufficient to satisfy luxury's demands' ('quae si vera sunt spes est numquam defutura luxuriae').

The problem of identifying marbles of Greek, Anatolian (Asia Minor), and Italian origin is extremely complex.[55] Visual identification, based on macroscopical examination of colour, brightness, grain, and other physical characteristics can be far from conclusive.[56] Samples from the same quarry are often morphologically different, while samples from different quarries may have similar characteristics. Mineralogical and petrographical identification, however, are somewhat more reliable, while geochemical analysis provides absolute evidence.

---

[55] Healy, 'Mines and Quarries', 789f. See further M. Coleman and S. Walker, 'Stable Isotope Identification of Greek and Turkish Marbles', *Archaeometry*, 21/1 (Feb. 1979), 107–12.

[56] L. Lazzarini *et al.*, A Contribution to the Identification of Italian, Greek and Anatolian Marbles through a Petrological Study and Evaluation of Ca/Sr Ratio', *Archaeometry* 22/2 (Aug. 1980), 173–82: table of results from samples from ancient and modern quarries, and from outcrops, pp. 177–9.

Minor and trace elements (iron, manganese, and strontium[57] allow marbles from Ephesus, Proconnesus (Marmara Island), and Luna to be distinguished. Sodium and manganese content separates certain Anatolian marbles from Attic. Analysis based on the isotopic ratios of carbon (C13/12) and oxygen (O18/16)[58] helps to identify marbles from quarries in Asia Minor, at Marmara and Aphrodisias and, perhaps, Afyon. It is also possible to characterize Pentelic and Proconnesian marble. The evidence of calcium/strontium ratios is limited, but Ephesian, or Docimean (Phrygian), can be isolated from other types.

The combination of the evidence of petrology and geochemical data, examined within the framework of their historical and archaeological context, can therefore provide 'prints' of the main types of marble and their sources, although much further research remains to be carried out.

Pliny (36. 14), mentioning the use of Parian marble, writes: 'All [these sculptors] employed only white marble from the island of Paros, a stone that eventually became known as "lychnites", because according to Varro, it was quarried in galleries by the light of oil-lamps [*lychni*]. Subsequently many other varieties of white marble have been found, some indeed only recently in the quarries at Luna. There is a bizarre story that in the quarries on Paros, once, when a single block was split open with wedges, it was found to contain an image of Silenus.' ('omnes autem candido tantum marmore usi sunt e Paro insula, quem lapidem coepere lychniten appellare, quoniam ad lucernas in cuniculis caederetur, ut auctor est Varro, multis postea candidioribus repertis, nuper vero etiam in Lunensium lapicidinis. sed in Pariorum mirabile proditur, glaeba lapidis unius cuneis dividentium soluta, imaginem Sileni intus extitisse.') The story of the statue is obviously an imaginative fiction: all that this implies is that within every

---

[57] L. Lazzarini *et al.*, A Contribution to the Identification of Italian, Greek and Anatolian Marbles through a Petrological Study and Evaluation of Ca/Sr Ratio', *Archaeometry* 22/2 (Aug. 1980), 173–82: table of results from samples from ancient and modern quarries, and from outcrops, pp. 177–9.

[58] L. Manfra *et al.*, 'Carbon and Oxygen Isotope Ratios of Marbles from Some Ancient Quarries of Western Anatolia and their Archaeological significance', *Archaeometry*, 17/2 (July 1975), 215–21.

block of marble there may be said to be a 'potential' statue. Pliny (36. 44) also observes that Thasian marble rivals Parian.

It has been suggested that *'lychnites'* which literally means 'lamplike', refers to the appearance of crystalline marble and might be translated as 'luminous'.[59] However, an inspection of the Parian quarries,[60] which can still be explored, seems to me to confirm the explanation of Varro, namely that the marble was, unusually, quarried by lamplight. Many large pieces of marble remain scattered about the area, their surfaces now brown in colour through weathering: when broken they reveal the gleaming, crystalline marble within.

Pliny refers to marble from other sources but gives no information other than the colour, as, for example, in the case of Chian (36. 46): 'In my opinion, the first specimens of our favourite marbles, with their particoloured markings, appeared from the quarries of Chios when the people of that island were building their walls' ('primum, ut arbitror, versicolores istas maculas Chiorum lapicidinae ostenderunt, cum exstruerent muros'). Chian marble, as Pliny rightly observes, occurs in a variety of colours: grey and red are common and the stone is delicately veined.

It is interesting to note that marble from Proconnesus (36. 47) was also used in the sixth century AD, when Hagia Sophia[61] was rebuilt under the emperor Justinian, to replace the basilica of Constantine the Great that had twice been destroyed by fire.

Occasionally, Pliny (36. 86) misidentifies limestone as marble although he is clearly not happy about the identification: 'There is a feature of the Egyptian Labyrinth which I, for my

---

[59] Eichholz, Pliny, *NH 36–37*, p. 12 note *a*.

[60] To reach the quarries, one takes the Naoussa road from Paroikia and, after about two miles (3.2 km.), the track to the deserted Monastery of Ayios Minas. The quarries—there are numerous excavations—lie on the slopes of Mt. Profitis Ilias. The largest is 330 feet long and nearly 30 feet wide (100 m. by 9 m.), with a chamber on either side of the central passage. Still plainly visible are the marks of the wedges used by the ancient quarrymen, who, entering the mountainside by adits, virtually 'mined' the marble. The quarries were left abandoned until the 19th cent. when, in 1844, marble from here was used for Napoleon's tomb in Les Invalides (Paris).

[61] After 1453, Hagia Sophia was converted into a mosque. In the 16th cent. Sinan, the famous architect of Suleiman the Magnificent, strengthened the buttresses. It remained in use as a mosque for about 500 years, but, in 1935, Kemal Atatürk established it as a museum.

part, find surprising, namely an entrance and columns made of Parian marble. The rest of the structure is of granite from Aswan'. ('Aegyptius [sc. labyrinthus] quod miror equidem, introitu lapidibus e Paro columnisque, reliqua e syenite molibus compositis.'[62])

In addition to the sculptural and architectural use of marble, Pliny (36. 46) records its use in veneers: 'The art of cutting marble into thin slabs may possibly have been invented in Caria. The earliest instance, so far as I can discover, is that of the palace of Mausolus at Halicarnassus, the brick walls of which were decorated with marble from Marmara Island' (secandi in crustas nescio an Cariae fuerit inventum. antiquissima, quos equidem inveniam, Halicarnassi domus Mausoli Proconnesio marmore exculta est lateri1iis parietibus'). Pliny is clearly uncertain as to the origin of marble veneers: elsewhere (36. 51) he comments, 'Whoever first discovered how to cut marble and carve up luxury into many portions, was a man of misplaced ingenuity' ('sed quisquis primus invenit secare luxuriamque dividere importuni ingeni fuit.')

Finally, Pliny (33. 68) also uses *marmor* in the technical meaning, 'quartz', when he states that 'channelled'—that is, reef—gold is found sticking to quartz:[63] 'canalicium vocant, alii canaliense,...marmoris glareae inhaerens.' *Marmoris glareae* is quartz gangue.

---

[62] The Egyptian Labyrinth: for a description of this remarkable structure, see further *NH* 36. 87 ff. Pliny was, of course, right to be surprised. The stone was not Parian marble but exceptionally white limestone, found, for example, in the Tura quarries east of the Nile, near Memphis. Theophrastus (*Lap.* 7) compares the physical characteristics of this stone ($\pi \acute{o} \rho o s$—the Greek name for limestone) with Parian marble. Pliny (36. 132) likewise states, 'Porus is similar to Parian marble in whiteness and hardness, only not so heavy' ('[porus] Pario similis candore et duritia, minus tantum ponderosus'). In his numerous references to marble, Pliny identifies the following types: Augustean (36. 55), marble from Carystus (36. 48–9), Chian (36. 46), Hymettan (36. 7, 114), Lacedaemonian (36. 55, 37. 73), Lesbian (36. 44), Lucullan (36. 6, 49), Luna (36. 18, 48), Numidian (36. 49), Parian (36. 14, 44, 86, 132, 135, 158). Proconnesian (36. 47), Synnadic (35. 3), Taenarian (36. 135, 158), Thasian (36. 44), Tiberian (36. 55).

[63] C. Domergue, 'Les Techniques minières antiques de *Le De Re Metallica* d' Agricola', in Domergue, *Mineria y Metalurgia*, ii. 77. 11: 'Marmor (marbre) doit désigner ici une gangue calcaire (calcite?), de même que dans le texte de Pline qui suit. Il est plus difficile de traduire *glarea*, qui peut désigner la structure friable de la gangue, comme c'est parfois le cas dans les filons.'

Minerals 209

calcium bicarbonate Pliny (31. 29) discusses waters with petrifying properties: 'At Perperena, a spring turns to stone any land it irrigates, and the thermal waters of Adepsus, in Euboea, have the same effect. Whatever rocks the stream reaches increase in size. At Eurymenae, garlands thrown into a spring turn to stone. Bricks thrown into the river at Colossae[64] are found to be of stone when retrieved. At the mine on Scyros all the trees washed by the river are petrified, branches and all.' ('in Perperenis fons est quamcumque rigat lapideam faciens terram, item calidae aquae in Euboeae Adepso. nam quae adit rivus saxa in altitudinem crescunt. in Eurymenis deiectae coronae in fontem lapideam fiunt. in Colossis flumen est quo lateres coniecti lapidei extrahuntur. in Scyretico metallo arbores quaecumque flumine adluuntur saxeae fiunt cum ramis.')

Some of the land in the region of the ancient site of Colossae (ancient Phrygia) has been affected as Pliny describes, and is covered by calcium deposits. Near the site of Hierapolis[65] (modern Pamukkale) the water from thermal springs[66] is supersaturated with calcium bicarbonate. In contact with the atmosphere it releases its carbon dioxide while trying to regain equilibrium: the mineral salt is precipitated as in the following equation:[67] $Ca(HCO_3)_2 = CaCO_3 + CO_2 = H_2O_2$. The reaction, which is affected by weather conditions, ambient temperatures, speed of flow of the water, and flow duration, has created a number of pools on terraces, at different levels,

---

[64] The classical town of Colossae was founded on the south side of the Cadmos (Honaz) mountain, on the banks of the river Lycus (Aksur). It was one of the most important centres within Phrygia, and Xenophon (*Anab.* 1. 2. 6) describes it as a 'flourishing and great city' (Κολοσσάς, πόλιν οἰκουμένην εὐδαίμονα καὶ μεγάλην). Colossae was prosperous under the Persian empire, but, from the 2nd cent. BC, after the foundation of Hierapolis and Laodicea, its importance began to decline. It was devastated by an earthquake during the principate of Nero. In the later Roman period, the inhabitants migrated to Hierapolis and Laodicea. (*NH* 5. 145). Here, St Paul delivered his Epistle to the Colossians. See further H. Huseyin Bayasal, Ali Ceylan, *et al.*, *Guide to Denizli: Pammukale* (museum guide [n.d., *c.*1996]), 53 ff.

[65] Strabo, 13. 4. 14, refers to Hierapolis as being in Phrygia. It is also mentioned by Pliny, 2. 208, and Vitruvius, *De arch.* 8. 3. 104.

[66] There are in this region no less than seventeen thermal springs which have a temperature range of between 35 °C and 100 °C.

[67] Baysal, Ceylan, *et al.*, *Guide to Denizli*.

and petrified 'waterfalls'. These spectacular formations are known as 'travertines'.[68]

calcium hydroxide (lime) Lime is a loose term embracing a variety of materials containing 'lime' in different forms: calcium oxide (CaO), calcium carbonate ($CaCO_3$), and calcium hydroxide ($CaOH_2$).[69]

'Another kind of white marl,' writes Pliny (17. 45), 'a variety also used for cleaning silver, is obtained from a considerable depth in the ground, usually from pits made a hundred feet deep, with a narrower mouth but with the shaft expanding in the interior, as is the practice in mines. This chalk is chiefly used in Britain. Its effect lasts for eighty years and there is no case of anyone having scattered it on the same land twice in his lifetime. ('alterum genus albae creta argentaria est; petitur ex alto, in centenos pedas actis plerumque puteis, ore angustiore, intus ut in metallis spatiante vena. hac maxime Britannia utitur. durat annis LXXX, neque est exemplum ullius qui bis in vita hanc eidem iniecerit.')

Slaked lime was also regularly used in ancient times in the building trade, especially for plastering and for mortar, but does not seem to have been much used on the land. Lime from kilns had, in Pliny's time, been applied with success to olive cultivation.

The addition of lime to the soil is beneficial in two ways: it not only makes up for deficiencies in one of the most important

---

[68] This term was coined, in Roman times, apropos the great travertine deposits of Tivoli. Precipitation continues until the carbon dioxide in the thermal waters reaches equilibrium with the carbon dioxide in the atmosphere. Measurements made at the source of the springs find atmospheric levels of 725 mg/l of carbon dioxide. By the time this water has flowed across the travertines, this figure has, fallen to 145 mg/l. Likewise, calcium carbonate falls from 1,200 mg/l to 400 mg/l and calcium from 576.8 mg/l to 376.6 mg/l. From these results it is calculated that 499.9 mg of calcium carbonate is deposited on the travertine for every litre of water. This means that for a flow rate of 1 l/s of water 43,191 grams are deposited daily. The average density of a travertine is 1.48 g/cm^3, which implies that it can whiten 13,584 m^2 a day, but in practice this area of coverage is difficult to attain. These theoretical calculations indicate that up to 4.9 km^2 can be covered with a white deposit of 1 mm thickness.

[69] Tylecote, *Metallurgy*, 25, describes the pyrological processes involved in the making of lime.

Minerals 211

mineral elements, but corrects undue acidity which may adversely affect the growth of many plants. Pliny (*NH* 17. 44) also states, 'White marl,[70] if found near springs, has unlimited fertilizing properties... but if scattered in excessive quantities burns up the soil' (... si inter fontes reperta est, ad infinitum fertilis... si nimia iniecta est, exurit solum.').

calcium sulphate Pliny (36. 182 f.) next discusses gypsum,[71] which is an evaporite mineral, $CaSO_4 \cdot 2H_2O$, found in clays and limestones—a sediment resulting from the evaporation of saline water (normally bodies of sea water).

Pliny (36. 182 f.) observes

an affinity between lime and gypsum, a substance of which there are several varieties. For it can be produced from a heated mineral, as in Syria and at Thurii: it can also be dug from the earth, as in Cyprus and Perrhaebia. The gypsum of Tymphaea is stripped from the earth's surface. The mineral that is heated ought to be like onyx marble, or crystalline limestone. In Syria the hardest stones are selected for the purpose and are heated along with cow-dung to accelerate the burning. However, it has been discovered that the best kind is prepared from selenite, or from stone that flakes in the same way. Gypsum, when moistened, should be used instantly, since it coheres with great rapidity. There is nothing to prevent it from being pounded and reduced again to a fine powder. Gypsum is used for white stucco and, with pleasing effects, for plaster-cast figures and decorative mouldings on buildings.

cognata calci res gypsum est. plura eius genera. nam e lapide coquitur, ut in Syria ac Thuriis, et e terra foditur, ut in Cypro ac Perrhaebia; e summa tellure et Tymphaicum est. qui coquitur lapis non dissimilis alabastritae esse debet aut marmaroso. in Syria durissimos ad id eligunt cocuntque cum fimo bubulo, ut celerius urantur. omnium autem optimum fieri compertum est e lapide speculari squamamve talem habente. gypso madido statim utendum est, quoniam celerrime coit; tamen rursus tundi se et in farinam resolvi patitur. usus gypsi in albariis, sigillis aedificiorum et coronis gratissimus.

Pliny derives this account, in the main, from Theophrastus (*Lap.* 64–5 and 69), who does not clearly distinguish gypsum,

---

[70] See further K. D. White, *Roman Farming* (London, 1970), 138 f.
[71] Whitten–Brooks, *Geology*, 219 and appendix.

that is hydrous calcium sulphate, from the dehydrated form of gypsum, namely plaster of Paris. The latter is obtained by heating the mineral so as to cause it to lose three-quarters of its water content. The gypsum of Cyprus, Perrhaebia, and Tymphaea is the actual mineral, while that from Syria and Thurii is plaster of Paris.

Bailey[72] erroneously identified the 'gypsum' of Tymphaea as fullers' earth, incorrectly asserting that Theophrastus said that it was used for 'cleaning clothes'—a purpose for which it would be entirely unsuitable. Theophrastus in fact says it is used 'for cloaks', later adding that fullers *sprinkled* gypsum on cloaks—to *whiten*, not to clean them, for which it would, of course, have been suitable.

Pliny also records the quick-setting property of plaster of Paris when mixed with water and mentions its use with moulds to produce cast figures. He further observes (34. 183) that there is nothing to prevent plaster of Paris, when set, from being pounded and reduced again to a fine powder. In practice, however, such reuse is not very effective.

cassiterite A widely distributed ore-mineral of tin[73] (SnO_2), with a specific gravity of 6.8–7.1; it is often found in the form of rolled pebbles in placer deposits known as 'stream tin', as Pliny (34. 157) describes.[74] 'It is also found in gold mines called 'alutiae', through which a stream of water passed that washed out black pebbles of tin, mottled with white spots (cassiterite) and of the same weight as gold. ...('invenitur et in aurariis metallis quae alutias vocant, aqua inmissa eluente calculos nigros paullum candore variatos, quibus eadem gravitas quae auro. ...')

It has been noted as an original constituent of igneous rocks and pegmatites, but is more commonly found in high-temperature hydrothermal veins, in, or near, granitic rocks.

Stannite[75] (Cu_2FeSnS_4), with a specific gravity of 4.45, occurs in tin-bearing veins associated with cassiterite.

cerussa See Ch. 15, s.v. 'psimithium'.

---

[72] *Chemical Subjects*, ii. 276 f.
[73] Whitten–Brooks, *Geology*, 72 and appendix.   [74] Cf. Diod. Sic. 5. 22. 2.
[75] Healy *Mining*, 38; Tylecote, *Metallurgy*, 36.

chalcitis The mineral referred to as *chalcitis*[76] (34. 2) is chalcopyrite, or copper pyrites in the process of decomposition.

Chalcopyrite[77] ($CuFeS_2$) is the most widely occurring ore-mineral of copper and one of the most important sources of that metal in the ancient world. Most sulphide ores contain some chalcopyrite but the most important economically are the hydrothermal vein and replacement deposits. It is also the principal primary copper mineral in 'porphyry-copper' deposits.

Pliny (34. 117) refers thus to *chalcitis*:

The ore from which copper and *cadmea* [zinc silicate and zinc carbonate] are obtained by smelting is called copper pyrites. *Cadmea*[78] is quarried from rocks above ground, while copper pyrites is mined underground and immediately crumbles into small pieces, being of a soft consistency resembling matted down. Copper pyrites further differs from *cadmea* in that it contains three kinds of mineral—that is copper, copper pyrites [*misy*], and decomposing marcasite[79] [*sori*]. ... the veins of copper in it are oblong in shape. Top-quality copper pyrites is honey-coloured, streaked with fine veins and liable to crumble but it is not stony.

chalcitim vocant ex quo et ipso aes coquitur. distat a cadmea, quod illa super terram ex subdialibus petris caeditur, haec ex obrutis, item quod chalcitis friat se statim, mollis natura, ut videatur lanugo concreta. est et alia distinctio, quod chalcitis tria genera continet, aeris et misyos et soreos. ... habet autem aeris venas oblongas. probatur mellei coloris, gracili venarum discursu, friabilis nec lapidosa.

Later (34. 121), Pliny adds further details about copper pyrites (*misy*). 'Some authorities have recorded that copper pyrites is made by burning copper ore in trenches, its fine yellow powder[80] mixing itself with the ash of the burnt pine-wood.

---

[76] Arist. (*HA* 552^{b}10) records that χαλκῖτις was found in Cyprus, and Plutarch (*Mor.* 434a) adds Euboea.
[77] Healy, *Mining*, 37 f. Cf. Hurlbut, *Mineralogy*, 251 ff.
[78] See above, s.v.
[79] Marcasite (FeS_2) is less stable than iron pyrites and is frequently found with lead and zinc ores. It is easily decomposed (as here described). Marcasite most frequently occurs as replacement deposits in limestone and often in concretions embedded in clays, marls, and shales.
[80] In other words, the yellow powder is the oxidation product of the preliminary roasting of copper sulphide—possibly chalcocite, or copper glance which is an important source of copper.

But, although obtained from copper pyrites, it is an integral part of the mineral and separated from it only by force. The best kind of copper pyrites is extracted in copper-works in Cyprus. When broken, its fragments sparkle like gold; when ground it has a sandy appearance. A mixture of copper pyrites is used by gold refiners.[81] ('misy aliqui tradiderunt fieri exusto lapide in scrobibus, flore eius luteo miscente se ligni pineae favillae. re vera autem e supra dicto fit lapide, concretum natura discretumque vi, optimum in Cypriorum officinis, cuius notae sunt friati aureae scintillae et, cum teratur, harenosa natura sine terra, chalcitidi dissimilis. hoc admiscent qui aurum purgant.')

chalcopyrite See above, s.v. 'chalcitis'.

chert A cryptocrystalline silica[82] which may be of organic, or inorganic, origin. It occurs as bands, or layers of nodules, in sedimentary rocks. It can be shown that it is sometimes a primary deposit, sometimes formed by the confluence of disseminated silica in a rock, and sometimes a secondary replacement material.

Flint is the variety of chert occurring primarily in the Upper Cretaceous and as detrital pebbles in the Tertiary. It has a conchoidal fracture, as opposed to the flat fracture of chert.

Lyddite is a dense variety of chert, formerly used as a touchstone.[83]

Chia terra candicans This substance (35. 194), 'white earth of Chios, is also among medicines; its effect is the same as that of Samian earth. It is specially used as a cosmetic for the skin of women' ('est in medicamentibus et Chia terra candicans, effectus eius idem qui Samiae; usus ad mulierum maxime cutem').

This must surely be a form of china clay?

---

[81] Healy, *Mining*, 159f. The result of the roasting of the sulphide was the production of copper oxide as: $Cu_2S + 2O_2 = 2CuO + SO_2$. The copper oxide was subsequently reduced with charcoal as: $CuO + CO = Cu + CO_2$.

[82] Whitten–Brooks, *Geology*, 76ff.

[83] See s.v. 'lapis Lydius' below. Pliny (36. 157) also refers to the touchstone as *basanites lapis* that is siliceous slate. In contrast, *basanites* (36. 58) is greywacke and, elsewhere (36. 147), *basanites lapis* is a variety of haematite (see below, s.v. 'haematite').

chrysocolla *Chrysocolla*[84] should not be confused with the modern mineral of that name. In the ancient world, the term was used of any bright-green copper mineral which occurred as an earthy incrustation. Dioscorides (5. 104) states that the best kind of *chrysocolla* was of a leek-green colour. The name referred mainly to malachite, or green copper carbonate, when in an earthy form, and also to the amorphous green copper silicate (still called chrysocolla).

The description of Pliny (33. 86) confirms the attribution to malachite.[85]

Cinnabar Mercury[86] (Hg) is found in its native state as a liquid, metallic element,[87] and in cinnabar, or mercury sulphide (HgS), its most important ore-mineral. Cinnabar occurs in veins and as impregnations where volcanic activity has taken place.[88] It was first extracted in the third millenium BC at Suplja Stena near Belgrade.

From the sixth century BC cinnabar was used as a pigment for colouring statues and architectural mouldings and in the planning of white-ground lekythoi. However, the earliest literary reference to cinnabar (τὸ κιννάβαρι) is found in Aristotle (*Mete.* 3. 378^a).

Some minerals in the ancient world were discovered accidentally—cinnabar was found in the search for silver. Pliny (33. 113 f.) provides a direct, but slightly abridged, translation of Theophrastus' account (*Lap.* 58 f.) of the mineral's origin, location, and circumstances of discovery. (This is a good example of his unquestioning acceptance of secondary sources.)

---

[84] Theophr. *Lap.* 26, Pliny *NH* 33. 4 and elsewhere.
[85] Cf. also E. R. Caley and J. F. Richards (trans. and comm.), *Theophrastus on Stones* (Columbus, Oh., 1956), 105. [86] See Ch. 17. 12 below.
[87] Healy, *Mining*, 227.
[88] Ibid. 42, cf. p. 66. For physical charactcristics see Whitten–Brooks, *Geology*, appendix on cinnabar. Cinnabar: hardness 2.0/2.5; sp.gr. 8.1; streak, scarlet; perfect cleavage. See also A. Sherratt, 'Resources, Technology and Trade: an Essay in Early European Metallurgy', in G. de G. Sieveking, I. H. Longworth, and K. E. Wilson (eds.), *Problems in Economic and Social Archaeology* (London, 1976). 566. The area in north-west Yugoslavia was to become an important mining area for mercury in later times.

... it was discovered by an Athenian named Callias[89] some ninety years before the archonship of Praxibulus at Athens [that is, in 405 BC]:[90] Callias had hoped that gold could be extracted, by smelting, from the red sand found in silver mines. A hard, sandy type of cinnabar was already known in Spain and in Colchian territory on a certain inaccessible rock from which the local people dislodged it by throwing spears. This cinnabar, however, was of an impure quality; the best occurs in the region of the Cilbi, to the east of Ephesus where the sand is the red colour of a cochineal beetle [*kermes*]. The powder from the ground-up sand is washed and the sediment rewashed. Skill makes a difference in these operations: some workers produce cinnabar after one washing, while, with others, the cinnabar is rather weak and is improved by a second washing. [He continues (33. 115),] The importance of the colour comes as no surprise to me, for in Trojan times red ochre was valued highly. Homer [*Il*. 2. 637] commends it as a distinguished colour for ships, although he otherwise rarely alludes to colours and paintings. The Greek name for it is $\mu\acute{\iota}\lambda\tau\sigma\varsigma$[91] and they call *minium* cinnabar.

Theophrastus LXXXX annis ante Praxibulum Atheniensium magistratum—quod tempus exit in urbis nostrae CCCXLVIIII annum—tradit inventum minium a Callia Atheniense initio sperante aurum excoqui harenae rubenti in metallis argenti; ... reperiri autem iam tum in Hispania, sed durum et harenosum, item apud Colchos in rupe quadam inaccessa, ex qua iaculantes decuterent; id esse adulterum, optimum vero supra Ephesum Cilbianis agris harena cocci colorem habente, hanc teri, dein lavari farinam et quod subsidat

---

[89] Callias is generally thought to have been a friend of Socrates and to have had a very extravagant lifestyle. See Zehnacker, *NH 33*, p. 201 n. 3: 'Quoiqu' il en soit, ce Callias avait sans doute pris à ferme une partie des mines du Laurion; malheureusement pour lui, l' exploitation minière fut abandonnée à partir de 412, consécutivement aux succès de sparte (Thuc. VI, 91 et VII, 27). Il est probable qu' il émigra alors en Asie, avec l'espoir d'y continuer ses activités. Le hasard d'une prospection lui fit découvrir le minium.'

[90] On the chronology, see further Eichholz, *Lap.*, p. 127.

[91] This is red ochre (Fe_2O_3), which is found mixed with silica and hydrated aluminium oxide. Cf. Hdt. 4. 191, 7. 69; Ar. *Eccl.* 378, and elsewhere. Roman authors often confused *minium* and cinnabar because of the similarity of colour. Whereas *minium Hiberum* (Propertius 2. 2. 21) is cinnabar, elsewhere (e.g. *NH* 33. 118; Verg. *Ecl*. 10. 27; Suet. *Caligula* 18) *minium* commonly means 'red lead'. Pliny (33. 116) adds a further bizarre reason for the confusion. The Greeks gave the name 'Indian cinnabar' to the gore of a snake crushed by the weight of dying elephants, when the blood of each animal gets mixed together. In fact this substance (*sanies draconis*) was really an exudation (still called 'dragon's blood') from a species of the oriental plant *Dracaena*, or *Pterocarpus*.

iterum lavari; differentiam artis esse, quod alii minium faciant prima lotura, apud alios id esse dilutius, sequentis autem loturae optimum.... auctoritatem colori fuisse non miror. iam enim Troianis temporibus rubrica in honore erat Homero teste, qui naves ea commendat, alias circa pigmenta picturasque rarus. milton vocant Graeci miniumque cinnabarim.

It is possible that Callias emigrated to Asia Minor in the hope of continuing his entrepreneurial activities. There, by chance, he discovered a source of cinnabar. With regard to the existence of this mineral near Ephesus, Zehnacker,[92] in his commentary on this passage, draws attention to the fact that Pliny disregards the evidence of Vitruvius,[93] that cinnabar was no longer exploited in the vicinity of Ephesus in his time. The Cilbian plain mentioned by both Vitruvius and Pliny, is located to the east of the Cayster plain, the site of Ephesus.

It is strange that Theophrastus describes the cinnabar found near Ephesus as very hard since it is only 2.5[94] on the Mohs scale—approximately the hardness, that is, of rock salt. The earthy varieties, however, are soft. Theophrastus is, perhaps, merely emphasizing the pronounced difference in hardness between the crystalline and earthy varieties. The impure kind, from near Ephesus, was probably the latter, whereas the so-called natural kind, from Spain and Colchis, was probably a pure crystalline, or massive, variety.

Pliny (33. 114) again describes the standard method of separating a mineral from its accompanying gangue. In the case of cinnabar, the specific gravities are 8.1 and 3.0, respectively. The cinnabar settles more rapidly and can be poured off with the water by a skilled operator. The method of gravity separation had been very successfully used at Laurion[95] and Callias, as an Athenian, would certainly have known about this technique.

---

[92] Zehnacker, *NH 33*, p. 202.
[93] *De arch.* 7. 8. 1: 'ingrediar nunc minii rationes explicare. id autem agris Ephesiorum Cilbanis primum esse memoratur inventum.' Also *De arch.* 7. 9. 4: 'quae autem in Ephesiorum metallis officinae, nunc traiectae sunt ideo Romam, quod id genus venae postea est inventum Hispaniae regionibus.'   [94] Healy, *Mining*, 42.
[95] Conophagos, *Laurium*, 213 ff.

The evidence of geology does not warrant any claim that cinnabar was ever found in southern Attica.

Pliny (33. 118) continues:

Juba[96] records that cinnabar is also produced in Carmania,[97] and Timagenes,[98] that it occurs in Ethiopia as well: but it is not exported to us from either place and indeed we only get cinnabar from Spain. The most famous cinnabar mine which provides revenue for the Roman people is that of Sisapo, in Baetica. The security precautions are second to none. Smelting and refining of cinnabar are not allowed to be carried out locally, but as much as 2,000[99] pounds of crude ore a year are sent to Rome under seal and there purified. To prevent the price going sky-high, it is fixed by law at seventy sesterces for about a pound.

Iuba minium nasci et in Carmania tradit, Timagenes et in Aethiopia, sed neutro ex loco invehitur ad nos nec fere aliunde quam ex Hispania, celeberrimo Sisaponensi regione in Baetica miniario metallo vectigalibus populi Romani, nullius rei diligentiore custodia. non licet ibi perficere id excoquique; Romam adfertur vena signata, ad bina milia fere pondo annua, Romae autem lavatur, in vendendo pretio statuta lege, ne modum excederet HS LXX in libras.

Sisapo—Almaden[100] in New Castile—is celebrated for native mercury as well as cinnabar. The deposit at Almaden rests on quartzite of Lower Silurian date and consists of three

---

[96] Zehnacker, *NH 33*, p. 205, n. ad loc.

[97] Carmania (Kerman) is a mountainous region of Iran, to the east of Chiraz (cf. references in *NH* 6. 84–152). Rare and valuable gems and minerals were found there, including onyx (*NH* 36. 59 and 61), fluorspar (37. 21), and a variety of other precious stones (37. 110, 131, 132, and 134).

[98] Zehnacker *NH 33* p. 205 n. 3. Timagenes of Alexandria, the son of a royal official, was taken prisoner in 55 BC by Gabinius and transported to Rome. There he became a professor of rhetoric and was a friend of Asinius Pollio. For some time he advised Octavius on oriental matters. Timagenes' historical work (see Quint. *Inst.* 10. 1. 75) treated the East and included geographical and ethnographical elements. Suidas' *Lexicon* also credits him with a *Periplus* (in five books).

[99] Bailey, *Chemical Subjects* i. 220 (note to 33. 118), preferred the reading 'dena milia', namely 10,000. Cf. Zehnacker, *NH 33*, p. 207 n. 6.

[100] Thirteen miles (21 km) from Mirobriga, on the way to Laminium. *CIL* x. 3964 records a 'vilicus sociorum Sisaponensium'. Strabo 3. 2. 2 also refers to silver-mines at Sisapo.

lenses on top of each other, 650–825 feet wide and 20 feet thick (200–250 m. by 6 m.). The ore body is penetrated by slaty gangue and consists mainly of cinnabar (HgS) impregnations between quartz grains. There are also prills of native mercury (Hg) which collect in dips and channels. Compared with other areas such as Idria (Istria) and Turkey, the ore body is comparatively rich and averages 6–7 per cent in mercury content. There seems to be no indication of a decrease in grade with depth. Crystallographic examination shows that the cinnabar had formed before the hardening of the gangue which contradicts the epigenetic hypothesis.

Creta Cimolia A form of fuller's earth derived from the island of Cimolus. Pliny (35. 195) describes two varieties: 'One is bright white, the other, inclining to purple. Both are used in medicine' ('Cimoliae duo ad medicos pertinentia, candidum et ad purpurissimum inclinans').

Pliny (35. 198) mentions a further use, namely as a test for dyed cloth.

The cloth is first washed with earth of Sardinia (*Sarda*),[101] then fumigated with sulphur and, afterwards, scoured with Cimolian earth, provided that the dye is fast. If it is coloured with inferior dye, this is detected and turns black and its colour is spread by the action of the sulphur. Genuine and valuable colours, however, are softened and brightened up with a sort of brilliance by Cimolian earth when they have been made more sombre by the sulphur. The 'rock' kind is more serviceable for white garments, after the application of sulphur, but is very detrimental to colour. In Greece they use Tymphaean gypsum instead of Cimolian earth.

primum abluitur vestis Sarda, dein sulpure suffitur, mox desquamatur Cimolia quae est coloris veri. fucatus enim deprehenditur nigrescitque et funditur sulpure, veros autem et pretiosos colores emollit Cimolia et quodam nitore exhilarat contristatos sulpure.

---

[101] Rackham, *NH 33–35*, p. 406, note *a*, identifies *Sarda* as a mixture 'of strong clacium montmorillonite, Umbrian earth, some kaolin and *saxum*'. They are kinds of fuller's earth. Bailey, however, *Chemical Subjects*, ii. 244, thinks that *saxum* is quicklime. This is a further example of Pliny's imprecise terminology.

candidis vestibus saxum utilius a sulpure, inimicum coloribus. Graecia pro Cimolia Tymphaico utitur gypso.

creta Eretria (Eretrian earth) See Ch.15, s.v. 'creta Eretria'.

crystallum Rock-crystal (known to the Greeks as κρύσταλλος) is a clear variety of quartz (SiO_2), a common mineral in all kinds of rocks, whether igneous, metamorphic, or sedimentary.[102] Characteristic features of quartz are the absence of cleavage and its hardness, 7 on the Mohs scale.

The perfectly limpid, colourless quartz is exceptional among all other minerals by reason of its clarity and transparency, in which it often surpasses even that of the diamond, although it is not comparable with the latter in lustre, or play of colours. It commonly occurs in fine crystals, the prism faces[103] of which are, almost without exception, largely developed so that the habit of the crystals is columnar.[104] Some crystals may be attached to one or to both ends of the matrix and show considerable variation in size. Pliny observed the habit of quartz but was puzzled by its cause!

Pliny (37. 23 ff.) states that the most sought after rock-crystal occurs in India and is also found in the Near East, Cyprus, and the Alps. He quotes Juba (ibid. 24) as his authority that it occurs on an island (Necron) in the Red Sea, facing Arabia. 'There', he claims, 'one of Ptolemy's officers, named Pythagoras, dug up a piece measuring about eighteen inches in length. Cornelius Bocchus mentions that rock-crystal of surprising weight was found in Lusitania in the Ammaeensian mountains when wells were being sunk to water-level.' ('et in ea [Necron]...cubitalem effossam a Pythagora, Ptolemaei praefecto; Cornelius Bocchus et in Lusitania perquam mirandi ponderis in Ammaeensibus iugis depressis ad libramentum aquae puteis.')

Pliny, later (37. 27) adds, 'The largest mass of rock-crystal ever seen by us is that which was dedicated in the Capitol by

---

[102] Healy, *Mining*, 20. [103] Healy, 'Pliny and Ancient Mineralogy', 172.
[104] Ibid. 173, fig. 5.

Livia, the wife of Augustus: this weighs about 150 pounds. Xenocrates also records that he saw a vessel that could hold almost seven gallons[105] and some authorities mention one from India with a capacity of four pints.' ('magnitudo amplissima adhuc visa nobis erat quam in Capitolio Livia Augusti dedicaverat, librarum circiter CL. Xenocrates idem auctor est vas amphorale visum, et aliqui ex India sextariorum quattuor.')

Quartz, in the form of sand, was extensively used in glass-making[106] and, as an abrasive, for 'sawing' marble (36. 51). The cutting of marble is achieved ostensibly by iron, but, in reality, by sand. The saw merely presses the sand on a finely traced line and the passage of the instrument, owing to the rapid movement to and fro (friction),[107] is in itself enough to cut the stone: 'harena hoc fit et ferro videtur fieri, serra in praetenui premente harenas versandoque tractu ipso secante.'

Finally, quartz was stained to counterfeit gemstones, as, for example, beryls of which there were good forgeries in the first century AD.[108]

diamond See above, s.v. 'adamas' (and 'Crystal Systems', Ch. 13.4 above).

emerald See below, s.v. 'smaragdus'.

fluorspar (or fluorite) See below, s.v. 'myrrha'.

galena Lead sulphide (PbS), specific gravity 7.4–7.6, is found in veins associated with sphalerite, pyrite, and other minerals. When it occurs in hydrothermal veins, galena[109] is frequently argentiferous and is the main source of silver in the ancient world.[110]

---

[105] Cf. *NH* 36.59 ('vas amphorale'). [106] See Ch. 18.2 below.
[107] Ch. 12.5.1 above.
[108] E. H. Warmington, *The Commerce between the Roman Empire and India*[2] (Cambridge, 1974), 251.
[109] Whitten–Brooks, *Geology*, 196 and appendix.
[110] Healy, *Mining*, 38; cf. also 157 ff.

Cerussite ($PbCO_3$) like galena, is found at Laurium;[111] it is a widely distributed secondary ore-mineral of lead formed by the action of carbonated waters on galena.

gypsum See above, s.v. 'calcium sulphate'.

haematite Haematite[112] (Fe_2O_3), specific gravity 4.7, is widely distributed in rocks of all ages and forms the most abundant and important ore-mineral of iron. It may occur as a sublimation product in connection with volcanic activities, in contact metamorphic deposits, and as an accessory mineral in feldspathic igneous rocks such as granite. It is found from microscopic scales to enormous masses in regionally metamorphosed rocks where it may have orginated by the alteration of limonite, siderite, or magnetite. Haematite is found also in red sandstones as the cementing material that binds the quartz grains together.

Like limonite, it may be formed in irregular masses and beds as the result of the weathering of iron-bearing rocks. The oolitic ores are of sedimentary origin and may occur in beds of considerable size. Varieties of haematite include: specular iron ore, a micaceous form with metallic lustre and crystals often splendent, and kidney ore. The latter is often discovered in association with limonite.

Pliny (36. 144 f.) writes, 'A close relationship exists between *schistos* and *haematites*. The latter is found in mines and, when roasted, reproduces the colour of red lead. ... it is sometimes counterfeited,[113] the points of distinction being the red veins of ore and its friable nature' ('schistos et haematites cognationem habent. haematites invenitur in metallis, ustus minii colorem imitatur.... adulteratur haematites; discernunt venae rubentes et friabilis natura').

---

[111] Conophagos, *Laurium*, 126.
[112] Healy, *Mining*, 39 ff.; Hurlbut, *Mineralogy* 281 ff.
[113] Cf. A. Rosenfeld, *The Inorganic Raw Materials of Antiquity* (London, 1965). 143 ff. Under conditions of strong chemical weathering, magnetite is converted into haematite, or limonite; limonite itself, when heated to about 400 °C for short periods, is converted to Fe_2O_3, identical in composition to haematite. It is possibly to this phenomenon that Pliny refers in his use of the term 'counterfeiting'.

He continues (36. 146 f.) 'Among the oldest authorities, Sotacus [early third century BC] records five kinds of haematite, in order of quality: [1] Aethiopian,[114]... [2] *andromas* [specular iron ore],[115] black in colour and conspicuous for its weight and hardness... [3] Arabian, ... [4] *hepatites* [kidney ore], and [5] *schistos* [limonite].'

Limonite is the omnibus term used for a range of mixtures of hydrated iron oxides and iron hydroxides. It consists of various amorphous and cryptocrystalline constituents and also includes various amorphous iron hydroxides. Limonites occur widely as the weathering product of all iron-containing ore-minerals. The most characteristic property of the mineral is its yellow, or brownish-yellow, streak.[116]

Among well-known types of limonite are deposits of gossan red and yellow ochres,[117] bog iron ore, and laterite, deposited in swamps and lakes probably by bacterial action; they may have a low content of phosphorus and sulphur and can yield very pure iron.[118]

Oolitic limonites[119] constitute the most important iron ore deposits in Europe. Pliny (34. 142) observes that 'in Cappadocia alone it is merely a question whether the presence of iron is to be credited to water, or to earth, as that region supplies iron from the furnaces when the earth has been flooded by the river Cerasus, but not otherwise.' ('in Cappadocia tantum quaestio est, aquae an terrae fiat acceptum, quoniam perfusa Ceraso fluvio terra neque aliter ferrum a fornacibus reddit.')

Pliny concludes that there are numerous varieties of iron, depending on the kind of soil, or climate (and other factors): 'differentia ferri numerosa... in genere terrae caelive.'

iris Pliny (37. 136) describes this mineral, otherwise known as the 'rainbow stone': 'It is quarried on an island in the Red Sea,

---

[114] This variety cannot be identified with any certainty.
[115] Whitten–Brooks, *Geology*, 423. The popular name for haematite.
[116] Ibid. 269.
[117] Ibid. 211. The leached and oxidized near-surface part of a vein containing sulphides, especially iron-bearing ores—for example iron pyrites.
[118] Whitten–Brooks, *Geology*, 56. [119] Healy, *Mining*, 41.

sixty miles from the city of Berenice' ('effoditur in quadam insula Rubri maris quae distat a Berenice urbe LX p.'). It has hexagonal faces and is similar to rock-crystal.[120] It can act as a prism in refracting light[121] to produce the colours of the rainbow, hence its name (*iris*).

iron pyrites Iron pyrites[122] (FeS_2) is the most common and widespread of the sulphide minerals. Although associated with many minerals, it is found frequently with chalcopyrite,[123] sphalerite,[124] and galena.[125] It was not much exploited for iron because of the detrimental effect of the presence of sulphur.

Pliny does not distinguish, or discuss, iron pyrites.

lapis Lydius This stone is a siliceous schist (flinty slate) of a black colour, the best examples being of uniform deep black, without light-coloured veins or spots, and of a fine grain.[126] Pliny's description (33. 126) is as follows: 'Some people call it Heraclean stone and others, Lydian: the pieces are of moderate size, not exceeding four inches in length and two in breadth' ('alii Heraclium, alii Lydium vocant. sunt autem modici, quaternas uncias longitudinis binasque latitudinis non excedentes').

He repeats the statement of Theophrastus (*Lap.* 46 f.) that, formerly, it was not usually found anywhere but in the river Tmolus (he clearly means Pactolus).

The *lapis Lydius* was used as a touchstone for making an estimate of the quality of gold.[127]

lapis specularis (selenite)

Pliny (36. 160) gives an account of selenite,[128] much valued because of the utilization of its property of cleavage. It has an

---

[120] See above, s.v. 'crystallum'. [121] See further Ch. 12. 1. 2 above.
[122] Healy, *Mining*, 39, and Hurlbut, *Mineralogy* 262 ff.
[123] See above, s.v. 'chalcitis'.
[124] Whitten–Brooks, *Geology*, 423 and appendix.
[125] See above, s.v. [126] Healy, *Mining*, 205.
[127] D. T. Moore and W. A. Oddy, 'Touchstones: Some Aspects of their Nomenclature, Petrography, and Provenance', *Journal of Archaeological Science*, 12 (1985), 59–80.
[128] Although *lapis specularis* has been identified with mica (Cf. Bailey, *Chemical Subjects*, i. 202), it is more likely to have been selenite.

affinity with the Siphnian[129] stone and a soft stone found in Gallia Belgica that can be cut with a saw (36. 159 f.): 'hi quidem sectiles sunt.'

Selenite, although it is called a rock, is far more tractable and can be split into plates of any desired thinness. At one time it was produced only in Hither Spain, not throughout the whole province, but only within a radius of a hundred miles from the city of Segobriga. Today it also comes from Cyprus, Cappadocia, Sicily, and—a recent discovery—Africa. All these latter types are, however, inferior to the Spanish kind. Cappadocia produces the largest pieces but they are opaque.[130] Moreover, in the region of Bologna in Italy, small streaks occur tightly embedded in hard rock; and yet they are large enough for their essential similarity to the rest to be unmistakable.

specularis vero, quoniam et hic lapidis nomen optinet, faciliore multo natura finditur in quamlibet tenues crustas. Hispania hunc tantum citerior olim dabat, nec tota, sed intra C̄ passuum circa Segobrigam urbem, iam et Cypros et Cappadocia et Sicilia et nuper inventum Africa, postferendos tamen omnes Hispaniae, Cappadocia amplissimos magnitudine, sed obscuros. sunt et in Bononiensi Italiae parte breves maculae complexu silicis alligatae quarum tamen appareat natura similis.

Pliny continues (36. 161) with his description of the mining and quarrying of the mineral.

In Spain, selenite is mined to a great depth by sinking shafts; it is also found just below the surface, embedded in the prevailing rock of the locality, from which it is torn, or cut away. However, for the most part it is in a form that can be dug up since it occurs free-standing, like rough blocks in quarries; no block, to date, has been found more than five feet in length. It is evident that selenite comes from a liquid that, like rock-crystal, has been frozen and hardened into stone by an exhalation within the earth; for when wild animals fall down these shafts, the marrow in their bones assumes the form of selenite after one winter.[131]

---

[129] Cf. Theophr. Lap. 42. This is probably soapstone (steatite), a variety of talc; similarly, the stones found in Italy and Gallia Belgica are soapstone.

[130] Bailey, *Chemical Subjects*, ii. 133, translates 'obscuros' as 'are marred by their dark colour', but 'opaque' is more in keeping with the nature of the mineral.

[131] There is no evidence to support this belief. On the phenomenon of petrification, however, see further on the sarcophagus stone, below, s.v.

puteis in Hispania effoditur e profunda altitudine, nec non et saxo inclusus sub terra invenitur extrahiturque aut exciditur, sed maiore parte fossili natura, absolutus in se caementi modo, numquam adhuc quinque pedum longitudine amplior. umorem hunc terrae quadam anima crystalli modo glaciari et in lapidem concrescere manifesto apparet, quod cum ferae decidere in puteos tales, medullae in ossibus earum post unam hiemem in eandem lapidis naturam figurantur.

'Sometimes [36. 162] a black variety of selenite[132] is discovered; this is bright and, although noted for its softness, has the remarkable property of withstanding heat and cold; it does not deteriorate providing it is not damaged. This also applies to blocks of many other kinds of stone. Men have found yet another purpose for selenite; it is spread as flakes during the Games to give the surface of the Circus Maximus a much-praised bright appearance.' ('invenitur et niger aliquando, sed candido natura mira, cum sit mollitia nota, perpetiendi soles rigoresque, nec senescit, si modo iniuria absit, cum hoc etiam in caementis multorum generum accidat. invenere et alium usum in ramentis squamaque, Circum maximum ludis Circensibus sternendi ut sit in commendatione candor.')

Selenite was also employed in the ancient world as a substitute for glass in windows[133] (*specularia*), a use that continued well into the eighteenth century.

lignite See Appendix B below, s.v. 'coal'.

lime See above, s.v. 'calcium hydroxide'.

lychnis Tourmaline[134]—hexagonal trigonal—is perhaps the best known six-membered ring silicate. Many tourmalines display colour zoning and the mineral is both pyroelectric and piezoelectric.[135]

---

[132] Bailey, *Chemical Subjects*, ii. 267, identifies this as a black variety of mica—such as lepidomelane.

[133] Sen. *Ep.* 86. 11, 90. 25; *Prov.* 4. 9; *QN* 4. 13. 7; The Younger Pliny, *Letters* 2. 17. 4 and 21; Mart. *Epigr.* 8. 14; Juv. *Sat.* 4. 21; cf. *NH* 19. 64 (of cold frames).

[134] Whitten and Brooks, *Geology*, 115 f.   [135] See Ch. 12.2.1b–c.

Pliny (37. 103) describes tourmaline: 'To the same class of fiery stones belongs the *lychnis*, so called from the kindling of lamps [reading; 'lucernarum accensu'], because at that time it is exceptionally beautiful. *Lychnis* is found around Orthosia and throughout Caria[136] and the neighbouring regions, but occurs at its finest in India.[137] ... I find that there are two other varieties as well, one of which has a purple, and the other a scarlet, sheen.'[138] ('ex eodem genere ardentium est lychnis appellata a lucernarum accensu, tum praecipuae gratiae. nascitur circa Orthosiam totaque Caria ac vicinis locis, sed probatissima in Indis. ... et alias invenio differentias; unam quae purpura radiet, alterum quae cocco.') Pliny (ibid.) is the first authority to draw attention to the pyroelectric property of tourmaline.

It is interesting to note that the adjective *lychnites* is used to refer to Parian marble, which was mined by lamplight and not quarried (36. 14).

lyddite See above, s.v. 'lapis Lydius'.

magnetite Magnetic iron ore[139] (Fe_3O_4), specific gravity 5. 18, is a common ore-mineral found disseminated as an accessory through most igneous rocks. In certain types, because of magmatic segregation, it becomes one of the chief constituents and may thus form large ore bodies. Magnetite is most commonly associated with crystalline metamorphic rocks and, frequently, with rocks which are rich in ferromagnesian minerals such as diorite, gabbro, and peridotite. It is also found in immense beds and lenses, enclosed in old metamorphic rocks, and, similarly, in the black sands of the seashore. Magnetite is often intimately associated with corundum, forming emery. The properties of magnetite were well known throughout the ancient world, as

---

[136] This stone was the red garnet (cf. 37. 92).
[137] Indian *lychnis* may have included rubies.
[138] These are violet-red and rose-red tourmaline respectively. (There is also a variety known as lyncurium, which, from Pliny's description (37. 53), appears to be brown tourmaline.
[139] Healy, *Mining*, 40. On the magnetic properties of this mineral, see further Ch. 12. 3 above.

numerous literary references attest.[140] Pliny (36. 127) writes, 'The magnet is called by the Greeks by another name, "iron-stone" [σιδηρῖτας] and by some of them the "stone of Heracles". According to Nicander, it derives its name from Magnes, its discoverer, who found it on Mt. Ida.' The name, however, probably refers to the locality of Magnesia ad sipylum (north of Izmir).[141]

malachite See above, s.v. 'chrysocolla'.

marble See above, s.v. 'calcite'.

melinum See Ch. 15 s.v.

mercuric sulphide See above, s.v. 'cinnabar'.

mica See above s.v. 'lapis specularis'.

minium See above, s.v. 'cinnabar'.

myrrha (fluorspar) One of the most colourful minerals that Pliny describes (33. 5) is *myrrha*,[142] although he never uses this term[143] in the *Natural History*. 'Out of this same earth we dug "myrrhine" [fluorspar] and rock-crystal which owe their value to their very brittleness' ('murrina ex eadem tellure et crystallina effodimus, quibus pretium faceret ipsa fragilitas').

Myrrhine ware was brought to Rome after Pompey's defeat of Mithridates and his settlement of the East. (37. 18) 'He

---

[140] For example, Pl. *Ion* 535d–e, and Ti. 80c; Theophr. *Lap.* 29; Lucr. *DRN* 6. 908; Pliny, as here, 4. 147, and elsewhere.

[141] *NH* 36. 128, where Sotacus is recorded as describing another variety of magnetite.

[142] *Myrrha* is employed in two senses: (1) a resinous gum from an Arabian tree: cf. Pliny 12. 66, 'murra in iisdem silvis permixta arbore nasci tradidere aliqui'—this resin was used to flavour wine and also as an unguent in the ancient world; (2) the mineral fluorspar: cf. Mart. *Epigr.* 10. 80. 1, 'maculosae pocula murrae'.

[143] Pliny employs the adjective *myrrhina* (sc. *vasa*) when referring to the mineral from which they are made, namely fluorspar. So Juv. *Sat.* 6. 156 and elsewhere.

was... the first to dedicate fluorspar bowls and cups from his triumph to Capitoline Jupiter. Such vessels immediately passed into everyday use and even display stands and tableware were eagerly sought'. ('eadem victoria primum in urbem myrrhina invexit, primusque Pompeius capides et pocula ex eo triumpho Capitolino Iovi dicavit. quae protinus ad hominum usum transiere, abacis etiam escariisque vasis expetitis; et crescit in dies eius luxuria.')

Pliny (37. 21) follows the long-accepted theory about the formation of the mineral: 'The substance is thought to be liquid which is solidified underground by heat. In size the pieces are never larger than a small display stand, while in bulk they rarely equal the drinking-vessels we have discussed.' ('umorem sub terra putant calore densari. amplitudine numquam parvos excedunt abacos, crassitudine raro quanta dicta sunt potoria.')

The evidence of colour is significant for the identification of the mineral.[144] Pliny (37. 21) writes:

The [vessels] shine, but without intensity; indeed it would be truer to say that they glisten rather than shine. Their value lies in their varied colours: the veins as they revolve, repeatedly vary from purple to white or a mixture of the two, the purple becoming fiery, or the milk-white becoming red as though the new colour was passing through the vein. Some people particularly appreciate the edges of a piece, where colours may be reflected such as we observe in the inner part of a rainbow, Others prefer thick veins (any trace of transparency or fading is always a fault) and also specks and spots.... The smell of the substance is also its merit.

splendor est iis sine viribus nitorque verius quam splendor. sed in pretio varietas colorum subinde circumagentibus se maculis in purpuram candoremque et tertium ex utroque, ignescente veluti per transitum coloris purpura aut rubescete lacteo. sunt qui maxime in iis laudent extremitates et quosdam colorum repercussus, quales in

---

[144] So Mart, *Epigr.* 10. 80. 1 (cited at above), n. 142 in the sense of 'variegated', or 'striped', rather than 'spotted'. Bailey, *Chemical Subjects*, i. 177, also observes, incorrectly, that agate, with its undulating, many coloured-bands of chalcedony, carnelian, jasper, amethyst and rock-crystal, corresponds very well with Pliny's description. In identifying the mineral, however, other, more significant factors must be taken into consideration, for example hardness values.

caelesti arcu spectantur imi. aliis maculae pingues placent—tralucere quicquam aut pallere vitium est—itemque sales verrucaeque . . . aliqua in odore commendatio est.

Pliny (36. 198) describes glass painted to look like myrrhine ware:[145] 'There is, furthermore, opaque white glass and others that reproduce the appearance of fluorspar, blue sapphires, or lapis lazuli, and indeed glass exists in any colour' ('fit et album et murrina aut hyacinthos sappirosque imitatum et omnibus aliis coloribus').

The reason for the variegated colours of fluorspar, as for example, in the case of 'Blue John',[146] as it is currently known, has been the subject of much research and speculation. A detailed study by Hazeldine, suggests that, in the case of Treak Cliff (Derbyshire), it arose from its special geological history whereby radioactivity from ores of uranium type, thousands of years ago, displaced some of the calcium atoms from their normal positions in the lattice of the colourless calcium fluoride crystals. The calcium atoms displaced in this way form colloidal groups of calcium atoms which aggregate preferentially in the defects or irregularities found in the lattice. The blue colour is formed by the scattering and by the absorption of light in the aggregates of colloidal atoms as light passes through the crystals of calcium fluoride.

The colour of fluorspar can be modified by heat. The technique of firing, as used today, is a process whereby a very dark specimen, such as is obtained in the so-called Bull Beef vein in the 'Blue John' mines in Derbyshire, is lightened in colour. The mineral is heated in a fire until the craftsman judges that the piece has the best colour: opaque blue-black fluorspar is turned to a translucent purple colour with dark veins. The success of the operation depends on the skill and experience of the craftsman. Too long an exposure to heat results in an opaque off-white colour being produced.

---

[145] On the colouration of glass, see Partington, *Chemistry*, 374.

[146] See generally A. E. Ollerenshaw, R. J. Harrison, and D. Harrison, *The History of Blue John Stone: Methods of Mining and Working, Ancient and Modern*² (Scarborough [n. d.]), and A. E. Ollerenshaw, *Blue John Cavern and Blue John Mine: Castleton via Sheffield* (Scarborough [n.d.]). The term is a corruption of the French *bleu jaune*.

The interpretation of the verse in Propertius (4. 5. 26), 'murreaque in Parthis pocula cocta focis', has long presented a problem to commentators. The rendering 'cups of porcelain baked in Parthian kilns' is in no way tenable. The phrase 'in focis' retains its basic meaning of 'on the hearth', that is 'in front of a fire'. In the light of the technical use of *coquere* the verse may, therefore, refer to the practice of modifying the colours of the fluorspar cups by heat, or, simply, to warming their contents. Wine was certainly enhanced by warmth as Martial[147] and other poets confirm.

The difficulty of mining and working fluorspar, because of its fragility, its mineral associations, and the comparative rarity of its sources—it was mainly found in the East, in the kingdom of Parthia—explains the high value of objects. Pliny (37. 21) adds that the finest specimens exist in Carmania: 'Oriens myrrhina mittit. inveniuntur ibi pluribus locis... maxime Parthici regni, praecipua tamen in Carmania.'

Myrrhine vases were collectors' items and extremely expensive. Pliny (37. 18) writes about an ex-consul who

drank from a cup for which he had paid 70,000 sesterces, although its capacity was only three pints. He was so enamoured of it that he used to chew the rim. Yet this damage increased its value and no item of myrrhine today bears a higher price-tag on it. The amount of money squandered by this same man upon other articles of this material in his possession can be gauged from their number, which was so great that, when Nero took them away from the man's children and displayed them, they filled the private theatre in Nero's garden across the Tiber, a theatre which was large enough to satisfy even Nero's desire to sing before a full house at the time when he was rehearsing for his appearance in Pompey's theatre.

a myrrhino L̄X̄X̄ HS empto, capaci plane ad sextarios tres calice, potavit... anus consularis, ob amorem adorso margine eius, ut tamen iniuria illa pretium augeret; neque est hodie myrrhini alterius praestantior indicatura. idem in reliquis generis eius quantum voraverit, licet aestimare ex multitudine, quae tanta fuit ut auferente liberis eius Nerone exposita occuparent theatrum peculiare trans Tiberim in hortis quod a populo impleri canente se, dum Pompeiano proludit, etiam Neroni satis erat.

---

[147] Cf. Mart. *Epigr.* 14. 113 and elsewhere.

The nameless consul who gnawed the rim of his priceless cup further confirms the identification of myrrhine. The brittle mineral, capable of being damaged by the teeth[148] (4+ on the Mohs scale of hardness) is clearly more likely to have been fluorspar (Mohs 4) than agate (Mohs 7).[149] This is the first recorded example of someone whose teeth enjoyed the benefits of fluoride!

Pliny (37. 20) also refers to two further outrageously expensive pieces, 'When the ex-consul Petronius was facing death, to spite Nero, and to deprive the emperor's dining-room table of this legacy, he broke a myrrhine ladle that had cost him 300,000 sesterces. Nero, however, as was proper for an emperor, outdid everyone by paying 1,000,000 sesterces for a single bowl.' ('T. Petronius consularis moriturus invidia Neronis, ut mensam eius exheredaret, trullam myrrhinam HS CCC emptam fregit; sed Nero, ut par erat principem, vicit omnes HS |X| capidem unam parando.')

Modern commentators, including Zehnacker[150] and Eichholz,[151] although generally accepting that the mineral of the vases which Pliny describes, is fluorspar, still mistakenly refer to the possibility that they may have been carved from agate,[152] as originally suggested by Bauer.[153] Some others believe

---

[148] I am indebted to J. M. Heathcote, BDS, for advice on this matter.
[149] Whitten–Brooks, *Geology*, 221 f.
[150] Zehnacker, *NH 33*, pp. 121 f., quite properly accepts the conclusion of A. I. Loewenthal and D. B. Harden, 'Vasa murrina', *Journal of Roman Studies*, 39 (1949), 31–7, that the mineral is fluorspar. He is incorrect, however, when he claims that Bailey was wrong in suggesting that heating can change the colours in fluorspar. E. de Saint-Denis (ed.), *Pline l'Ancien: Histoire Naturelle, Livre 37* (Budé: Paris, 1972), ad. loc. also accepts the identification.
[151] Eichholz, *NH 36–37*, p. 178, although he agrees that Pliny's description suits fluorspar, commenting on Dionysius, *Periplus Maris Erythraei* 49, (see next paragraph), suggests that the mineral is agate, which, like fluorspar, was also used for cups. Some agate, he observes was burnt, as is the case today, so as to modify its colours. He adds that Propertius, 4. 5. 26 ('murreaque in parthis pocula cocta focis', cited in text above), is probably confusing fluorspar from Persia with burnt agate from India.
[152] So P. M. Green, *Juvenal: The Sixteen Satires* (Harmondsworth, 1970), where *Sat.* 6. 155–6, 'grandia tolluntur crystallina maxima rursus | myrrhina', is rendered 'She goes on a shopping-spree: huge crystal vases, outsize myrrh-jars of finest *agate*'. He appears to follow J. D. Duff, *Juvenal* (Cambridge, 1909): 'Some good authorities hold that *murra* was a variety of agate containing shades of red and purple.'
[153] M. Bauer, Precious Stones, 2 vols. (New York, 1968), i. 530 f.

that Propertius (4. 5. 26) confuses these two very different minerals.

The identification of myrrhine is an excellent example of the collation of philological, archaeological, and scientific evidence which has resolved a number of problems and advanced our understanding of Pliny's text.

Although the evidence of Dionysius Periegetes (*Periplus Maris Erythraei* 49) is inconclusive when he joins ὀνυχίνη λιθεία καὶ μουρρίνη ('onyx and myrrhine')—which have both been identified with agate—the brittle property of myrrhine, with its low hardness value, its variegated colours, and the effect of heat on these, proves beyond any reasonable doubt that this mineral is fluorspar, or calcium fluoride (CaF).

A fine example of an artefact made from fluorspar, is the Crawford Vase[154] (of the first century BC or AD), found near the Turkish–Syrian border.

'Blue John' (fluorspar) has been exploited in Derbyshire, since the Roman period, in the vicinity of Castleton (near the Roman settlement of Anuvio). It is found in veins[155] of an average thickness of three inches (7.5 cm.), or in nodular form, lining the inner wall of cavities in the carboniferous limestone of the region. The nodules are spherical, fungus-like growths, composed of concentric bands of blue, purple, and white or yellow fluorite which radiate from a central point. Massive varieties seem to be composed of inter-penetrating cubic crystals. Fluorite was probably discovered by the Romans in their search for lead.

Present-day methods used in the preparation and working of Blue John are as follows. Mined fluorspar is scrubbed, dried out for about a year, and then rough sawn to the shape of the intended article. After this initial sawing, all the pieces of the mineral are subjected to carefully controlled heat, either in an oven, or on a hotplate, and American resin, or less commonly shellac, is applied. Another method, but more costly, is to

---

[154] Now in the British Museum. Cf. Sen. *Ep.* 119. 3: 'poculum murreum'.
[155] See further Ollerenshaw, Harrison, and Harrison, *History of Blue John*, 3 and 9. It is difficult to separate the fluorspar, which often occurs as a gangue mineral, or in veins completely surrounded by limestone, thus needing a much more complicated process of removal.

immerse the Blue John in a bath of molten resin. The resin melts and impregnates the outer layers of the crystals, thus binding the whole more securely together. In this stabilized state the fluorspar is ready to be worked, but once the impregnated layer is removed, the resining process must be repeated.[156] Even then, sawing must be carried out with great care, because the heat generated by friction would be sufficient to crack and break the mineral.

Evidence from the smell of the wine is, at best, inconclusive. Pliny (*NH* 14. 92) observes that 'the finest wines in early days were those spiced with scent of myrrh, as appears in the plays of Plautus, although, in *The Persian*, he recommends the addition of sweet-reed [calamus] also' ('lautissima apud priscos vina erant murrae odore condita, ut apparet in Plauti fabulis, quamquam in ea quae *Persa* inscribitur et calamum addi iubet'). Pliny continues, with a quotation from Fabius Dossennus: '"I sent them a fine wine, one flavoured with myrrh" ("mittebam vinum pulchrum, murrinam"), and, in his *Acharistio*, "bread and pearl-barley and wine spiced with myrrh" ("panem et polentam, vinum murrinam"). I also observe that Scaevola, Lucius Aelius, and Ateius Capito thought the same: so we find in [Plaut.] *Pseudolus* [740–1]: "But if he has to bring out a sweet wine from that same cellar, does he have one?" "Have one? [I should say so.] Myrrh-wine, raisin-wine and boiled-down must and honey..." ("quod si opus est ut dulce promat indidem, ecquid habet?" "rogas? murrinam, passum, defrutum, mella..."), which shows that myrrh-wine was counted not only among wines but among sweet wines.'

Martial's observation (*Epigr.* 14. 113) may support the view that the wine was flavoured by contact with the cup: 'If you drink your wine warm, myrrhine suits the burning Falernian and better flavour comes therefrom to the wine' ('si caldum potas ardenti murra Falerno | convenit et melior fit sapor inde mero').

---

[156] The crucial importance of the resining process suggests that the Romans too must have had some means of neutralizing the friability of this mineral. However, no conclusive laboratory tests to confirm the impregnation of ancient fluorspar with resin have, to my knowledge, been carried out.

In conclusion, although this evidence is circumstantial, modern practice in the working of Blue John[157] does appear to suggest the use of a stabilizing agent also in the ancient world, to counteract the natural brittleness of the fluorspar. The literary evidence, however, mainly, but by no means exclusively, supports the view that the distinctive flavour of the wine was achieved by the use of *myrrha* as an additive. One is, of course, reminded of Greek retsina.

onyx marble See below s.v. 'phengites'.

orpiment Orpiment, or yellow sulphide of arsenic (As_2S_2), is found in the oxidized zones of arsenic minerals, in veins, and in hot-spring deposits.[158] It is lemon-yellow in colour and of lamellar structure.[159]

Pliny in describing the mineral uses the Latin name *auripigmentum* (33. 79) and the Greek term *arrhenicum* (34. 178).[160] There is also an artificially produced kind of yellow ochre known as *ochra* (35. 30).

Pliny erroneously believes that orpiment[161] is a source of gold as well as a pigment. But when he writes, 'it is found on the surface of the earth and is of a gold colour, easily broken up like selenite',[162] his description is accurate in all respects.

He continues (33. 79): 'The emperor Caligula,[163] who was extremely covetous for gold, gave orders for a great weight of orpiment to be smelted. As a matter of fact, it did produce excellent gold, but so small a weight that he found himself a loser by his avarice, although orpiment was selling for 4 denarii a pound. No one subsequently repeated the experiment.'

---

[157] I am indebted to the guides at the Blue John Mine for their explanation of modern working practices, which help to throw light on the problems connected with this mineral.
[158] Whitten–Brooks, *Geology*, 325 and appendix.
[159] Zehnacker, *NH 33*, ad loc. (33. 79. 3).
[160] ἀρρενικόν (ἀρσενικόν) Arist Pr. 966^{b}28; Theophr. *Lap.* 40; Dioscorides 5. 104; Strabo 15. 2. 14. This term includes realgar, cf. Celsus 5. 5 and Vitr. *De arch.* 7. 7. 5. See further Dioscorides 5. 104–5, and H. Le Bonniec (ed. and trans.) and Gallet de Santerre (comm.), *Pline l'Ancien: Histoire Naturelle, Livre 34* (Paris, 1953), ad. loc.
[161] See Ch. 15 below, s.v.   [162] *NH* 36. 160 ff.
[163] Caligula's main interest in orpiment was as a poison, cf. F. d'Erce, 'La Mort de Germanicus et les poisons de Caligula', *Janus*, 56 (1969), 123–48.

('invitaveratque spes Gaium principem avidissimum auri, quam ob rem iussit excoqui magnum pondus et plane fecit aurum excellens, sed ita parvi ponderis, ut detrimentum sentiret propter avaritiam expertus, quamquam auripigmenti librae X IIII permutarentur. nec postea temptatum ab ullo est.')

Pliny was clearly deceived by this story since there is, of course, no way in which gold can be produced from arsenic sulphide. Zehnacker draws attention to Bailey's[164] imaginary scenario in which a quantity of orpiment might have been seeded by an unscrupulous merchant to make it appear like auriferous earth. This is less likely than his own suggested explanation:[165] 'Il est beaucoup plus simple de penser que l'orpiment en question avait été extrait d'une terre aurifère à laquelle il restait mêlé.' There does not seem to be any evidence for this association, although Pliny elsewhere (34. 177) states that realgar[166] is found in gold and silver mines.

paraetonium See Ch. 15 below, s.v.

phengites Pliny (36. 163) describes this unusual stone: 'In the principate of Nero there was discovered, in Cappadocia, a stone as hard as marble, white, and, even where deep yellow veins occurred, translucent. Because of its appearance, it was called "phengites"[167] [or the "luminary stone"].' ('Nerone

---

[164] *Chemical Subjects*, i. 202 (note to 33. 79).

[165] Although at tourist attractions today, where visitors are able to pan for gold, Nature is, more often than not, assisted by the seeding of the 'placer' with grains of gold.   [166] See Ch. 15 below, s.v. 'sandaraca'.

[167] From the Greek φεγγίτης which is equated with σεληνίτης (Dioscorides 5. 141). In LSJ, it is incorrectly translated as foliated sulphate of lime. The stone was known as 'moonstone' by the Greeks because it was thought to 'wax' and 'wane' with the moon. See Bailey, *Chemical Subjects*. ii. 268 (note to 36. 163). The only other reference is in Suetonius, *Domitian* 14, where the emperor is terrified: 'porticuum in quibus spatiari consuerat, parietes phengite lapide distixit, e cuius splendore per imagines quidquid a tergo fieret, provideret. quoddam est candidum sed ei fulvae incidunt venae, quod quia ubique translucet, ex argumento lapis phengites vocatur.' ('Domitian decorated the walls of the colonnades, where he used to walk, with a "shining stone" [this may have been selenite, or a form of onyx marble] so that from the reflection which resulted from its brilliance, he could see what was happening behind him. Some stone is white, but striated with yellow veins running through it, and because it shines everywhere, it is called the "shining stone" in token of its appearance.') This passage of Suetonius was clearly known to Pliny, as the nature of the stone and the wording of the reference to the derivation of the name both confirm. The 16th-cent. writer Georgius Agricola, *De Natura Fossilium* (Basel, 1546), vii. 307, considers *phengites* to be a marble.

principe in Cappadocia repertus est lapis duritia marmoris, candidus atque translucens etiam qua parte fulvae inciderant venae, ex argumento phengytis appellatus.') He continues (ibid.): 'Nero rebuilt the temple of Fortune, known as the shrine of Sejanus,[168] of this stone.... this was incorporated by Nero in his Golden House.[169] Thanks to this stone, in the daytime it was as light as day there even when the doors were shut; but the effect was not that of windows of selenite,[170] since the light was, so to speak, trapped within rather than allowed to penetrate from without. According to Juba there exists in Arabia too a stone that is transparent, like glass, and is used as window panes.' ('hoc construxit aedem Fortunae quam Seianio appellant.... amplexus aurea domo; quare etiam foribus opertis interdiu claritas ibi diurna erat alio quam specularium modo tamquam inclusa luce, non transmissa. in Arabia quoque esse lapidem vitri modo tralucidum, quo utantur pro specularibus, Iuba auctor est.')

The stone mentioned by Juba may well be a form of selenite, but Pliny's description of *phengites* undoubtedly fits that of marble rather than selenite. Elsewhere (36. 59) onyx marble is called by the generic term *alabastrites*.[171]

porus πῶρος, πώρινος πόρος (cf. Theophr. *Lap.* 7) is basically a stone used for building. Pliny (36. 53 and 36. 132) describes *porus* as 'like Parian marble in colour and solidity, but lighter'. It is also used of a local (ἐπιχώριος) conglomerate, as found at Olympia (Pausanias 5. 10. 2). The term is generally applied to any form of soft, brown limestone, or calcareous tufa, the material of archaic sculptures, for example early pediments (Hdt. 5. 62).

psimithium See Ch. 15 below, s.v.

---

[168] Sejanus, prefect of the Praetorian Guard, under Tiberius, was executed in AD 31.
[169] R. B. Bandinelli, *Rome the Centre of Power: Roman Art to AD 200* (London, 1970), 130 ff. The Golden House (Domus Aurea) covered a vast area of land between the Caelian and Esquiline hills in Rome.
[170] See above, s.v. 'lapis specularis'.
[171] Cf. also 36. 60 and 158, 37. 73 Cf. and 144. Onyx marble was generally used for unguent containers and, of course, in Egypt, for Canopic jars.

pumex Pyroclastic rocks consist of fragmental volcanic material that has been blown into the atmosphere by explosive activity. They are usually produced from volcanoes whose lava is of a more viscous type, and are of a number of different types.[172]

Pumice is a highly vesicular material derived from acidic lavas and produced in large quantities. The vesicles are produced by bubbles of gas trapped during the solidification of lava. The term is also used by ancient authors in the more general sense of a type of porous rock.[173]

Pliny (36. 154f.) states that 'the finest quality occurs in Melos, Nisyros, and the Aeolian Islands. The test of its quality is that it should be white, very light in weight, extremely porous, dry, and easy to grind without being sandy, when rubbed.' ('laudatissimi [pumices] sunt in Melo, Nisyro, et Aeoliis insulis. probatio in candore minimoque pondere et ut quam maxime spongiosi aridique sunt teri faciles nec harenosi in fricando.') His description of pumice is accurate, except in respect of the colour which is usually light grey, rather than white.

Pliny 36. 154 makes a brief reference to some uses of pumice. 'This name is given to the hollowed rocks in the buildings called by the Greeks "Homes of the Muses", where such rocks hang from the ceilings so as to create an artificial imitation of a cave' ('appellantur quidem ita erosa saxa in aedificiis, quae musaea vocant, dependentia ad imaginem specus arte reddendam'). 'Pumice is used as a depilatory for women and, nowadays, also for men' ('pumices...in usu corporum levandorum feminis, iam quidem et viris'). Catullus[174] (22. 8) reminds us that pumice was employed for smoothing the surface of papyrus (for books), 'pumice omnia aequata', and (1. 1–2) the ends (*frontes*) of the roll (*volumen*).

Pumice, when calcined, quenched in white wine, and ground to a powder, is used in a variety of medicinal compounds (*NH* 36. 155ff.), but mainly for eye-salves

---

[172] Whitten–Brooks, *Geology*, 368f.
[173] Cf. Ov. *Met.* 3. 159, 8. 561; *Fast.* 2. 315; Verg. G. 4. 44 and *Aen.* 5. 214, and elsewhere.
[174] See further C. J. Fordyce, *Catullus* (Oxford, 1968), 148f. and 84ff.

('oculorum medicamentis'). Tooth powder[175] (*dentifricia*) is compounded from pumice (36. 156), and pumice can be added to new wine, because its cooling properties stop it bubbling (ibid.):[176] 'tantamque refrigerandi naturam esse ut musta fervere desinant pumice addito.'

Pliny elsewhere (36. 53) refers to its use, like sand from Thebes and powder produced from limestone, for polishing marble: 'Thebaica polituris accommodatur et quae fit e poro lapide aut e pumice.'

In the second half of the nineteenth century pumice from Santorini[177] in the Cyclades was imported by Egypt for use in consolidating, or stabilizing, the channel of the Suez Canal. Pumice is still mined on the island—the only major exporter of this commodity in the region; ship-loading chutes can be seen projecting into the sea, not far from Skala Fira.

pumice See above, s.v. 'pumex'.

quartz See above, s.v. 'crystallum'.

realgar See Ch. 15 below, s.v. 'sandaraca'.

sarcophagus stone Pliny (36. 131) refers to 'the "sarcophagus" stone, found at Assos in the Troad, which splits along a

---

[175] Elsewhere, Pliny (28. 178) states that tooth powder is made from the ash of a deer's horns, compounded with myrrh and, similarly, from the ash of burnt eggshells (29. 46).

[176] Pliny (36. 156) adds a curious piece of information—allegedly quoting Theophrastus—namely that 'drinkers competing in drinking contests first take a dose of powdered pumice'; but adds that 'they run a grave risk unless they fill themselves with wine at a single draught'. ('Theophrastus auctor est potores in certamine bibendi praesumere farinam eam [pumicem], sed, nisi universo potu inpleantur, periclitari.')

[177] Santorini (Thera), originally a volcano of marble and metamorphic schist, with its main crater (caldera) forming what is now the roadstead for ships, has, from earliest times, changed as a result of eruptions. The largest, *c*.1500 BC, measured 6 on the Volcanic Explosivity Index. This was the most violent eruption in the Mediterranean region within 10,000 years. It left the island covered with a layer of pumice and volcanic ash up to 140 feet thick. See further M. T. Greene, *Natural Knowledge in Preclassical Antiquity*, (Baltimore, 1992) 58, and T. Simkin *et al.*, *Volcanoes of the World: A Regional Directory, Gazetteer and Chronology of Volcanism during the Last 10,000 Years* (Stroudsburg, Pa. 1981), pp. vii, 1–3.

line of cleavage[178] ['fissili...scinditur']. It is established that corpses buried in it are, with the exception of the teeth,[179] consumed within a period of forty days. Mucianus[180] is the authority who adds also that mirrors, strigils, clothes, and shoes buried with the corpses are also petrified. There are rocks of the same type in Lydia and in the East, which, when attached even to living persons, eat away their bodies.' ('in Asso Troadis sarcophagus lapis fissili vena scinditur. corpora defunctorum condita in eo absumi constat intra XL diem exceptis dentibus. Mucianus specula quoque et strigiles et vestes et calciamenta inlata mortuis lapidea fieri auctor est. eiusdem generis et in Lydia saxa sunt et in oriente, quae viventibus quoque adalligata erodunt corpora.')

No stone, however, could consume a body in so short a time, or petrify objects under these conditions. Bailey[181] made the suggestion that this was probably a fissile limestone. Lime may also have been thrown into the sarcophagus to accelerate decomposition. Water seeping into it may have deposited calcium carbonate on the objects, or the lime may have become slaked with moisture from the decomposing body and have been deposited on the objects, gradually forming a hard crust.[182] The evidence provided does not allow any certain identification of this mineral.

---

[178] See Ch. 13.3.2.

[179] Pliny (7. 69 f.) also incorrectly alleges that teeth are not destroyed by the fire when bodies are cremated: 'dentes autem in tantum invicti sunt ignibus ut nec cremantur cum reliquo corpore.'

[180] Governor of Syria in AD 69, who compiled a volume of *Mirabilia*.

[181] *Chemical Subjects*, ii. 251 f. Bailey favours the identification of this mineral with limestone. He writes: 'Theophrastus, *De Igne* 46, refers to coffins made of this stone which reduced the body to ashes. He says that it acted by virtue of its intrinsic heat, a proof of which is τὸ γίνεσθαι κονίαν ἐξ αὐτοῦ. What κονία here means is uncertain, but, as the word indicates an alkaline powder in Aristophanes, *Lys.* 470, where it was employed as a soap, and is used of lime-plaster in Eustathius, 382, 36, it seems likely that it here means "lime". That Theophrastus adduces the production of lime as a proof of the intrinsic heat of the stone is probably due to the observation of the evolution of heat when lime is wetted. This would make sarcophagus lapis a limestone.' In the course of time, water soaking through a limestone coffin might deposit sufficient calcium carbonate on objects within to cause petrifaction. Cf. stalactites and stalagmites. There is, of course the further possibility that lime, quicklime, or other corrosive chemicals were thrown into the sarcophagus to accelerate decomposition of the body.

[182] Stanley Smith, in Eichholz, *NH 36–37*, p. 105, note *e*.

schistos See above, s.vv. 'alumen' and 'haematite'.

selenite See above, s.v. 'lapis specularis'.

smaragdus Pliny (37. 76) correctly implies that emeralds[183] and beryls[184] are similar, but does not understand that their hexagonal crystals are naturally formed.[185]

Pliny (37. 62 ff.) describes the emerald which he rates as third in value of the precious stones. (For some reason he puts pearls in second place.)

No colour is more delightful in appearance. For, although we enjoy looking at plants and leaves, we regard emeralds with all the more pleasure because, compared with them, there is absolutely nothing that is more intensely green. Moreover, emeralds, alone of precious stones, satisfy our gaze without causing a surfeit of colour. And, after straining our eyes by looking at other objects, we can restore our vision to normal by gazing at an emerald. Lapidaries find this the most agreeable means of refreshing their eyes. So soothing is the mellow green of the gem for tired eyes. Apart from this property, emeralds appear larger when they are viewed at a distance because they reflect their colour upon the air around them.[186]

tertia auctoritas smaragdis perhibetur pluribus de causis, quippe nullius coloris aspectus iucundior est. nam herbas quoque silentes frondesque avide spectamus, smaragdos vero tanto libentius, quoniam nihil omnino viridius comparatum illis viret. praeterea soli gemmarum contuitu inplent oculos nec satiant. quin et ab intentione alia aspectu smaragdi recreatur acies, scalpentibusque gemmas non alia gratior oculorum refectio est: ita viridi lenitate lassitudinem mulcent. praeterea longinquo amplificantur visu inficientes circa se repercussum aera.

Later (37. 64), Pliny continues: 'Emeralds are generally concave in shape, so that they concentrate one's vision. Because of

---

[183] Healy, 'Pliny and Ancient Mineralogy', 177.
[184] Ibid. 174, figs. 10 and 11. [185] See above, s.v. 'beryllus'.
[186] cf. *NH 33*. 128, where the reflection of images is ascribed to the air 'rebounding' from a mirror. Here the reflection of colour is assumed to be caused in the same way. These are true emeralds. Few precious stones are harder than emeralds and so engraving, in any case, would have been difficult.

these properties, mankind has decreed that emeralds must be preserved in their natural state and has forbidden them to be engraved. In any case, those of Scythia and Egypt are so hard as to be unaffected by blows. When emeralds that are tabular in shape are laid flat, they reflect objects just like mirrors.[187] The emperor Nero used to watch fights between gladiators which were reflected by the surface of an emerald.' ('idem plerumque concavi, ut visum conligant. quam ob rem decreto hominum iis parcitur scalpi vetitis quamquam Scythicorum Aegyptiorumque duritia tanta est ut non queant volnerari. quorum vero corpus extentum est, eadem qua specula ratione supini rerum imagines reddunt. Nero princeps gladiatorum pugnas spectabat in smaragdo.'

It is improbable that Pliny means that a concave emerald could be used as a lens;[188] a green gemstone would hardly serve this purpose. Although magnifying mirrors were known in the ancient world, there is no certain evidence that anything was known of magnifying lenses, although engravers must have had some means of magnification when making coin dies, or engraving gems. His observations about the soothing effects of a green gemstone are a reasonable assumption. Nero's use of a reflecting emerald should, perhaps, be understood in such a context; this would have neutralized the glare!

Pliny (37. 65 ff.) records twelve kinds of emerald.

The most notable is the Scythian, so called from the nation in whose territory it is found.[189] No kind is deeper in colour, or more free from defects: it differs as widely in quality from other emeralds as they do from other gems. Next to this in esteem, as in locality, is the Bactrian. These stones are said to be gathered by the natives in the fissures of rocks when the Etesian winds[190] blow. For at this season the ground is uncovered and the stones glitter here and there because the sands of the desert are shifted violently by these winds. These stones,

---

[187] See Ch. 12.1.1. above.
[188] Eichholz, *NH 36–37*, pp. 212 f., note *c*.
[189] Ibid., p. 214, note *c*: it is suggested that these were from workings near the river Takovaya, in the Urals, about 50 miles (80 km.) east of Sverdlovsk. From there the stones could have reached the river Don and the Black Sea.
[190] These are the northerly winds which blow over the Aegean in the late summer. Theophrastus (*Lap.* 35) uses the term of similar seasonal winds blowing over the deserts of Persia (or, as he thought, of Bactria).

however, are said to be much smaller than the Scythian. Third in order come those of Egypt, which are dug near Coptos a city of the Thebaid, from mines in the hills.[191] The other kinds are found in copper-mines and so the first place among them is held by stones of Cyprus. Their special asset is their colour, which is limpid without being at all faint. On the contrary it combines body and clarity and wherever one peers through the stones, reproduces the transparency of sea water, the stones being in equal degree translucent and brilliant. In other words, they dissipate their colour and also allow the sight to penetrate within.

nobilissimi Scythici, ab ea gente in qua reperiuntur appellati. nullis maior austeritas nec minus vitii; quantum smaragdi a gemmis distant, tantum Scythicus a ceteris smaragdis. proximam laudem habent, sicut et sedem, Bactriani, in commissuris[192] saxorum colligere eos dicuntur etesiis flantibus; tunc enim tellure deoperta internitent, quia iis ventis harenae maxime moventur. sed hos minores multo Scythicis essse tradunt. tertium locum Aegyptii habent, eruuntur circa Copton, oppidum Thebaidis, collibus excavatis, reliqua genera in metallis aerariis inveniuntur, quapropter principatum ex iis optinent Cyprii. dos eorum est in colore liquido nec diluto, verum ex umido pingui quaque perspicitur imitante tralucidum maris, pariterque ut traluceat et niteat, hoc est colorem expellat, aciem recipiat.

Pliny, however (37. 70 ff.), includes other varieties of green stones under the heading *smaragdi* which are not emeralds—for example, Attic stones found in the silver-mines at Thoricus.[193] These are often marred by leadlike inclusions ('frequens est in his et plumbago') and are probably calamine (zinc carbonate). Median stones that show a great variety of tints and, on occasion, are even blended to some extent with

---

[191] These are the same as Ethiopian emeralds, the provenance of which Pliny describes more accurately in 36. 69. He quotes Juba as his source. The mines from which the emeralds are obtained were those of Gebel Zubara and Gebel Sikait, lying close to the caravan route which led from Coptos to the Red Sea port of Berenice, on the Ethiopian border. The district is about 180 miles from Coptos.

[192] Eichholz, *NH 36–37*, p. 215, note *e*. This passage ('commissuris... moventur') is a garbled version of Theophrastus (*Lap.* 35), 'in commissuris saxorum... being a mistranslation of εἰς τὰ λιθοκόλλητα χρῶνται. Pliny, or his source, gratuitously assumed that Theophrastus was referring to *smaragdi*, whereas he was alluding to one of the blue stones used by the Persians in inlay-work, probably the blue turquoise. The Bactrian *smaragdus* is a product of Pliny's imagination. [193] Conophagos, *Laurium*, 160 f.

lapis lazuli, are malachite ('Medici plurimum habent varietatis, interdum aliquid et e sappiro')—that is, malachite (green copper carbonate) intergrown with azurite (blue copper carbonate) which resembles lapis lazuli in colour.

Pliny continues (37. 73): 'There is a mountain known as Smaragdites [or Emerald Mountain] near Chalcedon, on which emeralds used to be gathered.[194] Juba states that a *smaragdus*, known as "chlora", or green stone, is used as an inlay in decorating houses in Arabia[195] ... Several of our most recent authorities mention not only Laconian *smaragdi*[196] which are dug out of Mt. Taygetus and resemble the Median variety, but also others that are found in Sicily.'[197] ('mons est iuxta Calchedonem, in quo legebantur, Smaragdites vocatus. Iuba auctor est smaragdum quam chloram vocent in Arabia aedificiorum ornamentis includi. ... complures vero e proximis et Laconicos in Taygeto monte erui Medicis similes et alios in Sicilia').

Pliny (37. 74 f.) goes on to cite further authorities on *smaragdi*:

Theophrastus [*Lap*. 24], writes that in Egyptian records are to be found statements to the effect that to one of the kings, a king of Babylon once sent as a gift a *smaragdus* measuring 6 feet in length and 3 feet in breadth;[198] and that there existed in Egypt, in a temple of Jupiter, an obelisk made of four *smaragdi* and measuring 60 feet in height and 6 feet in breadth at one extremity and 3 feet at the other.[199] He states, moreover, that at the time when he was writing there existed in the temple of Hercules, at Tyre, a large square pillar of *smaragdus*[200] unless this was to be regarded as a 'false *smaragdus*'; for according to him there is another variety that is found. He mentions that there was once discovered in Cyprus a stone of which half

---

[194] These were malachite crystals from the copper-mines of Demonesus (Chalke), one of the Prinkipo Islands, in the Sea of Marmara, not far from Chalcedon. Cf. Theophr. *Lap*. 25–6.
[195] The inlay possibly consists of panels of green porphyry.
[196] Not emeralds, but small pieces of green porphyry.
[197] This stone may be identified as green jasper.
[198] Possibly a block of malachite.
[199] Theophrastus' text is corrupt. Obelisks are *monoliths* and are usually in pairs, so that it is virtually certain that Theophrastus is describing not *one*, but *four* obelisks, each made of a single *smaragdus*, which may have been green basalt, or green schist. The temple was that of Amon, at Thebes.
[200] Cf. Hdt. 2. 44. This was possibly malachite, or green jasper, but Theophrastus may have supposed it to have been green porphyry. The temple belonged to Melkart.

was a *smaragdus* and half a *iaspis*,[201] because the liquid matter had not yet been completely transformed.[202] Apion,[203] surnamed Plistonices [or 'the Cantankerous'], has lately left on record the statement that there still exists in the Egyptian Labyrinth[204] a large statue of Serapis, thirteen and a half feet high, and made of *smaragdus*.

Theophrastus tradit in Aegyptiorum commentariis reperiri regi eorum a rege Babylonio muneri missum smaragdum quattuor cubitorum longitudine ac trium latitudine, et fuisse apud eos in Iovis delubro obeliscum e quattuor smaragdis quadraginta cubitorum longitudine, latitudine vero in parte quattuor, in parte duorum, se autem scribente esse in Tyro Herculis templo stelen amplam e smaragdo, nisi potius pseudosmaragdus sit, nam et hoc genus reperiri, et in Cypro inventum ex dimidia parte smaragdum, ex dimidia iaspidem, nondum umore in totum transfigurato. Apion cognominatus Plistonices paulo ante scriptum reliquit esse etiam nunc in labyrintho Aegypti colosseum Serapim e smaragdo novem cubitorum.

smithsonite See above, s.v. 'cadmea'.

stibnite Stibnite was the most important ore-mineral of antimony (Sb_2S_3), variously known to the ancient world as στίβι, στίμμι, *stimmi*, or *stibium*. It is found in low temperature hydrothermal veins.[205]

Many European countries, including Britain,[206] have deposits, mainly in association with other minerals. Dionysius Periegetes mentions a stibnite trade from East Africa to India in the first century AD.[207]

For Pliny and his contemporaries the mineral posed problems of differentiation and there is no evidence to suggest

---

[201] Jasper is a green-coloured stone, an opaque, impure form of quartz. Other varieties are red, yellow, or brown. So Mart. *Epigr.* 5. 11. 1, 9. 60. 20, and Verg. *Aen.* 4. 261.

[202] Eichholz, *NH 36–37*, p. 224, note *b*. Theophrastus seems to have thought that gemstones were formed of liquid matter impregnated with earthy particles held in solution. As for the stone, the uncertainties of ancient nomenclature make identification impossible. It may, for example, have been a malachite crystal joined to a piece of ordinary malachite, or a green jasper joined to a dull grey jasper, or a plasma joined to a dull chalcedony.

[203] A scholar of Alexandria, who later lived in Rome under Tiberius and Claudius. He wrote on Egypt. (See Josephus, *Against Apion*.)

[204] See further *NH* 36. 84.

[205] Whitten–Brooks, *Geology*, 427 and appendix.   [206] Healy, *Mining*, 67.

[207] Cf. Forbes, *Ancient Technology*, ix. 165.

that the Greeks, or Romans refined stibnite to produce pure antimony.

From Pliny's confused description (33. 103 ff.), the 'male' stibnite would appear to be antimony sulphide (SbS) and the 'female', metallic antimony (Sb). Dioscorides (5. 84) considers the bright variety of *stimmi*, as the best: he does not, however, use the terms 'male' and 'female'.[208]

sulphur The physical characteristics of sulphur[209] are as follows: its hardness value is 2.0, specific gravity 2.07, and melting-point 113 °C.

Sulphur and sulphurous fumes[210] are universally associated with thunder and thunderbolts in Greek[211] and Roman[212] sources. Lightning is occasionally referred to as *sulpur aetherium*[213] because of the acrid fumes which accompany it. Similarly, the hostile environment of Lake Avernus,[214] the entrance to Hades, avoided by birds, is mainly the result of the emission of sulphurous gases.[215]

The non-metallic, chemical element sulphur is widely distributed in nature both in its free state and in combination. Various forms of native (*vivum*) sulphur occur in areas of volcanic activity, around hot springs, and in sedimentary rocks together with gypsum and limestone. Sulphur deposits are clearly visible in the Aeolian or Lipari Islands (3. 93 f.), an archipelago off the north-east coast of Sicily, originally named after Aeolus, god of the winds, whose kingdom was thought to have been in this region.[216] The island of Vulcano is still

---

[208] On the processing of antimony sulphide, see further below, Ch. 17. 11.
[209] Whitten–Brooks, *Geology*, 434 and appendix (Sulphur).
[210] Cf. Cato, *RR* 39. 1; Columella 8. 5. 11; Pers. *Sat.* 2. 25. 3.
[211] Hom. *Il.*14, 415; *Od.* 12. 417, and elsewhere.
[212] Lucr. *DRN* 6. 219–21: 'quod superest ⟨quali⟩ natura praedita constent | fulmina, declarant ictus et inusta vaporis | signa notaeque gravis halantis sulpuris auras.' ('what, then, of the nature and composition of thunderbolts? The places where they have struck, the scorch marks and traces of sulphur giving off noxious fumes, provide clear evidence.')
[213] Lucan, *BC* 7. 160.   [214] Verg. *Aen.* 6. 236 ff.
[215] Lucr. *DRN* 6. 747 f. 'is locus est Cumas apud, acri sulpure montes | oppleti calidis ubi fumant fontibus aucti.' ('There is such a spot, near Cumae, where hills give off acrid fumes of sulphur.')
[216] Verg. *Aen.* 1. 52 ff.

a smoking mass of lava: steam and gases issue through the fumaroles or vents. In the vicinity of volcanoes there is often a strong smell of sulphur. According to Pliny the most famous source of sulphur is the island of Melos.

Volcanic sulphur usually occurs as a sublimate on the walls of vents, probably as a result of action between hydrogen sulphide and sulphur dioxide. Free sulphur may also result from the weathering of pyrites.

Pliny also mentions sulphurous gases encountered when digging wells (31. 49) and Seneca (*QN* 3. 15. 5) refers to a vein of sulphur. Inexplicably, sulphur is not included by Theophrastus in his *De lapidibus* and Pliny is the first to attempt to give any detailed account of this element (35. 174): 'Among other kinds of minerals, the one with the most remarkable properties is sulphur which exercises a great power over many other substances' ('mira natura est sulpuris, quo plurima domantur'). Sulphur is found in the Aeolian Islands between Sicily and Italy, round Naples, and in Campania (18. 114), on the hills called Leucogaei.[217] It is dug out of mine galleries and purified by fire.

Pliny is clearly unaware of the allotropic forms of sulphur[218] and restricts his description to native sulphur which is related to rhombic. There is no mention of monoclinic sulphur which does not exist in nature. Pliny characterizes the use of sulphur (35. 175).

There are four kinds [of sulphur]: the first is free sulphur—the Greek name for which is ἄπυρον [that is 'untouched by fire']: this alone forms a solid mass. All other kinds are liquid and treated by boiling in oil. Free sulphur is obtained by mining and is of a green, translucent colour: this is the only kind used in medicine. The second kind is called 'clod' sulphur [*glaeba*], generally only found in fullers' shops. The third kind, called *egula*, is employed only for bleaching[219] woollens that are held

---

[217] Chalk is found in these hills—hence the name—as well as sulphur. Pliny (31. 12) also mentions fountains of the same name between Puteoli and Naples.   [218] Partington, *Chemistry*, 689–90.
[219] K. W. Whitten, K. D. Gailey, and R. E. Davis, *General Chemistry with Qualitative Analysis* (Philadelphia, 1992), 890. Gaseous sulphur dioxide and aqueous solutions of sulphuric acid and sulphur trioxide ions are used as bleaching agents. Cf. Celsus 4. 5, 'lana sulphurata'.

over it—this imparts whiteness and softness; a fourth kind [not specifically named by Pliny] is employed for making lamp-wicks.

genera IIII: vivum, quod Graeci apyron vocant, nascitur solidum solum—cetera enim liquore constant et conficiuntur oleo incocta—vivum effoditur tralucetque et viret. solo ex omnibus generibus medici utuntur. alterum genus appellant glaebam, fullonum tantum officinis familiare. tertio quoque generi unus tantum est usus ad lanas suffiendas, quoniam candorem mollitiamque confert, egula vocatur hoc genus, quartum autem ad ellychnia maxime conficienda.

The terms *vivum*,[220] *vivax*,[221] and *lurida*[222] are common epithets for sulphur. It is always contaminated with clay, limestone, or gypsum and its purification depends on melting it and running it off from the accompanying gangue. This process is known as 'liquation'[223] and has been practised in Sicily since early times. Pliny's reference to boiling oil is significant, since oil, like naptha, pitch, benzine, and similar hydrocarbons,[224] is a solvent. The term 'clod' sulphur (*glaeba*) is puzzling and certainly does not signify an allotrope of sulphur. Elsewhere, *glaeba*[225] is found in the sense of a lump, or mass, and Pliny seems only to be describing its appearance. *Egula* is probably sulphur dioxide, the product of burning sulphur, and does have bleaching properties. Later (35. 198), in a further discussion of fulling, Pliny writes, 'The cloth is first washed with earth of Sardinia, then fumigated with sulphur dioxide, and, finally, scoured with Cimolian earth, provided that the dye is fast. If it is coloured with inferior dye, this is detected and turns black and the colour runs as a result of the sulphur. Whereas, if it is genuine and valuable, the colours are softened and brightened up with a sort of brilliance by Cimolian earth, when they have been made more sombre by

---

[220] This term is applied to any substance that is natural, for example rock. Cf. Lucr. *DRN* 6. 221; Verg. *Aen.* 1. 167; Livy 39. 12. 12; Ov. *Fast.* 4. 739.
[221] Ov. *Met.* 3. 374.   [222] Ibid. 14. 791 and 15. 351.
[223] A mass of ore was placed in a kiln and melted, either by setting fire to a proportion, or by external heating. The molten sulphur was collected in wooden moulds.   [224] See Ch. 14 App. B below.
[225] In this meaning see: Lucr. *DRN* 1. 887, 6. 553; Verg. *G.* 1. 94; *Cic. Verr.* 2. 28; Livy 4. 11. Cf also Vitruvius (*De arch.* 7. 9. 4), referring to cinnabar: 'id genus venae postea est inventum Hispaniae regionibus e quibus metallis glaebae portantur ('that variety of cinnabar was later found in Spain and lumps of the mineral were brought out of the mines').

the action of the sulphur. The "rock" kind is more serviceable for white garments that have been treated by sulphur, but is very detrimental to colour. In Greece they use Tymphaean gypsum instead of Cimolian earth.' Celsus (4. 5) refers to similarly treated wool as 'lana sulphurata'.

Pliny (35. 177) observes, ... 'no other substance is more easily ignited, showing that it contains a powerful abundance of fire' ('neque alia res facilius accenditur, quo apparet ignium vim magnam et inesse'). His suggestion, however, that a fourth kind of sulphur was used for lamp-wicks is highly unlikely, since the fumes from the combustion of sulphur would have resembled those from a 'fumite' bomb. On the other hand, his explanation (35. 177) that sulphur also has a place in religious ceremonies, for the purpose of purifying houses is more acceptable:[226] 'habet et in religionibus locum ad expiandas suffitu domos.'

Sulphur was also widely used, as Pliny records (35. 176 ff.), in medicine, as it is today, as an alternative treatment for certain skin disorders, in addition to other applications.

Finally, sulphur was a material originally employed by the Greeks in siege warfare[227] and, later, in imperial times by the Romans, as Martial (*Epigr.* 10. 3. 3)[228] states, for the production of 'matches' ('sulphuratum ramentum').[229]

Although his treatment is somewhat cursory, Pliny nevertheless makes a unique contribution to our knowledge of sulphur.

tourmaline See above, s.v. 'lychnis'.

travertine See above, s.v. 'calcium bicarbonate'.

zinc carbonate See above, s.v. 'cadmea'.

zinc hydrosilicate See above, s.v. 'cadmea'.

---

[226] Hom. *Il.* 16. 228; *Od.* 22. 481. Sulphur is here used in purificatory rites.
[227] Thuc. 2. 77: ἐμβαλόντες πῦρ ξὺν θείῳ καὶ πίσσῃ ἦψαν τὴν ὕλην ('they lit the wood by throwing on to it sulphur and pitch'). Cf. also Thuc. 4. 100.
[228] Also *Epigr.* 1. 41. 4, 12. 57. 14, and elsewhere.
[229] Martial (*Epigr.* 1. 4. 42) also uses the term 'sulfurata fila'.

Appendix A: Amber

The yellow, translucent fossil resin, known as amber, intrigued the Greeks and Romans. Although it is correctly described as an exudation from pine-trees, in their appraisal of this substance legend and fact are often closely linked.

Aeschylus, Philoxenus, Euripides, Nicander, and Satyrus explain its origin through the legend of Phaethon. When he was struck by the thunderbolt, his sisters, because of their grief, were transformed into poplar trees and every year by the banks of the river Eridanus (the Po), shed tears of amber, known to the Greeks as electrum.[1] Pliny (37. 31 f.) states that this story is false and adds, 'Amber provides an opportunity for exposing the false accounts of the Greeks. My readers should bear with me patiently, since it is important to realize that not everything handed down by the Greeks merits admiration.' ('occasio est vanitatis Graecorum detegendae; legentes modo aequo perpetiantur animo. cum hoc quoque intersit vitae sciré, non quidquid illi prodidere mirandum.') Here, as elsewhere, Pliny is critical and not prepared to accept such far-fetched stories (*vanitas*) from Greek sources.

The existence of islands called the Electrides[2] is a figment of the imagination. Likewise, the location of the Eridanus is uncertain. (Aeschylus says it is in Iberia and that it is called the Rhône, while Euripides and Apollonius assert that the Rhône and the Po meet on the coast of the Adriatic!) Such assertions only serve to underline the imprecise character of ancient geography, also the weakest aspect of the *Natural History*. According to Pliny (37. 33),

Other more cautious but equally misguided authorities have made similarly false assertions to the effect that on inaccessible rocks at the head of the Adriatic there are trees which, at the rising of the Dog Star, shed this gum [*cummis*], that is amber. Theophrastus states that amber is dug up in Liguria,[3] while Chares says that Phaethon died in Ethiopia on an island, the Greek name of which is the Isle of

---

[1] Amber (ἤλεκτρον) first occurs in Hom. *Od.* 15. 460, where a necklace of pieces of amber is mentioned. It is also apparently used of the alloy of gold and silver—which, it was generally thought, resembled the colour of amber—as in *Od.* 4. 73 and *Hes. Shield*, 142. See also below, n. 10.

[2] Although islands of this name are also mentioned by Strabo (5. 1. 9), there is no independent evidence for their existence, or for any other islands which might have been within easy reach of anything carried down to the sea by the river Po.

[3] No amber has been found in Liguria, but some resin occurs in the region to the north-east of the Apennines.

Ammon, and that here is his shrine and oracle and the source of amber. Philemon states that amber is a mineral which is dug up in two regions of Scythia, in one of which it is of a white, waxy colour and is called 'electrum', while in the other it is tawny[4] and known as 'sualiternicum.'

modestiores, sed aeque falsum, prodidere in extremis Hadriatici sinus inviis rupibus arbores stare quae canis ortu hanc effunderent cummin. Theophrastus effodi in Liguria dixit, Chares vero Phaëthontem in Aethiopia Ἄμμωνος νήσῳ obisse, ibi et delubrum eius esse atque oraculum electrumque gigni. Philemon fossile esse et in Scythia erui duobus locis, candidum atque cerei coloris quod vocaretur electrum, in alio fulvum quod appellaretur sualiternicum.

Demostratus[5] calls amber 'lyncurium',[6] or 'lynx-urine', a common misunderstanding in the ancient world (cf. Zenothemis, Sudines, and other authorities).[7]

Many locations for amber are cited by Pliny (37. 38 ff.), following Theomenes, Ctesias, Mithridates,[8] Xenocrates,[9] and others. Xenocrates adds that amber in Italy is known not only as 'sucinum', but also as 'thium', while, in Scythia, it is called 'sacrium'. He states that others suppose it is produced from mud in Numidia. Pliny is outraged by Sophocles'[10] explanation that amber is formed in the

---

[4] Most succinite is transparent to translucent, and yellow, red, or brown in colour, but some is opaque and of a pale straw colour because innumerable bubbles are present in it.

[5] A Roman senator and historian of the early 1st cent. AD.

[6] See J. H. Whatmough, in Eichholz, *NH 36–37*, p. 204, note *a*. *Lyncurium*, identified as brown tourmaline, is pyroelectric. The term may have been extended to hessonite, which resembles it, although it does not possess the same electrical properties. The form 'lyncurium' is probably due to a false etymology. λιγγούριον, cited by Strabo (4. 6. 2) as a Ligurian term for amber, may be a more correct form. It is suggested that the term originally meant 'the Ligurian stone' and was applied to amber (*NH* 37. 33 f.). Later the term may have been applied to one or more kinds of hard gemstone similar to amber in their colour and electrical properties. [7] *NH* 37. 34.

[8] Mithridates VI, king of Pontus (120–63 BC).

[9] Xenocrates, of Aphrodisias, was a physician at the time of Nero and the Flavians (AD 54–96). His περί τῆς ἀπὸ τοῦ ἀνθρώπου καὶ τῶν ζῴων ὠφελείας is full of superstitious means of treatment, borrowed largely from previous works such as Ps.-Democritus, λιθογνώμων—a lexicon of gems. For frs. see M. Wellmann (ed.), *Quellen und Studien zur Geschichte der Naturwissenschaften und der Medizin* (Berlin, 1935).

[10] This passage is not extant. The only Sophoclean reference to ἤλεκτρον is now *Antigone* 1037 f., where it refers, not to *amber*, but to the mixture of gold and silver ('white' gold)—commonly miscalled 'electrum'—from Sardis, in Lydia.

lands beyond India, from tears shed for Meleager by the birds known as Meleager's daughters—this is a gross insult to man's intelligence ('summa hominum contemptio est').

Pliny (37. 42) correctly observes: 'It is now well known that amber occurs on islands in the Northern Ocean and is called by the Germans "glaesum".[11] Accordingly, one of these islands, the local name for which is Austeravia, was nicknamed by our troops Glaesaria [or Amber Island] when Germanicus Caesar was involved in naval operations there'. ('certum est gigni in insulis septentrionalis oceani et ab Germanis appellari glaesum, itaque et ab nostris ob id unam insularum Glaesarium appellatam, Germanico Caesare res ibi gerente classibus, Austeraviam a barbaris dictam.')

Pliny goes on to explain (37. 42) that amber 'is formed from a liquid seeping from the interior of a species of pine, just as the gum in a cherry-tree, or the resin of a pine bursts forth when the liquid is excessively abundant' ('nascitur autem defluente medulla pinei generis arboribus, ut cummis in cerasis, resina in pinis erumpit umoris abundantia'). He adds the classic proof of the origin and nature of amber (37. 46), namely that 'it occurs as a distillation, as is shown by the presence of certain objects, such as ants, gnats, and lizards that are visible inside it. These must certainly have stuck to the fresh sap and have remained trapped inside it as it hardened.' ('liquidum id primo destillare argumento sunt quaedam intus tralucentia, ut formicae culicesque et lacertae, quae adhaesisisse musteo non est dubium et inclusa durescente eodem remansisse'.) Furthermore, he states (ibid.) that, according to Archelaus,[12] some amber brought from India[13] in its rough state has bark adhering to it: 'Archelaus ... illinc pineo cortice inhaerente tradit advehi rude.'

Pliny (37. 42 f.) is not sure, however, why it solidifies and ascribes this process to frost, moderate heat, or the sea. 'At any rate, the amber is washed up on the shores of the mainland, whirling along in such a way that it seems to be suspended in the water and not to settle on the sea bed. Even our ancestors believed that amber was a distillation from a tree and so named it "sucinum".' Pliny continues (ibid. 43): 'The fact that it smells of pine when rubbed and burns like a pine-torch with its smoky smell, are indications that the tree belongs to a species of pine.'

---

[11] Modern German *das Glas*. English derivatives include 'glass' and 'glaze'.
[12] King of Cappadocia and possibly the nominee of Antony.
[13] It has been suggested that the amber came from Myanmar (Burma), or, possibly, was another substance.

Amber was mainly exported to Pannonia and, later, to the Veneti, close neighbours of the Pannonians who live round the Adriatic. Pliny deduces that the association of amber with the region of the Po, in his time, is due to the fact that the peasant women of Transpadane Gaul were still wearing pieces of amber, both as an adornment and because of its medical properties.

Pliny (37. 45) adds that 'a Roman knight—still alive—was commissioned to procure amber by Iulianus,[14] when the latter was in charge of a display of gladiators, given by the emperor Nero. He traversed both the trade route and the coasts and brought back so plentiful a supply that the nets used for keeping beasts away from the parapet of the amphitheatre were knotted with pieces of amber.... the largest lump brought back weighed thirteen pounds.' ('vivit eques Romanus ad id comparandum missus ab Iuliano curante gladiatorium munus Neronis principis. qui et commercia ea et litora peragravit, tanta copia invecta ut retia coercendis feris podium protegentia sucinis nodarentur.... maximum pondus is glaebae attulit XIII librarum.') Pliny lists several kinds of amber (ibid. 47)—pale, tawny, and 'Falernian': its colour can be varied by tinting.[15]

At the present time, Baltic amber (succinite) comes not from an island but from the Samland Peninsula, north-west of Gdansk, where it is partly worked in open pits, partly mined, and partly dredged from the sea. It is washed up on other parts of the Baltic coast and occasionally carried to the east coast of Britain. It is a fossil resin derived from a now extinct species of pine (*Pinus succinifera*). Succinite contains up to 6 per cent of succinic acid and a considerable amount of this acid is present in most of the ancient amber artefacts found in Italy and Greece that have been chemically analysed. The northern origin of the amber of which most of these objects were made is thus confirmed. As for Southern Europe, Sicilian amber contains very little succinic acid, Romanian, some, but it is harder than succinite: there is, however, no evidence that this amber was known in the ancient world.

Amber is best known for its well-documented pyroelectric properties.[16]

---

[14] Iulianus is otherwise unknown.

[15] Pliny (37. 48) explains that the root of alkanet ('anchusae radice') was used: alkanet, also commonly employed as rouge in antiquity, would have reddened the amber. It was, similarly, in Pliny's time, coloured with purple dye. The modern technique for tinting is to open a fissure, introduce the colourant and then heat the amber.   [16] See Ch. 12. 2.16 above.

Appendix B: Hydrocarbons

The term hydrocarbon minerals[17] is usually taken to include the naturally occurring solid varieties of carbon and hydrogen (sometimes also with oxygen) compounds—that is, the natural, 'organic' minerals. Two groups may be recognized (1) coal, bitumens, waxes, and resins, and (2) oil and natural gas.

Coal

Coal is the general name given to stratified accumulations of carbonaceous matter derived from vegetation, under a process of compaction and slight heating, in which peat is converted to the familiar black mineral. The Romans used coal, especially in Britain, where it was mined in the Mendips and in north-east England.[18] There is no specific reference to coal in the *Natural History*. Pliny (37. 99) does, however, refer to a stone 'called "anthracitis" which is dug up in Thesprotia and resembles charcoal'; and he observes, 'Statements that it is found in Liguria I consider to be false, unless it is a fact that it was discovered there at the time of the statements.' ('est et anthracitis appellata, in Thesprotia fossilis, carbonibus similis. falsum arbitror quod et in Liguria nasci prodiderunt, nisi forte tunc nascebatur.') This appears to be giving a garbled account of lignite.[19]

Bitumen

Asphalt[20] and its relatives and the natural waxes appear as residues from the volatilization of the lower boiling-point liquid hydrocarbons from a natural oil deposit. Seepages of asphalt may form asphalt lakes (as in Trinidad and Jamaica)[21] or tar pits, but the material also occurs occupying joints and fissures in porous sandstones, and even in lenses in sediments. The natural waxes all appear to be oil residues. The resins include amber[22] and gum copal.[23]

---

[17] Whitten–Brooks, *Geology*, 229 f.

[18] M. Cunnington, 'Mineral Coal in Roman Britain', *Antiquity*, 7 (1933), 89, Coal is thought to have been used in lead smelting.

[19] Healy, *Mining*, 149. See also, *Theophr. Lap.* 16, and Caley and Richards, *Theophrastus on Stones*, p. 88. Fibrous lignite, in its natural state, contains as much as 20% water and thus cannot be readily ignited, although it is combustible when properly dried out.

[20] ἄσφαλτος: Hdt. 1. 179, 6. 119; Theoc. *Id.* 16. 10; Arist. *Mir.* 842^{b}15 (ὀρυκτή, 'dug out'); Dioscorides 1. 72. 1 (ὑγρά, 'raw').

[21] Cf. J. Henderson, *The Caribbean and Bahamas* (London, 1997), 99. Pitch Lake, near San Fernando in Trinidad covers an area of about 47 hectares (116 acres) of smooth surface, resembling caked mud, but which really is black tar; it is 41 metres (135 ft.) deep.

[22] The yellow, translucent fossil resin derived from extinct coniferous trees. See App. A above.

[23] A hard aromatic resin obtained from various tropical trees and used in making varnishes and lacquers.

Many hydrocarbons are also produced from the distillation of coal tar, or petroleum, and so are part of our industrialized society. Pliny's acquaintance with hydrocarbons, however, is superficial and descriptive. Moreover his nomenclature is imprecise and he generally refers only to varieties of bitumen or asphalt.

In the *locus classicus* (2. 235) he writes:

In Samosata, the capital of Commagene, there is a marsh that exudes an inflammable mud called fossil pitch (*maltha*). When it touches anything solid, it sticks to it: also, when touched, it adheres to people trying to escape from it.[24] So the people of Samosata defended their city walls with this substance when Lucullus attacked them [74 BC]. The army was repeatedly burnt by its own weapons. Water feeds the flames. Experiments have shown that the flames can be extinguished only by earth. Naptha[25] has similar properties—this is the name of the substance that flows out, like liquid bitumen, in the vicinity of Babylon and in Parthia, near Astacus. It has an affinity with fire, which leaps out immediately it sees it, wherever it is. This is how Medea,[26] according to the story, burnt her rival when her wreath caught fire after she had gone to the altar to make a sacrifice.

in urbe Commagenes Samosata stagnum est emittens limum (maltham vocant) flagrantem cum quid attigit solidi, adhaeret; praeterea tactus et sequitur fugientes. sic defendere muros oppugnante Lucullo, flagrabatque miles armis suis. aquis etiam accenditur; terra tantum restingui docuere experimenta. similis est natura napthae: ita appellatur circa Babylonem et in Astacenis Parthiae profluens bituminis liquidi modo. hic magna cognatio ignium, transiliuntque in eam protinus undecumque visa. ita fertur a Medea paelicem crematam postquam sacrificatura ad aras accesserat, corona igne rapto.

Pliny (5. 72) writing of the Dead Sea, states, 'Its only product is bitumen, the Greek word[27] for which gives it its name' ('Asphaltites nihil praeter bitumen gignit, unde et nomen').

Bitumen is discussed in more detail elsewhere (35. 179):

Bitumen has similar properties to sulphur. In some places it resembles mud, in others a mineral. It issues with the consistency of mud from the Dead Sea, as I have said [5. 72], and is found as a mineral in

---

[24] A forerunner of the modern napalm—petrol gelled with aluminium soaps used in flame-throwers.
[25] Pliny's naptha is not to be confused with the modern substance, a distillation product from coal tar, or petroleum, used as a solvent and in petrol. Pliny equates naptha with *maltha*, 'tar' (35. 179 and cf. 2. 235).
[26] The daughter of King Aeëtes of Colchis and his wife Eidyia, granddaughter of Helios and niece of Circe.
[27] So Diod. Sic. 19. 98 (λίμνη ἀσφαλτίτης), and Strabo 7. 5. 8 (βῶλος), referring to tar.

Syria round about the coastal town of Sidon. Both varieties solidify, thickening to a dense consistency. There is, however, a liquid form of bitumen such as that from Zacynthus and the kind imported from Babylon; at the latter place it also occurs with a white colour. The bitumen from Apollonia is similarly a liquid and all these liquid kinds are known to the Greeks as 'pissasphalton'[28] because of their similarity to a vegetable pitch.

et bituminis vicina natura est. aliubi limus, aliubi terra est, limus e Iudaeae lacu, ut diximus, emergens, terra in Syria circa Sidonem oppidum maritimum. spissantur haec utraque et in densitatem coeunt. est vero liquidum bitumen, sicut Zacynthium et quod a Babylone invehitur; ibi quidem et candidum gignitur. liquidum est et Apolloniaticum, quae omnia Graeci pissasphalton appellant ex argumento picis et bituminis.

He continues (35. 179) with what is perhaps one of the most interesting references to bitumen:

There is also a bitumen of the consistency of oil; it is found in a spring at Agrigentum in Sicily and pollutes the spring's waters.[29] The inhabitants collect it on bunches of reeds, to which it quickly adheres, and use it instead of lamp oil and also as a cure for scab in beasts of burden. Some authorities include naptha, about which I have spoken in my second book [2. 235], among the types of bitumen, but its burning property and liability to ignition is far removed from any practical use. The test of bitumen is that it should be extremely brilliant, massive, and with an oppressive smell. When quite black, its brilliance is moderate, as it is commonly adulterated with vegetable pitch. In its properties it is like sulphur: it solidifies, it separates, it collects together, it is viscous.

gignitur et pingue oleique liquoris in Sicilia Acragantino fonte, inficiens rivum. incolae id harundinum paniculis colligunt, citissime sic adhaerescens, utunturque eo ad lucernarum lumina olei vice, item ad scabiem iumentorum. sunt qui et naptham, de qua in secundo diximus volumine, bituminis generibus adscribant, verum eius ardens natura et ignium cognata procul ab omni usu abest. bituminis probatio ut quam maxime splendeat sitque ponderosum, graveolens. atrum modice, quoniam adulteratur pice. vis quae sulpuri: sistit, discutit, contrahit, glutinat.

---

[28] Dioscorides 1. 73, a compound of asphalt and pitch (πίσσα), cf. Hom. *Il.* 4. 277; Herodotus (4. 195) refers to sources of pitch—lakes on the islands of Cyranis and Zacynthus.

[29] Modern surveys for oil and natural gas have been carried out in this region, and Gela currently has the largest petrochemical refining plant in Europe.

Another kind of asphalt, known as *ampelitis, or 'vine'* earth (35. 194)—not readily identifiable—is, observes Pliny, 'very like bitumen. The test for it is whether it dissolves when oil is put in it, like wax, and whether, when roasted, it retains its blackish colour' ('bitumini simillima est ampelitis. experimentum eius, si cerae modo accepto oleo liquescat et si nigricans colos maneat tostae').

Pliny (16. 56) mentions another substance, called *zopissa*, which is a mixture of pitch and wax scraped from the bottom of ships in the process of careening: 'non omittendum apud eosdem zopissam vocari derasam navibus maritimis picem cum cera.'

Finally, the term 'pitch' (*pix*) has a variety of meanings. It signifies any of various heavy, dark, viscid substances obtained as a residue from the distillation of tars. In addition, the term pitch may be loosely applied to any naturally occurring substance of a similar type, such as asphalt. Pitch can also be crude turpentine, obtained as sap from pine-trees. Pliny (16. 52) writes: 'In Europe, pitch is obtained from the torch-pine by heating it. ...('pix liquida in Europa e taeda coquitur: navalibus muniendis multosque alios ad usus'.) It is used too for coating ships' tackle (as in this passage), caulking,[30] in warfare,[31] and for many other purposes.

Amber, the well-known fossil mineral of the ancient world is discussed separately.

APPENDIX C: METEORITES

Pliny (2. 147f.) records celestial phenomena,[32] including the story that in 54 BC it rained iron with the appearance of sponges: 'effigies quo pluit ferri spongearum similis fuit.' This reference to the texture of iron is extremely interesting since it accords with the general description of bloomery iron. Elsewhere (34. 146), Pliny uses the same image to describe smelted blooms of iron: 'It is remarkable that when a vein of ore is fused, the iron becomes liquid like water and, afterwards, acquires a spongy and brittle texture.' ('mirumque cum excoquatur vena, aquae modo liquari ferrum, postea in spongeas frangi').

---

[30] That is filling gaps in ships' timbers. This was a substitute for oakum, that is inferior flax (*NH* 19. 17).

[31] Especially in ancient warfare as, for example, at the siege of Delium, described by Thucydides (4. 100): ἡ δὲ πνοὴ ἰοῦσα στεγανῶς ἐς τὸν λέβητα, ἔχοντα ἄνθρακάς τε ἡμμένους καὶ θεῖον καὶ πίσσαν, φλόγα ἐποίει μεγάλην καὶ ἧψε τοῦ τείχους. ('The wind blowing through a *tuyère* into the cauldron, which contained red-hot embers, sulphur, and pitch, produced a great blaze and set the wall on fire.')

[32] On meteorites see Whitten–Brooks, *Geology*, 288 f. A meteorite is extraterrestrial material which falls to the surface of the earth when captured by the earth's gravitational field. Much of the material is in the form of fine dust and

In book 2, Pliny refers (2. 149) to Anaxagoras' prediction[33] that a rock (meteorite) would fall from the sun. This happened in 467 BC, a fact attested by other sources. 'The Greeks declare that Anaxagoras, in the second year of the seventy-eighth Olympiad, through his knowledge of literature relating to astronomy, prophesied that, within a certain number of days, a stone would fall from the sun and this happened during the day at Aegospotami, in Thrace. The stone [of siderite] is on show even now and is as big as a cart and brown[34] in colour. There was also a blazing comet at night in the same place.' ('celebrant Graeci Anaxagoran Clazomenium Olympiadis septuagesimae octavae secundae anno praedixisse caelestium litterarum scientia quibus diebus saxum casurum esset e sole idque factum interdiu in Thraciae parte ad Aegos flumen (qui lapis etiamnunc ostenditur magnitudine vehis, colore adusto) comete quoque illis noctibus flagrante.')

---

only exceptionally do large fragments survive the passage through the atmosphere. Four classes of meteorite may be identified: (1) meteorites composed entirely of metal (nickel-iron)—siderites; (2) those consisting of both metal and silicate—stony-irons, siderolites, pallasites; (3) those composed dominantly of silicate material—stones, aerolites; (4) glassy meteorites—tektites.

[33] Anaxagoras' best known doctrine was that the sun is an incandescent stone larger than the Peloponnese. He held that the stars and planets were also stones, flung off the earth by centrifugal force and ignited by their motion (friction). Indeed he believed that the whole sky was built of stones and that these were kept aloft by centrifugal force. Comets also were basically made of stone, being aggregates of two, or more planets throwing off flames. Meteors were not solid bodies, but merely sparks produced from the revolution of the firmament. He clearly did not associate stones falling from heaven with ordinary meteors, but with the more permanent bodies found in the celestial regions.

[34] This confirms that the composition of the meteor was basically siderite ($FeCO_3$). Anaxagoras' precise description of the great comet (next sentence), implies that he was an eyewitness of this. See M. L. West, 'Anaxagoras and the Meteorite of 467 BC', *Journal of the British Astronomical Association*, 70/8 (1966), 368–9, who concludes that the prophecies (other stones fell at Abydos and at Potidaea) were made on theoretical grounds and vindicated by the fortuitous fall of a large meteorite. This may, or may not, have been connected with the comet in fact; what matters is that Anaxagoras may have found, in some aspect of the comet's apparition, reason to think that some of the furniture of heaven was slipping. If he thought that, then according to his cosmological beliefs, he naturally expected a large stone to arrive on earth within a matter of days, and luckily for his fame it did. The comet appears in Chinese records and has usually been identified with Halley's comet. But see D. J. Shove, 'Halley's Comet and Kaminski's Formula', *Journal of the British Astronomical Association*, 66 (1956), 131–9, esp. 138.

15

Minerals as Pigments

Pliny (35. 29 ff.) gives a brief account of minerals used as pigments by artists.[1] However, his claim that painting began with single colours ('monochromata') is not borne out by the extant evidence.

Colours are defined as brilliant (*floridi*), or sombre (*austeri*). In the former category are cinnabar (*minium*),[2] *armenium* or azurite, dragon's blood (*cinnabaris*), *chrysocolla*, indigo (35. 46), and bright purple, that is earth stained with Tyrian purple[3] (35. 44). All the rest are classed as sombre.

Among the *natural colours* are *sinopis*, a brown-red ochre, or red oxide of iron, found in Pontus (35. 31 f.) There are three kinds (35. 31): red, faintly red, and intermediate—the last being *rubrica* (35. 33), ruddle, or red ochre. Other natural colours are *paraetonium* (a white chalk), melinum (*a white marl*), Eretrian earth, and orpiment.

The *artificial colours* include yellow ochre (*ochra*), *cerussa*, realgar, *sandyx*, Syrian colour, and *atramentum*.

The following catalogue of mineral-pigments is set out alphabetically, in a similar manner to Chapter 14.

armenium (azurite) *Armenium*, or copper carbonate[4] ($CuCo_3$ $Cu(OH)_2$), is the alteration product of copper sulphides under the oxidizing influence of surface waters. It is less common than malachite but with the same origin and mineral associations: they are often combined in a spectacular colour arrangement.

---

[1] On pigments see also Healy, *Mining*, 252 f. For the main treatment of minerals, see Ch. 14 above. On art and artists see generally, G. Isager, *Pliny on Art and Society: The Elder Pliny's Chapters on the History of Art* (London, 1991).
[2] See Ch. 14 above, s.v. 'cinnabar'.   [3] See *NH* 9. 125 ff.
[4] Healy, *Mining*, 36. It has the same origin and mineral associations as malachite.

Pliny (35. 47) writes: 'Armenia sends us *armenium* (azurite), named after the country of origin. This is also a mineral that is dyed like malachite and the best is that which most closely approximates to that substance, the colour also partaking of the blue.' ('Armenia mittit quod eius nomine appellatur. lapis est, hic quoque chrysocollae modo infectus, optimumque est quod maxime vicinum et communicato colore cum caeruleo.') He does not mention any of the physical properties of azurite and, in particular, is unaware of the efflorescence[5] that may be found on its surface.

atramentum This black pigment, ferrous sulphate,[6] 'although listed among the artificial colours, is also derived from the earth in two ways (35. 41). It either exudes from the earth, like the brine in salt pits, or actual earth of a sulphur colour is approved of for the purpose.' ('atramentum quoque inter facticios erit, quamquam est et terrae, geminae originis. aut enim salsuginis modo emanat, aut terra ipsa sulpurei coloris ad hoc probatur.')

All other black pigments are carbon based.

creta Eretria Pliny (35. 30) lists Eretrian earth,[7] as a pigment. It is named after its country of origin and may possibly be identified with magnesite, that is magnesium carbonate ($MgCO_3$), found in irregular veins in serpentine and formed by the replacement of dolomite and limestone.

melinum *Melinum* (35.37) is a white marl—the best comes from Melos—but it is also dug up on Samos.

orpiment Orpiment[8] (As_2S_2) was used as a pigment in the ancient world and is still part of the artist's palette today. 'It is mined for painters on the surface in Syria and is the colour of gold', ('in Syria foditur pictoribus in summa tellure auri colore'). An artificial kind of yellow ochre is also listed by Pliny (35. 30).

---

[5] See Ch. 11.7 above.
[6] F. Greenaway, 'Chemical Tests in Pliny', in French–Greenaway, *Roman Science*, 156 f. [7] Whitten–Brooks, *Geology*, 279.
[8] See also Ch. 14 above, s.v.

paraetonium 'This is a white chalk [35, 36] that derives its name from Paraetonium in Egypt. People say it is sea-foam hardened with mud, which explains why tiny shells are found in it. It also occurs on the island of Crete and at Cyrene.'

psimithium (*cerussa*) *Psimithium*, also known as *cerussa* (ceruse), is lead acetate, from shavings of lead placed over a vessel of very sour vinegar and so made to drip down (34. 175). This pigment, white lead, was not only part of the artist's palette,[9] but was also used, ill-advisedly, by women for making their skin white. Although a poisonous substance, it was used in medicine.

Pliny (35. 37) lists it as the third of the white pigments included in his discussion. He states that there was also once a native ceruse earth found on the estate of Theodotus at Smyrna, which was employed in olden times for painting ships. At the present time all ceruse is manufactured from lead and vinegar: 'tertius e candidis colos est cerussa. ... fuit et terra per se in Theodoti fundo inventa Zmyrnae, qua veteres ad navium picturas utebantur. nunc omnis e plumbo et aceto fit.'

'Burnt ceruse', observes Pliny (35. 38), 'was discovered by accident, when some was burnt up in jars in a fire at Piraeus. ... it is also called purple ceruse and costs six denarii a pound. It is made at Rome by calcining yellow ochre, which is as hard as marble, and quenching it with vinegar.'[10] ('usta casu reperta est in incendio Piraeei cerussa in urceis cremata. ... quae et purpurea appellatur. pretium eius $\bar{X}$ VI fit et Romae cremato silc marmaroso et restincto aceto.')

sandaraca (realgar) Realgar[11] (34. 177) is arsenic sulphide (AsS), found in association with orpiment, in hydrothermal sulphide veins and as a deposit from hot springs and vulcanicity.[12] This pigment 'occurs in both gold- and silver-mines: the redder it is, the more it gives off the poisonous smell of sulphur, and the purer and more friable it is, the better the quality' ('invenitur et in aurariis et in argentariis metallis, melior quo

---

[9] Pliny (35. 38) states that this is indispensable for representing shadows.
[10] See further Ch. 17.8.
[11] Strabo 12. 3. 40 Realgar was mined at Sandaracurgium, from which it derived its name—*sandaraca*. [12] Whitten–Brooks, *Geology*, 379.

magis rufa quoque magis virus sulpuris redolens ac pura friabilisque'. 'According to Juba [*NH* 35. 39] sandarach [or realgar] and ochre are products of the island of Topazus in the Red Sea, but they are not imported from those parts to us. ... an adulterated sandarach is also made from ceruse, boiled in a furnace. It ought to be flame-coloured. ('Sandaracam et ochram Iuba tradidit in insula Rubri maris Topazon nasci, sed inde non pervehuntur ad nos. ... fit et adulterina e cerussa in fornace cocta. color esse debet flammeus.')

Realgar, like orpiment, is used by artists.

sandyx Pliny (35. 40) describes a method of making the pigment called *sandyx*: 'If ceruse[13] is mixed with red ochre in equal quantities and burnt, it produces *sandyx* [or vermillion]'; but he adds that Virgil (*Ecl.* 4. 45) held the view that *sandyx* was a plant. ('haec [cerussa] si torreatur aequa parte rubrica admixta, sandycem facit, quanquam animadverto Vergilium existimasse herbam id esse.')

Pliny adds (35. 40) that Syrian colour, which is used as an undercoating for cinnabar and red lead, is made by mixing *sandyx* and *sinopis* (a kind of red ochre).

---

[13] See above, s.v. 'psimithium'.

16

Gems and Precious Stones

In the final book of the *Natural History* (37) Pliny discusses precious stones, or gems.[1] Here (37. 1), Nature's grandeur is gathered together within the narrowest limits. ...very many people find that a single gemstone alone is enough to provide them with a supreme and perfect aesthetic experience of the wonder of Nature.' ('in artum coacta rerum naturae maiestas. ... ut plerisque ad summam absolutamque naturae rerum contemplationem satis sit una aliqua gemma.')

'The question of which stones are supreme has been settled by a decree of our women-councillors [37, 85], whereas men tend to indulge their whims when choosing gemstones. The emperor Claudius wore an emerald, or a sardonyx but many years previously, according to Demostratus[2] the elder Africanus was the first to adopt a sardonyx and hence arose the esteem which this gemstone enjoys at Rome.' ('hactenus de principatu convenit mulierum maxime senatus consulto minus certa sunt de quibus et viri iudicant; singulorum enim libido pretia singulis facit praecipueque aemulatio, velut cum Claudius Caesar smaragdos induebat vel sardonyches. Primus autem Romanorum sardonyche usus est Africanus prior, ut tradit Demostratus, et inde Romanis gemmae huius auctoritas.')

The following list (in order of appearance in book 37) includes a selection of those gems described by Pliny, chosen for their

---

[1] Pliny (37. 186 ff.) discusses methods of classification of gems, by comparison with (*a*) parts of the body, (*b*) names of animals, (*c*) inanimate objects, trees, and leaves of plants. For a modern definition of precious stones, see Whitten–Brooks, *Geology*, 198.

[2] A Roman senator and historian of the early 1st cent. AD.

interest, or importance. (The diamond, emerald, and beryl, which are among the finest of gemstones, are discussed in detail elsewhere,[3] in illustration of Pliny's understanding of their 'crystal systems.'[4])

opalus (opal) Writing on opals,[5] Pliny (37. 80 ff.) incorrectly states that 'India is the sole producer of these stones. They combine', he writes, 'the brilliant qualities of the most valuable gems and, above all other, defy description. Opals display the more subtle fires of the *carbunculus* [cf. ibid. 92–7], the flashing purple of the amethyst, and the sea-green tint of the emerald—combined together in incredible brilliance.' ('India sola et horum mater, atque ut pretiosissimarum gloria compositi gemmarum maxime inenarrabilem difficultatem adferunt. est in his carbunculi tenuior ignis, est amethysti fulgens purpura, est smaragdi virens mare, cuncta pariter incredibili mixtura lucentia.')

Opals are also known by the Greek name *paedaros* (παιδέρως), but this term is shared with a variety of amethyst (37. 123).[6] Nevertheless, Pliny's description of the *paederos* (ibid. 129) matches that of the opal. He is, however, obviously uncertain about this identification and the evidence of the text is ambiguous.

sardonyx The name sardonyx,[7] derived from sard, a carnelian (σάρδιον), and onyx, a finger-nail (ὄνυξ), 'encapsulates its own description [37. 86]. It is a stone with a layer of carnelian resting on a layer of white, that is like flesh superimposed on a human finger-nail, both parts of the stone being translucent.' ('sardonyches olim, sicut ex ipso nomine apparet, intellegebantur in candore sarda, hoc est veluti carne ungui hominis inposita, et utroque tralucido.')

The term *sarda*—derived from the Persian word *sered*, meaning 'yellowish-red', includes carnelian (clear red chalcedony)

---

[3] See Ch. 14 above, s.vv. 'adamas', 'smaragdus', and 'beryllus', respectively.   [4] See Ch. 13.4.   [5] Whitten–Brooks, *Geology*, p. 413.

[6] The places where this was found include Egypt (cf. also 37. 121) and this seems to suggest that the gemstone described in Pliny's source is an amethyst.

[7] Cyclosilicate mineral. See Whitten–Brooks, *Geology*, 413.

and sard (reddish-brown, brown, and yellow chalcedony). In the present context, Pliny is describing carnelian. 'The stone is a common one and was first discovered at Sardis—hence its name—but the most valuable specimens are found near Babylon. The stones come to light when certain quarries are being opened up' (37, 105).

Sardonyx is strictly a banded chalcedony[8] containing at least a layer of carnelian. Owing to the popularity of the genuine sardonyx, the word had come to be applied, no doubt by gem dealers, to the varieties of onyx containing a layer of red jasper ('imitante et ungue minium'), or sard ('melleae aut faeculentae', 37. 89).

onyx Onyx is the name given both to a stone (36, 59), namely onyx marble (see Ch.14, s.v. 'phengites'), and a gem—onyx[9] or banded agate (37. 90). Pliny (ibid.) records Sudines'[10] description. 'In onyx one finds a white band resembling a human finger-nail, as well as the colour of the chrysolith, or yellow sapphire [37. 126 below, 'chrysolithus'], the sard, or carnelian, and the *iaspis*, or green-chalcedony' ('in gemma esse candorem unguis humani similitudine item chrysolithi colorem et sarda et iaspidis').

carbunculus (ruby) Pliny (37. 92 ff.) continues with a categorization of gemstones by colour, beginning with those that are red (*carbunculi*), which include rubies,[11] red garnet[12] (pyrope and almandine), and, possibly, red spinel. He follows Theophrastus (*Lap.* 33). The term *carbunculi* derives from their resemblance to pieces of burning charcoal.[13] Ancient authorities, including Theophrastus (ibid. 19), claim that they

---

[8] See n. 7 above. [9] See n. 7 above.
[10] An astrologer (cf. 36. 59) who wrote on the mystic properties of stones, c. 240 BC. He lived at the court of King Attalus I of Pergamum.
[11] Whitten–Brooks, *Geology*, p. 395 and appendix. A red, transparent variety of corundum (Al_2O_3): it is found widely in metamorphosed shales, in certain kinds of metamorphic limestone veins, and in some undersaturated igneous rocks.   [12] Whitten–Brooks, *Geology*, 196 f.
[13] *Carbunculus* is a diminutive of *carbo*. Cf. Plaut. Rud. 2. 6. 48; Lucr. *DRN* 6. 802; Cic. *Offi*; 2. 7. 25; Hor. *Odes* 3. 8. 3. Pliny, *NH* 2. 82 and elsewhere.

are not affected by fire, but this only applies to rubies, since all (red) garnets *are* fusible.

The origin of the 'Carthaginian' *carbunculi*, however, is a mystery, since no African garnets are presently known.[14] Pliny (37. 96) correctly observes that male Carthaginian stones have a blazing star inside them: this may be seen when the garnet is fibrous.

The stone called *anthracitis* (37. 99) is in fact lignite[15] and should not be included in this section. Pliny explains how, when it is touched, its glow dies away and disappears, but, on the other hand, when it is soaked with water, it blazes forth again: 'tactu velut intermortuae extinguuntur, contra aquis perfusae exardescunt.'

lychnis (tourmaline) Pliny (37. 103) mentions two forms of *lychnis*, namely violet-red and rose-red tourmaline, respectively (see Ch. 14, s.v. 'lychnis').

topazos (peridot) Pliny begins his account of green stones (37. 107 ff.) with the *topazos*, that is olivine of gem quality. He records Juba's assertion (ibid. 108) that 'Topazos is the name of an island situated in the Red Sea, at a distance of some thirty-five miles from the mainland. Peridot was, it is said, brought from there as a gift for queen Berenice, the mother of

---

[14] Pliny, in his account of the geography of Africa (5. 34 and 35–7), refers to Ethiopian *carbunculi* and mentions the 'Cave-dwellers [Trogodytae] with whom our only intercourse is the trade in precious stones, imported from Ethiopia, which we call garnets (*carbunculi*). Before reaching them in the direction of the African desert, stated already to be beyond the lesser Syrtis, is Fezzan.' ('Trogodytae ... cum quibus commercium gemmae tantum quam carbunculum vocamus ex Ethiopia invectae. intervenit ad solitudines Africae supra Minorem Syrtim dictas versa Phazania.') 'Some way further is a town of the Garamantes called Thelgae [5. 36] ... finally comes Mt. Goriano where there is an inscription stating that it is an area where precious stones occur.' ('... mons Gyri in quo gemmas nasci titulus praecessit,') The Garamantes are obviously 'middlemen', and trade in garnets. Pliny (37. 104) refers to a Carthaginian stone (*Carchedonia*) which has pyroelectric properties akin to those of tourmaline. It is formed in the mountain country of the Nasamones (who live to the north-east of the Garamantes) and is the garnet.

[15] See Ch. 14, App. B, s.v. 'coal'. Lignite was said to have occurred in Thesprotia, south-west Epirus (Antigonus of Carystus, *Hist. Mir.* 120), and in Liguria (Theophr. Lap. 16). Furthermore, ancient writers observe that moisture causes the spontaneous combustion of this fossil fuel.

Ptolemy II, by his governor Philo... a statue of peridot (six feet tall), was made in honour of Arsinoe, Ptolemy's wife and consecrated in the shrine which was named after her, the Arsinoeum.' ('Iuba Topazum insulam in Rubro mari a continuenti stadii CCC abessse dicit... ex hac primum importatam Berenicae reginae, quae fuit mater sequentis Ptolemaei, ab Philone praefecto... inde factam statuam Arsinoae Ptolemaei uxori quattuor cubitorum, sacratam in delubro quod Arisinoeum cognominabatur.')

callaina (green turquoise) Pliny (ibid. 110) next includes a porous stone of similar appearance, called *callaina*: this is green turquoise. The Caucasus and Carmania are said to be the source. Collecting the gem is supposedly a hazardous task, although this warning may have been put out to deter collectors!

Among other green stones, are included, prase (ibid. 113), green jasper (*iaspis*), malachite, and unnamed varieties of chalecdony. *Nilios* (ibid. 114)[16] is incorrectly listed here.

lapis lazuli One of the best-known of the blue stones (37. 119 ff.) is Lapis lazuli [17] (of which *cyanus* is probably a form), the highest quality of which Pliny (ibid. 121) misguidedly states is found in Persia. The most important mines are found in northern Afghanistan.[18]

amethyst A gem, in colour like the perfect shade of Tyrian purple,[19] found in India and, according to Pliny (37. 121), also in that part of Arabia known as Petra.[20] There are five main varieties (ibid. 122). 'Of these a second kind of amethyst inclines towards the sapphire. Its colour is known to the

---

[16] See further below, n. 21.
[17] Whitten–Brooks, *Geology*, 174. This is the semi-precious variety of lazurite. It belongs to the group of feldspathoids.
[18] See Eichholz, *NH* 36–37, p. 262, note *c*. Mines in the upper Koksha valley, in Badakshan, have been worked for thousands of years.
[19] Cf. *NH* 9. 137 ff., on purples.
[20] The capital of the Nabataeans. Situated in a ravine, it was, in the 1st cent. AD, on the main overland caravan route from the *East* to the Mediterranean. Its importance diminished, however, as commercial sea routes developed.

Indians as 'socos' and the variety of gem as 'socondios'. A fainter variety of the same stone is called 'sapenos' and, also in the districts adjacent to Arabia, 'pharanitis', after the name of a tribe. The fourth kind, however, has the colour of red wine [and is, probably a garnet], while a fifth degenerates nearly into rock-crystal, since its purple fades away towards colourlessness.' ('alterum earum genus descendit ad hyacinthos; hunc colorem Indi socon vocant, talemque gemmam socondion. dilutior ex eodem sapenos vocatur, eademque pharanitis in contermino Arabiae, gentis nomine, quartum genus colorem vini habet. quintum ad vicina crystalli descendit albicante purpurae defectu.')

sapphire (hyacinthus) 'There is a considerable difference [in colour] between the amethyst and the sapphire [37. 125].... The difference lies in the fact that the violet brilliance that is characteristic of the amethyst is here diluted with the tint of the "hyacinth flower".' ('multum ab hac distat hyacinthos.... differentia haec est, quod ille emicans in amethysto fulgor violaceus diluitur hyacintho.') The gem to which Pliny assigns the name *Nilios*—derived from the Sanskrit word *nila* meaning 'dark blue'—is likely to have been a blue sapphire.[21]

The *asteria* (37. 131), or 'star stone', with an inclusion, may be a pale star-sapphire, although this identification is not certain.

chrysolithus Finally, there are yellow stones, among them the *chrysolithus* (37. 126) and the golden amber (hessonite) and pale yellow corundum, *leucochrysi*, and others.

After his account of coloured gems, Pliny (ibid. 132) turns to consider other precious stones without colour, among them the moonstone (*astrion*) which closely resembles rock-crystal and occurs in India and on the coasts of Patalene.[22] It has inside, at the centre, a star shining brightly like the full moon: 'similiter candida est quae vocatur astrion, crystallo propinqua, in India nascens et in Patalenes litoribus. huic intus a centro stella lucet fulgore pleno lunae.'

---

[21] See Eichholz, *NH 36–37*, p. 256, note *c*.
[22] E. H. Warmington, *The Commerce between the Roman Empire and India*[2] (Cambridge, 1974), 248.

It is said that the best variety is found in Carmania and that no gem is less liable to possess defects. He also describes inferior kinds of moonstone.[23]

iris (rainbow stone) The *iris* rock-crystal (37. 136); see Ch. 14 above, s.v. 'crystallum', 'iris') occurs naturally as a prism, and its properties are described elsewhere.[24]

agate Agate[25] is dismissed by Pliny (37. 139) as a stone once held in high esteem, but, in his time, enjoying none: 'achates magna fuit auctoritate, nunc in nulla est.' It was first discovered in Sicily near the river Achates. Agate is credited with remarkable properties—from being able to counteract the bites of spiders and scorpions, to allaying the thirst, if placed in the mouth!

apsyctos A curious stone, the identification of which is uncertain, is described by Pliny (37. 148): 'The *apsyctos*, or uncooled stone, retains its warmth for seven days if it is thoroughly heated in a fire and is black, heavy, and marked with red veins. It is thought to counteract cold.' ('apsyctos septenis diebus calorem tenet excalefacta igni, nigra ac ponderosa, distinguentibus eam venis rubentibus. putant prodesse contra frigora.') It has been tentatively identified with lignite but this seems far from likely.

coral One further stone, called *Gorgonia*, or 'Gorgon's stone' (37. 164) is of interest: 'it is, in fact, coral. The reason for its name is that it is transformed into the hardness of stone after being softened in the sea.' ('Gorgonia nihil aliud est quam coralium. nominis causa, quod in duritiam lapidis mutatur emollitum in mari.')

Pliny concludes (37. 195) his account: 'There is no end to the names given to precious stones and I have no intention of listing them in full, innumerable as they are, thanks to the

---

[23] The omission of Sri Lanka as a source of the moonstone is surprising since the region was well known to the Romans of the 1st cent. AD as an important source of pearls.   [24] See Ch. 12.1.2 above.
[25] Whitten–Brooks, *Geology*, 17.

conceit and imagination of the Greeks. Now that I have mentioned the precious stones and also some, indeed, that are common, I must be content with having given emphasis to the rarer varieties that deserve notice.' ('cum finis nominum non sit—quae persequi non equidem cogito, innumera ex Graeca vanitate—indicatis nobilibus gemmis, immo vero etiam plebeis, rariorum genera digna dictu distinxisse satis erit.')

In book 37, Pliny lists a great number of precious, or semi-precious stones, very many of which cannot be identified with any certainty. In addition, there is evidence of repetition and confusion in nomenclature. Finally, varieties of gems are not always unequivocally distinguished.

In spite of some obvious shortcomings, Pliny deserves credit for being more detailed and comprehensive in his treatment than Theophrastus.

17

Metals

17.1 Introduction

A metal possesses all, or many, of the following properties: it is solid (except for mercury)[1] at ordinary temperatures, opaque, a good reflector of light when polished, and a good, or fairly good, conductor of heat. When melted it solidifies as a compact mass of crystals generally invisible to the eye but clearly visible in metallographic analyses.

17.2 Physical Characteristics

Pliny's knowledge of the physical characteristics of metals is superficial and, for obvious reasons, cannot include references to melting-points or boiling-points, nor does he have any understanding of thermal expansion. He merely describes the colour[2] of metals, which, with the exception of gold and copper, are a greyish colour varying from bluish-grey (lead) to white (silver). Pliny is aware of their varying weights although not of their densities, that is the relationship of a volume of metal to an equal volume of water.

17.3 Mechanical Properties

Pliny is, on the whole, better informed about the mechanical properties of metals although he sometimes confuses, or conflates them.

---

[1] See Healy, *Mining*, 227.
[2] As, for example, in the case of tin (*NH* 34. 157).

17.3.1 Ductility

Ductility[3] is the property of a metal which enables it to be given a considerable amount of mechanical deformation (especially stretching) without cracking. This was well known to Aristotle (Mete. 4. 386^b), who gives an accurate definition: 'Ductile things (ἑλκτά) are those whose surface will extend in the same plane, for to be drawn out is to have the surface extended in the direction of the motive force without breaking.'

Pliny (33. 59) mentions the weight and malleability of gold, but, erroneously, rates these properties as second to those of lead:[4] 'nec pondere aut facilitate materiae praelatum est ceteris metallis, cum cedat per utrumque plumbo.' Yet, later (33. 61) he correctly observes: 'No other metal is as malleable as gold or able to be divided into so many portions: thus an ounce of gold can be beaten into upwards of 750 leaves, four inches square'[5] ('nec aliud laxius dilatatur aut numerosius dividitur, utpote cuius unciae in septingenas quinquagenas pluresque bratteas quaternum utroque digitorum spargantur').

Similarly, copper is highly ductile and malleable, when annealed, as Pliny (34. 94) explains, 'In the case of the copper of Cyprus, "chaplet copper" is made into thin leaves' ('in Cyprio coronarium tenuatur in lamnas'); and later, 'Bar copper also is produced in other mines and likewise fused copper. The difference between them is that the latter can only be fused, as it breaks under the hammer. Whereas, bar copper, otherwise called ductile, is malleable, which is the case with all Cyprus copper. But also in the other mines, this difference of bar copper from fused copper is produced by treatment; for all copper, after impurities have been rather carefully removed by fire and smelted out of it, becomes bar copper.' ('regulare et in aliis fit metallis itemque caldarium. differentia quod caldarium funditur tantum, malleis fragile, quibus regulare obsequitur ab aliis ductile appellatum, quale omne Cyprium est. sed et in ceterius metallis cura distat a caldario; omne enim diligentius

---

[3] Healy, *Mining*, 225.
[4] The specific gravities are 19.3 and 11.3 respectively.
[5] The thickness was approximately 0.0002 mm., a limit which stood until the 18th cent. See further E. D. Nicholson, 'The Ancient Craft of Gold Beating', *Gold Bulletin*, 12/4 (1979), 161–6.

purgatis igni vitiis excoctisque regulare est.') Pliny (34. 5, 46, 97) makes many references to the fusibility of copper and other metals and adds (34. 97) that 'all copper and bronze fuses better in very cold weather' ('aes omne frigore magno melius fundi'). It has a tenacity which is second only to that of iron.

17.3.2 Hardness

Hardness is the property of a metal which causes it to resist deformation from the application of external forces. Although hardness was experienced and known, in practice, in the ancient world, neither the Greeks nor the Romans were able to quantify this measurement, as today, when hardness figures for individual metals, or alloys, may be given as HV (Vickers Hardness), or B (Brinell).

17.3.3 Tenacity

Iron is malleable, ductile, and the most tenacious[6] of all metals: it is smelted to give hardness to a blade (34. 144): 'excoquitur [in fornacibus] ad indurandam aciem.' Pliny (34. 156) confuses tin and lead, calling the former 'white lead' ('plumbum candidum'). Tin is detected by its weight: 'pondere tantum ea deprehenditur.' It is fairly soft and easily flattened. Almost devoid of tenacity, and elastic within only narrow limits, when bent it emits a characteristic sound known as its 'cry'.[7]

Black lead (34. 164) is used to make pipes and sheets: 'nigro plumbo ad fistulas lamnasque utimur.' Pliny (34. 165) further comments, without any justification: 'It is remarkable that vessels made of lead will not melt if they have water put in them, but if one adds a pebble, or a bronze quadrans, the fire burns through the vessel.' ('mirum et addita aqua non liquescere vase e plumbo, eadem, si in aquam addantur calculus vel aereus

---

[6] Healy, *Mining*, 227 n. 18. Iron is affected by a whole range of impurities which may be present to a greater or lesser extent in any ore. Phosphorus, for example, affects the hardness and strength of pure iron, air-cooled from 950°C. The hardness increases from 75 HV (no phosphorus present) to 165 HV (with 0.7%).   [7] Ibid.

quadrans, peruri.') This is untrue, but suggested emendations to the text are unsatisfactory.

17.3.4 Magnetic Properties

Iron is attracted by a magnet, and temporary magnetization may be induced by contact with a magnet, as in the example of the chain of suspended rings, described by Plato and Lucretius.[8]

17.3.5 Fusibility

The natural properties of metals may be modified, or altered, in three significant ways: (1) by alloying; (2) by mechanical deformation; and (3) by heat treatment—all of which processes are described by Pliny. Certainly in the case of bronze, its most important property was its fusibility,[9] namely that it could be melted and cast, especially in the production of statues by the *cire perdue* process.

MAIN METALS (AND ALLOYS) IN USE IN THE ANCIENT WORLD

The main metals exploited by the Greeks and Romans were gold (and its alloys, white gold and electrum), silver, copper (bronze, brass), lead (pewter), tin, mercury, and iron. Zinc was known only in the form of calamine, or zinc carbonate, its ore-mineral.

17.4 GOLD

17.4.1 Introduction

Gold is chemically unchangeable and permanent and this, together with its other unique properties, and symbolism, led to its being the most highly sought-after metal in the ancient world.

Gold occurs in its native state,[10] in placer deposits, and in four main mineral groups: (1) sulphides[11] in which gold is the

---

[8] See Ch. 12.3 above. [9] Healy, *Mining*, 226 f.
[10] A. B. Edwards, *Textures of Ore-minerals and their Significance* (London, 1954), 168.
[11] From the 1st cent. AD, the Romans began to obtain gold as a by-product from the processing of sulphides and pyrites.

only valuable constituent; (2) sulphides containing valuable base metals; (3) tellurides[12]—tellurium and possibly selenium are the only elements which are combined with gold in nature; (4) sylvanite,[13] which occurs associated with native gold and other gold and silver tellurides, with pyrite, sulphides, quartz, calcite, and fluorite.

The tellurides were, for technological reasons, not able to be exploited in the ancient world.[14]

17.4.2 Native Gold and Alluvial Deposits

Although Strabo (13. 1. 23) refers to the existence of gold-mines at Astyra (in the Troad) in early times, gold was found mainly as the native metal, or in the form of placer deposits in the preclassical period.

Alluvial deposits[15] are formed by grains of gold, silver, and copper, together with traces of platinum group elements—the product of weathered parent lodes[16]—which have been transported to their locations by rivers, such as the Pactolus and Hermus,[17] as Herodotus (5. 101) explains.

---

[12] Hurlbut, *Mineralogy*, 268 ff. The main tellurides are calaverite, $AuTe_2$, and petzite $(Ag,Au)Te_2$. See further O. Davies, *Roman Mines in Europe* (Oxford, 1935), 56. Some tellurides were found in Dacia in Roman times and, to a much lesser extent, on Andros.

[13] Hurlbut, *Mineralogy*, 269 f. The name sylvanite is derived from Transylvania, where it was first found and in allusion to sylvanium, one of the names originally proposed for the element tellurium $(Au,Ag)Te2$.

[14] Healy, *Mining*, 36.

[15] Ibid. 31 ff. and 74 ff. See further S. V. Griffith, *Alluvial Prospecting and Mining*² (Oxford, New York, and Paris), 1 ff. There are five main types: (1) eluvial (or residual); (2) river and stream; (3) river terrace, or bench; (4) beach, or marine; (5) deep leads (or buried placers).

[16] J. H. Kirsch (trans. K. A. Jones), *Applied Mineralogy for Engineers, Technologists and Students* (London, 1968), 76. Cf. Hdt. 1. 93. 1, where he mentions Mt. Tmolus, in this connection.

[17] G. M. A. Hanfmann, *Letters from Sardis* (Cambridge, Mass., 1972), 141 ff. In Hanfmann, the final report from the Turkish geologist Mustafa Saydamer sets a wider framework for the question of Lydian resources in gold: 'Between the towns of Turgutlu and Allahiye, gold is found in the alluvium of several torrents which flow into the Hermus river from the south. There is a close relation between the abundance of quartz formations and the gold content of the alluvial deposits. The Sart Çay (our Pactolus) and the Tabak Çay (which flows east of the citadel) have both large amounts of quartz pebbles.'

Sheepskins (fleeces) were pegged out in the rivers to trap the grains.[18] Later, this technique was improved by the Soanes[19] who put the fleeces in troughs through which the river water was diverted. It is highly probable, therefore, that the quest for the Golden Fleece[20] is a legendary reference to this method of collection. The laden fleeces were hung up to dry and beaten to release the gold dust. It is interesting to recall that, when Jason went in search of the Golden Fleece, he found it 'hanging upon a huge oak-tree, like to a cloud that blushes red with the beams of the rising sun'.[21] The colour description may well have been a poetic allusion to a metallurgical fact, namely that the presence of copper[22] could have given the mixture such an appearance—in addition to the reflection of the colour of the rays of the sun at dawn.

Strabo (3. 2. 8) explains how, in the case of the auriferous sand in the waterless districts of Turdetania, the inhabitants diverted water to the area to make the grains of gold glitter and stand out, since they could not otherwise be readily distinguished. Yet a further method of collecting gold was employed in Turdetania. Soil was washed in pits, or carried along in streams and washed in nearby troughs—a primitive form of 'sluicing'. Strabo (ibid.) adds that by his time 'gold washeries' (χρυσοπλύσια) were more numerous than actual gold-mines. (Gold washing had been practised by the Egyptians from as early as the Twentieth dynasty.) In Lusitania the ground 'effloresced' with white gold (λευκὸν χρυσίον), and a form of 'panning'[23] was employed in which the women scraped up the

---

[18] See J. F. Healy, 'Greek White Gold and Electrum Coin Series', in D. M. Metcalf (ed.), *Metallurgy in Numismatics*, i (Royal Numismatic Society, Special Publication, 24; London, 1980), 194.   [19] Strabo 11. 2. 19.
[20] Healy, *Mining*, 75 ff.   [21] Ap. Rhod. *Argon*. 4. 123–6.
[22] Healy, *Mining*, 153.
[23] Griffith, *Alluvial Prospecting*, 33: 'The pan (or container) is filled with "dirt", placed in water, and the dirt thoroughly wetted and stirred by hand to break up the lumps of clay and the like, large stones at the same time being well washed, picked out and discarded, after being examined to see that they do not contain any gold, or other mineral matter. The pan, still under water, is then given a circular, or gyratory motion in a horizontal plane, which causes the heavy particles to settle and the lighter ones to come up to the surface; at frequent intervals the pan is gently tipped forward and the lighter material at the surface washed off. This process is continued until nothing but gold, or other minerals, and a little heavy sand is left. The pan is then dried and the mineral separated from the sand by gently blowing the sand away.'

gold-bearing sand with shovels and washed it in sieves woven like baskets; this is a well-known technique employed to the present day by primitive communities.[24]

17.4.3 Gold-mining

As the placers became worked out, the Greeks were forced to seek other sources of gold (χρυσὸς μεταλλευτός). Mines on the islands of Siphnos[25] and Thasos[26] are well documented, but the richest sources of gold in the Greek world were found in the Macedonia –Thrace region and, by 358 BC, Philip II was exploiting the mines of Mt. Pangaeus.[27] There is evidence to support the belief that the Bessi, from Paeonia also exploited placers in Macedon during the fourth century BC, as [Aristotle,] *De mirabilibus auscultationibus* 45 states, in so far as some gold staters of Philip II contain traces of platinum group elements (PGE)[28] not normally associated with reef gold.

A number of mines continued to be exploited by the Romans, especially under the empire, but new sources of gold were developed, particularly in Spain, a region rich in mines, as Pliny (3. 30)[29] records: 'Nearly the whole of Spain is rich in lead-, iron-, copper-, silver-, and gold-mines' ('metallis plumbi ferri aeris argenti auri tota ferme Hispania scatet').

---

[24] Forbes, *Ancient Technology*, vii. 222.
[25] Hdt. 3. 57. 2. These mines were already being exploited in the 6th cent. BC. See G. Weisgerber, 'Recent Research on Prehistoric and Ancient Mining in Greece', in Domergue, *Mineria y Metalurgia*, i. 197 ff. (on Siphnos and Thasos).
[26] On Thasos see further: G. A. Wagner, E. Pernicka, W. Gentner, and H. Gropengiesser, 'Nachweis antiken Goldbergbaus auf Thasos: Bestätigung Herodots', *Naturwissenschaften*, 66 (1979), 613, and G. A. Wagner, E. Pernicka, G. Gialoglou, and M. Vavelidis, 'Ancient Gold Mines on Thasos', *Naturwissenschaften*, 68 (1981), 263.   [27] Healy, *Mining*, 46.
[28] J. Ogden, 'Platinum Group Inclusions in Ancient Gold Artefacts', *Journal of the Historical Metallurgical Society*, 11 (1977), 53–72, esp. 56, fig. 2. See also, J. F. Healy, 'Greek and Roman Gold Sources: The Literary and Scientific Evidence', in Domergue, *Mineria y metalurgia*, ii. 15.
[29] See also Strabo (3. 2. 8), who makes the same observation.

In his main account of gold, Pliny (33. 66 ff.) writes: 'Leaving aside tales of Indian gold obtained by ants,[30] or the gold dug up by griffins in Scythia,[31] gold is obtained in our part of the world in three ways. First, it can be obtained from placers in rivers—for example, the Tagus[32] in Spain, the Padus[33] in Italy, the Hebrus[34] in Thrace, the Pactolus[35] in Asia Minor, and the Ganga[36] in India; this is gold in its most perfect state, thoroughly polished by the friction induced by the current. A second method is to sink shafts. And, thirdly, it is sought for in the debris of collapsed mountains.' ('aurum

---

[30] The story is first found in Hdt. 3. 102–5. Pliny (11. 111) again refers to ants which 'carry gold out of caves in the earth in the region of the northern Indians called the Dardae. These creatures are the colour of cats and the size of Egyptian wolves [some ants!] The gold that they dig up in winter-time the Indians steal in the hot weather of summer when the heat makes the ants hide in burrows [or their mines?—*cuniculos*]. Nevertheless they are attracted by the Indians' scent and fly out and sting them repeatedly although retreating on very fast camels: such speed and ferocity do these creatures combine with their love of gold.' ('aurum hae e cavernis egerunt cum terra in regione septentrionalium Indorum qui Dardae vocantur. ipsis color felium, magnitudo Aegypti luporum. erutum hoc ab iis tempore hiberno Indi furantur aestivo fervore, conditis propter vaporem in cuniculos formicis, quae tamen odore sollicitatae provolant crebroque lacerant quamvis praevelocibus camelis fugientes: tanta pernicitas feritasque est cum amore auri.') An interesting modern reference to 'ants', only this time to (human) miners (*garimperos*) who are locally nicknamed 'formigas' (ants), comes from Serra Pelada (Brazil), 700 miles (1,127 km.) south of Belem, where, in 1979, a gold nugget weighing 137 pounds (62.14 kg.) and 80% fine was found. This led to the discovery of the largest alluvial deposit in the world. Within fifteen days some 15,000 prospectors arrived and their number grew to 80,000 at the peak period of exploitation.

[31] Hdt. 3. 116, 4. 13 and 217. Cf. Pliny (7. 10), who, mentioning the Arimaspi, states: 'Many authorities, the most distinguished being Herodotus and Aristeas of Proconnesus, write that these people wage a continual war around their mines with the griffins, a kind of wild beast with wings, as commonly reported, that digs gold out of mines; these creatures guard the gold which the Arimaspi try to pillage. Both griffins and Arimaspi display remarkable greed.' ('quibus adsidue bellum esse circa metalla cum grypis, ferarum volucri genere, quale vulgo traditur, eruente ex cuniculis aurum, mira cupiditate et feris custodientibus et Arimaspis rapientibus, multi sed maxime inlustres Herodotus et Aristeas Proconnesius scribunt.') See further R. Schilling, *Pline l'Ancien: Histoire Naturelle, Livre 7* (Budé: Paris, 1977), note ad. loc.

[32] Although Pliny refers to this and the following rivers in his sections on geography, only the Tagus is said to have gold-bearing sands, for which it is famous (4. 115): 'Tagus auriferis harenis celebratur.' Cf. also Livy 21. 5. 8; Ov. *Am.* 1. 15. 34, *Met.* 2. 251; Lucan, *BC* 7. 755; Mart. *Epigr.* 1. 50. 15 and 10. 96. 3.

[33] *NH* 3. 117. [34] *NH* 4. 43. [35] *NH* 5. 119. [36] *NH* 6. 60 f

invenitur in nostro orbe, ut omittamus Indicum a formicis aut apud Scythas grypis erutum tribus modis: fluminum ramentis, ut in Tago Hispaniae, Pado Italiae, Hebro Thraciae, Pactolo Asiae, Gange Indiae, nec ullum absolutius aurum est, ut cursu ipso attrituque perpolitum. alio modo puteorum scrobibus effoditur aut in ruina montium quaeritur.')

He continues[37] (ibid. 67): 'Gold prospectors begin by gathering earth that indicates the presence of gold [*segullum*].[38] The auriferous material of the deposit is washed and from the sediment an analysis of the parent ore body [*tenor*][39] can be made. Sometimes, by a rare stroke of luck, a pocket is found on the surface, such as the one recently found in Dalmatia during Nero's principate which yielded fifty pounds of gold in a day. Gold occurring in this manner is called an outcrop [*talutium*][40] if there is also auriferous earth beneath. The otherwise parched and barren mountains of Spain, in which nothing else is produced, are compelled to yield a harvest of this commodity.' ('aurum qui quaerunt, ante omnia segullum tollunt; ita vocatur indicium. alveus hic est harenae, quae lavatur, atque ex eo, quod resedit, coniectura capitur. invenitur aliquando in summa tellure protinus rara felicitate, ut nuper in Dalmatia principatu Neronis, singulis diebus etiam quinquagenas libras fundens. ....')

'Gold mined by means of shafts is called "channelled",[41] or "trenched" [33. 68]: it is found adhering to quartz, not in the way it shines in lapis lazuli in the East, or in the granite of Thebes, or in other precious stones, but glistening as nodules in a quartz matrix. Traces of veins appear here and there along the walls of underground galleries—this is the derivation of its

---

[37] See generally P. R. Lewis and G. D. B. Jones, 'Roman Gold-mining in North-west Spain', *Journal of Roman Studies*, 60 (1970), 169 ff.

[38] Cf. the Spanish term *segullo*.

[39] A. M. Bateman, *Economic Minerals*² (London, 1950), 24, and Whitten–Brooks, *Geology*, 441. Cf. Forbes, *Ancient Technology*, ix. 6 ff.

[40] Cf. C. Domergue, 'Apropos de Pline, *Naturalis Historia* 33. 70-8', *Archivo Español de Arte y Arqueologia*, 45–7 (1972–4), 502. See further Zehnacker, *NH 33*, p. 174 ad loc. (33. 67). O. Bloch (ed.), W. von Wartburg (rev.), *Dictionnaire étymologique de la langue française*, s.v. 'talus', takes this to be a Gallic word derived from *talo*, 'front'. Zehnacker rightly concludes that, in this context, *talutium* signifies 'la couche de terre aurifère' and is a Spanish term.

[41] *Canalis* normally refers to water-pipes, conduits, or channels, whether open, or closed. In isolation this word would have posed a problem, but here Pliny provides his own definition.

name—and the overburden is supported by wooden props.' ('quod puteis foditur, canalicium vocant, alii canaliense, marmoris glareae inhaerens, non illo modo, quo in oriente sappiro atque Thebaico aliisque in gemmis scintillat, sed micans amplexu marmoris. vagantur hi venarum canales per latera puteorum et huc et illuc, inde nomine invento, tellusque ligneis columnis suspenditur.')

The third method [33. 70 f.] used for extracting gold rivals the achievements of the giants. By the light of lamps, long galleries are cut into the mountain. Men work in long shifts, measured by lamps, and may not see daylight for many months on end. The local people call all these mines deep-vein. The roofs of these are liable to give way and crush the miners, which makes diving for pearls,[42] or getting purple-fish from the depths[43] seem comparatively safe. So much more dangerous have we made the earth. Arches are left at frequent intervals to support the mountains above. [In opencast and deep-vein mining] masses of flint are encountered. These can be split by fire-setting[44] which involves the use of vinegar. Fire-setting in galleries usually makes them suffocatingly hot and smoke-filled. Instead, therefore, the rocks are split by means of crushers which carry 150 pounds of iron. The miners then carry the ore out of the workings on their shoulders, each man forming part of a human chain working in the dark; only those at the end of the line see daylight. If the bed of

---

[42] *NH* 9. 110.   [43] *NH* 9. 125 ff.

[44] The practice of fire-setting is well attested in ancient literature, as, for example, in Diodorus Siculus (3. 12. 1–13): 'The gold-bearing earth which is hardest they first burn with a hot fire and when they have crumbled it in this way, they continue working it by hand.' In particular, Livy (21. 37. 2–3) is the first to refer to the additional use of vinegar by Hannibal to disintegrate a fall of rocks during his crossing of the Alps. B. W. Holman, 'Heat Treatment as an Agent in Rock Breaking', *Transactions of the Institute of Mining and Metallurgy*, 36 (1926–7), 21–9, has confirmed by experiment that ores rich in quartz, if heated to a temperature of between 560 and 600 °C and properly quenched, undergo a remarkable physical (molecular) change. They may be rendered so friable as to be reduced to powder by rubbing with the fingers. Cf. G. Boon, 'Aperçu sur la production des métaux non-ferreux dans la Bretagne romaine', *Apulum IX* (Alba Iulia, 1971), 483 n. 93. The efficiency of methods employing fire, water, or vinegar have recently been re-examined by R. Shepherd, 'Hannibal the Rockbreaker', *Minerals Industry International*, 1008 (Sept. 1992), 39–47: 'Results from laboratory tests showed soured wine to be the most effective, but acetic acid, even at its weakest strength in solution, produced more efficient rock breakage than water. Based on these results ... tests were carried out at a suitable limestone quarry embracing the application of water and acetic acid solution on larger pieces of rock which had

flint seems too long, the miners bypass it. Yet flint is considered relatively easy to work.

tertia ratio opera vicerit Gigantum. cuniculis per magna spatia actis cavantur montes lucernarum ad lumina; eadem mensura vigiliarum est, multisque mensibus non cernitur dies. arrugias id genus vocant. siduntque rimae subito et opprimunt operatos ut iam minus temerarium videatur e profundo maris petere margaritas atque purpuras. tanto nocentiores fecimus terras! relinquuntur itaque fornices crebri montibus sustinendis. occursant in utroque genere silices; hos igne et aceto rumpunt saepius vero, quoniam id cuniculos vapore et fumo strangulat, caedunt fractariis CL libras ferri habentibus egeruntque umeris noctibus ac diebus per tenebras proximis tradentes; lucem novissimi cernunt. si longior videtur silex, latus sequitur fossor ambitque. et tamen in silice facilior existimatur opera.

Gold is obtained by hydraulicking,[45] as Pliny (33. 74 ff.) explains:

Another equally laborious and even more expensive task involves the feat of bringing streams along mountain-ridges—often a distance of a hundred miles[46]—to wash away the debris from mining operations. The miners call these water-channels 'corrugi', a term derived from the word 'confluence' [*conrivatio*] and they involve countless problems. The incline must be steep to produce a *surge* rather than a steady *flow* of water and consequently high-level sources are required. Gorges and crevasses are bridged by aqueducts. Elsewhere, impassable rocks are cut away to allow space for hollow wooden troughs. The workmen cutting out the rock hang suspended by ropes, so that viewed from a distance the operation seems to involve not so much a species of strange animals as of birds. Most hang suspended as they take levels and mark out the route—man leads rivers to run

---

been subjected to heating in a wood fire.' Shepherd's experiments clearly demonstrate that fire-setting is a viable technique for splitting rocks. Vinegar, or acetic acid, dissolved calcium carbonate, but not necessarily other substances and this might limit its use to the breakage of limestone-type rocks, the derived limestone schists (marble, etc.) and other rocks containing significant amounts of calcium. Pliny in this passage specifically refers to flints (*silices*). Similarly, Vitruvius (*De arch*. 8. 3) states that 'flints, when heated in a fire and sprinkled with "acid", fly asunder and are dissolved'. Flints can, in fact, be broken by the process of fire-setting and quenching with water; the use of vinegar does not necessarily increase the effectiveness of the technique.

[45] Healy, *Miniere*, 100 f.
[46] Domergue, 'Apropos de *NH* 33. 70–8'. 509–11, suggests a distance of about 33 miles (53 km.) rather than this obviously exaggerated figure.

where there is no place for him to plant his own footsteps. The washing of the ore is spoilt if the current of the stream brings salt, or silt, as this residue is called. To avoid this, the water is guided over flint, or pebbles.

alius labor ac vel maioris inpendii: flumina ad lavandam hanc ruinam iugis montium obiter duxere a centesimo plerumque lapide; corrugos vocant, a conrivatione credo. mille et hic labores: praeceps esse libramentum oportet, ut ruat verius quam fluat; itaque altissimis partibus ducitur. convalles et intervalla substructis canalibus iunguntur. alibi rupes inviae caeduntur sedemque trabibus cavatis praebere coguntur. qui caedit, funibus pendet, ut procul intuenti species ne ferarum quidem, sed alitum fiat. pendentes maiore ex parte librant et lineas itineri praeducunt, quaque insistentis vestigiis hominis locus non est, amnes trahuntur ab homine. vitium lavandi est si fluens amnis lutum inportet; id genus terrae urium vocant. ergo per silices calculosve ducunt et urium evitant.

Pliny continues (33. 75): 'On the ridge above the head of the mine reservoirs are excavated, measuring more than 200 feet each way and 10 feet deep. Five sluices each about three feet square, are constructed in the walls. When the reservoir is full, the sluices are knocked open so that the violent down-surge of water is sufficient to sweep away the rock debris.' ('ad capita deiectus in superciliis montium piscinae cavantur ducenos pedes in quasque partes et in altitudinem denos. emissaria in iis quina pedum quadratorum ternum fere relinquuntur, ut repleto stagno excussis opturamentis erumpat torrens tanta vi ut saxa provolat.')

'The next process takes place on level ground [33. 76]. Water conduits—the Greek name for which is *agogae* [ἀγωγαί[47]]—are cut in steps and their floors covered with gorse, a plant resembling rosemary, which is rough and so traps the gold particles. The conduits are boarded with planks and carried on arches over steep pitches. Thus the tailings flow down to the sea and the shattered mountain is washed away. It is because of this that Spain has now encroached substantially on the sea.' ('alius etiam in plano labor. fossae per quas profluat, cavantur—agogas

---

[47] (ὕδατος ἀγωγαί) also used of aqueducts: *IG* xii (5), 872 (Tenos). Cf. Dion. Hal 3. 67.

vocant; hae sternuntur gradatim ulice. frutex est roris marini similis, asper aurumque retinens. latera clauduntur tabulis, ac per praerupta suspenduntur canales. ita profluens terra in mare labitur ruptusque mons diluitur, ac longe terras in mare his de causis iam promovit Hispania.')

[33. 77 f.] Gold mined from deep veins does not require heat treatment but is pure. Nuggets are found in such mines and similarly in pits; some weigh more than ten pounds. These are called 'palagae', or 'palacurnae', while gold dust is also known as 'balux'. The gorse is dried and burnt and its ash washed on a bed of grassy turves so that the gold is deposited on these. According to some authorities Asturia and Callaecia and Lusitania produce in this way 20,000 pounds weight of gold a year, Asturia supplying the largest amount.

aurum arrugia quaesitum non coquitur, sed statim suum est. inveniuntur ita massae, nec non in puteis, et denas excedentes libras; palagas, alii palacurnas, iidem quod minutum est balucem vocant. ulex siccatur, uritur, et cinis eius lavatur substrato caespite herboso, ut sidat aurum. vicena milia pondo ad hunc modum annis singulis Asturiam atque Callaeciam et Lusitaniam praestare quidam prodiderunt, ita ut plurimum Asturia gignat.

17.4.4 Gold-refining

Deep-vein, or reef, gold[48] needs only mechanical treatment to release it from its quartz matrix, but alluvial[49] deposits need to be refined. Pliny (33. 80) erroneously states that '*all* gold contains silver in various proportions, a tenth part in some cases, an eighth in others. In one mine only, that of Callaecia called the Albucrara mine, the proportion of silver found is one thirty-sixth and consequently this is more valuable than all the others.' ('omni auro inest argentum vario pondere, aliubi decuma parte, aliubi octava. in uno tantum Callaeciae metallo, quod vocant Albucrarense, tricensima sexta portio invenitur; ideo ceteris praestat.')

Most alluvial gold contains not only silver, but small amounts of copper and traces of platinum group elements (PGE).[50] The silver and copper could be removed by a cementation process

---

[48] Healy, *Mining*, 33 and elsewhere.   [49] Ibid. 153 f.
[50] Healy, 'Greek and Roman Gold Sources,' 15 ff.

although the PGE remained intact through all heat, or mechanical, treatments. There is ample evidence to show that the Greeks, like the Egyptians, were able to refine gold from an early date, since they distinguished 'white gold' (λευκὸς χρυσός) from purified gold (ἄπεφθος χρυσός).[51] Few references to refining survive, and Agatharcides[52] gives the most complete account:

> In the last stage, skilled workmen receive the ore which has been ground and take it for its final processing; they rub the crystalline rock, reduced to a fine consistency, on the surface of a wide board with a slight incline and pour water over this throughout the operation. Then the gangue is separated by the water and flows down the board; the part of the ore which contains the gold, however, remains on the wood by reason of its weight. They do this a number of times, first of all rubbing it gently with their hands then pressing it lightly with loose-textured sponges. They absorb the porous material and gangue until only pure gold dust remains.
>
> On completion of this, other skilled workmen take what has been recovered and put it by measure and weight into earthen pots. They mix with this a lump of lead according to the mass, lumps of salt, a little tin, and barley bran. They put on a closely fitting lid carefully smearing it with mud and heat it in a furnace for five days and nights continuously; they allow the pots to cool and find no residual impurities in them; the gold they recover in a true state with little wastage. This processing of gold is carried on round about the most distant boundaries of Egypt.

This method would appear to be a conflation of two processes, namely cupellation and salt cementation.

Archaeological and numismatic evidence from Croesus' refineries at Sardis[53] shows that the Lydians were able to recover gold by such a technique, certainly by the mid-sixth century BC, which led to the introduction of bimetallic coin

---

[51] See Healy, 'Greek White Gold', 195.

[52] Quoted by Diod. Sic. 3. 14. 1–4. See generally J. F. Healy, 'Mining and Processing of Gold Ores in the Ancient World', *Journal of Metals* (American Institute of Metallurgical Engineers), 31/8 (1979), 11–16, and 'Greek Refining Techniques and the Composition of Gold–Silver Alloys'; *Revue belge de Numismatique*, 120 (1974), 25 ff. Cf. also Z. Szabó, 'Az aranyfinomításról' [Gold-refining], *Múzeumi Műtárgyvédelem*, 2 (1975), 105–19.

[53] Hanfmann, *Sardis*, 230 ff. and figs. 176–7.

series[54] in place of the earliest issues struck in the mixture known as 'white gold'. Excavations of gold and silver workshops in the vicinity of the Artemisium, uncovered gold lumps, globules of gold, and pitted foil. The surface pitting indicates that a salt cementation technique had been employed.[55] Pászthory[56] convincingly reconstructs the process at Sardes: 'Gold dust from Pactolus was melted and hammered into thin plates (to give the maximum surface exposure) prior to being heated with common salt in an earthenware pot. In addition, further confirmatory evidence is provided by electron microscopy which characterizes the structures of the electrum/silver interface, showing the growth of sodium–silver chloride crystals.'

Modern experiments[57] have simulated the method described by Agatharcides with the twofold aim of confirming the method and assessing its effectiveness. In the absence of samples of alluvial gold, commercial nine-carat gold was used. Three main groups of experiments were conducted: (1) salt and alloy were smelted under a variety of conditions; (2) salt, brick dust, and alloy were subjected to the traditional cementation processes; (3a) tin and lead were added to these, and (3b) tin, lead, and carbon. All charges were subjected to temperatures of 800 °C in sealed sillimanite crucibles for five days. The reaction was seriously inhibited in 3b by the presence of tin which formed a covering of oxide and recovered between 51 and 76 per cent gold. The best result, however, yielded a high gold content of 93.6 per cent, the silver and impurities forming a 'vitreous' (lead-based) slag.

A similar method of cementation is described by Strabo (2. 2. 8): 'They say that nuggets weighing half a pound are sometimes found in the gold dust; these they call "palae" and need little refining. ... the product of smelting the gold and

---

[54] P. R. Franke and M. Hirmer, *Die griechische Münzen* (Munich, 1964), 123.
[55] See Hanfmann, *Sardis*, 236 and fig. 178.
[56] 'Investigations of the Early Electrum Coins of the Alyattes Type', in D. M. Metcalf (ed.), *Metallurgy in Numismatics*, i (Royal Numismatic Society, special publication, 24; London, 1980), 153.
[57] J. F. Notton, 'Ancient Gold Refining', *Gold Bulletin*, 7/2 (1974), 50–6. See also Healy, 'Greek Refining Techniques', 19–33.

refining it with a kind of "styptic" earth[58] (στυπτηριώδης γῆ) is electrum. They further smelt this mixture of gold and silver; the silver is burned away and the gold remains.... the gold is preferably smelted by a fire fuelled by chaff since the flame is gentle and suitable for an alloy which yields and fuses easily; a charcoal fire, on the other hand, consumes a lot since it overmelts the gold and volatilizes it because of its intensity.' This is basically a salt cementation process except that styptic earth is a mixture of other salts of sodium and not just sodium chloride.

From this account it appears that Strabo does not understand the technique and confuses the role of styptic earth in the process of separation.

Pliny (33. 84) briefly describes the refining of gold: 'Gold is first heated with twice its weight of salt, three times its weight of copper pyrites,[59] and then again with two parts of salt and one of alum. This process removes the impurities when the other substances have been burnt away in an earthenware crucible; the gold itself is left behind pure and uncorrupted.' ('torretur et cum salis gemino pondere, triplici misyis ac rursus cum II salis portionibus et una lapidis quem schiston vocant. ita virus trahit rebus una crematis in fictili vase, ipsum purum et incorruptum.')

Pliny (33. 80) also refers to electrum[60] defining it as an 'ore' in which the proportion of silver is at least one-fifth: 'ubicumque quinta argenti portio est, *electrum* vocatur.'

17.4.5 Amalgamation

Pliny (33. 99), like Vitruvius (*De arch.* 7. 8. 5), incorrectly refers to the amalgamation[61] of gold with mercury as a method

---

[58] The natural assumption is that the mineral occurred locally, or at least elsewhere in North Africa, possibly in lower Egypt. If this is correct, then 'styptic earth' should be identified as the complex compound natron which comprises more than 50% sodium salts and would have been a suitable flux for a cementation process. It is possible that the Roman equivalent is alumen. Cf. J. F. Healy, 'Problems in Mineralogy and Metallurgy in Pliny the Elder's Natural History', *Atti del Convegno di Como—Technologia, economia, e società nel mondo romano* (Como, 1980), 191 n. 122. For an early account of ancient and modern natron, see further A. Lucas, *Ancient Egyptian Materials and Industries* (London, 1959), 548 ff. The mineral has also been incorrectly identified as ferrous sulphate.   [59] See Ch. 14, s.v. 'chalcitis'.

[60] The term is now used for the artificial alloy of gold and silver. See Healy, *Mining*, 207 and elsewhere.   [61] Tylecote, *Metallurgy*, 240.

of purifying gold. He does not realize that a further process of distillation[62] would have been needed to recover the gold. He writes: 'All substances float on its surface except gold which is the metal that it attracts to itself. Mercury is thus very good for refining gold, since, if the two are repeatedly shaken together in earthenware vessels, the mercury draws out all the impurities in the gold. After this, the separation of the mercury from the gold is achieved by pouring both on to well-dressed hides: the mercury is exuded through the hides like a kind of sweat and the gold is left pure.' ('omnia ei innatant praeter aurum: id unum se trahit. ideo et optime purgat, ceteras eius sordes exspuens crebro iactatu fictilibus in vasis, ita vitiis eiectis ut et ipsum ab auro discedat, in pelles. subactas effunditur, per quas sudoris vice defluens purum relinquitur aurum.')

17.4.6 *Gold-coating*

In a passage relating to mirrors, Pliny (33. 130) describes a process whereby glass was coated with gold to provide a reflective surface:[63] 'The best mirrors known to our ancestors were made at Brindisi of a mixture of *stagnum* [also *stannum*—an alloy of silver and lead] and copper. Silver mirrors have come to be preferred: they were first made by Pasiteles in the time of Pompey the Great. But it has recently come to be believed that a more reliable reflection is given by applying a layer of gold to the back of the glass.' ('optima [specula] apud maiores fuerant Brundisina, stagno et aere mixtis. praelata sunt argentea; primus fecit Pasiteles Magni Pompei aetate. nuper credi coeptum certiorem imaginem reddi auro opposito vitris.')

Whether the reading accepted here is 'aversis', or 'vitris', Pliny's meaning is, in my opinion, clear. The gold is surely placed on the reverse side of the 'mirror'. For this to reflect, the substance of the mirrors can only be glass. The sense of 'aversus' is confirmed by Martial (*Epigr* 8. 62. 1): 'Picens writes epigrams on the reverse side of the paper' ('Scribit in aversa Picens epigrammata charta'). This practice is also

---

[62] Ibid. 47 f.
[63] A modern application of gold coating is seen in astronauts' visors.

alluded to by Juvenal (*Sat.* 1. 6), 'liber scriptus in tergo'. Zehnacker[64] retains the reading 'aversis', but on the grounds that the evidence does not justify its emendation. The reading, however, does not modify the meaning Pliny intended.

17.4.7 Gold Leaf

The malleability[65] of gold, known to Pliny (33. 61), is one of its most important physical properties. For at least 5,000 years craftsmen have exploited this unique feature, by hammering gold into leaf of extraordinary thinness which then, because of its beauty and durability (especially its resistance to corrosion) could be employed to ornament and protect that to which it was applied.

The Egyptians appear to have been the earliest practitioners of the art of gold-beating, and the illustrations on tombs at Saqqara and Thebes[66] show the gold-beaters working together with gold-founders and smiths. It seems likely that the thickness of the leaf employed varied considerably as knowledge of the art increased: the Luxor samples (late Eighteenth Dynasty) were some $0.3\,\mu m$ thick. Homer, in the *Odyssey*, refers to the gilding of the horns of sacrificial oxen. Persian officers are said to have slept in gilded beds before the battle of Plataea.[67]

Lucretius (*DRN* 4. 724 ff.) compares 'films from the surface of objects encountering one another... with gossamer, or gold leaf' ('rerum simulacra vagari... tenuia... ut aranea bratteaque auri').

Pliny (33. 61 f.) is well aware of the technique of gold-beating: 'no other metal... can be divided into so many portions; thus an ounce of gold can be beaten into upwards of 750 leaves four inches square. The thickest kind of gold leaf is called Praenestian, retaining a name derived from the superbly gilded statue of Fortune at Praeneste. The next in thickness is called Quaestorian.' ('nec aliud laxius dilatatur aut numerosius dividitur utpote cuius unciae in septingenas quinquagenas pluresque bratteas quaternum utroque digitorum spargantur.

---

[64] *NH 33*, p. 215, note ad. loc.   [65] Tylecote, *Metallurgy*, 84.
[66] See, generally, Nicholson, 'Ancient Craft of Gold Beating', 161–6 and illustration, p. 161.   [67] Ibid. p. 162.

crassissimae ex iis Praenestinae vocantur, etiamnum retinentes nomen Fortunae inaurato fidelissime ibi simulacro. proxima brattea quaestoria appellatur.') The thickness here indicated is of the order of 0.00034 mm., that is a little less than four times that of the finest modern gold leaf.[68]

The cult statue in the temple of Fortuna is one of a number of similarly gilded statues, as, for example, that of the young Alexander which Nero had gilded (*NH* 34. 63) and of Castor (Juv. Sat. 13. 152). Pliny's use of the term 'Quaestorian' is not readily explicable. It has been compared by Zehnacker[69] to a possible parallel in Martial, *Epigrams* 8. 33. 1 ff.:

> De praetorica folium mihi, Paule, corona
> mittis et hoc phialae nomen habere iubes.
> hac fuerat nuper nebula tibi pegma perunctum
> pallida quam rubri diluit unda croci.

You send me a gold leaf from the crown given at the Games by the praetor, and bid it have the name *phiale* [shallow dish]. Recently a raised platform had been finely covered by you with gold leaf from which the pale water, mixed with [essence of] red saffron, drains the colour.

With these two properties (resistance to corrosion and intrinsic malleability), as Tylecote explains, it was possible to use gold sparingly, since it could be beaten to thin foils only 1 μm thick which would stay bright and not lose thickness due to corrosion. Many striking pieces, such as the 'Mask of Agamemnon',[70] are of very thin foil. The inherent weakness can be offset by judicious alloying, but this will, of course, decrease its malleability.

A gold-beater can today reduce the thickness of an ingot of gold from 13 mm. to 0.05 μm. A 'book' of 256 gold leaves, about 3 inches (80 mm.) square, weighs only one-third of a gramme: it is this fact that has made it economic to use gold leaf so widely. (Egyptian lead gold leaf, used from 2300 to 1700 BC, had a thickness of between 0.3 and 1.0 μm.)[71]

---

[68] Zehnacker, *NH 33*, p. 170 ad. loc. (33. 61).
[69] Ibid.
[70] Tylecote, *Metallurgy in Archaeology*, 239.
[71] Tylecote, *Metallurgy*, 84.

Nutting and Nuttall,[72] in their discussion of the properties of gold leaf, conclude that the reason for its remarkable capacity of deformation is not due to gold being more ductile than other metals with a similar crystal structure, but to the absence of an oxide film which normally prevents the escape of dislocations from a foil when the foil thickness approaches the sub-grain size.

17.4.8 Gilding

One of the many controversial topics in the *Natural History*, until modern laboratory experiments resolved the problem, has been Pliny's account of the gilding[73] of copper substrates. He writes (33. 64):

The regular way to gild copper would be to use natural, or at all events, artificial quicksilver, concerning which a method of adulteration has been devised, as I shall relate in describing the nature of these substances. The copper is first subjected to fire; then, when it is red hot, it is quenched with a mixture of brine, vinegar, and alum, and, afterwards, put to a test, its brilliance of colour showing whether it has been sufficiently heated; then it is again dried in the fire, so that, after a thorough polishing with a mixture of pumice and alum, it is able to take the gold leaf laid on with quicksilver. Alum has the same cleansing property here that we said is found in lead.

aes inaurari argento vivo aut certe hydrargyro legitimum erat, de quis, ut dicemus illorum naturam reddentes, excogitata fraus est. namque aes cruciatur in primis accensumque restinguitur sale, aceto, alumine, postea examinatur an satis recoctum. sit, splendore deprehendente, iterumque exhalatur igni, ut possit, edomitum mixtis pumice et alumine inductas accipere bratteas, alumen et in purgando vim habet qualem esse diximus plumbo.

It was long thought that Pliny's description of gilding referred to the technique called 'hot-mercury' gilding, or 'fire-gilding'.[74] This implied that he, like Vitruvius (*De arch.* 8. 9), who also describes the method, omits to mention the final

---

[72] 'The malleability of Gold: An Explanation of its Unique Mode of Deformation', *Gold Bulletin*, 10/1 (1977), 2–8.
[73] O. Vittori, 'Pliny the Elder on Gilding: A New Interpretation of his Comments', *Gold Bulletin*, 12/1 (1979), 35–9.   [74] Ibid. 35.

stage of the process, which consists of heating the piece in order to evaporate the mercury.

A study of the surface characteristics of the Golden Horses of St Mark's Basilica, Venice,[75] led to a wider, detailed examination of gilding techniques used in the ancient world. Craddock[76] suggests that the choice of pure copper for some gilded artefacts of the classical age is the result of difficulties encountered by craftsmen in mercury-gilding substrates made of lead, or tin-rich copper alloy—a view which Theophilus seems to support.

Pliny (33. 100) continues his account: 'Consequently, when also things made of copper are gilded, a coat of quicksilver is applied underneath the gold leaf and keeps it in its place with the greatest tenacity but, if the gold leaf is put on in one layer, or is very thin, it reveals the quicksilver by its pale colour. Consequently, persons intending this fraud adulterated the quicksilver used for this purpose with white of egg; later they also falsified hydrargyrus, artificial quicksilver, about which I will speak in its proper place.' ('ergo et cum aera inaurentur, sublitum bratteis pertinacissime retinet, verum pallore detegit simplices aut praetenues bratteas. quapropter id furtum quaerentes ovi liquore candido usum eum adultavere, mox et hydrargyrum de quo dicemus suo loco.' Later (33. 125) he writes: 'At the present time, silver is almost the only substance that is gilded with artificial quicksilver, though really a similar method ought to be used in coating copper' ('hydrargyro argentum inauratur solum nunc prope, cum et in aera simili modo duci debeat').

Pliny refers to gold leaf applied directly to the substrate and the following operations are involved, as Vittori[77] explains.

(1) Mercury is rubbed on the surface of the copper substrate when it is cold. Some copper dissolves in the mercury and forms a very thin layer of copper–mercury amalgam. At room temperature, copper

---

[75] O. Vittori and Anna Mestitz, 'Artistic Purpose of Some Features of Corrosion on the Golden Horses of Venice', *Burlington Magazine*, 117 (1975), 132–9. See also Healy, 'Problems in Mineralogy', 182 and fig. 10 (p. 183).
[76] 'The Golden Horses of San Marco', *Journal of Archaeological Science*, 4 (1977), 103–23.   [77] 'Pliny on Gilding', 36 n. 59.

does not become soft in the presence of mercury. (2) Any excess of mercury is mechanically removed. This operation leaves the surface of the object shining and smooth as a mirror. (3) Gold leaf is then pressed on the surface, it absorbs a little of the mercury from the copper, but does not soften and does not form a conventional amalgam. Even if the gold leaf absorbs all the mercury from the copper–mercury amalgam, it does not disintegrate, but keeps its original leaf form and bonds very firmly to the substrate. This method, best referred to as 'cold-mercury' gilding, does not appear to require special operator skills since the amount of mercury involved is independent of the procedure followed by the craftsman and the mechanics of the operation are defined exclusively by the solubility of copper in mercury. An additional advantage is that the consumption of mercury is minimal.

Vittori[78] confirms these conclusions by a series of laboratory tests using pure copper, or 7 per cent tin bronze as the substrate. In his experiments the copper surface was polished until it had a mirror-like finish and the mercury was spread on to this to act as an adhesive for the gold leaf. Any mercury that contaminates the outer surface causes a characteristic paleness in the colour of the gold leaf, a fact accurately recorded by Pliny.

Vittori also observes an interesting aspect of cold-mercury gilding, which concerns the finishing of the gilt surface. While gilding applied on pure copper can be easily finished with a burnishing tool, this is not the case with a bronze substrate. On the latter alloy, burnishing destroys the gilded layer. Alloying of gold with tin easily takes place in the presence of mercury at the gold/bronze interface (mercury amalgamates well with tin) and the brittleness of the products must be the cause of the observed fragility of the gild.[79] Even in hot cladding, where mercury is not used, tin can migrate from the bronze into the gold leaf at high temperature. As might be expected in terms of the interpretations above, cladding in which the use of the burnishing tool is of fundamental importance gave good results (in experiments) on pure copper but was impossible on bronze.[80]

---

[78] Ibid. 37 f.
[79] *Enciclopedia della Chimica Guareschi* (Turin, 1910), 428.
[80] See Vittori, 'Pliny on Gilding', 39 n. 17.

The association of gilding with pure copper in ancient artefacts from the classical period is confirmed by the nature of the metal objects found on Roman ships recovered from the bottom of Lake Nemi.[81] A list of these artefacts classifies those of interest in the present connection as either of 'pure copper' or of 'bronze'. All the former are reported to be gilt, while of the latter items only one is described as gilt.

In his detailed examination of Pliny's account of gilding, Vittori[82] gives a fair assessment of the position with regard to the technological information found in the *Natural History*. The vicissitudes to which the manuscripts have been exposed through the centuries explain the difficulties experienced in the interpretation and understanding of the original text of the work. But for the experiments carried out here, the whole technique decribed by Pliny might well have been dismissed as not viable and cited as a further example of his incompetence as a 'scientist'.

Vittori's research yet again demonstrates the importance of the 'marriage of science and philology'.

17.4.9 Tests

17.4.9a Touchstone (chert, lyddite)

Touchstones[83] are small pieces of rock which have been used by goldsmiths for at least 2,500 years to determine the approximate purity of gold, and, to some extent, silver. They include any black, fine-grained rock (chert, lyddite, or basalt) which is sufficiently abrasive to produce a streak when a gold object is rubbed across the surface and the assay is made by comparing the colour of the streak with alloys of known composition. They were used without acid reagents in the ancient world.

In a detailed petrographic examination of a number of examples, Moore and Oddy[84] discuss the identification of touchstones which have, over the years, been erroneously identified with a number of minerals. Partington, for example,

---

[81] G. Ucelli, *Le Navi di Nemi* (Rome, 1940).
[82] See 'Pliny on Gilding', 36. [83] Tylecote, *Metallurgy*, 80 f.
[84] D. T. Moore and W. A. Oddy, 'Touchstones: Some aspects of their Nomenclature, Petrography and Provenance', *Journal of Archaeological Science*, 12 (1985), 59–80.

described the 'Lydian' stone as a variety of black basalt. This would in any case be unlikely, since the Menderes Massif, in which region early examples are said to have been found, is a predominantly Precambrian or Palaeozoic metamorphic massif, where quartzites, phyllites, marble, and gneisses are known to occur. Volcanic rocks are not mentioned in accounts of that region, neither are they shown on geological maps. The terms basanite, basalt, black flint, black marble, clay and flinty slate, bituminous quartzite, and lyddite are all more or less defined in terms of their mineralogy and texture, and none should be used to describe touchstones unless a petrographic examination has shown that they apply.

Recent research on a series of touchstones would appear to confirm the early findings of Werner, Pinkerton, and Dana.[85] Nearly all the specimens were siliceous and many showed a degree of secondary silification. Indeed the specific gravities of most of the touchstones suggested that few mafic minerals were present but that they were made principally from rocks such as chert, consisting of quartz and secondary silica.

Ancient references to the touchstone are found in Greek literature as early as the sixth century BC when the term βάσανος first occurs in Theognis,[86] a gnomic poet born in Megara. Pindar (*Pyth.* 10. 67) also specifically refers to the testing of gold by the touchstone in an analogy relating to the proof of an upright mind: πειρῶντι δὲ καὶ χρυσὸς ἐν βασάνῳ πρέπει καὶ νόος ὀρθός ('Just as gold shows its nature to one who tests it by the touchstone, so does an upright mind reveal itself when put to the test').

Theophrastus (*Lap.* 45 ff.),[87] however, gives the first definitive account:

The nature of the stone that tests gold is remarkable, for it seems to have the same power as fire, by which it can also be assayed. On that account some people are puzzled about this, but without good reason,

---

[85] D. T. Moore and W. A. Oddy, 'Touchstones: Some aspects of their Nomenclature, Petrography and Provenance', *Journal of Archaeological Science*, 12(1985), 59–80.

[86] Thgn. 417: ἐς βάσανον δ᾽ ἐλθὼν παρατρίβομαι ὥστε μολύβδῳ χρυσός ('I am put to the test like gold beside lead [on the touchstone]'); cf. Thgn. 450 and 1105.

[87] E. R. Caley and J. F. Richards (trans. and comm.), *Theophrastus on Stones* (Columbus, Oh., 1956), 157.

for the stone does not test in the same way. Fire works by changing and altering the colours; the stone works by friction. For it seems to have the power of picking out the essential quality of each metal. They say that a much better stone has now been found than the one used before; for this not only detects purified gold, but also gold and silver that are alloyed with copper, and it shows how much is mixed in each stater. And the indications are obtained from the smallest weight. All such stones are found in the river Tmolus. They are smooth in nature and like pebbles, flat, not round and, in size, twice as big as the largest pebble. The top part which has faced the sun differs from the lower surface in its testing power and tests better than the other. This is because the upper surface is drier, for moisture prevents it from picking out the metal. Even in hot weather the stone does not test well, for it gives out a moisture which causes slipping. This happens also to other stones including those from which statues are made, and this is supposed to be a peculiarity of the stone.

Theophrastus here makes an exaggerated claim that the stone can give a 'quantative analysis' of what is in effect a 'ternary' alloy. Such accuracy is, even now, only able to be achieved by chemical analysis, spectrography, or sophisticated non-destructive techniques![88]

The reference to Tmolus[89] as a river is incorrect, since it is the name of a mountain in Lydia lying between the rivers Hermus (in the north) and Cayster (to the south). Caley and Richards, in their commentary, consider that Theophrastus could have meant the river Pactolus, which rises on Mt. Tmolus: this is a reasonable assumption. The name 'Lydian'— perhaps inevitably, but not convincingly—is associated with the fact that early 'white gold' coinage (all too often mistakenly referred to as 'electrum') was first struck by the Lydians, in the seventh century BC in the vicinity of Sardis.[90] It is possible, but not proven, that the use of the touchstone was, like so many discoveries in the ancient world, accidental.

---

[88] Healy, 'Greek White Gold', 197 ff. These include XRF, NAA, XPS, PIXE, PAA, and EPMA (X-Ray Fluorescence spectroscopy, Neutron Activation Analysis, X-Ray photoelectron spectroscopy, Proton Induced X-Ray Emission spectroscopy, Photon Activation Analysis, and Electron-Probe Microanalysis).
[89] Pliny refers to Tmolus as a mountain in Lydia in 5. 110 f. and 118, 6. 215, and 7. 159. The present designation is clearly a slip on Pliny's part. Mt. Tmolus is part of the Menderes Massif.
[90] See, generally, Hanfmann, *Sardis*, 141 ff.

Pliny (33. 126) mainly follows Theophrastus' account of the touchstone. 'Some people call it 'Heraclean'[91] stone and others, 'Lydian'.[92] The pieces are of a medium size, not exceeding four inches in length and two in breadth. The side that has been exposed to the sun[93] is better than that in contact with the soil. Experts use this touchstone like a file to take a scraping[94] from an *ore*; they can tell at once to a scruple[95] how much gold, silver, or copper is present, and their amazing calculation is absolutely without error.' ('alii Heraclium, alii Lydium vocant. sunt autem modici, quaternas uncias longitudinis binasque latitudinis non excedentes. quod a sole fuit in iis, melius quam quod a terra. his coticulis periti cum e vena ut lima rapuerunt ramentum, protinus dicunt quantum auri sit in ea, quantum argenti vel aeris, scripulari differentia, mirabili ratione non fallente.')

In the case of a less complicated, binary alloy of gold and silver,[96] however, a remarkable degree of accuracy can be achieved by an experienced operator using a touchstone. Table 17.1 below, an analysis of four specimens of gold by four independent 'assayers', shows the range of results that may be obtained.[97]

---

[91] This and the following stone are incorrectly linked by Theophrastus (*Lap*. 4) and by Pliny who follows his text implicitly. The 'Heraclean' stone, however, is not an alternative name for a touchstone. In *NH* 36. 127, Pliny correctly identifies this: 'The magnet is called by the Greeks by another name, the ironstone, and by some of them the stone of Heracles' ('sideritim ob id alio nomine vocant, quidam Heraclian').

[92] This is the original name, found, for example, in Bacchylides, fr. 10 (Jebb) Λυδία λίθος cf. also Theoc. Id. 12. 36, Λυδία πέτρη. See also n. 100 below. Surprisingly, however, in Sophocles, fr. 800 (Nauck/Snell), it is the term used for a magnet.)

[93] Pliny is again following Theophrastus' observation. Contact between the specimen and the surface would naturally be better if the surface were dry.

[94] Reading 'ramentum', not 'experimentum', as Zehnacker, *NH 33*, p. 211 n. 4. Cf. *NH* 33. 127, 'ramentum'.

[95] See Zehnacker, *NH 33*, p. 211 n. 1, who follows Eichholz, *Lap*., comm. ad loc. He bases his interpretation on the degree of accuracy here specified—namely 1/24. Cf. *NH* 34. 94: 'added in the proportion of six scruples of gold to the ounce' ('in uncias additis auri scripulis senis').

[96] Healy, *Mining*, 205. In the testing of electrum, results are not as accurate when substantial amounts of copper are present. See further, Healy, 'Greek White Gold', 204 ff.

[97] W. A. Oddy and F. A. Schweizer, 'A Comparative Analysis of Some Gold Coins', in Hall–Metcalf, *Chemical and Metallurgical Investigation*, 179, and Healy, *Mining*, 206. See also Christiane Eluère. 'A Prehistoric Touchstone from France', *Gold Bulletin*, 19/2 (1986), 58 ff. Touchstones are still in use

TABLE 17.1

Specimen number	Assayers				Neutron activation	Chemical analysis	
	1	2	3	4		(a)	(b)
(1)	>90	>90	>90	>90	91.6	98.49	97.57
(2)	<30	<30	<30	<30	28.0	21.1	22.99
(3)	65.5	64.5	65.5	64.5	67.5	64.4	64.27
(4)	58.5	58.5	58.5	58.5	58.8	58.7	58.30

The strength of the reagent used as standard is as follows: (a) for 18 ct. and finer samples: nitric acid—40 cc., hydrochloric acid—1 cc., distilled water—15 cc., (b) for 10–18 ct. samples: nitric acid—30 cc., hydrochloric acid—0.5 cc., distilled water—10 cc.; (c) for samples below 10 ct.: equal volumes of concentrated nitric acid and distilled water. The touch-needles are made of gold-silver alloys of known composition.[98] To test a sample needles which seem to correspond most closely with the gold, or gold alloy, to be examined are chosen. The needle is rubbed firmly up and down the touchstone to make a streak about 3 cms. long and 0.5 cm. wide. Next a similar streak is made with the unknown metal as close to the first one as possible. After this a glass rod is dipped in

---

today. For the degree of accuracy that may be achieved see E. Walchi, 'Touching Precious Metals', *Gold Bulletin*, 14/4 (1981), 154–8. Differences of 10–20 parts per thousand in gold content may be established, using appropriate reference standards and acids to discolour the contaminant metals.

[98] I am indebted to Mr A. H. Westwood, formerly Assay Master at the Birmingham Assay Office, for this account of the use of the touchstone. 'The stone employed is a siliceous schist (flinty slate) of a black colour, the best stones being of a uniform deep black, without light-coloured veins or spots, and of a fine grain. The stone must be ground so that the surface is completely flat, otherwise acid will accumulate in indentations and cause uneven reactions on the touch streaks. The surface must also be matt and not polished, or the touch will not adhere to the stone. ... In the ancient world, in the absence of suitable acid reagents, comparisons were made directly between streaks of alloy for analysis and those from standard, made-up alloys' (personal communication). The accuracy achieved is likely to have been surprisingly high in the case of gold alloyed with silver, or copper in the range of 50–75% gold content, as modern experiments confirm. See n. 97 above.

the reagent and drawn evenly across the two streaks so that a strip about 0.5 cm. wide is uniformly covered with acid. From the action of the acid on the two streaks it will be ascertained which is of the higher quality. Another touch needle is then selected and so on, until two streaks are made which react equally with the acid.

An interesting suggestion by Moore and Oddy[99] is that the appropriate general term for such minerals should be 'touchstone', or 'Lydian' stone! The dense variety of chert is currently called lyddite.

At the ingot stage, however, the purity of gold may be checked by weight against a standard of equal volume[100]—the latter being able to be calculated by linear measurement.

17.4.9b Colorimetry

A much simpler method of estimating gold content is based on colorimetry. At Serra Pelada,[101] in Brazil, the refined gold (*bombril*) is compared with a range of prepared samples of known composition.

17.4.9c Gold Trial

Finally, Pliny (33. 59) observes that gold is the 'only [metal] that loses nothing by contact with fire.... Indeed the opposite is the case: the quality of gold is enhanced the more it is subjected to fire.' ('rerum uni nihil igne deperit.... quin immo quo saepius arsit, proficit ad bonitatem.') This view prevailed through the centuries and was held as late as Sir Isaac Newton's day,[102] as is shown by the Gold Trial of 1710,

---

[99] See further, 'Touchstones', p. 293, n. 84, above.

[100] Healy, *Mining*, p. 207. On density measurements generally, cf. W. A. Oddy and M. J. Hughes, 'A Reappraisal of the Specific Gravity Method for the Analysis of Gold Alloys', *Archaeometry*, 12/1 (1970), 1–11, and C. Webster, 'Some Practical Aspects of Specific Gravity Determinations', *Numismatist*, 89 (1976), 991–1000.

[101] Auriferous material is placed in packets, together with borax as a flux, and melted at 1500 °C by a blow torch, if less than 100 gms. in weight, or, if a larger amount, in a high-temperature oven. When the blackened, fused borax is broken open, lumps of refined gold of varying size are found within. I am indebted to Mr R. Saunders of BBC TV and Dr I. P. de Martins of the Brazilian Embassy for information and advice about operations at Serra Pelada.

[102] E. G. V. Newman, 'The Gold Metallurgy of Isaac Newton', *Gold Bulletin*, 8/3 (1975), 90–5, esp. 93 ff.

following the Scottish coinage issued after the Act of Union. It was found that the trial plate was purer than any coin. There was, however, clearly no idea of 'absolute purity', since Newton claimed that, by re-heating, gold might be refined to half a carat finer than 24 carat.

17.5 SILVER

Pliny's detailed treatment of gold is not matched by his description of the mining and metallurgy of silver. Silver (33. 95) 'is found only in deep shafts and does not advertise its existence, having no shiny particles such as are seen in association with gold' ('non nisi in puteis reperitur nullaque spei sui nascitur, nullis, ut in auro, lucentibus scintillis'). His account of silver, which is obtained from argentiferous lead ('plumbum argentarium'), that is the mineral galena, relies heavily on secondary sources. He continues (33. 96): 'Silver is found in very nearly all the provinces but the best comes from Spain, where, together with gold, it occurs in barren ground and even in the mountains' ('reperitur in omnibus paene provinciis, sed in Hispania pulcherrimum, id quoque sterili solo atque etiam montibus'). Pliny does not add any further commentary, unaware that the lack of vegetation indicates the presence of poisonous minerals in the subsoil, such as arsenopyrite.[103]

Significantly, Pliny does not make any reference to silver production at Laurium, since Spain was a more important source in his day. Laurium[104] had, originally, been one of the largest of all ancient silver-mining complexes; outcrops were exploited from Dipileza to Cape Sunium.[105] With the exception of the Peloponnesian War, mining continued until the Slave Revolt in 103 BC. By the first century AD, however, only some tailings were being reworked and operations eventually ceased until the resumption of working in the fourth century.

Pliny (33. 95) is vague in his description of the processing of argentiferous galena: 'Its ore is sometimes red, sometimes the

---

[103] Healy, *Mining*, 71; B. Boky (trans. J. Scott), *Mining* (1967), 19.

[104] Healy, *Mining*, 78 ff. See, generally, Conophagos, *Laurium*; R. J. Hopper, 'The Laurion Mines: A Reconsideration', *Annual of the British School at Athens*, 63 (1968), 293 ff., and 'The Attic Silver Mines in the Fourth Century BC', *Annual of the British School at Athens*, 48 (1953), 200 ff.

[105] Healy, *Mining*, 78.

colour of ash.[106] It can only be smelted with lead, or with the lead mineral called "galena", mixed with veins of silver. When cupelled ['eodem opere ignium'], part of the ore precipitates as lead, while the silver floats on the surface like oil on water.' ('terra est alias rubra, alias cineracea. excoqui non potest nisi cum plumbo nigro aut cum vena plumbi—galenam vocant—quae iuxta argenti venas plerumque reperitur. et eodem opere ignium discedit pars in plumbum, argentum autem innatat superne, ut oleum, aquis.')

The extraction of silver[107] involves roasting, cupellation, and reduction. The argentiferous galena was first brought, by roasting, to its metallic state—an almost self-effecting process even in the simplest form of furnace. When the partly roasted ore, which consists of lead oxide and unchanged sulphide, was heated to a relatively low temperature, the oxide and sulphide reacted together and metallic lead was released and carried silver, antimony, and other impurities with it. By cupellation the argentiferous lead was next converted into a 'scum' of oxide which was removed by being scraped over the lip of the furnace until only pure silver was left. Finally the lead oxide, thus 'purified', was converted into metallic lead by reduction with charcoal.

The charge for the furnace consisted of alternate layers of charcoal and argentiferous galena, suitably concentrated by dressing (that is grinding, sieving, and washing), to maximize the surface exposed for roasting. The gangue accompanying the metalliferous fraction varied. In the north of the mining area, it was ferruginous, in the centre, chalk, and, in the south, fluorspar.

The furnaces at Laurium were constructed of mica schist obtained locally and of trachyte, from Melos, both eminently suitable refractory materials. Strabo (3. 2. 8) describes similar

---

[106] Zehnacker, *NH 33*, Ch. 95, note 1. Minerals containing a substantial tenor of silver are, for the most part, grey (or black), or red. Among the former, is silver sulphide (Ag_2S), or argyrite, and, in the latter category, is argyrose (AgS_2), pyrargyrite (Ag_3SbS_3), and proustite (Ag_3AsS_3). Pliny could be thinking of certain poor minerals such as blende (ZnS), which is black, or brown. On lead generally and the terminology of its ore-minerals, see Healy, *Mining*, 38 f.

[107] Conophagos, *Laurium*, 305 ff., and Healy, *Mining*, 157 ff.

furnaces in Turdetania: 'They built their silver-smelting furnaces with large chimneys so that the gas from the ore might be carried high into the air; for it was heavy and deadly.' The upper part of the chimney was used for the smelting operation, while the lower part was used for the oxidization of the lead before cupellation. The many chimneys belching fumes must have led to Xenophon's unfavourable comments (*Vect*. 3. 6. 12) about the general atmospheric pollution in the Laurium region.

17.6 COPPER

A substantial proportion of book 34 is devoted to the topic of copper and bronze, although Pliny's main interest appears to lie in the use of the metal and its alloys[108] rather than in the mining and processing of the ores of copper.[109] He introduces his subject as follows (34. 2 ff):

Copper is produced from an ore that the Greeks call 'cadmea',[110] a highly reputed variety. *Cadmea* comes from overseas; formerly, it was found in Campania and now, it occurs in the region of Bergamum, on the very border of Italy. It has also recently been reported in the province of Germany. In Cyprus, where copper was first discovered, it is obtained from *chalcitis* [chalcopyrite, or copper pyrites];[111] this ore is of a particularly low quality and better sources, especially of *aurichalcum* [copper pyrites] were found in other countries. ... The highest reputation is now enjoyed by Marian copper, also called Corduba copper.

fit et e lapide aeroso, quem vocant cadmean, celebri trans maria et quondam in Campania, nunc in Bergomantium agro extrema parte Italiae; ferunt nuper etiam in Germania provincia repertum. fit et ex alio lapide, quem chalcitim appellant, in Cypro, ubi prima aeris inventio, mox vilitas praecipua reperto in aliis terris praestantiore maximeque aurichalco. ... summa gloria nunc in Marianum conversa, quod et Cordubense dicitur.

'The ore from which copper and *cadmea* are obtained by smelting is called *chalcitis* ['copper-stone']. *Cadmea* is quarried from rocks above ground, while *chalcitis* is worked

---

[108] Healy, *Mining*, 209 ff.   [109] Ibid. 158 ff.   [110] See Ch. 14, s.v.
[111] See Ch. 14, s.v. 'chalcitis'.

underground and immediately crumbles into small pieces, being of a soft consistency, resembling matted down.' 'chalcitim vocant, ex quo et ipso aes coquitur. distat e cadmea, quod illa super terram et subdialibus petris caeditur, haec ex obrutis, item quod chalcitis friat se statim, mollis natura, ut videatur lanugo concreta.')

Pliny (34. 94 ff.), while briefly mentioning the various forms of copper and its blends, refers to refining:

Bar copper also is produced from the other mines and likewise fused copper. The difference between them is that the latter can only be fused, as it breaks under the hammer, whereas bar copper, otherwise called ductile copper, is malleable,[112] which is the case with all Cypriote copper. But also in the other mines, this difference of bar copper from fused is produced by treatment; for all copper, after the impurities have been rather carefully removed by fire and melted out of it, becomes bar copper. Among the remaining kinds of copper the prize goes to bronze of Campania, which is most esteemed for utensils. There are several ways of preparing it. At Capua it is smelted in a fire of wood, not of charcoal, and then poured into cold water and cleaned in a sieve made of oak and this process of smelting is repeated several times. At the last stage Spanish silver lead is added to it in the proportion of ten pounds to a hundred pounds of copper; this treatment renders it pliable and gives it an agreeable colour of a kind imparted to other sorts of copper and bronze by means of oil and salt.

Bronze resembling the Campanian is produced in many parts of Italy and the provinces, but there they add only eight pounds of lead and do additional smelting with charcoal, because of their shortage of wood. The difference produced by this is noticed especially in Gaul, where the metal is smelted between stones heated red hot, as this roasting scorches it and renders it black and friable. Moreover, they only smelt it again once whereas to repeat this several times contributes a great deal to the quality. It is also not out of place to observe that all copper and bronze fuses better in very cold weather.

regulare et in aliis fit metallis, itemque caldarium. differentia quod caldarium funditur tantum, malleis fragile, quibus regulare obsequitur ab aliis ductile appellatum. quale omne Cyprium est. sed et in ceteris metallis cura distat a caldario; omne enim diligentius purgatis igni vitiis excoctisque regulare est. in reliquis generibus palma

---

[112] Healy, *Mining*, 227.

Campano perhibetur utensilibus vasis probatissimo. pluribus fit hoc modis. namque Capuae liquatur non carbonis ignibus, sed ligni, purgaturque roboreo cribro profusum in aquam frigidam ac saepius simili modo coquitur, novissime additis plumbi argentarii Hispaniensis denis libris in centenas aeris. ita lentescit coloremque iucundum trahit, qualem in aliis generibus aeris adfectant oleo ac sale.

fit Campano simile in multis partibus Italiae provinciisque, sed octonas plumbi libras addunt et carbone recocunt propter inopiam ligni quantum ea res differentiae adferat, in Gallia maxime sentitur, ubi inter lapides candefactos funditur; exurente enim coctura nigrum atque fragile conficitur. praeterea semel recoquunt quod saepius fecisse bonitati plurimum confert. id quoque notasse non ab re est, aes omne frigore magno melius fundi.

It is interesting to review Pliny's account of copper smelting in the light of modern experiments carried out under varying conditions.[113] These showed that an open, hot-wood fire gave a temperature of 600–700 °C, while malachite (copper carbonate) requires a temperature of between 700–800 °C for smelting.

1. A charcoal fire was exposed to natural wind draught, in March, and attained a temperature high enough to smelt malachite. The atmosphere within, however, was not sufficiently reducing—that is, the carbon monoxide to carbon dioxide ratio was not high enough ($2CO + O_2 = 2CO_2$)—so that the malachite was only calcined to copper oxide and no pure copper was produced. But, with just sufficient air, charcoal burns to carbon monoxide and reduces the malachite as in the following two equations:[114] (i) $2C + O_2 = 2CO$ (production of carbon monoxide-reducing agent); (ii) $CuCO_3 + CO = Cu + 2CO_2$.

2. Malachite was put into a flat pottery dish, with an upturned porous pot over it, to make a type of 'kiln' or covered crucible. With the same heat source as the first experiment, after several hours, a well-reduced copper sponge was obtained which appeared to have been melted.

---

[113] H. H. Cohglan, 'Prehistoric Copper and Some Experiments in Smelting', *Transactions of the Newcomen Society*, 20 (1939), 49–65.

[114] Tylecote, *Metallurgy in Archaeology*, 26 f.

3. Ground malachite under the same conditions as in 2 (above) produced a bead of refined copper.

The result of these experiments led Coghlan to the conclusion that copper was first accidentally produced in a pottery kiln, perhaps from a pigment with a copper-mineral base used on pots when they were being fired.

Because of the high melting-point of fine copper (1083 °C),[115] copper was usually smelted into ingots formed within the smelting furnace. Early smelting furnaces undoubtedly consisted of a small hollow, in a depression in the ground, above which was placed the fire containing charcoal and oxidized ore, assisted with bellows. The copper was reduced from the ore and dripped into the hollow below.

From an open-hearth type of furnace, the next significant development was the bowl furnace which needed a charge of ore and charcoal in layers. It would, therefore, have been natural to extend the furnace upwards to accommodate a supply of fuel and ore following the principle on which the modern 'gravity feed' boiler is based. Such an extension led to the development of the shaft furnace.

Copper produced from sulphides—as chalcocite—had first to be roasted. The minerals were calcined[116] to remove, as soon as possible, the arsenic and antimony compounds present as impurities and the sulphur. The process required little fuel as it developed sufficient heat to be self-perpetuating until the completion of the roasting. The result of this roasting was the production of copper oxide from the original sulphide as $Cu_2S + 2O_2 = 2CuO + SO_2$. The copper oxide was then put with charcoal to reduce it: $CuO + CO = Cu + CO_2$.

The second main method of smelting sulphide ores resembles present-day techniques. The sulphide ore was washed and smelted in a furnace to separate the gangue from the copper sulphide. The sulphide underwent no change in this operation and was left as a cake in the bottom of the hearth. A slag, similar in composition to primitive iron-smelting slag, but containing some copper, was run off and the cake of sulphide left

---

[115] Healy, *Mining*, 226.   [116] Forbes, *Ancient Technology*, ix. 18.

to cool. This was then broken up, made into balls, and roasted to oxide. The oxide was then resmelted in the original furnace with charcoal but this time was reduced to cakes of metal weighing about four to five pounds and some slag which solidified on the surface of the ingot. The copper so produced was often impure and needed further refinement in a separate shaft furnace, or was refined in the course of melting.

Of particular interest are furnaces which produced the Roman 'bun' ingots, which vary in size from thirty to fifty pounds.[117] The charge was a mixture of oxide, or roasted sulphide ore, and charcoal, from which the slag would have been tapped from the surface at intervals. Only at the end of the process, when the amount of copper reached a suitable weight, was the *whole* metallic content tapped into a mould to form the characteristic ingot shape. (Copper, with its high melting-point, unlike lead, could not have been allowed to escape slowly, since it would have tended to solidify before reaching the mould.)

The surfaces of Roman copper ingots suggest that they were cooled in an oxidizing atmosphere, and that many of the stamps had been impressed before the ingots were completely cold, while the copper was still soft. Analyses of a limited number of ingots show that, mainly because of their low sulphur content, the ingots from Wales were the result of skilful smelting of pure oxidized ores.

Much of the evidence for Roman copper smelting is derived from Rio Tinto,[118] where oxide, or roasted sulphide ores, continue to be smelted.

17.6.1 *Bronze*

Bronze is basically a mixture of copper and tin and was the earliest and most extensively used alloy in the ancient world. Pure copper, as, for example, native copper, is soft, but can be hardened by the addition of alloying elements such as tin and

---

[117] Healy, *Mining* 160; Tylecote, *Metallurgy in Archaeology*, 31 ff. and table 9.
[118] B. Rothenberg and A. Blanco, *Ancient Mining and Metallurgy in South-West Spain* (London, 1981).

arsenic. It is easy to add tin by stirring in cassiterite (SnO_2) with plenty of charcoal on the surface. Arsenic exists in a native form or as arsenical pyrites (FeAsS), and when mixed with copper, under reducing conditions, the arsenic goes into the copper, and the iron and sulphur form iron sulphide which floats on the surface and can be later oxidized to form a slag.

17.6.2 Arsenical Copper/Bronze

Arsenic[119] occurs in copper from the Early Bronze Age and its presence modifies the physical properties of the bronze produced from it.

The maximum solubility of arsenic in copper is of the order of 8 per cent and this does not change markedly with the temperature of the melt. This is in contrast with tin bronzes where the solubility of tin falls significantly with temperature, from a maximum of 15 per cent to a low figure at room temperature. The chemical effect of the presence of arsenic improves its casting quality, but brittleness increases rapidly after 1 per cent.[120]

Maréchal[121] has shown that in the cold-worked state in particular the hardness of the two materials is very similar and that the arsenic-containing alloys soften less readily with reheating, re-emphasizing the effect on the recrystallization temperature. It is thus clearly established that, even with up to 8 per cent arsenic, the alloys are readily worked hot or cold, and can give strength and hardness equivalent to tin bronze and that the presence of arsenic is particularly beneficial in maintaining the workability of copper which would, otherwise, contain copper oxide.

Experiments[122] have also shown that, whereas the effect of arsenic on the strength of copper in the cast (Brinell hardness 79) and cold-worked (B 132) condition is relatively high, as little as 1.04 per cent arsenic will raise the strength of

---

[119] N. H. Gale and Z. A. Stos-Gale, 'Some Aspects of Early Cycladic Copper Metallurgy', in Domergue, *Mineria y metalurgia*, i. 21–37.
[120] Forbes, *Ancient Technology*, ix. 168.
[121] J. R. Maréchal, *Zur Frühgeschichte der Metallurgie* (Lammersdorf über Aachen, 1962). [122] Healy, *Mining*, 228.

hammered copper to the equivalent of Brinell 124–77. Thus, the very significant effect of arsenic on the hardness of a hammered cutting edge would not fail to have been noticed and efforts would have been made to reproduce the compositions that gave this result. The presence of antimony and silver[123] also contribute to the hardness of copper and it must have been mortifying to early smelters when they found that, by improving their technique, and so obtaining pure copper, they had reduced its strength. The alloys were produced as a result of observation and repetition of successful ore combinations.

Pliny lists a number of copper-based alloys (*aes*).[124] '*Delian* [34. 9] bronze was the earliest to become famous, the whole world thronging to the markets in Delos; and hence the attention paid to the processes of making it.' ('antiquissima aeris gloria Deliaco fuit, mercatus in Delo celebrante toto orbe, et ideo cura officinis.') Tantalizingly there is no record of the composition of this bronze. *Aeginetan* was celebrated because of the production of the alloy in its foundries. '... officina temperatura nobilitata'.

17.6.3 Copper–Gold–Silver

Corinthian bronze described by Pliny (34. 6 ff.), and other authorities, was a ternary alloy of copper–gold–silver blended in varying proportions.[125] He states that the alloy was accidentally produced when Corinth was captured and burnt down in 146 BC ('hoc casus miscuit Corintho, cum caperetur incensa'), adding that there are three varieties: (1) where the silver predominates, (2) where the gold predominates, and (3) where all three metals are blended in equal proportions. ('tria genera: candidum argento nitore quam proxime accedens, in quo illa mixtura praevaluit; alterum in quo auri fulva natura; tertium in quo aequalis omnium temperies fuit').

---

[123] See further F. C. Thompson, 'Hardness and Brittleness in Silver–Copper Alloys', in Hall–Metcalf, *Chemical and Metallurgical Investigation*, 67–8.
[124] Healy, *Mining*, 212 ff.
[125] P. T. Craddock, 'Gold in Antique Copper Alloys', *Gold Bulletin*, 15/2 (1982), 69–72, esp. 69–70.

Hepatizon: Pliny (34. 8) records another mixture 'for which the formula cannot be given, although it is manufactured. But the bronze valued in portrait statues and others for its peculiar colour, approaching the appearance of liver and consequently called by a Greek name *hepatizon* meaning 'liverish', is a blend produced by luck; it is far behind the Corinthian variety, yet a long way in front of the bronze of Aegina and that of Delos, which long held the first rank.' ('praeter haec est cuius ratio non potest reddi, quamquam hominis manu est, at fortuna temperatur in simulacris signisque illud suo colore pretiosum ad iocineris [cf. *NH* 22. 80; celsus 2. 8] imaginem vergens, quod ideo hepatizon appellant, procul Corinthio, longe tamen ante Aegineticum atque Deliacum, quae diu optinuere principatum.')

17.6.4 Copper–Tin–Lead

Pliny (34. 94 ff.) also records a number of copper-based alloys of copper–tin–lead.

Campanian was produced at Capua in the proportions 100:5:5, and, elsewhere in Italy and the provinces in the proportions 100:4:4.

Statue bronze (also used for tablets) is possibly a variety of *Delian*, or *Aeginetan* (34. 97), since Pliny (34. 96) also observes that Myron and Polycleitus used these bronzes for their statues.

Graecanic: Pliny describes (34. 97 ff.) the proper blend for making statues and tablets.

At the outset the ore is melted and then there is added to the melted metal a third part of scrap copper, that is copper, or bronze, that has been bought up after use. This contains a peculiar seasoned quality of brilliance that has been subdued by friction and, so to speak, tamed by habitual use. Silver-lead is mixed with the fused metal in the proportion of 1:8. There is also, in addition, what is called *mouldblend* of bronze of a very delicate consistency, because a tenth part of black lead is added and a twentieth of silver-lead; and this is the best way to give it the colour called Graecanic 'after the Greek'.

massa proflatur in primis, mox in proflatum additur tertia portio aeris collectanei, hoc est ex usu coempti. peculiare in eo condimentum attritu domiti et consuetudine nitoris veluti mansuefacti.

miscentur et plumbi argentarii pondo duodena ac selibrae centenis proflati. appellatur etiamnum et formalis temperatura aeris tenerrimi, quoniam nigri plumbi decima portio additur et argentarii vicesima, maximeque ita colorem bibit quem Graecanicum vocant.

Pot bronze: 'The last kind is called pot bronze, taking its name from the vessels made of it; it is a blend of three or four pounds of silver-lead with every hundred pounds of copper' ('novissima est quale vocatur ollaria, vase nomen hoc dante, ternis aut quaternis libris plumbi argentarii in centenas aeris additis').

The addition of lead to Cyprus copper produces the purple colour seen in the bordered robes of statues: 'Cyprio si addatur plumbum, colos purpurae fit in statuarum praetextis.'

17.6.5 Copper–Antimony

Bronze containing antimony[126] occurs from an early date, but, in most cases, the amount present is so low that it may be safely explained as an impurity in the copper ores and not as a conscious addition. Antimony is found in early Cycladic Greek bronzes.[127]

17.6.6 Copper–Silver

These alloys were mainly, if not exclusively, used for imperial Roman coin series. Some 'silver' coinages were in effect 'argentiferous bronzes'. Under the empire, the Romans issued denarii of different alloys of silver with copper varying from 1:12 to 15:12[128]—the latter as a result of the great debasement by Septimius Severus in AD 193. The varying degree of debasement of such alloys provides an important economic indicator.

---

[126] See section 17.11 below.
[127] See Gale and Stos-Gale, 'Cycladic Copper Metallurgy'.
[128] L. H. Cope, 'The Metallurgical Analyses of Roman Imperial Silver and Aes Coinages', in Hall–Metcalf, *Chemical and Metallurgical Investigation*, 44 ff.

17.6.7 Brass

Two references from the fourth century BC appear, superficially, to be related to the production of brass. According to [Aristotle,] *De mirabilibus auscultationibus* 62, 'They say that the copper of the Mossynoeci is very brilliant and light in colour, though tin is not mixed with it. They add that the discoverer of the mixture did not instruct anyone else, so that the copper objects formerly produced in these regions are superior, whereas those made subsequently are no longer so.'

Similarly, Theophrastus (*Lap*. 49) writes: 'The earth mixed with copper is most unusual; for in addition to melting and mixing with the metal, it also has the remarkable power of improving the beauty of its colour.' Caley[129] sums up the arguments against this being brass. The 'unusual earth' may not have been calamine[130] but perhaps an arsenical compound such as realgar,[131] or orpiment:[132] of some significance perhaps is the adjective (λευχότατον) used to describe the appearance of the metal (or *alloy*). It may be translated 'very white', or 'very light' in colour, instead of 'very pale' and, if it has this meaning, then the metal could not have been brass. However, it could have referred to a copper–arsenic alloy, since such alloys, even when they contain a relatively small proportion of arsenic are white in colour, not yellow, like brass or bronze. Alloys of copper and arsenic were known in the Aegean region from very early times as has been shown by chemical analysis.[133]

Zinc ores were used directly by the Romans in the manufacture of brass. Pliny (34. 2 ff.) introduces the mineral *cadmea*,

---

[129] In Caley and Richards, *Theophrastus*, pp. 162 ff. See also E. R. Caley, *Orichalcum and Related Alloys: Origin, Composition, and Manufacture, with Special Reference to the Coinage of the Roman Empire* (American Numismatic Society, Notes and Monographs, 151; New York, 1964).

[130] See Ch. 14, s.v. 'cadmea'. [131] See Ch. 15, s.v. 'sandaraca'.

[132] See Ch. 14, s.v.

[133] J. A. Charles, 'Early Arsenical Bronzes—a Metallurgical View', *American Journal of Archaeology*, 71 (1967), 21–6. The presence of arsenic and the actual amount in copper arsenical bronzes also has a significant bearing on their origin. The naturally occurring impurity may be distinguished from the deliberately added arsenic (reckoned as amounts in excess of 2%).

Metals 311

but his account of its use is limited.

The metal [copper] is also obtained from a coppery stone called by the Greek name 'cadmea' (καδμεία), a kind in high repute coming from overseas and also, formerly, found in Campania and, at the present day, in the territory of Bergamo on the farthest confines of Italy. It is also reported to have been found recently in the province of Germany. In Cyprus, where copper was first discovered, *cadmea* is additionally obtained from another stone also, called 'chalcitis'. This, however, was afterwards of exceptionally low value when a better copper was found in other countries, and especially gold–copper [*aurichalcum*] which long maintained an outstanding quality and popularity, but which for a long time has not been found, since the ground has been exhausted.

fit et e lapide aeroso, quem vocant cadmean, celebri trans maria et quondam in Campania, nunc et in Bergomatium agro extrema parte Italiae; ferunt nuper etiam in Germania provincia repertum. fit ex alio lapide quem chalcitim, in Cypro ubi prima aeris inventio, mox vilitas praecipua reperto in aliis terris praestantiore maximeque aurichalco, quod praecipuam bonitatem admirationemque diu optinuit nec repetitur longo iam tempore effeta tellure.

Cadmea includes two groups of mineral:[134] (1) natural *cadmea*, that is calamine, which can be zinc carbonate ($ZnCO_3$), or hydrated zinc silicate (Zn_4 ($Si_2O_7(OH)$, H_2O) and (2) artificial *cadmea*, that is zinc oxide (ZnO). Pliny continues (34. 4), with a brief, but accurate, account of the chemistry involved: 'The highest reputation has now gone to Marius copper, also called Cordova; next to the Livia variety, this kind *most readily absorbs cadmea* and reproduces the excellence of *aurichalcum* [brass] in making sesterces and double-*as* pieces, the single *as* having to be content with its proper Cyprus copper.' ('summa gloriae nunc in Marianum conversa, quod et Cordubense dicitur. hoc a Liviano cadmean maxime sorbet et aurichalci bonitatem imitatur in sestertiis dupondiariisque, Cyprio suo assibus contentis.')

Elsewhere (34. 100 ff.) *cadmea* is said to be

produced in furnaces where silver is smelted, this kind being whiter and not so heavy; it is, however, by no means to be compared with

---

[134] See Rackham, *NH 33–35*, p. 126, note *c*. Cf. Whitten–Brooks, *Geology*, appendix (smithsonite).

that from copper. There are several varieties; for while the mineral itself, from which the metal is made, is called 'cadmea', which is necessary for the fusing process, but is of no use for medicine,[135] so again another kind is found in furnaces which is given a name indicating its origin. It is produced by the thinnest part of the substance being separated out by the flames and the blast, and becoming attached, in proportion to its degree of lightness, to the roof chambers and the side walls of the furnaces—the thinnest being at the very mouth of the furnace, which the flames have belched out; this is called 'smoky' *cadmea* from its burnt appearance and because it resembles hot white ash in its extreme lightness. The part inside is best, hanging from the vaults of the roof chamber and this consequently is designated 'grape-cluster' *cadmea*; this is heavier than the preceding kind but lighter than those that follow—it is of two colours, the inferior kind being the colour of ash, and the better, the colour of pumice—and it is friable and extremely useful for making remedies for eye conditions. A third sort is deposited on the sides of furnaces, not having been able to reach the vaults because of its weight; this is called in Greek 'placitis' ['caked residue', πλακίτης], in this case by reason of its flatness.

fit sine dubio haec et in argenti fornacibus, candidior ac minus ponderosa, sed nequaquam comparanda aerariae. plura autem genera sunt, namque ut ipse lapis, ex quo fit aes, cadmea vocatur, fusuris necessarius, medicinae inutilis, sic rursus in fornacibus exsistit alia, quae originis suae nomen recipit. fit autem egesta flammis atque flatu tenuissima parte materiae et camaris lateribusque fornacium pro quantitate levitatis adplicata. tenuissima est in ipso fornacium ore quam flammae eructarunt, appellata capnitis, exusta et nimia levitate similis favillae. interior optuma, camaris dependens et ab eo argumento botryitis (βοτρυίτης) nominata, ponderosior haec priore, levior secuturis—duo eius colores, deterior cinereus, pumicis melior—friabilis, oculorum medicamentis utilissima. tertia est in lateribus fornacium quae propter gravitatem ad camaras pervenire non potuit. haec dicitur placitis et ipsa ab argumento planitiei.

To sum up, the Romans produced brass by a cementation process in which a zinc-rich mineral was added to copper under reducing conditions in a crucible. In this reaction zinc vapour is given off at quite low temperatures and it is best to

---

[135] Contrast *NH* 34. 119 and elsewhere: Pliny states that *cadmea* and *chalcitis* are compounded as a drug.

granulate the copper and carry out the process just below the melting-point so that the zinc vapour is absorbed by the large area of solid copper. Pliny's description, as we have observed, matches this. The technique would have been developed by repeating successful ore combinations rather than as a result of any understanding of the metallurgical principles involved.

Roman copper-based alloys usually also contain some tin and zinc.[136] There is a tendency for the wrought alloys to contain more zinc than tin, while, in cast alloys, the reverse is the case. Lead is often added to copper alloys (34. 98 ff.).

The following figures provide a brief indication of the average content of major elements found in Roman copper-based alloys.[137]

Element	Cast alloys	Wrought alloys
Zinc	2.7% (11 examples)	6.1% (12 examples)
Lead	7.1% (11)	0.4% (13)
Tin	13.3% (13)	5.7% (15)

The quality of brass produced for Roman coin series[138] was clearly controlled within close limits, but there is no independent evidence, literary, or otherwise, to indicate how this was achieved.

The physical properties of brass are interesting. When progressively increasing amounts of zinc are alloyed with copper, the brass becomes gradually harder. At 36 per cent zinc content, the properties are sharply altered and further additions of zinc bring about a more rapid hardness than before (HV 90, compared with HV 53 for pure copper). However, brass containing more than 64 per cent copper was malleable, which increased its usefulness. Modern brass is of very different composition from known ancient samples: when used for casting, the zinc content is 40 per cent, while for working, it is between 10 and 30 per cent.

---

[136] Tylecote, *Metallurgy in Archaeology*, 53 ff.
[137] Healy, *Mining*, 214. Tylecote, *Metallurgy in Archaeology*, 54, and table 17.
[138] Cope, 'Metallurgical Analyses' 44 ff.

The cementation method of brass production, as described by Pliny, continued to be used until the eighteenth century. It was only then, in the 1750s, that Champion[139] invented a method of recovering pure zinc from its ores by means of a vertical crucible, downward condensation process.

From Pliny's incomplete and fragmented account of copper, arsenical bronzes, and copper-based alloys, one may conclude that he possessed little more than a layman's general knowledge of the main ore-minerals of copper and the operations involved in their smelting. As in other metallurgical contexts, with the notable exception of gold, he certainly did not understand the underlying chemistry. Nevertheless, Pliny does include some surprising details, such as an example of hydrometallurgy.[140]

17.7 TIN

There are two main sources of tin, alluvial (stream) and vein ore. Traders along the Mediterranean were carrying tin by 2000 BC,[141] first from Galicia and, later, from Cornwall.[142] Recent evidence confirms that tin was also available from Turkey from an early period.[143]

Greek sources refer to the Cassiterides, or 'Tin Islands',[144] the location of which has never been firmly established.[145] According to legend (Pliny 34. 156 f.), traders used to fetch

---

[139] Tylecote *Metallurgy*, 144. Champion's method was based on the Indian process described by P. T. Craddock *et al.*, 'Early Zinc Production in India', *Mining Magazine* (Jan. 1985), 45–52.   [140] See Ch. 11.5 above.

[141] Tylecote, *Metallurgy*, 21.

[142] See, generally R. D. Penhallurick, *Tin in Antiquity: Its Mining and Trade throughout the World, with Particular Reference to Cornwall* (London, 1986). Alluvial placers were worked in Cornwall from *c*.500 to 43 BC, but subsequently abandoned because of the volume of tin being produced in Spain, until the 3rd cent. AD. At that time Pytheas visited the miners of Belericum (Land's End) and their tin deposits at Ictis (St Michael's Mt.). See Davies, *Roman Mines*, 143.

[143] Evidence of the exploitation of tin ores, from an early period, has recently been discovered in Anatolia, Turkey.

[144] Herodotus (3. 115) disclaims knowledge of the Tin Islands. Contrast Diod. Sic. 5. 38. 4.

[145] Among numerous locations suggested for the 'Cassiterides', Plymouth, Devon, has been put forward.

tin from the islands in the Atlantic and brought it back in wicker containers covered with stretched hides: 'fabuloseque narratum in insulas Atlantici maris peti vitilibusque navigiis et circumsutis corio advehi.' Tin (erroneously referred to as 'plumbum candidum'), continues Pliny (ibid.),

is now found in Lusitania and Gallaecia. ... It is extracted from surface strata of a sandy black colour. Tin is detected only by its weight[146] [34. 157]. Tiny pebbles of tin [cassiterite] appear sporadically, especially in the dried up beds of what were once fast-moving rivers, The workmen wash the sand and heat the residue in furnaces, tin is also found in gold mines called 'alutiae', through which a current of water is sent to wash out the black pebbles with their white spots of tin. The pebbles weigh the same as the gold and so remain with it in the bowls in which they are collected. Afterwards the pebbles are separated in the furnaces and fused into tin.

nunc certum est in Lusitania gigni et in Gallaecia ... summa tellure, harenosa et coloris nigri. pondere tantum ea deprehenditur; interveniunt et minuti calculi, maxime torrentibus siccatis. lavant eas harenas metallici et, quod subsedit, cocunt in fornacibus. invenitur et in aurariis metallis, quae alutias vocant, aqua inmissa eluente calculos nigros paullum candore variatos, quibus eadem gravitas quae auro, et ideo in catillis quibus aurum colligitur, cum eo remanent; postea caminis separantur conflatique in plumbum album resolvuntur.

Cassiterite (SnO_2)[147] is a widely distributed ore-mineral in small amounts being often found in the form of rolled pebbles in placer deposits known as stream tin. It has been noted as an original constituent of igneous rocks and pegmatites, but is more commonly found in high-temperature hydrothermal veins in, or near, granitic rocks. There is a further source, namely stannite[148] (Cu_2FeSnS_4): this is found in tin-bearing veins associated with cassiterite, chalcopyrite, tetrahedrite, and pyrite, which often occur as inclusions.

---

[146] Pliny is simply observing that the ore is heavier than the accompanying earthy material, which would have been washed away by the current. (Specific gravities: tin, 7.29; gold, 10.3; quartz, 2.65.)
[147] Healy, *Mining*, 38; Whitten–Brooks, *Geology*, 72.
[148] Healy, *Mining*, 38.

17.7.1 Tin-streaming

It is most probable that, as in Cornwall, the first ore exploited was alluvial, occurring in stream beds, where the ground had been broken up and washed along by running water, as Pliny (34. 157) records. As with gold, the movement of the water was an effective way of concentrating finely comminuted ore-mineral: the heavy tin ore, being more difficult to move than the light gangue, tended to settle in pockets in the stream bed, or at any irregularities, such as a bend or boulder where the eddy currents could swirl up the rock particles and promote the conditions needed for the (gravity) separation of the mineral. Searching along the beds was the earliest form of tin-streaming. The method used was to find a favourable site and dig the deposit which was then agitated and stirred carefully in a stream of clear water. The heavy ore was settled as much as possible and the waste material gently raked off. The cassiterite that remained was then retrieved.

An improvement of this technique was made by digging out the mineral into a stream of water, diverted along a channel provided especially for the dressing process, and slowly raking up against the current of water. The ore was carried farthest against the stream because of its relatively high density and, over a period, a concentrate was formed and collected. The concentration of the original deposit of stream tin was such that the ore was comparatively easy to work up to a grade fit for smelting.[149]

Diodorus Siculus (5. 38. 4), discussing tin found in Iberia, states: 'Tin is not found on the surface of the earth, as certain writers continually repeat in their histories, but is dug out of the ground and smelted in the same manner as silver and gold. For there are many mines of tin in the country above Lusitania and on the islets that lie off Iberia, or in the Ocean, and, because of that, they are called Cassiterides.'

Although at first sight this description might seem to suggest the mining of vein ore (shode) and not alluvial deposits, it is almost certain that Diodorus means the latter. In some districts of Cornwall, the alluvial gravels are not on the surface,

---

[149] Ibid. 89.

but, in one place for example, they are beneath fifty feet (15 m) of sand and silt and, in another, they are covered with peat, gravel, and sand to a depth of twenty feet (6 m). Such deposits are, in fact, buried placers, not veins of ore. At Montebras[150] (Creuse) there is evidence of extensive opencast mining: some twenty trenches, running at right angles to each other for prospecting and exploitation, are 98–130 feet (30–40 m.) wide and 26–33 feet (8–10 m.) deep.

By the first century AD, the mines in Gaul had to close down and the main producer of tin was Spain: the sources were mainly placers. Strabo (15. 2. 10) mentions Turdetania, in addition to Lusitania and Gallaecia.

Pliny (34. 156) erroneously confuses tin with a form of lead ('plumbum candidum') and here and later (34. 161) describes it, like Caesar (*BG* 5. 12), as 'plumbum album' when discussing its properties: 'White lead has more dryness, whereas black lead is entirely moist. Consequently, tin cannot be used for anything without an admixture of another metal, nor can it be employed for soldering silver because the silver melts before the tin.' ('albi natura plus aridi habet, contraque nigri tota umida est. ideo album nulli rei sine mixtura utile est neque argentum ex eo plumbatur, quoniam prius liquescat argentum.') Pliny is in error here. Silver does not melt sooner than tin: the melting-points of silver and tin are 961 °C and 232 °C respectively.[151]

17.7.2 *Smelting*

Few accounts of the smelting of tin appear in ancient sources and our knowledge is largely derived from the interpretation of the evidence of slag heaps. Diodorus Siculus (5. 22. 2) writes in very general terms, clearly conflating some of the processes which he does not logically separate: 'They it is who work the tin, treating the bed that bears it in an ingenious manner. The bed, being like rock, contains earthy seams and in them the workers quarry the ore, which they then melt down and cleanse of its impurities. They then work the tin

---

[150] Davies, *Roman Mines*, 83.
[151] Healy, *Mining*, 226, and G. W. C. Kaye and T. H. Laby, *Tables of Physical and Chemical Constants*[13] (London, 1966), 118 ff.

into pieces the size of knuckle bones and convey it, to an island which lies off Britain and is called Ictis.'

Fortunately, archaeological evidence[152] allows a reconstruction of furnace methods employed. Simple, clay-lined holes, or trenches, were dug in the ground and wood was piled into them and lighted. If it burned fiercely, charges of ore and charcoal (or wood) were thrown in alternately and the slag was tapped from the furnace into a second hole until enough tin had been produced to be ladled out. The function of the trench was merely to hold the embers and contain the metal. Many of these slag hearths, as they are known, have been found in Cornwall.[153] It was gradually recognized that cassiterite required fairly high reduction temperatures so that, for example, gold would melt from the mixture before the tin ore was reduced. This high temperature ensures good fluidity of the slag but involves a tendency for the tin to enter the slag. Some slags in Gaul contain no less than 21 per cent tin.[154] To obviate this loss of tin and to ensure the maximum recovery rate from the ore, a careful control of the temperature is necessary.

An unpublished tin ingot of the fourth century AD, recovered from the river Thames, contained 94 per cent tin and 4.59 per cent lead,[155] comparable in composition with other known fabricated objects, but different from other ingots found at Port Vendres,[156] in southern France: the latter contain less than 0.02 per cent lead.

---

[152] On tin-smelting techniques, see Tylecote, *Metallurgy*, 140 ff. Evidence of slag from Cornish tin-smelting confirms that it was smelted from the 15th cent. BC. We know that it is possible to smelt high-purity cassiterite in a crucible, as this is the basis of a tin assay technique. This is done today in a clay-graphite crucible with excess carbonaceous reducing agent, usually anthracite, and the tin globules, formed by heating to 1000 °C, are poured out of the crucible, separated from the excess anthracite and ash, and agglomerated by melting in another crucible. To do this effectively, the ore has to be well dressed to give over 60% tin (stannous oxide (SnO) = 79% tin (Sn)).

[153] Forbes, *Ancient Technology*, ix. 138 f.   [154] Healy, *Mining*, 178.

[155] I am indebted to the British Museum Research Laboratory for these figures, obtained from analyses by atomic absorption spectrometry. The ingot contained, in addition to the major elements of tin and lead, the following trace elements: silver, 0.0018%; iron, 0.187%; nickel, 0.0027%; copper 0.115%; and antimony, 0.090%. Its composition is very similar to that of an ingot from Corbridge. (See further J. A. Smythe, 'Notes on Ancient and Roman Tin and its Alloys with lead, *Transactions of the Newcomen Society*, 18 (1937), 255–66.)

[156] D. Colls, C. Domergue, F. Lanbenheimer, and B. Lion, 'Les Lingots d'étain de l'épave, Port Vendres', *Gallia*, 33 (1975), 61–94.

Vein ore could not be directly refined because of its quartz matrix, as at Vaulry,[157] and had first to be calcined to eliminate this. The preliminary roasting also removed sulphur and arsenic compounds. After this treatment, the ore was crushed and washed a number of times until the cassiterite had been sufficiently concentrated. One important fact should be borne in mind, however, when considering the efficiency of the refining processes, namely that neither the Greeks nor the Romans, demanded a high degree of purity in their tin, which was mainly used in the production of alloys (bronze and pewter).

17.7.3 Tinning

Finally, Pliny (34. 162) refers to an interesting process used in tinning copper or bronze artefacts. A method discovered in the Gallic provinces is to dip bronze articles in white lead (tin) so as to make them almost indistinguishable from silver. Objects thus treated are called 'tinned'. The actual process is not described by Pliny, who was clearly unaware of the technical details.[158]

There are two main methods of 'tinning':[159] (1) where the artefact is immersed in a crucible full of molten tin, or (2) 'wiping' hot bronze with a stick of tin, or with pellets.

Pliny (34. 160) adds: 'When copper vessels are coated with *stagnum* [tin], the contents have a more agreeable taste and the formation of destructive verdigris is prevented and, what is remarkable, the weight is not increased' (stagnum inlitum aereis vasis saporem facit gratiorem ac compescit virus aeruginis, mirumque pondus non auget').

There is also evidence that tin sheet was applied to fourteenth- and fifteenth-century BC Mycenaean vases[160] to simulate the appearance of silver. This is now fully oxidized to tin oxide and cassiterite, but there is little doubt that it was

---

[157] Healy, *Mining*, 160 f.

[158] Pliny (ibid.) also states that, in the town of Alesia, they plated silver in a similar manner, particularly ornaments for horses, pack animals, and yokes of oxen. He adds that the distinction of developing this method belongs to the Bituriges.   [159] Tylecote, *Metallurgy*, 238 f.

[160] Sara A. Immerwahr, 'The Use of Tin on Mycenean Vases', *Hesperia*, 35 (1966), 381–96.

applied to pottery to make effective substitutes for the real thing—silver—for funerary purposes.

17.7.4 Pewter

Pewter[161] is an alloy of tin and lead mixed by the Greeks and Romans in the proportions of 2:1. The modern equivalent is 4:1.

In metallurgical terminology, freezing is synonymous with solidification. Some alloys solidify progressively and the temperatures within which solidification takes place constitute the freezing range. This occurs in most alloys but is not found in pure metals, in metallic or chemical compounds, or in some special alloy compositions that melt and freeze at fixed temperatures.

Tin–lead alloys[162] provide an interesting example of such a special composition. Lead melts at 327 °C and tin at 232 °C. If lead is added to molten tin and the alloy is then cooled, the freezing-point of the alloy is found to be lower than the freezing-point of both metals. When a molten alloy containing 2:1 (tin/lead) is cooled, the mixture reaches a temperature of just under 200 °C before it begins to solidify and finishes only when the temperature has fallen to 183 °C.

One particular type of pewter containing 62 per cent tin and 38 per cent lead,[163] melts and solidifies entirely at 183 °C. Similar effects may occur in many other alloy systems; the special composition of each series that has the lowest freezing-point and which entirely freezes at that temperature, is called the *eutectic alloy*, and 183 °C is the *eutectic temperature*.

17.8 LEAD

At 33. 106, Pliny writes:

In the same mines [silver],[164] occurs the mineral called 'litharge'[165] [PbO, that is lead monoxide]. There are three kinds with Greek

---

[161] Forbes, *Ancient Technology*, ix. 159. [162] Healy, *Mining*, 214 f.
[163] Ibid. 214.
[164] See Conophagos, *Laurium*, generally for the mining and processing of argentiferous galena.
[165] Pliny employs the term *spuma argenti* for litharge.

names: *chrysitis*, 'golden'; *argyritis*, 'silver'; and *molybditis*, 'leaden'. Generally speaking, all these colours[166] are found in the same sample. The Attic kind is the most sought after, then the Spanish. The 'golden scum', is obtained directly from the mineral ore, the 'silver' from the silver, and the 'leaden' from smelting the lead itself, which is processed at Puteoli, hence its name 'argyritis Puteolana'. Each kind of litharge is made by heating the ore until it liquefies; it then flows from an upper into a lower vessel and is taken out of that by means of small iron spits. It is then directly exposed to the flame on a spit to make it of moderate weight.[167]

fit in isdem metallis et quae vocatur spuma argenti. genera eius tria: optima quam chrysitim vocant, sequens quam argyritim, tertia quam molybditim. et plerumque omnes hi colores in isdem tubulis inveniuntur, probatissima est Attica, proxima Hispaniensis. chrysitis ex vena ipsa fit, argyritis ex argento, molybditis e plumbi ipsius fusura—quae fit Puteolis—et inde habet nomen. omnis autem fit exocta sua materia ex superiore catino defluens in inferiorem et ex eo sublata vericulis ferreis atque in ipsa flamma convoluta vericulo, ut sit modici ponderis.

Litharge presents two allotropic forms,[168] different in colour, namely yellow and red. They derive from two different arrangements of their atoms. Yellow litharge is orthorhombic and stable below 489 °C. Red litharge is tetragonal in structure and similarly stable below the same temperature. If, at 910 °C, the litharge drops into water it remains yellow because it cools quickly without changing its crystalline structure. The red is the slower to cool. This is the reason why in the same 'tube' of litharge these two colours exist side by side, according to the speed of the cooling.

Pliny's treatment of lead is confused because of his *mélange* of the ore-minerals of silver and lead (and even tin!). The method of cupellation here described is, according to Conophagos,[169] exceptional. The technology is as follows. When lead is melted, a crust of suboxide (PbO) forms on the surface; on heating this, one obtains amorphous lead monoxide, also called 'massicot'. According to differential cooling times, it

---

[166] The colours refer to the mineral ore and not the product of refining.
[167] Litharge and lead have different densities, namely 9. 36 and 11. 3 respectively.   [168] Conophagos, *Laurium*, 327 f.   [169] Ibid. 326.

crystallizes into litharge with a fawn (Pliny's yellow) or reddish colour. At Laurium, fragments of litharge have been found in the form of truncated cones pierced by a canal. These cones are called *tubuli*,[170] and result from the lifting of the litharge by spits, as Pliny explains. He also refers to these (33. 108) as 'scierytis',[171] 'reumenen',[172] and 'molybdaena' (lead sulphide).

In his discussion of silver Pliny identifies three kinds of litharge according to the way in which they are recovered from the melt. Similarly, Dioscorides (5. 87) defines his categories according to (1) the method of production, (2) the place of production, and (3) the then current metallurgical terminologies. Pliny (34. 173) also mentions 'molybdaena',[173] which he elsewhere (33. 95) 'refers to as galena, that is a mineral compound of silver and lead. This is better the more golden its colour and the less leaden. It is friable and of moderate weight. When boiled with oil, it acquires the colour of liver. *Molybdaena* is also found adhering to furnaces in which gold and silver are smelted: in this case it is called metallic [sulphide of lead]'. ('est et molybdaena, quam in alio loco galenam appellavimus, vena argenti plumbique communis. melior haec, quanto magis aurei coloris quantoque minus plumbosa, friabilis et modice gravis. cocta cum oleo iocineris colorem trahit. adhaerescit et auri argentique fornacibus; hanc metallicam vocant.') *Molybdaena* is here identifiable as red monoxide of lead, not lead sulphide, which is black in colour.

---

[170] Conophagos, *Laurium*, 327 f. and figs 12. 19 and 12. 20. See also R. Halleux, 'Les Deux Métallurgies du plomb-argentifère dans Conophagos, *Laurium*, 327 f. 'Histoire Naturelle de Pline, *Revue de Philologie*, 49 (1975), 73 ff.

[171] Ibid. 83–6. In Dioscorides 5. 87, Wellmann reads σκαλανθρῖτις (*vericulum*, 'spit').

[172] This reading is adopted by Zehnacker, *NH 33*, pp. 196 f. (ad loc.): 'A la variété qui se présente sous la forme de couches de litharge figées autour des ringards, s'oppose la variété qui a débordé de la coupelle à l'état liquide; ni peumenen ni pneumenen, ne sont donc acceptables. On choisira la participe de ῥέομαι, la confusion entre Π (grec) et P (latin) étant facile.'

[173] Halleux, 'Les Deux Métallurgies', 86–8, following Dioscorides, states that *molybdaena* is the third kind of litharge. J. Ramin, 'Les Connaissances de Pline l'Ancien en matière de métallurgie', *Latomus*, 36 (1977), 148–9 and 154, sees in galena a generic term for *all* lead minerals. In this context, however, *molybdaena* is specifically lead sulphide (PbS).

In his main discussion of lead, Pliny (34. 156 ff.) utilizes two lost sources[174]—namely the works of Cornelius Bocchus (34. 156 ff.), and of Sextus Niger (33. 106–8 and 34. 173). In the data derived from the former, he confuses lead and tin! 'There are two kinds of lead—black and white. White lead [that is tin] is the most valuable.'

Black lead [34. 158 ff.] does not occur in Gallaecia, but the neighbouring country of Biscaya has large quantities of black lead only; and white lead yields no silver although it is obtained from black lead. ... There are two different sources of black lead; it is either found in a vein, without associated substances, or it forms together with silver and is smelted with the two veins mixed together. From this mixture, the liquid which melts first in the furnaces is called 'stagnum'[175] [a crude alloy of silver and lead], the second, argentiferous lead, and the residue left in the furnace is impure lead. This represents about 30 per cent [by weight] of the original ore. When this is again fused, it yields black lead, having lost two-ninths of its bulk.

non fit in Gallaecia nigrum, cum vicina Cantabria nigro tantum abundet, nec ex albo argentum, cum fiat ex nigro. ... plumbi nigri origo duplex est; aut enim sua provenit vena nec quicquam aliud ex sese parit aut cum argento nascitur mixtisque venis conflatur. huius qui primus fuit in fornacibus liquor stagnum appellatur; qui secundus, argentum; quod remansit in fornacibus, galena, quae fit tertia portio additae venae; haec rursus conflata dat nigrum plumbum deductis partibus nonis II.

Forbes suggested that the Romans knew and used an early version of the Pattinson process[176] in refining their lead. In such a context, *stagnum* and *argentum* would signify 'crude lead' and galena 'purified lead'. The process[177] by which precious metals are nowadays concentrated, utilizes a peculiar physical property of lead–silver mixtures. If melted and cooled again, the first crystals formed consist of pure lead and the remaining solution will, therefore, become richer in silver. This formation of lead crystals, will go on until the remaining lead contains about 2.4 per cent of silver. Then the rest of the

---

[174] See Halleux, 'Les Deux Métallurgies'.
[175] Tylecote, *Metallurgy in Archaeology*, 75 ff. See also Forbes, *Ancient Technology*, viii. 228.
[176] Forbes. *Ancient Technology*, viii. 230 ff. Contrast Davies, *Roman Mines*, 180, who disagrees with Forbes's view. [177] Healy, *Mining*, 180.

molten metal will solidify all at once. By pouring off the molten metal before this happens, the silver is concentrated as far as possible and the lead, thus enriched, can be desilvered by cupellation. Finally the lead is recovered from the litharge by reduction. Conophagos understandably challenges the ability of Roman metallurgists to have used this method: 'Mais tout metallurgiste du plombe peut dire que ce procédé ne peut pratiquement pas être appliqué sans thermomètres et sans de grandes cuves convenablement chauffées.' The Pattinson process, which was not actually discovered until 1829, is based on the fact that an alloy of lead and silver with 2.5 per cent silver, is a 'eutectic' with a melting point of 300 °C.

The Greeks were able to desilver lead to 0.02 per cent silver content, while the Romans were even more successful (0.01 per cent, or less). Roman ingots, when desilvered, carried the inscription EX ARG, or EX ARGENT, and can often be precisely dated from their legends.[178]

Pliny (34. (164) briefly lists the sources of lead: 'The lead that we use in the manufacture of pipes and sheets is mined with considerable labour in Spain and throughout the Gallic provinces.[179] In Britain, however, it is found on the surface in such large quantities that there is a law limiting production. The following are the names of the different sources of lead—Oviedo, Caparia, and Oleastrum. There is, however, no intrinsic difference between them provided that the slag is carefully removed by smelting.' ('nigro plumbo ad fistulas lamnasque utimur, laboriosus in Hispania eruto totasque per Gallias, sed in Britannia summo terrae corio adeo large, ut lex interdicat ut ne plus certo modo fiat. nigri generibus haec sunt nomina: Ovetanum, Caprariense, Oleastrense, nec differentia ulla scoria modo excocta diligenter.') He adds (34. 165): 'It is a remarkable fact in the case of these mines only that, when they have been abandoned, they replenish themselves and become more productive.[180] This appears to be due to air infusing itself to saturation through the open orifices, just as a miscarriage seems to make some women more prolific.' ('mirum in his solis metallis, quod derelicta fertilius revivescunt. hoc videtur

---

[178] Healy, *Mining*, 180.   [179] Ibid. 61f.   [180] Ibid. 19.

facere laxatis spiramentis ad satietatem infusus aer, aeque ut feminas quasdam fecundiores facere abortus.') Elsewhere, Pliny (36. 125) mentions the common belief that marble quarries regenerate spontaneously.

The term 'bright lead' ('plumbum argentarium') has been interpreted by many scholars as a mixture of tin and lead, or just tin. Both views are incorrect, as Rottländer[181] has shown—the term refers to lead itself. This was a by-product of the silver industry when lead oxide was reduced with charcoal to make the pure metal which remains clean and bright for a long time since it is not very inclined to oxidize in the air.

Lead has many practical uses and its sulphide is the basis for some medical preparations (33. 110). Pliny (33. 108) describes how it is treated:

To make this available for use, it is boiled a second time after the 'cones' have been broken up into pieces the size of finger rings.[182] After being heated up with bellows to separate the cinders and ash from it, the lead sulphide is washed with vinegar or wine and cooled down in the process. In the case of the silvery kind [*argyritis*], in order to give it brilliance the instructions are to break it into pieces the size of a bean and boil it in water in an earthenware pot with the addition of wheat and barley wrapped in new linen cloths, until the sulphide is cleaned of impurities. Then they grind it in mortars for six days, washing the product three times a day with cold water, and, when they have ceased operations, with hot; they add salt[183] from a

---

[181] R. C. A. Rottländer, 'The Pliny Translation Group of Germany', in French–Greenaway, *Roman Science*, 16. The lead ingots so described (particularly those from England) have a relatively high silver content. Rottländer offers as a possible, but not convincing, explanation that either it was not worth while to extract the silver, or that the ancient analysis of the lead ore was erroneous.

[182] The manuscripts universally favour the reading 'anulorum', rather than 'abellanarum'—'filbert', or 'hazel-nut' (cf. *NH* 15. 88), which is paralleled by Dioscorides 5. 87 (κατακόψας εἰς καρύων μεγέθη). Pliny is obviously well aware of the passage in Dioscorides, since he later uses the phrase 'magnitudine fabae'. (εἰς μεγέθη κυάμων). But as Bailey, *Chemical Subjects*, i. 206 (note to *NH* 33. 108), observed (even before the discovery of actual *tubuli* at Laurium), this shape would naturally break into rings.

[183] The salt has two roles: (1) to whiten the litharge by forming chlorides of the type $5PbO, PbCl_2$ (yellow) and finally $PbCl_2, PbO, H_2O$ (white); (2) the salt can eliminate traces of silver oxide (black) by forming silver chloride which dissolves with the salt and is removed by washing.

salt-mine—an obol to a pound of lead sulphide.[184] On the last day they store it in a lead vessel.

spuma, ut sit utilis, iterum coquitur confractis tubulis ad magnitudinem anulorum. ita accensa follibus ad separandos carbones cineremque abluitur aceto aut vino simulque restinguitur. quodsi sit argyritis, ut candor ei detur, magnitudine fabae confracta in fictili coqui iubetur ex aqua addito in linteolis tritico et hordeo novis, donec ea purgentur. postea VI diebus terunt in mortariis, ter die abluentes aqua frigida et, cum desinant calida, addito sale fossili in libram spumae obolo. novissimo die dein condunt in plumbeo vase.

In many ways Pliny's treatment of lead, derived from a conflation of secondary sources, well illustrates his limited understanding of metallurgical problems, although it contains much of interest.

17.9 Iron

By the time of Homer[185] the Greeks had some knowledge of iron, but Herodotus[186] seems to imply that ironworking is a relatively unfamiliar sight. The Romans considered iron an extremely important metal, although they had an equivocal attitude to it. Pliny (34. 138) writes: 'It is the best and worst of life's materials' ('optimo pessimoque vitae instrumento est').[187]

Pliny (34. 142) correctly observes: 'Deposits of iron are found almost everywhere... and there is very little difficulty

---

[184] It is difficult to estimate the proportion that this represents, without knowing what relationship Pliny accepts between the drachma and the pound. Zehnacker, NH 33 p. 198 examines three possibilities: '(1) Si libra est une transcription de μνᾶ (et le texte de Dioscoride, citée *supra*, incite à le croire), le rapport est de 1/600. (2) Si Pline pense à la livre romaine (96 drachmes), le rapport est de 1/576 (solution retenue par Bailey). Enfin, (3) si Pline songeait à la prétendue livre attique de 75 drachmes (qui n'a jamais existé, mais qu'ont mentionnée les textes métrologiques), le rapport serait 1/450.' As Zehnacker observes, possibility no. 3 is a very low proportion compared with those indicated by Dioscorides (for example, 5 drachmas for a μνᾶ).

[185] Although bronze is the predominant metal in Homeric epic, iron is mentioned: Hom. Il. 7. 473; *Od.* 1. 184, and elsewhere. Cf. Hes *Theog.* 862–4.

[186] 1. 67. 3 also 1. 68. 1 f. See further, Healy, *Mining*, 193 f.

[187] Pliny (33. 1) comments with reference to iron that, 'amid warfare and slaughter, it is even more prized than gold' ('ferrum auro etiam gratius inter bella caedesque').

in recognizing them as they are indicated by the actual [red] colour of the earth' ('ferri metalla ubique propemodum reperiuntur... minimaque difficultate adgnoscuntur colore ipso terrae manifesto').

Modern estimates record that no less than 5.06 per cent of the earth's crust is formed of iron, or its compounds; only silicon and aluminium exceed this volume,[188] although many rich sources of iron ore are inaccessible for exploitation even today. Neither Greeks, nor Romans were able to mine chromite ($FeCr_2O_4$).[189]

17.9.1 Sources

Iron exists in three basic forms:[190] (1) meteoric,[191] (2) telluric, and (3) mineral ores. Meteoric iron was well known to the ancient world. So Pliny (2. 147) states: 'It is entered in the records that it rained milk and blood. ... similarly that it rained iron in the district of Lucania in the year before Marcus Crassus was killed by the Parthians [53 BC]. ... the shape of the iron that fell resembled sponges.' ('relatum in monumenta est lacte et sanguine pluisse. ... item ferro in Lucanis anno antequam M. Crassus a Parthis interemptus est. ... effigies quo pluit ferri spongearum similis fuit.') A little later, Pliny (2. 150) describes a stone which Anaxagoras[192] had predicted would fall from the sun, in the Goat's River district, 'as big as a cart and brown in colour' ('lapis... magnitudine vehis, colore adusto'), and others at Abydos, and Cassandria. Pliny claims to have seen a stone that had recently come down in the territory of the vocontii (in Gaul). All are likely to have been meteorites and the colour of Anaxagoras's stone suggests that it was siderite ($FeCO_3$).[193]

---

[188] Healy, *Mining*, 181.
[189] Forbes, *Ancient Technology*, ix. 182, and Healy, *Mining*, 41. This is a common consituent of peridotites and of serpentines derived from them. On iron ores generally, see further Tylecote, *Metallurgy*, 47 ff.
[190] Cf. Healy, *Mining*, 181.   [191] See further Ch. 14 above, App. C.
[192] For a critical account of Anaxagoras' prediction, see M. L. West, 'Anaxagoras and the Meteorite of 467 BC, *Journal of the British Astronomical Association*, 70/8 (Oct. 1960), 368–9. Healy, *Mining*, 41.
[193] Hurlbut, Mineralogy, 323.

Telluric iron was not exploited in antiquity because of the extreme difficulty of extracting iron from ores that had a high nickel content.

The two main categories of iron ores are oxides and carbonates.[194] (Figures in brackets indicate their practical iron content.) Oxides: Magnetite (60–68%), Haematite (40–66%), Limonite (25–58%), including two varieties—Oolitic (24–46%) and Bog-ore (35–55%). Carbonates: Siderite—Spathic iron ore (30–44%), Sphaerosiderite (23–40%), and Blackband ores (36–40%).

Pliny (34. 143) was well aware that there are many different kinds of iron, but credits the determinant factors to the nature of the deposits and the climate. He observes (34. 145) that 'the quality of the iron depends on the tenor of the lode, as at Noricum,[195] the method of working, as at Sulmona, and, in other places, the water' ('aliubi vena bonitatem hanc praestat, ut in Noricis, aliubi factura, ut Sulmone, aqua aliubi').

17.9.2 Smelting

Iron ores varied in regard to difficulty of smelting. Those from the mines of the Chalybes,[196] near Amisus, on the southern shores of the Black Sea, were difficult to smelt on account of the quantity of clay contained in them and were softened only by raising them to a great temperature. At Noricum the Romans preferred limonites,[197] while rejecting spathic iron ores. Unknown to the Romans one of these ores contained titanium, an excellent component for making steel. Lake iron contains phosphorus and varying amounts of manganese. The

---

[194] Tylecote, *Metallurgy*, 48 f.
[195] Healy, *Mining*, 64, and Forbes, *Ancient Technology*, ix. 268 f. On Noricum generally, see G. Alföldy, *History of the Provinces of the Roman Empire: Noricum* (London and Boston, 1974).
[196] Virgil (*Aen.* 10. 174) also mentions inexhaustible mines of 'Chalybean' iron on *Elba*.
[197] Limonite is an omnibus term used for a range of mixtures of hydrated iron oxides and iron hydroxides (FeO·OH, or Fe_2O_3). Limonite consists of various amorphous and cryptocrystalline constituents and also includes amorphous iron hydroxides. Oolitic limonites (24–46% iron) constitute the most important iron deposits in Europe.

best ore for smelting is not necessarily that with the highest iron content. The ore should be low in silica, alumina, lime, and magnesia.

Pliny (34. 142) mistakenly states that the method of melting out the veins is the same as in the case of copper ('ratio eadem excoquendis venis'). The technology is, however, totally different. Four stages[198] may be considered: (1) the use of meteoric iron; (2) the production of iron as a by-product of gold-refining; (3) the reduction of iron ores in bloomeries; (4) cementation processes used in the manufacture of steel.

It was suggested, many years ago, that one of the earliest sources of iron may have been the gold placers of Nubia,[199] where the gravels contained a high percentage of magnetite. Iron objects began to appear in Egypt after $c.2000$ BC.[200] According to Hume,[201] iron was discovered in a bath of molten gold; it was identified with meteoric iron. This is extremely unlikely, however, since, as Tylecote[202] observed, for this to happen, the ratio of magnetite to sand would have to be very high, greater, in fact, than two to one and conditions would have to be very reducing, which would not be necessary for the melting of gold out of the ore.

To refine copper, or lead, all that is essential is a mixture of sulphide and oxide minerals and heat.[203] With iron the situation is quite different. No amount of heat alone will reduce a mixture of iron oxide and iron sulphide to usable metal. To produce iron the metallurgist must employ iron oxide, together with both heat and carbon, the latter performing an essential function in the process. Iron in comparison with tin oxide, for example, is difficult to reduce and needs much more strongly reducing conditions.[204]

Some metals are produced by smelting below their melting-point, others, above. Fortunately iron may be reduced from

---

[198] Healy, *Mining*, 182.
[199] Healy, 'Greek and Roman Gold Sources', 10.
[200] A. Lucas, *Ancient Egyptian Materials and Industries*⁴ (London, 1962), 241 ff.
[201] E. W. Hulme, 'Iron Smelting with Lake and Bog-iron Ores', *Antiquity*, 11 (1937), 222.   [202] *Metallurgy in Archaeology*, 185.
[203] Healy, *Mining*, 182 f.   [204] Ibid.

pure iron oxide at about 800 °C, considerably below its melting-point of 1535 °C. Since it is not practicable to separate the mineral from its gangue by washing, a large amount must be removed by 'slagging' in the smelting process. The slags consist mostly of a compound formed from iron oxide and silica. As much as possible of these slags has to be removed from the iron by liquation; the smelting process, therefore, must take place at, or above, the temperature at which the slags become sufficiently fluid to drain away the solid iron. This is about 1150 °C, well above the minimum temperature at which iron oxide can be reduced to iron, but substantially below its melting-point. The iron is produced in the solid state as a sponge, or raw *bloom* from which the slag partly drains away. The rest is removed by hammering while the slag is still in a fluid state. While it is comparatively easy to bring about strongly reducing conditions at a moderate temperature (up to 800 °C) in a primitive furnace, this becomes more and more difficult as the temperature increases, since more air is needed and therefore conditions tend to become more oxidizing. To offset this tendency a larger proportion of charcoal is required and some way must be found of keeping the whole mass of charcoal and iron hot without consuming too much charcoal for this purpose. In other words the heat must not be allowed to escape too easily and the furnace must therefore be well lagged.

Iron ores vary considerably in composition but the majority of those used in early times have ferric oxide, or haematite, contents exceeding 60 per cent when ready for the furnace.

Preliminary roasting[205] has a number of aims, among them the following: (1) to drive out carbon dioxide from carbonate ores; (2) to remove water from limonite containing bog-ore and excess water; (3) to make ores more porous and so more easily reducible; (4) to enable some very hard ores to be broken up, thereby facilitating the entry of carbon monoxide into the resultant pieces.

The principles of roasting differ from those of smelting. In roasting, a moderate temperature is required and, very often,

---

[205] Healy, *Mining*, 183.

oxidizing conditions. Therefore, there is no need to keep out excess air, and a platform in the open is quite sufficient. Raw wood can be used and a lattice-like construction made on which the ore can spread. It is quite unlikely that primitive peoples used bowl hearths of furnaces for roasting as these would be an unnecessary complication and relatively inefficient.

The roasting reactions of carbonate ores are: $FeCO_3 \rightarrow FeO + CO_2$ (decomposition), and $4FeO + O_2 \rightarrow 2Fe_2O_3$ (oxidation).

The drying, or decomposition, of hydrated ores of limonite, or bog-ore type, proceeds according to the equation: $Fe_2O_3 \cdot H_2O \rightarrow Fe_2O_3 + H_2O$ (drying).

17.9.2a Bloomery Iron

The Greeks and the Romans initially produced iron in the form of a bloom, that is a rough, spongy mass of iron, or 'steel', obtained directly from the iron ores. Possibly the first reference to such a bloom occurs in Homer (*Il.* 23. 826): 'Then the son of Peleus set forth a mass of rough-cast iron' (σόλον αὐτοχόωνον).

The early bloomery[206] was either a shallow trench about two feet deep, with sides of turf or stone built several feet above the ground (basically, that is, a sheltered hole), or a hearth fire to accommodate the ore and charcoal (the charge);[207] lignite was, occasionally, also used as fuel. At first, a natural draught was used to raise the furnace temperature, and furnaces were often built on high ground to take advantage of prevailing winds. In the course of time, the use of bellows, already known in Homer (*Il.* 18. 468), replaced or supplemented natural draught. Over a period of eight to twelve hours, the ore itself did not become molten but, gradually, coagulated into the bloom. The slag trickled to the bottom of the furnace whence it was in turn displaced by the heavier iron. Infusible material remained at the top. On completion of the smelting

---

[206] Forbes, *Ancient Technology*, ix. 186.
[207] Healy, *Mining*, 185, fig. 26

the furnace was broken open and the bloom removed. The bloom, at this stage, still contained a number of impurities and was reheated in another furnace and hammered, or forged: this operation was repeated a number of times to drive out the impurities, or to dislodge the adhering slag. The final product was wrought iron. This, as the evidence of analysis shows, often contained much slag.[208]

Diodorus Siculus (5. 13. 1–2) records iron–working on Elba: 'The island possesses a great amount of iron ore. ... they crush the rock that has been quarried and in certain ingenious furnaces they smelt the lumps of iron ore by means of a great fire and form them into pieces of moderate size which in their appearance are like large sponges.' Similarly, Strabo (5. 2. 6) observes: 'I saw the people who work the iron that is transported from Aethalia [Elba]; for it cannot be brought into complete coalescence by heating in the furnaces on the island. It is carried immediately from the mine to the mainland.' Strabo here indirectly refers to the problem of the high temperatures needed to bring the iron from a brittle spongy mass to a soft texture.

17.9.2b Shaft Furnaces

Bloomery hearths were common in the ancient world and continued to be used even after $c.500$ BC, when shaft furnaces first began to be developed in central Europe.[209] Although Pliny does not specifically refer to this development, Wynne and Tylecote's[210] account of their experiments with a reconstructed

---

[208] Tylecote, *Metallurgy*, 248 f.: 'Wrought iron starts life as a smelted bloom of iron particles and slag, and the first process of smithing is to weld the iron particles together and expel the slag. The higher the temperature, the greater the difference in the physical properties of pure iron and slag; as the melting-point of the iron is $1540\,°C$ and that of a fayalitic slag ($2FeO. SiO_2$) is about $1200\,°C$, therefore, the smithing temperature should be as near to $1200\,°C$ as possible. ... What is clear is that the welding of iron is implicit in the smithing operation and the early smith must have known that pieces of roughly smithed iron could be welded together.'

[209] Forbes, *Ancient Technology*, ix. 2203 ff.

[210] 'An Experimental Investigation into Primitive Iron-smelting', *Journal of the Iron and Steel Industry*, 190 (1958), 339–48. See further Tylecote, *Metallurgy*, 50 and 52.

shaft furnace is included to complete the picture of iron-smelting technology in Roman times. In the shaft furnace the ore and fuel travel down together, although they are usually added layer by layer. In the upper levels water is removed and, as the temperature of the ascending gases reaches 500 °C, any iron carbonates are decomposed. Then, lower down, reduction starts at about 750 °C, at first converting the higher oxides (magnetite and haematite) to the lower oxide (FeO), providing that the carbon monoxide content of the atmosphere is high enough.

Furnace remains suggest that a considerable amount of iron carbide was produced as: $3Fe+3Fe+C \rightarrow Fe_3C$. At 900 °C the following equations illustrate the process: (1) $Fe_3C+O_2 \rightarrow 3Fe+CO_2$ and (2) $2Fe+O_2+SiO_2 \rightarrow 2FeO \cdot SiO_2$.

17.9.3 Treatments of Iron

One of the main problems of the Greeks and the Romans was the hardening of iron, or steel. This was achieved by quenching and carburization.

17.9.3a Quenching

The process by which highly heated iron, or steel, is suddenly cooled in water, or other liquids, to fix into the cold structure the various properties acquired by them at different temperatures, is known as quenching.[211] The cementite structure is rapidly carried through the dangerous zone of disintegration and more or less stabilized at low temperatures. Slowly cooled steel has a characteristically different structure.

Pliny (34. 144) continues his account of iron with the processes to which it is subjected to modify its physical properties. 'The chief difference [that is, in *hardness*] depends on the water in which, at intervals, the red-hot metal is plunged'

---

[211] Healy, *Mining* 233 f. On the confusion of the terms *tempering* and *quenching*, see R. Aiano, 'The Roman Iron and Steel Industry at the Time of the Empire (MA diss., Aberystwyth, 1975), 43 ff.

('summa autem differentia in aqua, cui subinde candens immergitur'). 'It is the custom [34. 146], however, to quench smaller iron forgings in oil for fear that the water might harden them and make them brittle. And it is remarkable that, when a vein of ore is fused, the iron becomes liquid like water and afterwards acquires a spongy and brittle texture.' ('tenuiora ferramenta oleo restingui mos est, ne aqua in fragilitatem durentur. mirumque cum excoquatur vena, aquae modo liquari ferrum postea in spongeas frangi.')

17.9.3b Carburization

One of the major problems facing the Greeks and the Romans in iron technology was the enhancement of the low carbon content of wrought iron which made it unsuitable for the manufacture of tools and weapons.

The process, known as carburization[212] (cementation), was effected during the reheating and hammering of wrought iron: in this it picked up additional carbon from the charcoal. The process was clearly empirical, since there is no evidence to suggest that the smelters understood the principle of carburization. They would doubtless have observed in practice that iron, when reheated and forged, improved in quality and toughness and subsequently that quenching would complete the improvement.

[Aristotle,] *Meteorologica* 4. 383[a–b], in a puzzling account of wrought iron, claims that it will melt and grow soft and then solidify again: 'This is the way "steel" is made. The dross sinks to the bottom and is removed from below and, by repeated subjection to this treatment, the metal is purified and the "steel" produced. They do not repeat this process often, however, because of the great wastage and loss of weight in the iron which is purified. But the better quality of the iron, the smaller the amount of impurity. Pyrimachos stone will also melt and form drops and become fluid: when it solidifies after having been fluid, it regains its former hardness. Millstones,

---

[212] Tylecote, *Metallurgy*, 271 f. and Healy, *Mining*, 231 ff.

too, melt and become fluid and, when they solidify again, they are black in colour but like lime in texture.' This is a controversial passage, as Lee[213] points out: εἰργασμένος σίδηρος is clearly the bloom which has been turned into wrought iron; μαλάττεται means 'grows soft' not 'melt', since this would be impossible granted the furnace conditions; στόμωμα must surely mean 'hardened iron', that is iron resulting from carburization and, possibly, quenching, as the later evidence of Plutarch suggests.

The account in the passage contains the sort of confusion one might expect from a layman's limited understanding of the processes involved, or from an abbreviated version of a more complicated description. The reference to millstones may be included as a result of the author's recollection of their use in ore-dressing.

In the same passage [Aristotle] refers to a 'fire-resisting' stone (πυρίμαχος λίθος) which also melts, forms drops, and becomes fluid. When this solidifies, after having been fluid, it regains its former hardness. This is repeated by Theophrastus (*Lap.* 9). Caley,[214] and others have seen in these references—ostensibly in connection with iron-refining—evidence for the addition of acidic limestone fluxes to the charge when smelting. Attractive as this interpretation is, however, such fluxes are not likely to have been employed, as Tylecote[215] observes, for, in practice, additional lime (from limestone or chalk) cannot be absorbed in large pieces by fayalite slags at such a low temperature as 1200 °C. It takes an appreciable time for a modern open-hearth steel-smelting furnace working at 1600 °C to absorb limestone into the slag. However, if the limestone was ground into powder and absorbed in small quantities the result would only be to reduce the melting-point by about fifty degrees C, which would not much benefit the process. Large

---

[213] H. D. P. Lee (ed. and trans.), *Aristotle: Meteorologica* (Loeb: London and Cambridge, Mass., 1952), comm. ad loc., pp. 324 ff.

[214] In Caley and Richards, *Theophrastus*, pp. 76 f. But contrast Eichholz, *Lap.* comm. on p. 94, for the more acceptable view that these were the building materials of lime kilns and the hearths of iron furnaces, where they were subjected to a fluxing agent (lime) 'the matter heaped upon them'.

[215] *Metallurgy*, 271 f.

quantities would raise the melting-point, clog the furnace, and be detrimental to the smelting process.

17.9.3c Nucleus ferri

Pliny (34. 144) refers to a special process by which iron is smelted to give hardness to a blade: 'All of these [different kinds of iron] are called "edging ores" [*stricturae*] a term not used in the case of other metals; it is, as assigned to these ores, derived from the term "to draw out a sharp edge" [*stringere aciem*].' Later, in the same passage, he uses the somewhat cryptic term 'nucleus ferri' in connection with the hardness of a blade—adding, 'and, by another process, they give solidity to anvils, or the heads of hammers' ('stricturae vocantur hae omnes, quod non in aliis metallis, a stringenda acie vocabulo inposito. ... *nucleus*que quidam *ferri* excoquitur ... ad indurandam aciem, alioque modo ad densandas incudes malleorumque rostra').

The phrase *stringere aciem* must, originally, have meant 'to draw out a sharp edge', but Schaaber[216] (a specialist in the hardening of iron to form steel) argues that it had, by Pliny's time, become a technical term, meaning 'forming a point of steel'. Such a distinction, between iron and steel, is perhaps also supported by archaeological evidence. Vetters[217] points out that there are known from the La Tène period, and in Roman times, bipyramidal ingots of iron, of which one of the two points has been left as soft iron, as can be shown by bending it, while the other, in contrast, has been hardened with carbon to provide a steel capable of being reworked with heat. Similarly, apropos the problematical term *nucleus ferri*, Vetters and Schaaber conclude that it is 'another technical term for iron with a high carbon content, suitable for forging cutting edges. ... again we take *acies* as steel and not just pure iron,

---

[216] See Rottländer, 'Pliny Translation Group', 13 f., and, generally, O. Schaaber, H. Müller, and I. Lehnert, 'Metallkundliche Untersuchungen zur Frühgeschichte der Metallurgie, *Archäologie und Naturwissenschaften*, 1 (Mainz, 1977), 221–68.

[217] Rottländer, 'Pliny Translation Group', 14.

firstly because ancient ideas about what is pure metal are quite different from ours.'

The new meaning of *acies*[218] is supported by frequent archaeological finds in which the cutting edge of the artefact is hard, while the rest of it is soft.

Wrought iron can also be hardened and, when high in phosphorus from its ore-mineral, can reach a degree of hardness as high as HV 300 in the cold-worked state. This exceeds the hardness of a carburized, but non-heat-treated steel.

Pliny's account of iron and the technology of iron smelting adds little to the descriptions of earlier authorities such as [Aristotle], Diodorus Siculus, and Strabo. He incorrectly assumes that the processes are identical to those employed in the smelting of non-ferrous ores and so omits any further explanatory detail.

However, one important contribution of Pliny to our knowledge of iron technology is his reference to carburization and its effect on improving the quality and toughness of this widely used metal.

17.10 ZINC

Although a small piece of zinc was discovered in the Agora at Athens[219] there is no independent evidence to confirm that the Greeks, or the Romans, had the appropriate technology to produce the *pure* metal from compounds of zinc.

Strabo (13. 1. 56) also described a mineral found in Asia Minor in terms which suggest that it may have been an ore of zinc: 'There is a stone in the neighbourhood of Andeira which, when burned, becomes iron, and then, when heated in a furnace with a certain "earth", distils "zinc" [ψευδάργυρος] and this, with the addition of copper, makes the mixture called by some brass[220] [ὀρείχαλκος].'

---

[218] Compare the French *acier* ('steel').
[219] M. Farnsworth, C. S. Smith, and J. L. Rodda, 'Metallographic examination of a Sample of Metallic Zinc from Ancient Athens', *Hesperia*, Suppl. 8 (1949), 126–9.
[220] See *Homeric Hymn* 6. 9; Hes. *Shield*, 122; Stesich. 88; Ibyc. *apud P.Oxy.* 1790. 42; Arist. *Mir.* 834^{b}25.

Zinc-rich ores, such as *cadmea*,[221] were relatively plentiful in the ancient world, but the fact that at its smelting temperature of 1000–1100 °C, its vapour pressure is above that of atmospheric pressure means that it starts its existence as a metallic vapour and has to be condensed before it comes into contact with air. The temperature, accordingly, has to be rapidly reduced to 430 °C, that is its melting-point.

Zinc is produced today by a retorting and condensing process.[222]

Minor Metals

The minor metals known in the ancient world include antimony, mercury, and (certainly known to the Egyptians, but perhaps also to the Greeks and Romans) platinum.

17.11 Antimony

Antimony was often present in both early and late coppers up to as much as 5.6 per cent, and it occurs as a native metal in the Caucasus and is associated with silver deposits in both Tuscany and Siphnos. It is quite possible that, like arsenic, it was never produced in its metallic form in Europe until the eighteenth century.[223]

Stibnite, or antimony sulphide[224] (Sb_2S_2), appears to have been known by Pliny (33. 101) who gives a garbled account of the mineral:

In the same mines, is found what may properly be called a rock that is composed of white and shiny, but not transparent, froth. Some people call this '*stimmi*', or '*stibi*', others '*alabastrum*', or '*labarsis*'. There are two kinds, male and female. The female variety is preferred, the male being more uneven and rough to the touch, as well as lighter in weight, less brilliant, and more like sand. The female,

---

[221] Ch. 14, s.v.
[222] Tylecote, *Metallurgy*, 143 f. See above, end of section 17.6.7 and n. 139 on Champion's vertical crucible downward condensation process.
[223] Tylecote, *Metallurgy*, 144 f.
[224] Whitten–Brooks, *Geology*, appendix: Stibnite.

however, is sparkling and brittle and splits into thin layers rather than into globules.

in isdem argenti metallis invenitur, ut proprie dicatur, spumae lapis candidae nitentisque, non tamen tralucentis; *stimmi* appellant, alii stibi, alii alabastrum, aliqui larbasim. duo eius genera, mas ac femina. magis probant feminam. horridior est mas scabriorque et minus ponderosus minus radians et harenosior, femina contra nitet, friabilis fissurisque, non globis, dehiscens.'

The male would seem to be antimony sulphide (Sb_2S_2) and the female, metallic antimony (Sb). Pliny is followed by Dioscorides (5. 84) who considers the bright variety of *stimmi* as the best: he does not, however, use the terms 'male' and 'female'.

The density of stibnite is 4.5–4.7[225] which means that it is less heavy ('minus ponderosus') than native antimony which is 6.6–6.8: the latter, as Pliny also correctly observes, has an easy cleavage habit.

The processing of antimony trisulphide is complicated, as Pliny (33. 103) explains:[226]

It is prepared by being smeared round with lumps of ox-dung and burnt in ovens, and then cooled down with women's milk and mixed with rain-water and pounded in mortars. And next the turbid part is poured off into a copper vessel after being purified with soda. The lees are recognized by being full of lead and they settle to the bottom of the mortars and are thrown away. Then the vessel into which the turbid part was poured off is covered with a cloth and left for a night, and the next day anything floating on the surface is removed with a sponge. The sediment on the bottom is considered the choicest part and is covered with a linen cloth and put to dry in the sun but not allowed to become very dry and is ground up a second time in the mortar and divided into small tablets. But above all, it is essential to limit the amount of heat applied to it, so that it may not be turned into lead. Some people do not employ dung in boiling it, but fat. Others pound it in water and strain it through three thicknesses of linen cloth and throw away the dregs and pour off the liquor that comes through, collecting all the deposit at the bottom, and this they use as an ingredient in plasters and eye-washes.

---

[225] Whitten–Brooks, *Geology*, appendix: Stibnite.
[226] Cf. Dioscorides 5. 84.

uritur offis bubuli fimi circumlitum in clibanis, dein restinguitur mulierum lacte teriturque in mortariis admixta aqua pluvia; ac subinde turbidum transfunditur in aereum vas emundatum nitro. faex eius intellegitur plumbosissima, quae subsedit in mortario, abiciturque. dein vas, in quod turba tranfusa sint, opertum linteo per noctem relinquitur et postero die quidquid innatet effunditur spongeave tollitur. quod ibi subsedit, flos intellegitur ac linteo interposito in sole siccatur, non ut perarescat, iterumque in mortario teritu et in pastillos dividitur. ante omnia autem urendi modus necessarius est, ne plumbum fiat, quidam non fimo utuntur coquentes, sed adipe. alii tritum in aqua triplici linteo saccant faecemque abiciunt idque, quod defluxit, transfundunt, quidquid subsidat, colligentes. emplastris quoque et collyriis miscent.

Bailey[227] explains this passage as an account of the oxidation of antimony trisulphide that releases sulphur dioxide (SO_2) and leaves antimony oxide (Sb_2O_3). The ox-dung (or paste), when burnt, becomes carbon, which acts as the reducing agent. Part of the antimony oxide is thus reduced to metallic antimony which the ancients took for lead because of its similar appearance:[228] this was deposited at the bottom of the mortar (faex plumbosissima).

Dioscorides (5. 84) describes the same process except that he advises covering the *stimmi* in a paste made of flour rather than using ox-dung. The result is again the production of carbon. He adds that if too much heat is applied all the antimony oxide is reduced, and becomes 'lead', as he thought.

Zehnacker[229] draws attention to the fact that στέαρ (*adeps*), used by Dioscorides and other authorities, signifies not only fat, but paste.[230]

Antimony was extensively used for eye make-up and as a cure for a number of eye disorders.[231]

17.12 HYDRARGYRUS (MERCURY)

Mercury,[232] in its native state, has a metallic lustre which is silver-white. At ordinary temperatures, it is a coherent, mobile

---

[227] *Chemical Subjects*, i. 213.
[228] Antimony has a silvery-grey colour, not unlike that of lead.
[229] *NH 33*, p. 194 n. 2.
[230] στέαρ as fat. see Hom. *Od.* 21. 178; Arist. *PA* 651ª26, and generally; as dough made from flour, see Theophr. *HP* 9. 20. 2; Strabo 17. 2. 5.
[231] *NH* 33. 102.   [232] Healy, *Mining* 191 f.

liquid.[233] Pliny (33. 99) gives a good description of mercury (*argentum vivum*): 'There is also a mineral found in these veins of silver, which contains a humour, in round drops, that is always liquid, and is called quicksilver. ... substances float on its surface except gold, which is the only thing that it attracts to itself' (amalgamation). ('est et lapis in iis venis, cuius vomica liquoris aeterni argentum vivum appellatur. ... omnia ei innatant praeter aurum id unum ad se trahit.')

Theophrastus (*Lap.* 60) was the first to record, in detail, the refinement of mercury from the mineral cinnabar.[234] His account is not only intrinsically interesting but is important in that it is the earliest reference to any method of isolating a metal from one of its compounds. His description is brief and to the point: 'Mercury, or quicksilver, is made when cinnabar, mixed with vinegar, is ground in a copper vessel with a pestle made of copper.'[235] This, explains Caley,[236] is not a mechanical method for the liberation of the metal from a natural mixture of mercury and cinnabar, but a true chemical process which depended on the displacement of the mercury from the cinnabar, by the more active metal placed in contact with it. Laboratory experiments confirm that this method is viable. The reaction may be accelerated by warming the mixture. The resultant products are copper sulphide and mercury.

Sisapo—Almaden, in New Castile—is celebrated for mercury as well as for cinnabar. Pliny (33. 119) describes another kind of *minium*[237]

---

[233] Ibid. 227. Cf [Aristotle], Mete. 378ª, who states that the poet Philippos explained the movement of a statue of Aphrodite by saying that the sculptor Daedalos poured quicksilver into it.

[234] See Ch. 14, s.v. 'cinnabar'.

[235] Bailey used cinnabar in his experiments simulating the methods described by Theophrastus and Pliny. The reaction formed copper sulphide and an amalgam of copper with mercury; finally, the mercury was isolated by further heating.

[236] In Caley and Richards, *Theophrastus*, p. 204.

[237] Bailey, *Chemical Subjects*, i. 220. This kind of *minium* was probably red lead (Pb_3O_4) manufactured from cerussite, or native lead carbonate ($PbCO_3$). The latter, as described in this section, often has a colour not unlike that of lead. When roasted, it first loses carbon dioxide (CO_2) and leaves litharge (PbO), which, on further roasting (oxidation), gives red lead. The red colour would, therefore, only be obtained after roasting.

found in almost all silver-mines and, similarly, lead-mines. This is produced by smelting rock that is veined with the metal and is not obtained from the rock that, when smelted, produces round drops called mercury,[238] but from other minerals found in association. ... In the cinnabar mines of Sisapo [33. 121] the vein of sand contains no silver. It is smelted like gold, and assayed by means of gold brought to red heat. If it has been adulterated, it turns black, but, if genuine, retains its colour.

namque est alterum genus omnibus fere argentariis itemque plumbariis metallis, quod fit exusto lapide venis permixto, non ex illo cuius vomicam argentum vivum appellavimus—is enim et ipse in argentum ⟨vivum⟩ excoquitur—sed ex aliis simul repertis. ... Sisaponensibus autem miniariis sua vena harenae sine argento. excoquitur auri modo; probatur auro candente, fucatum enim nigrescit, sincerum retinet colorem.

This passage is clearly intrusive and indicates lack of revision.

Pliny continues (33. 123) with what he incorrectly believes is a method of producing artificial quicksilver (mercury).[239]

It is made in two ways, [1] by pounding red lead in vinegar with a copper pestle in a copper mortar,[240] or [2] it is put in an iron shell in flat earthenware pans and covered with a convex lid sealed in with clay. Then a fire, lit beneath the pans, is kept constantly burning by means of bellows and so the surface moisture (with the colour of silver and the fluidity of water) which forms on the lid is wiped off. This moisture is also easily divided into drops and rains down freely with slippery fluidity.

fit autem duobus modis: aerei mortariis pistillisque trito minio ex aceto aut patinis fictilibus impositum ferrea concha, calice coopertum, argilla superinlinita, dein sub patinis accenso follibus continuis igni atque ita calici sudore deterso, qui fit argenti colore et aquae liquore. idem guttis dividi facilis et lubrico umore compluere.

---

[238] Pliny is in error; mercury clearly cannot be obtained from *minium*. Cf. above, n. 237.

[239] Pliny believes that mercury and quicksilver are two different substances. It seems that *argentum vivum* is native mercury (cf. 33. 99), while the less valuable quicksilver is mercury obtained from cinnabar.

[240] Pliny's description here follows Theophr. *Lap.* 60 (as above, cf. n. 235).

Mercury is produced from cinnabar by distillation.[241] It boils at 360°C, at atmospheric pressure, and has to be condensed. As the ore is relatively low grade, it is heated in sealed shaft furnaces with grates and the gases are fed into a condensing system from which the mercury (Hg) is recovered. The ore (mercury sulphide, HgS) when heated, oxidizes to yield mercury vapour and sulphur dioxide. In Ladik,[242] in Anatolia, the remains of condensers, but not furnaces, have been found. These clearly belong to the Graeco-Roman period and it is suggested that selected pieces of rich cinnabar were mixed with charcoal and fired and covered by an inverted pot. But it seems likely that the pot is part of the condensing system and that the actual furnace was not found.

17.13 PLATINUM

Apart from its use by the Egyptians,[243] in the seventh century BC, as a 'white metal' in the decoration of a box, platinum, as an independent metal, was not recognized in Europe until trial pieces of it were exported to Spain by the Spaniards in South America c.1745.[244]

Platinum and platinum-group elements (PGE),[245] however, were present in white gold—alluvial placers—found mainly in the rivers of Asia Minor[246] but, later, also in a stater of Philip II of Macedon.[247] The PGE inclusions were relatively

---

[241] Tylecote, *Metallurgy*, 147 f.
[242] J. W. Barnes et al., *Geology and Ore Deposits of the Sizma-Ladik Mercury District, Turkey* (Cento Rep., 1969).
[243] Lucas *Ancient Egyptian Materials*, 244.
[244] D. M. McDonald, *A History of Platinum* (Johnson Matthey: London, 1960). Cf., generally, D. M. McDonald and L. B. Hunt, *A History of Platinum and its Allied Metals* (Johnson Matthey: London, 1982).
[245] Healy, 'Greek and Roman Gold Sources', 15 f. See also J. Ogden, 'Platinum-group Metal inclusions in Ancient Gold Artefacts', *Journal of the Historical Metallurgical Society*, 11 (1977), 53–72; N. D. Meeks and M. S. Tite, 'The Analysis of Platinum-group Element Inclusions in Gold Antiquities', *Journal of Archaeological Science*, 7 (1980), 267–75.
[246] Healy, *Mining*, 31 f.
[247] Healy, 'Greek and Roman Gold Sources', 15.

insoluble, or had a high melting-point,[248] and so survived subsequent mechanical and heat treatment.

Pliny (37. 55) discusses a hard substance found in gold: *Adamas* 'was the name given to the "knot of gold"[249] found very occasionally in mines in association with gold and, so it seemed, formed only in gold. Our ancient authorities thought that it was found only in the mines of Ethiopia between the temple of Mercury and the island of Meroe[250] and state that the specimens discovered were no larger than a cucumber seed and not unlike one in colour.' ('ita appellabatur auri nodus in metallis repertus perquam raro comes auri nec nisi in auro nasci videbatur. veteres eum in Aethiopum metallis tantum inveniri existimavere inter delubrum Mercuri et insulam Meroen, dixeruntque non ampliorem cucumis semine aut colore dissimilem inveniri.')

Although attempts have been made to identify Pliny's substance with platinum, the physical characteristics—in particular, the colour—do not support such an identification.[251]

METALS: CONCLUSIONS

Pliny's treatment of the main metals in use in Roman times is inconsistent and depends on whether his information is derived from personal knowledge, obtained at first hand, as in the case of gold-mining, or from secondary sources. His account of gold is, therefore, technically sound; yet he can still include references to mining ants in India and griffins in Scythia, stories which appear in Herodotus! Furthermore, although his description of techniques used in gilding copper

---

[248] Tylecote, *Metallurgy*, 67.
[249] Pliny here translates the Greek term χρυσοῦ ὄζος in Plato (Ti. 59b), synonymous with ἀδάμας.
[250] Meroe is located in the northern Sudan.
[251] For a contrary view, however, see Ogden, 'Platinum-group Metal Inclusions', 61: 'The resemblance between Pliny's 'adamas' and platinoid grains might be purely fortuitous but some similarity is obvious. Pliny appears to have based much of his knowledge of adamas on Egyptian authorities. ... The name 'adamas' itself has been compared with the Akkadian 'algamisu', a very hard substance, which, in turn, has been connected with the Egyptian 'Irkbs'—a mineral included in a list of tributes from Nubia.'

substrates has been shown, by modern experimentation, to be sound, he credits 'electrum' with the power of detecting poisons!

Pliny is clearly less familiar with silver-mining and production but understands the well-established technique of cupellation although he sometimes appears to conflate processes; he is not likely to have had any experience of silver-mining.

The confusion in his discussion of lead reflects his dependence on two sources and two distinct metallurgical terminologies. There are, in addition, inaccuracies, some of which are readily explicable in terms of errors in manuscript transmission. Pliny, for example, refers to tin as a variety of 'white lead', although he describes the separation of cassiterite (a primary ore of tin) from its accompanying gangue by stream water. His explanation is based on observation, not on any understanding of the principle of gravity separation which depends on the difference in density of the ore-minerals involved. Although pewter was commonly produced by both Greeks and Romans, Pliny, inexplicably, does not refer to this tin-lead alloy.

The main ores of copper are well known to Pliny, but he is less concerned with the metallurgy of extraction than with the uses to which its alloys and compounds were put in the classical world. Pliny records details of bronze statues, many of which were subsequently destroyed, or lost, or cannot now be readily identified with any degree of certainty. He is a valuable source of information in the field of classical art. Copper compounds were also used in pharmacology.

The account of mercury and its discovery is almost wholly derived from Theophrastus. Apart from the mention of the health hazard and environmental implications of working with mercury, Pliny makes a minimal contribution.

Pliny's interest in iron reflects the growing importance of this metal for the Romans in agriculture, civil engineering, and, above all, warfare and weapons systems. The exploitation of iron may be said to have transformed the Roman way of life. He covers all aspects of iron ores and their individual properties including magnetism. He is well aware of the mechanical and thermal treatments of iron, including quenching and carburization.

Pliny discusses zinc, with particular reference to the use of its mineral compounds in the production of brass.

Finally he includes a brief account of antimony and of the processing of antimony trisulphide, the main use of which was in cosmetics.

18

Technology

18.1 Introduction

Some advances in technology in the ancient world were attributed to accidental discoveries, among them the basic metallurgical processes. Lucretius, in a well-known passage (*DRN* 5. 1255–64), ascribes the rise of civilization to its acquisition of the technology needed for smelting and casting of metals.[1] Similarly Pliny (36. 191 ff.) records the traditional account of the origin of glass.[2]

Lynn White,[3] with justification, claims that 'no Greek or Roman ever tells us, either in words, or in iconography, what he, or his society, wanted from technology, or why he wanted it. Pliny the Elder, however, gives us more information about the manual arts and crafts than does any other author of Antiquity, but always in terms of specificities, not of general considerations. ... the same lack is found also in the relatively few surviving works of Graeco-Roman engineers, except

---

[1] This section is by way of an *epilogos*, since Pliny, throughout the *Natural History*, refers to many examples of applied technology, which are discussed in context. Lucretius imagines a scenario in which a vast conflagration has roasted the earth, and writes: 'Out of the melted veins there would flow into hollows on the earth's surface a convergent stream of silver, gold, copper, and lead. Afterwards, when men saw these lying solidified on the earth and flashing with resplendent colour, they would be tempted by their attractive lustre and polish to pick them up. They would notice that each lump was moulded into a shape like that of the bed from which it had been lifted. Then it would enter their minds that these substances, when liquefied by heat, could run into a mould, or the shape of any object they might desire and could also be drawn out by hammering into pointed tips of any slenderness and sharpness.' So, in Hesiod, *WD* 109 ff. and other authorities, it is significant that, of the Five (mythological) Ages of Mankind, *four* are metals, namely gold, silver, bronze, and iron.   [2] See below, section 18.2.

[3] Lynn White Jr., 'Technological Development in the Transition from Antiquity to the Middle Ages', in *Atti del Convegno di Como—Technologia, economia e società nel mondo romano* (Como, 1980), 235 ff.

perhaps in Vitruvius' overblown and utopian description of the ideal education for an architect.' White continues:[4] 'Mythology—often so useful for understanding the values of a culture—seems biased against technology, most notably in the myth of Prometheus.[5] Ancient historians[6] mention innovations in vague terms but give us almost no analytical accounts of invention. Philosophers do not ponder it at length. To understand what the Ancients wanted from technology, one must look not at what they wrote—I repeat they were amazingly nebulous on the subject—but both at what they accomplished and what they did not accomplish even though, to modern eyes, the means lay at hand. The Greeks, for example, although they had complicated gearing mechanisms in the hellenistic period, never developed a clock.'[7]

Graeco-Roman technicians were ingenious, but less in their elaboration of ideas than in producing original ideas of admirable simplicity. The mechanical helix—the basis of the Archimedian screw[8]—shows an insight of genius. The development of military weapons, especially catapults,[9] is noteworthy. Philo, of Byzantium[10] (second century BC), describes in detail a catapult firing successive shots without reloading, that is a 'machine-gun' catapult. He says that it was invented in Rhodes. The feed mechanism that places a new arrow into the slot left vacant by the previous shot, is unambiguously, a groove cam, the first of its kind and the only one recorded from antiquity. The groove cam is important for the later development of automation. The organ,[11] Hero's[12] pneumatic devices,[13] and steam engine

---

[4] Ibid. 237.

[5] Prometheus, son of the Titan Iapetus and Clymene, stole fire from heaven, gave it to mortal men and taught them all the useful arts. For this he was punished by being chained to a rock in the Caucasus. An eagle devoured his liver every day and it regenerated each night.

[6] See, generally, R. Kellermann and W. Treue, *Die Kulturgeschichte der Schraube*² (Munich, 1962).   [7] See below, section 18.4.

[8] Healy, *Mining*, 95 ff.

[9] E. W. Marsden, *Greek and Roman Artillery*, 2 vols. (Oxford, 1969–71); *Ancient Engineering*, 99 ff. and 123–5; W. Soedel and V. Foley, Ancient catapults', *Scientific American*, 240 (March 1979), 156, 159.

[10] Landels, *Ancient Engineerig*, 99, 120 ff.

[11] Fleury, *La Mécanique de Vitruve* (Caen, 1993), 179 ff.

[12] Landels, *Ancient Engineering*, 199 ff.

[13] A. G. Drachmann, 'Ktesibios, Philon and Heron: A Study of Ancient Pneumatics', *Acta historica scientarum naturalium et medicinalium*, 4 (Copenhagen, 1948), 128.

are noteworthy inventions. Hero describes a small steam turbine.[14] The question often asked is, Why didn't he, or his successors, go on to produce a working steam engine? A simple answer is that ancient skills in handling iron were so rudimentary that no one could have produced boilers capable of controlling steam pressures. Bursting boilers continued to be a major hazard late into the nineteenth century.[15] Landels[16] examines the possible overall efficiency of such a device as Hero's which he puts as low as 1 per cent . If that is indeed so, then even if a large-scale model could have been built to deliver 1/10 hp and do the work of one man, its fuel consumption would have been enormous, about 25,000 BTU (26.8 × 10^6 joules) per hour. The labour required to procure and transport the fuel, stoke the fire, and maintain the apparatus would have been much more expensive than that of one man it might replace, and the machine would have been much less versatile. The failure of the Greeks and Romans to harness steam as a power source was, without doubt, one of the many factors which prevented industrialization in their society.

The Romans were remarkably slow about remedying technological defects and developing promising innovations. Virgil mentions a wheeled plough and, a century later, Pliny states that in the Italian Alps wheeled ploughs are drawn by eight oxen.[17] One of the most promising agricultural machines was the first-century AD Gallic harvester (*pecten*),[18] briefly mentioned by Pliny (18. 296) and attested by four surviving monuments, all from the same area of north-eastern Gaul, where open fields predominate. This consisted of a toothed frame, mounted on a pair of wheels, the motive power being supplied by a mule, or donkey, working in shafts and pushing the machine from behind,[19] the movement being controlled by a steersman, also from behind.

---

[14] Landels, *Ancient Engineering*, 28, fig. 7.
[15] J. G. Burke, 'Bursting Boilers and the Federal Power', in *Technology and Culture*, 7 (1966), 1–23.   [16] *Ancient Engineering* 29.
[17] See, generally, K. D. White, *Roman Farming* (London, 1970), 174 ff.
[18] Ibid. 182, figs 36–8 and pp. 448 f. See also K. D. White, 'The Economics of the Gallo-Roman Harvesting Machines', in *Hommage à Marcel Renard*, iii (Collection *Latomus*, 102; Brussels, 1969), 807–9.
[19] K. D. White, *Agricultural Implements of the Roman World* (Cambridge, 1967), 157 ff.

Lynn White[20] includes in his discussion of technological development a very curious example of both Roman inventiveness and conservatism in engineering, provided by the history of the arches used in bridges. Paintings found at Pompeii, Herculaneum, and elsewhere, dating from no earlier than the first century BC, sometimes show flattened arches, often in a context of fantasy—in theatrical scenery, for example. In surviving structures from the time of Augustus, flattened arches are frequent, embedded in masonry walls to relieve the vertical pressure on the horizontal lintel of a window or doorway. Thereafter, flattened arches and vaults, usually rather small, are found as hidden structural elements in buildings, in cellars, warehouses and catacombs, and in 'counter-cultural' sites like the Mithraeum beneath the present church of San Clemente in Rome. The only clear exception that I have noticed is the Domus Aurea built by Nero after the great fire of AD 64. Roman architectural experts must, however, have understood that, if the abutments were firm, segmental arches would be as strong as the semicircular sort. Moreover, they must have seen that such arches were particularly appropriate for bridges: they increased the length of spans in relation to height and reduced the number of piers needed. Thus they made river navigation easier; by reducing the volume of masonry in a bridge they decreased the cost of the construction.

The Roman attitude to water power is also revealing. Although the water-mill was apparently invented in the first century BC, Roman engineers failed to apply water power to any industrial operation other than the grinding of corn.[21] The first certain reference to such an application is in Ausonius'[22] *Mosella*, lines 362–4 (written *c*. AD 369), where he refers to water-driven saws on the banks of the Ruwar, a tributary of the Moselle. Pliny (36. 44) mentions that in Belgic Gaul there is a stone cut like wood with saws—but this cutting is no doubt achieved by human power.

In book 7 of the Natural History (191 ff.) Pliny provides a list of the various discoveries of different persons, covering the

---

[20] Lynn White, 'Technological Development', 243 ff.
[21] Ibid.   [22] Ausonius was born at Burdigala, *c*. AD 310.

Arts and Sciences. Many of the names, however, belong to the shadowy world of received tradition and add little to our appreciation of the Greek and Roman contribution to technology.

Pliny is, of course, more concerned with applied science, which is borne out by his accounts of mining, and the extractive processes involved in metallurgy,[23] described elsewhere in this assessment.[24] In addition, he briefly describes some technical details of operations involved in the transportation of obelisks (36. 70 ff.) and the building of the Pyramids (ibid. 81 ff.). He describes the Pharos at Alexandria (ibid. 83) and labyrinths (ibid. 83 ff.). His account of the temple of Diana at Ephesus (ibid. 95 ff.) is of particular interest in respect of the preparation of the foundations—with a damp course, laid on charcoal, to stabilize the marshy ground. The placement of the architraves was also achieved by an ingenious use of a ramp made from sandbags.

Pliny's account (36. 101 ff.) of the buildings of Rome is mainly descriptive and includes the Cloaca Maxima[25] (ibid. 105), Curio's (technically impossible) revolving theatre[26] (ibid. 116), the aqueducts[27] (ibid. 121 ff.), drainage schemes, and the harbour at Ostia (ibid. 70 and 124 f.).

Finally, as Pliny well understood, applied science brought many beliefs to mankind. Note, for instance, Pliny's observations (18. 273) apropos Democritus who was

the first person to understand and demonstrate the alliance between the sky and earth. When the wealthiest of his fellow citizens despised his attention to these studies, Democritus foresaw...the rising of the Pleiads would be followed by an increase in the price of oil which, at the time, was very cheap because of the anticipated crop of olives. He bought up all the oil in the whole of the region, to the surprise of those who knew that the things he most valued were poverty and peace and quiet for his studies. When his reason for his action became apparent and they had seen great wealth accrue to him, he gave back to the anxious and greedy and now repentant landlords, the

---

[23] J. F. Healy, *Miniere e metallurgia nel mondo greco e romano* (Rome, 1993), a revised and augmented edition of *Mining and Metallurgy in the Greek and Roman World* (London, 1978). [24] See above, Ch. 17 *passim*.
[25] The main sewerage system for Rome, originally constructed under the kings: it was covered by arches. [26] See further Ch. 12.6.3.
[27] Landels, *Ancient Engineering*, 34 ff., and A. T. Hodges, 'Aqueducts', in Barton, *Roman Buildings* 127 ff.

money paid him for the olives, being content to have given this proof that riches would easily be within his reach when he chose.

ferunt Democritum, qui primus intellexit ostenditque caeli cum terris societatem, spernentibus hanc curam eius opulentissimis civium, praevisa olei caritate futura ex vergiliarum ortu... magna tum vilitate propter spem olivae, coemisse in toto tractu omne oleum, mirantibus qui paupertatem quietemque doctrinarum ei sciebant in primis cordi esse, atque ut apparuit causa et ingens divitiarum concursus, restituisse mercedem anxiae et avidae dominorum poenitentiae, contentum ita probavisse opes sibi in facili, cum vellet, fore.

Three examples of ancient technology are treated below: Glass manufacture (18. 2) Paper (papyrus) (18. 3), and Time and time measurement (18. 4).

18.2 Glass Technology

Pliny devotes the concluding chapters of book 36 (191 ff.) to a discussion of silica (sand) and its practical use in the production of glass.[28] He repeats the traditional, but improbable, account of its accidental discovery: 'The beach extends for not more than half a mile, but for many years this area [Sidon] was the sole producer of glass. A ship belonging to traders in soda[29] once called here, so the story goes, and they spread out along the shore to make a meal. There were no stones to support their cooking pots so they placed lumps of soda from their ship under them. When these became hot and fused with the sand on the beach, streams of an unknown translucent liquid flowed

---

[28] I am indebted to John A. Kennedy, Head of Research, at Waterford Crystal Ltd., for very generously sharing his technical expertise and advising me with reference to Pliny's chapters on the production of glass. On glass cf. Bailey, *Chemical Subjects*, i. 280 ff. (note to 36. 191 ff.). The essential constituents of common (soft) glass are soda, lime, and silica and its composition often approximates to the formula $Na_2O_3 \cdot 3CaO \cdot 6SiO_2$, a compound that has a relatively low melting-point and which possesses, in a striking degree, the characteristic property of glass, the ability to pass from the liquid to the solid state without crystallization. Mixtures of very different composition and containing other constituents also furnish types of glass. See R. C. A. Rottländer, 'Glasherstellung bei Plinius dem Alteren', *Glastechnische Berichte*, 52 (1979), 265–70.

[29] *Nitrum*, commonly translated by the generic term 'soda', embraces washing soda, baking soda, and common salt. The merchants were probably trading in the latter.

and this was the origin of glass.' ('quingentorum est passuum non amplius litoris spatium, idque tantum multa saecula gignendo fuit vitro. fama est adpulsa nave mercatorum nitri, cum sparsi per litus epulas pararent nec esset cortinis attollendis lapidum occasio, glaebas nitri e nave subdidisse. quibus accensis, permixta harena litoris, tralucentes novi liquoris fluxisse rivos, et hanc fuisse originem vitri.') He continues (36. 192):

Soon, in keeping with man's inventive genius, he was no longer content to mix in only soda. He began to add magnesian limestone also, because of the belief that it attracts to itself molten glass no less than iron [Pliny here mistakenly equated *magnes lapis* with limestone]. Similarly, in a number of places, lustrous stones of many kinds were added to the melt, and then again shells and quarried sand. Some authorities claim that, in India, glass is made from broken rock crystal[30] and that, for this reason, Indian glass is second to none. Copper and soda, especially Egyptian,[31] are added to this mix that is melted by a fire made from light, dry wood. Glass, like copper, is processed in a series of furnaces and forms dark lumps. The molten glass is so sharp that before one can feel any pain, it cuts through to the bone any part of the body it splashes. The lumps of glass are fused again in the workshop and coloured. Some glass is shaped by blowing, some fashioned by a lathe, and some engraved like silver. Sidon was once renowned for its glass factories: glass mirrors, among other products, were invented there. This was the old method of glass manufacture.

Mox, ut est ingeniosa sollertia, non fuit contenta nitrum miscuisse; coeptus addi et magnes lapis,[32] quoniam in se liquorem vitri quoque ut ferrum trahere creditur. simili modo et calculi splendentes multifariam coepti uri, dein conchae ac fossiles harenae. auctores sunt in India et crystallo fracta fieri et ob id nullum comparari Indico. levibus autem aridisque lignis conquitur addito Cyprio ac nitro, maxime

---

[30] On glass and its discovery, see W. S. Ellis, 'Glass: Capturing the Dance of Light', *National Geographic Magazine*, 184/6 (1993), 37–69: See esp. p. 62, 'From Pliny to Tiffany and Beyond'.

[31] Egyptian soda, or natron, was found extensively in North Africa in both Egypt and Cyrenaica See J. F. Healy, 'Greek and Roman Gold Sources: The Literary and Scientific Evidence, in Domergue, *Mineria y metallurgia*, ii. 13–14.

[32] There is some confusion here, since Pliny elsewhere (cf. 36. 127) employs the term *magnes lapis* to indicate magnetite. Theophrastus (*Lap.* 41) uses μαγνῆτις λίθος of Magnesian limestone. As a source of lime the latter identification would appear to be more appropriate to the present context.

Aegyptio. continuis fornacibus ut aes liquatur, massaeque fiunt colore pingui nigricantes. acies tanta est quacumque ut citra sensum ullum ad ossa consecet quidquid adflaverit corporis. ex massis rursus funditur in officinis tinguiturque, et aliud flatu figuratur, aliud torno teritur, aliud argenti modo caelatur, Sidone quondam his officinis nobili, siquidem etiam specula excogitaverat. Haec fuit antiqua ratio vitri

The discovery of glass is one of the the earliest technological achievements of ancient civilization. The Egyptians were already making glass jewellery by *c*.6000 BC. However, the above-quoted traditional account of the production of the first glass is seriously flawed. It is unlikely, for example, that an open fire could have fused the materials, and the absence of lime, unless present as an impurity, would have led to the production not of glass, but of water glass, that is a solution of sodium silicate. Pliny, or his source, is unaware of the importance of lime, although the addition of what is, possibly, limestone and shells (a source of lime) is mentioned without any awareness of their significance in the process.

Glass is the product of the fusion of silica (sand), and alkali (soda) and an alkaline earth (lime). The use of rock crystal (quartz), pure silica, in India would certainly have resulted in glass of a superior quality, as Pliny states. Its much prized characteristic was its transparency: the legacy of fine glass continues in the trademark 'Waterford Crystal'.

The addition of copper, Egyptian soda, and lustrous stones would have had the following effects: opal, agate, and onyx, all silica minerals, would have been useful as glass formers. Sapphire and ruby, alumina minerals, would have helped to reduce the melting temperature of the mixture. Malachite and azure impart colour to the glass, from green to blue in various gradations.

Pliny's description of the actual manufacture of glass is confused. When he refers to a 'series of furnaces',[33] this can only be a conflation of processes, as seen elsewhere in his accounts of extractive metallurgy. In the refining of gold (33. 84), for example, he conflates cupellation by lead, and salt cementation.[34] It is just possible that he intended to distinguish

---

[33] There would be three processes, some or all of which might be carrried out in different furnaces: (1) the first firing (fritting, at a comparatively low temperature). (2) the second firing (melting), and (3) slow cooling (annealing).

[34] See Healy, 'Greek and Roman Gold Sources', 13–14.

annealing furnaces from those used in making the melt, as in modern glass technology. Today the initial mixture is melted in a low-temperature furnace, fired, and subsequently annealed in a kiln. Pliny's reference to 'dark lumps' is interesting since it is convincing evidence for his having witnessed glass-blowing[35] at first hand. Globules of molten glass, when taken out of the furnace, on the end of a pipe for blowing, at first appear reddish-yellow then soon cool to a dark, dullish red colour.[36] The lumps of molten glass need continual reheating while being worked.

The claim that no pain is felt when workers come into bodily contact with molten glass is simply untrue. On the other hand, slivers of cold glass will cut into the skin without causing immediate pain.

Pliny (36. 194 ff.) continues with an update on glass manufacture in more recent times:

> But now [first century BC] in Italy, a white sand occurs in the river Volturnus: it is found along six miles of the seashore between Cumae and Liternum. The softest of the sand is ground in a mortar, or mill.[37] It is then mixed with three-quarters of a part soda, either by weight or by volume. After fusion, it is taken in its molten state to other furnaces. There it becomes a lump called 'sand soda'; this is remelted and ends up as pure glass—a lump, that is, of clear glass. At the present time, sand is blended in a similar way in the provinces of Gaul and Spain. There is a story that, during the principate of Tiberius, a method of blending glass was invented to make it flexible [*flexile*], but the craftsman's workshop was completely destroyed for fear that this might detract from the value of such metals as gold, silver, and copper.[38] The story was current for a long time, thanks to frequent repetition.... that, when Nero was emperor a technique of

---

[35] Glass-blowing, in contrast to glass-moulding, was a comparatively late development, introduced about the middle of the 1st cent. BC, or even later.

[36] This change in colour is clearly observable in modern glass-working, as in the factories at Murano (Venice) and elsewhere. Pliny may well have seen glass being produced in Spain (cf. 36. 194). Cooling certainly affects the colour of the blob of molten glass. A less likely possibility that has been suggested is that a darker colour is produced when sulphur reduces any traces of metals contained in the glass melt. On glass of. Bailey *Chemical Subjects*, i. 280 ff (note to 36. 191 ff.) [37] See Healy, *Miniere*, 182.

[38] See F. D. Lazenby, 'A Note on *vitrum flexile*', *Classical Weekly*, 44 (1951), 102–3. Similar problems—namely the conflict of commercial interests—were faced with the attempted introduction of a multi-strike match and, of course, more recently, when CDs began to replace audio tapes.

glass-making was invented that produced two small cups called 'stoneware' [*petroti*]: these sold for 6,000 sesterces. ... The most highly prized glass is transparent and as close as possible to the appearance of rock-crystal [36. 199]. But, although glass has replaced gold and silver as the material from which drinking vessels are made, it cannot bear heat unless cold water is first poured in. Glass globules containing water become so hot, when they face the sun, that they burn clothes. When heated to a moderate temperature, pieces of broken glass can be stuck together. They cannot, however, be completely fused but melt into separate globules, as happens in the manufacture of glass pebbles called 'eyeballs': in some examples, these glass pebbles are multi-coloured and patterned. When glass is heated with sulphur, it coalesces into stone.

Iam vero et in Volturno amne Italiae harena alba nascens sex milium passuum litore inter Cumas atque Liternum, qua mollissima est, pila molave teritur. dein miscetur III partibus nitri pondere vel mensura ac liquata in alias fornaces transfunditur. ibi fir massa quae vocatur hammonitrum atque haec recoquitur et fit vitrum purum ac massa vitri candidi. iam vero et per Gallias Hispaniasque simili modo harena temperatur. ferunt Tiberio principe excogitato vitri temperamento, ut flexile esset, totam officinam artificis eius abolitam ne aeris, argenti, auri metallis pretia detraherentur, eaque fama crebrior diu.... fuit... Neronis principatu reperta vitri arte quae modicos calices duos quos appellabant petrotos HS VI venderet?... maximus tamen honos in candido tralucentibus, quam proxima crystalli similitudine. usus vero ad potandum argenti metalla et auri pepulit. est autem caloris impatiens, ni praecedat frigidus liquor, cum addita aqua vitreae pilae sole adverso in tantum candescant ut vestes exurant. fragmenta teporata adglutinantur tantum, rusus tota fundi non queunt praeterquam abruptas sibimet in guttas, veluti cum calculi fiunt quod quidam ab oculis appellant, aliquos et pluribus modis versicolores. vitrum sulpuri concoctum feruminantur in lapidem.

During the early imperial period, naturally occurring sand was finely ground and used in the production of glass. Pliny records the proportions of sand and soda required for the basic mixture but there is still no specific mention of lime as an essential ingredient. Experiments following Pliny's formula initially failed to produce glass, until it was realized that 'tres partes' meant 'three-quarters', not 'three parts'.[39]

---

[39] See further Rottländer, 'Glasherstellung'. Cf. also the experiments carried out by L. D. Pye, Director of the Centre for Glass Research at

Pliny also makes an interesting reference to the alleged invention of 'flexible' glass, only able to be produced in very recent times. Although glass is classified scientifically as a liquid, its atomic structure dictates that it is rigid and, by its very nature, fundamentally non-flexible. Pliny explains its disappearance in strangely familiar terms of its having been suppressed by rival manufacturing interests.

The material he describes as 'stoneware' poses a problem of identification: it has been suggested that this refers to the mixing of glass and ceramics, a technique that has been practised for many years. Whether it was known to the Romans is debatable.

Pliny, in this account, alludes to some properties of glass, correctly implying that a glass vessel will shatter if hot water is put into it without some cold having previously been put in. The same purpose is served by pouring hot water over a metal spoon to dissipate the initial heat. He completely misunderstands the effect of the sun's heat on water in a globe. There is little significant rise in temperature, but the water concentrates the sun's rays to form a sort of primitive magnifying glass, which could, in theory, cause combustion—but this is unlikely.

Pieces of shattered glass can be stuck together at temperatures above the annealing point of glass, that is about 500°C: their edges become rounded. Pliny is incorrect in stating that they can never fuse completely. Fusion is possible provided that the temperature is high enough to melt both pieces.

The references to 'eyeballs' is unclear: in his description they seem not unlike the multicoloured glass marbles with which children played in a less sophisticated age! They may, however, possibly refer to inserts made to fill the hollow eye sockets of bronze statues, but this seems equally unlikely in the light of the adjective 'versicolores', which would be inappropriate in describing eyes.

Pliny's statement that glass, when heated with sulphur, may coalesce into stone is likely to have been an extension of an

---

Alfred University, New York. Following the Phoenician recipe, he used natron and sand, fired at 1600°C for two hours and produced glass (personal communication).

earlier observation (31. 122) about salt and soda turning into stone when similarly heated with sulphur: 'sal nitrum[40] sulpuri concoctum in-lapidem vertitur'. There does not seem to be any justification for this equation.

18.3 PAPER

Arguably one of the most important discoveries in the ancient world was a method of making paper. Pliny (13. 74) outlines the basic process: 'Paper is manufactured from papyrus by splitting it with a needle into fine strips that are very thin but as long as possible. The quality of the papyrus is best at the centre of the plant and decreases progressively towards the outsides.' ('praeparatur ex eo charta diviso acu in praetenues sed quam latissimas philyras; principatus medio, atque inde scissurae ordine.')

'All paper', writes Pliny (13. 77), 'is 'woven' on a board, dampened with water from the Nile; the muddy liquid acts as glue. First an upright layer is smeared on the table—the whole length of the papyrus is used—and both its ends are trimmed; then strips are laid across and complete a criss-cross pattern, which is then squeezed in presses. The sheets are dried in the sun and then joined together; each succeeding sheet decreases in quality down to the worst. There are never more than twenty sheets on a roll.' ('texitur omnis madente tabula Nili aqua; turbidus liquor vim glutinis praebet. in rectum primo supina tabulae schida adlinitur. longitudine papyri quae potuit esse resegminibus utrimque amputatis, traversa postea crates peragit. premitur ergo prelis, et siccantur sole plagulae atque inter se iunguntur, proximarum semper bonitatis deminutione ad deterrimas. numquam plures scapo quam vicenae.')

'The surface roughness [13. 81] is smoothed out with a piece of ivory or a shell, but writing easily fades because the

---

[40] See Eichholz, *NH 36–37*, p. 157, note *d*, where Eichholz suggests that Pliny may have mistaken '*nitrum*' for '*vitrum*' in his source, but this does not resolve the inherent problem.

polished paper, with its shinier surface, does not absorb ink as well' ('scabritia levigatur dente conchave sed caducae litterae fiunt; minus sorbet politura charta, magis splendet').

Pliny (13. 82) describes a further method of making paper, from papyrus strips, by the use of a paste made from flour. 'A basic paste is made from fine-quality flour mixed with boiling water and a sprinkling of vinegar; carpenter's glue and gum are too brittle an adhesive. A still better method of making paste is to mix breadcrumbs in boiling water. This results in the least amount of paste at the joins and produces paper softer even than linen.... when pasted the paper is beaten thin with a mallet and its surface covered with paste, once more it has its creases removed by pressure and is flattened with a mallet. This process may enable written records to last a long time' ('glutinum vulgare e pollinis flore temperatur fervente aqua, minimo aceti aspersu, nam fabrile cummisque fragilia sunt. diligentior cura mollia panis fermentati colat fervente; minimum hoc modo intergerivi, atque etiam lini lenitas superatur.... postea malleo tenuatur et glutino percurritur, iterumque constricta erugatur atque extenditur malleo. ita sint longinqua monimenta.')

Modern experiments have followed Pliny's basic description (13. 77) except that no glue is needed to bond the strips which provide their own glutinous adhesive.[41] The resultant product is a highly successful 'paper'. The Nile water, mentioned by Pliny, merely prevents the strips from drying out. No simulation of the other technique, based on the use of an artificial adhesive, appears to have been attempted, as far as I am aware.

Finally, Pliny (13. 74f.) categorizes the various grades of paper, from the first quality—known as 'Augustan' (formerly, 'hieratic')—to 'packing paper', the equivalent of our 'brown paper' (*emporitica*).

---

[41] Demonstrations of the manufacture of papyrus can be seen at the Papyrus Institute in Syracuse (Sicily).

18.4 TIME AND THE MEASUREMENT OF TIME

Longitude affects the time of sunrise and sunset,[42] as Pliny (2. 181) explains:

> Although night and day are the same thing throughout the world, it is not night and day at the same time all over the world.... This fact is known as a result of many experiments—for example, with Hannibal's towers in Africa and Spain, and in Asia, where, because of the fear of pirates, similar watch-towers were built for protection; warning fires lit on these at midday were often found to have been seen by people furthest to the rear at 9 p.m. Philonides, one of Alexander's runners, completed the 136 miles from Sicyon to Elis[43] in nine hours from sunrise, but although the return journey was downhill, it took until 9 p.m. This happened often. The reason was that, on the way there, his route was with the sun, but, on the return, he passed the sun as it met him coming in the opposite direction. This is the reason why ships sailing westwards, even on the shortest day, beat the distance they sail at night because they travel with the sun.
>
> ideo nec nox diesque, quamvis eadem, toto orbe simul est. ... multis hoc cognitum experimentis in Africa Hispaniaque turrium Hannibalis, in Asia vero propter piraticos terrores simili specularum praesidio excitato, in quis praenuntios ignes sexta hora diei accensos saepe compertum est tertia noctis a tergo ultimis visos. eiusdem Alexandri cursor Philonides ex Sicyone Elin mille et ducenta stadia novem diei confecit horis, indeque, quamvis declivi itinere, tertia noctis hora remensus est saepius. causa, quod eunti cum sole iter erat, eundem remeans obvium contrario praetervertebat occursu. qua de causa ad occasum navigantes quamvis brevissimo die vincunt spatia nocturnae navigationis ut solem ipsum comitantes.

Pliny is in error, of course, since the distances involved are by no means great enough to involve any significant time change of the order he implies. Moreover his knowledge of the topography of the Peloponnese is faulty. Elis is higher above sea level than Sicyon it is true, but his route is only downhill in the latter stages.

---

[42] See in general, A. E. Samuel, 'Calendars and Time-telling', in Grant–Kitzinger, *Mediterranean Civilization*, 1, i. 389–95. Longitude is the distance in degrees, east or west of the prime meridian at 0° (Greenwich), measured by the angle between the prime meridian and that of the meridian through the point in question, or by the time difference; 15° represents one hour.

[43] The actual distance from Sicyon to Elis, in a direct line, is approximately 80 miles (130km.) Philonides' outward journey was accomplished at just over 15 miles (24km.) per hour.

The hours of daylight vary with latitude and Pliny (2. 186) continues:

So it happens because of the varied lengthening of daylight, the longest day lasts 12 and eight-ninths equinoctial hours at Meroe, 14 hours at Alexandria, 15 in Italy, and 17 in Britain, where the light summer nights confirm what theory urges us to believe, namely that, because the sun approaches nearer to the top of the world on summer days, owing to a narrow circuit of light, the parts of the earth that lie at the poles have continuous daylight for six months at a time and continuous night for six months when the sun has withdrawn in the opposite direction towards midwinter. Pytheas[44] of Massalia writes that this happens in the island of Thule, six days north of Britain and some say it also happens in the island of Mona which is about two hundred miles from the British town of Camulodunum.

sic fit ut vario lucis incremento in Meroe longissimus dies XII horas aequinoctiales et octo partes unius horae colligat. Alexandriae vero XIV horas, in Italia XV, in Britannia XVII, ubi aestate lucidae noctes haut dubie repromittunt id quod cogit ratio credi, solstiti diebus accedente sole propius verticem mundi angusto lucis ambitu subiecta terrae continuos dies habere senis mensibus, noctesque e diverso ad brumam remoto. quod fieri in insula Thyle Pytheas Massilienis scribit sex dierum navigatione in septentrionem a Britannia distante, quidam vero in Mona, quae distat a Camuloduno Britanniae oppido circiter ducentis milibus adfirmant.

18.4.1 Time measuring Instruments

Greeks and Romans, as earlier peoples, divided the periods of light and darkness into equal twelve-hour segments. Thus the length of an 'hour' was not fixed.[45]

The period that constituted a 'day' was differently interpreted by different peoples, as Pliny (2. 188) sets out:

The Babylonians reckon this as the interval between two sunrises, the Athenians that between two sunsets, the Umbrians that from midday

---

[44] The Greek explorer who circumnavigated Britain and sailed beyond to Thule. He calculated the latitude of Massalia and established bases for cartographic parallels through northern Gaul and Britain.
[45] On measuring-instruments see A. Le Bœuffle, *Le Ciel des Romains* (Paris, 1989), 29ff., and J. Soubiran (ed. and comm.), *Vitruvius*, ix (Budé: Paris), on 9. 2. 109. See also Samuel 'Calendars'; 395.

to midday, ordinary people everywhere from dawn to dark. Roman priests and authorities who fixed the 'civil day', and, likewise the Egyptians and Hipparchus, reckon the day from midnight to midnight. But it is clear that the interruption of daylight between sunset and sunrise is smaller near the solstice than at the equinoxes, because the position of the zodiac is more slanting around its middle points but straighter near the solstice.

Babyloni inter duo solis exortus, Athenienses inter duos occasus, Umbri a meridie ad meridiem, vulgus omne a luce ad tenebras, sacerdotes Romani et qui diem diffiniere civilem, item Aegypti et Hipparchus, a media nocte in mediam. minora autem intervalla esse lucis inter occasus et ortus solis iuxta solstitium quam aequinoctia apparet quia posito signiferi circa media sui obliquior est, iuxta solstitium autem rectior.

Sundials, hour-glasses, water-clocks, star clocks, and shadow-tables were used by the Egyptians and Babylonians long before the rise of Greek civilization, and Herodotus[46] claims that the Greeks learned the use of the sundial from the Babylonians.[47]

By Hellenistic times, the interest in astronomy had led to the development of precisely geared devices which operated to represent, in model form, the motions of celestial bodies. In the light of such inventions there is little doubt that the Greeks had the technological expertise to have produced mechanical clocks had they so desired.

At Rome (7. 212), however, 'in the Twelve Tables,[48] only sunrise and sunset were specified: a few years later, noon was also added, the consuls' aide announcing it, when from the Senate house he saw the sun between the Rostra and the Ambassador's house. When the sun sloped from the Maenian column to the Prison, he announced the last hour, but this only on clear days, down to the First Punic War.' ('XII tabulis ortus tantum et occasus nominatur, post aliquot annos adiectus est et meridies accenso consulum id pronuntiante cum

---

[46] Hdt 2. 109: πόλον μὲν γὰρ καὶ γνώμονα καὶ τὰ δυώδεκα μέρεα τῆς ἡμέρης παρὰ Βαβυλωνίων ἔμαθον οἱ Ἕλληνες ('The Greeks learned about the concave sundial, the gnomon, and the twelve divisions of the day from the Babylonians').

[47] See also below, n. 57.

[48] The surviving fragments of the twelve Tables reflect the condition of society in Italy in the 5th cent. BC. They embrace both public and private life.

a curia inter Rostra et Graecostasim prospexisset solem a columna Maenia ad carcerem inclinato sidere supremam pronuntiavit, sed hoc serens tantu, diebus usque ad primum bellum.'

'The theory of shadows and the art of making sundials was', writes Pliny (2. 187), 'discovered by Anaximenes,[49] of Miletus, the pupil of Anaximander:[50] he exhibited, at Sparta, the timepiece they call "Hunt the Shadow"'[51] ('umbrarum hanc rationem et quam vocant gnomonicen invenit Anaximenes Milesius, Anaximandri discipulus, primusque horologium quod appellant sciothericon Lacedaemone ostendit.')

Pliny (7. 213) writes about the introduction of sundials to Rome:

Fabius Vestalis records that the first sundial was set up eleven years before the war with Pyrrhus, at the temple of Quirinus, by Lucius Papirius Cursor when dedicating this temple which had been vowed by his father. Fabius Vestalis does not indicate the principle of the sundial, or the name of its maker, or where it originated,[52] or the name of the authority for this assertion. Marcus Varro[53] states that the first sundial was set up on a column along the Rostra, during the First Punic War after the capture of Catina, in Sicily, by the consul Manius Valerius Messala and that it was brought from Sicily thirty years later than the date assigned to Papirius' sundial, namely 263 BC.[54] The lines on the face did not match up with the hours, yet the

---

[49] Anaximenes (*fl. c.*546 BC) believed that air was the essence of all things and that the soul was composed of air. See further G. E. R. Lloyd, *Early Greek Science: Thales to Aristotle* (London, 1970), 1 ff.

[50] Anaximander (born *c.*610 BC) was one of the earliest philosophers of the Ionian school. He was the immediate successor of Thales its founder. He maintained that the infinite (τὸ ἄπειρον) was the primary source of all things.

[51] Cf. Strabo 2. 5. 24: σκιοθηρικοὶ γνώμονες. So Vitr. *De arch.* 1. 6. 6: σκιοθήρης; 2. Diog. Laert. 1; Plut. *Marc.* 19, and elsewhere. See further Le Bœuffle, *Le Ciel*, 30 f. The shadow-clock, or obelisk, was a vertical pillar sometimes mounted on a rectangular, or circular, stone base. The angle which the shadow of a vertical pole makes with the meridian is the sun's azimuth, and the azimuth angle through which the shadow rotates when the sun has turned through a given angle is not the same at all seasons. It depends on the sun's declination. So the length of a working-day, as recorded by the shadow-clock, was not the same fraction of a day at different times of the year.

[52] See n. 46 above, and n. 57 below.

[53] *RR.* 3. 5. 17. See also Cic. *Fam.* 16. 18. 2, and Vitr. *De arch.* 9. 8. 4–15.

[54] Apropos the chronology of the introduction of the first sundial to Rome, Pliny here gives two supposed dates, the second of which is attested by Varro (*RR* 3. 5. 17) and, therefore, to be preferred.

Romans followed its time for ninety-nine years until Quintus Marcius Philippus, who was censor with Lucius Paulus, set up a more carefully calibrated one next to it: this gift was one of the most welcome results of Philippus' works as censor. Even then, however, the hours were uncertain in cloudy weather.

princeps Romanis solarium horologium statuisse ante undecim annos quam cum Pyrrho bellatum est ad aedem Quirini L. Papirius Cursor, cum eam dedicaret a patre suo votam, a Fabio Vestale proditur; sed neque facti horologi rationem vel artificem significat nec unde translatum sit aut apud quem scriptum id invenerit. M. Varro primum statutum in publico secundum Rostra in columna tradit bello Punico primo a M' Valerio Messala consule, Catina capta in Sicilia, deportatum inde post XXX annos quam de Papirio horologio traditur, anno urbis CCCCLXXXXI. nec congruebat ad horas eius liniae, paruerunt tamen ei annis undecentum, donec Q. Marcius Philippus qui cum L. Paullo fuit censor diligentius ordinatum iuxta posuit; idque munus inter censoria opera gratissime acceptum est. etiam tum tamen nubilo incertae fuere horae. ...

The original sundial had, of course, been calibrated for a lower latitude, while Philippus' replacement was designed to function for the latitude of Rome (41° 53′ 33″ N).

From the second century BC there was a proliferation of sundials in public places and the homes of the wealthy. The poet Aquilius (apud Aul. Gell. *NA* 3. 3. 5) complained about this situation, pretending that the best timepiece was his stomach.

Pliny (36. 72) describes a gigantic sundial of Augustus:[55]

Augustus used the obelisk in the Campus Martius in a remarkable way—namely to cast a shadow and thus mark the length of days and nights. A paved area was laid out commensurate with the height of the monolith in such a way that the shadow at noon, on the shortest day, might extend to the edge of the paving. As the shadow gradually grew shorter and longer again, it was measured by bronze rods fixed in the paving. This device deserves study; it was the result of a brainwave of the mathematician Facundus Novius. Novius placed a gilded ball on the apex of the monolith so that the shadow would be concentrated at its tip; otherwise the shadow cast would have been very indistinct. He got this idea, so it is said, from seeing the shadow cast by a man's head.

---

[55] Cf. Le Bœuffle, *Le Ciel*, 29.

ei qui est in campo divus Augustus addidit mirabilem usum ad deprehendendas solis umbras dierumque ac noctium ita magnitudines, strato lapide ad longitudinem obelisci, cui par fieret umbra brumae confectae die sexta hora paulatimque per regulas quae sunt ex aere inclusae, singulis diebus descrescet ac rursus augesceret, digna cognitu res, ingenio Facundi Novi mathematici. is apici auratam pilam addidit, cuius vertice umbra colligeretur in se ipsam alias enormiter iaculante apice ratione, ut ferunt, a capite hominis intellecta.

This monument was constructed to commemorate Augustus' victory in Egypt in 30 BC. The base measured 492 feet by 246 feet (150 m × 75 m). The gnomon was an obelisk brought from Heliopolis to Rome in 10 BC; it was 72 feet (22 m.) high and weighed 226 tons (230 tonnes). The obelisk is represented on a bas-relief sculpture at the base of the Antonine column. Damaged by an earth tremor and fire in AD 80, the sundial was restored on the order of Domitian. Excavations carried out between 1976 and 1980 show that this was not merely a sundial but a complete calendar indicating, for example, the beginning of summer on 7 May. On the evening of 23 September, the anniversary of Augustus, the shadow of the gnomon pointed towards the Ara Pacis[56]—then part of the same complex, but now removed to a location by the Mausoleum of Augustus. The inscriptions and signs of the zodiac were in Greek.

Vitruvius (*De arch.* 9. 8. 1 ff.) describes various types of sundials[57] in which the shadow from the gnomon is cast not only on vertical, but also on horizontal and conical surfaces and concave hemispheres—variously orientated.

---

[56] Ibid. See also R. B. Bandinelli, *Rome the Centre of Power: Roman Art to AD 200* (London, 1970), 186 ff. The Ara Pacis was inaugurated in 9 BC to celebrate Augustus' victorious return from Spain and Gaul. Cf. the account of his campaigns set out in *Res gestae* 2. 37 f.

[57] The sundial was further developed by the Arabs, who made considerable progress in the study of spherical geometry, and is quite different from the shadow-clocks of antiquity. The pole of the shadow-clock was set along the earth's polar axis, with the style, or pointer, set so that you would see the pole star if your eye followed the line along its upper edge. The scale can be graduated in divisions corresponding with equivalent intervals at all seasons of the year, that is to say that the style of the sundial is set along the meridian and its edge is elevated at an angle equivalent to the latitude of the place at which it is sited. So it follows that a sundial which keeps time in one place will not be accurate in another.

Pliny (2. 182 f.) refers also to

travellers' sundials [which] are not the same for reference everywhere, because the shadows thrown by the sun as they alter, change the readings at every 34 miles, or, at farthest, every 57 miles. So, in Egypt, at midday on the day of the equinox, the shadow of the gnomon measures a little more than half its length, whereas, in the city of Rome, the shadow is a ninth shorter than the gnomon, at Ancona, a thirty-fifth longer, and, in the region of Italy called Venetia, the shadow is equal to the length of the gnomon at the same hours. Similarly, it is reported that at the town of Syene, 575 miles south of Alexandria, at noon in midsummer no shadow is cast and that in a well made for the the sake of this experiment, the light reaches the bottom, clearly showing that the sun is vertically above that place at that time. This is stated, in the writings of Onesicritus, also to occur at the same time in India, south of the river Hypasis. Moreover, it is said that in the Cave-dwellers' city of Berenice, and 554 miles away, at the town of Ptolemais in the same tribe, which was founded on the shore of the Red Sea for the earliest elephant hunts, the same thing occurs forty-five days before and the same number of days after midsummer and that, during that period of ninety days the shadows are thrown southwards.

vasa horoscopica non ubique eadem sunt usui, in trecenis stadiis aut ut longissime in quingentis, mutantibus semet umbris solis. itaque umbilici (quem gnomonem appellant) umbra in Aegypto meridiano tempore aequinocti die paulo plus quam dimidiam gnomonis mensuram efficit, in urbe Roma nona pars gnomonis deest umbrae, in oppido Ancona superest quinta tricesima, in parte Italiae quae Venetia appellatur iisdem horis umbra gnomoni par fit. simili modo tradunt in Syene oppido, quod est supra Alexandriam quinque milibus stadiorum, solstiti die medio nullam umbram iaci, puteumque eius experimenti gratia factum totum inluminari, ex quo adparere tum solem illi loco supra verticem esse; quod et in India supra flumen Hypasim fieri tempore eodem Onesicritus scribit. constatque in Berenice urbe Trogodytarum, et inde stadiis quattuor milibus DCCCXX in eadem gente Ptolemaide oppido, quod in margine rubri maris ad primos elephantorum venatus conditum est, hos idem ante solstitium, quadragenis quinis diebus totidemque postea fieri, et per eos XC dies in meridiem umbras iaci.

The precision of such portable 'sundials' was relative, as Seneca (Apocol. 2. 2.) ironically implies. 'As regards the time I cannot tell you exactly—the philosophers will sooner be in agreement than the timepieces. However it's between the sixth and the seventh hour!'

The invention of the sundial marked an important advance but, by its very nature it could not provide measurements on dull days. A more reliable means of measuring time, independent of the weather, was needed.

'Scipio Nasica, the colleague of Laenas, instituted the first water-clock,[58] which divided the hours of night and day equally; he dedicated this time-keeping device in a roofed building in 159 BC. For so long a time the hours of daylight had remained undivided for the Roman people.' ('Scipio Nasica, collega Laenatis, primus aqua divisit horas aeque noctium ac dierum, idque horologium sub tecto dicavit anno urbis DXCV: tamdiu populo Romano indiscreta lux fuit').

In the Roman Agora at Athens stands the so-called Tower of the Winds, more correctly referred to as the Horologium of Andronicus Cyrrhestes, built in the second or first century BC by the astronomer Andronicus of Cyrrhus. It served a triple purpose as sundial, water-clock,[59] and weather-vane. The tower, which survives, is in the form of a marble octagon, with a pyramidal roof held together by a round keystone. Each face marks a cardinal point and is adorned with a relief representation of

---

[58] *Sen. Ep.* 24. 19, Landels, *Ancient Engineering*, 188 ff. The clepsydra, or water-clock, was basically a vessel of fixed capacity with a hole at its base, of fixed diameter, through which water drained to mark the passage of time. It was extensively used in lawcourts to indicate the allocation of time for speeches. Ar. *Wasps* 93 and 857; [Arist.] *Ath. Pol.* 67. 2; *Arist. Pol.* 1451^{a}8, and elsewhere.

[59] G. E. R. Lloyd, *Greek Science after Aristotle* (London, 1973), 101, writes: 'Vitruvius (9. 8. 2 ff.) informs us that Ctesibius was one of the first to investigate the principles of construction of water-clocks; he attributes to him a constant-head water-clock and various devices by which the length of the hour—conceived by the ancients not as an absolute unit of time, but as a division of a period of sunlight—could be adjusted according to the season of the year.' Landels, *Ancient Engineering*, 188 f., adds: 'There may have been some attempt to standardize this unit of time, since the surviving remains of one clepsydra (in the Agora Museum, Athens) bear two marks which seem to indicate a check on its capacity by an inspector of weights and measures. A reconstructed model of this clock runs for about 6 minutes per filling. But this use of a time-measuring device was in the interests of fairness, not in the pursuit of science.' Landels adds the anecdote concerning a well-known prostitute in 4th-cent. Athens who was nicknamed 'The clepsydra', because, in order to ensure a fair distribution of her favours, she installed a water-clock in her boudoir, and timed her clients' visits by it! See also R. A. Cordingley (rev.), *Banistes Fletcher: A History of Architecture on the Comparative Method*[17] (London, 1961), 140.

the wind blowing from that direction. The tower, according to Vitruvius (*De arch.* 1. 6. 4), was originally surmounted by a revolving bronze Triton holding a wand, which pointed out the face corresponding to the prevailing wind. Attached to the south face is a reservoir in the form of a semicircular turret, from which water could be released in a steady stream to run the 'clock'. The Horologium was, in fact, a large-scale public clepsydra.[60]

18.4.2 Calendars

Le Bœuffle writes of the Romans: 'Ces conquérants d'empire terrestre ont aussi voulu saisir le ciel dans sa réalité totale: ils y ont vu tour à tour un calendrier, une horloge, une parade de mimes, un paradis de héros, surtout un modèle d'ordre de beauté et de sagesse.'[61]

The motions of the earth, moon, and sun create natural units of time that are apparent to observers at all places on the earth. The rotation of the earth creates the alternation between night and day, and a complete period, measured by twenty-four hours under our system. The circuit of the moon around the earth completed when the moon returns to the same line made between earth, moon, and sun, takes about 29.5 days. The motion of the earth around the sun, measured as the interval required for the earth to return to the same point in its orbit in relation to the sun is a year—or, more technically, a tropical year, namely about 365.25 days.

Greeks and Romans, like most ancient peoples, used all three natural units of time, but because they used the periods as they actually occur in the course of the celestial motions, their time measurement was complicated by the variability of the

---

[60] Le Bœuffle, *Le Ciel*, 30 n. 4 (and Soubiran, *Vitruvius*, ix, pp. 266 f.) adds that clepsydras were also adapted as alarm clocks: 'on avait aussi inventé des horloges à l'eau qui servaient de réveil-matin en projetant de petits cailloux sur un gong ou en produisant un sifflement.'

[61] *Le Ciel*, 137. In general, see also Samuel, 'Calendars', and Le Bœuffle, *Le Ciel*, 11 ff.

movements or incommensurability of the durations of one or another.

The perception of the earth changed from being a flat disc surrounded by water to being a solid, spherical body isolated in space, even if some philosophers continued to think of this as stationary in the centre of the universe. The sky was thought of as a vault to which the stars were fixed and, between these, were the five planets, Mercury, Venus, Mars, Jupiter, and Saturn: to these were added the sun and moon which Anaximander had distinguished from the stars.

In the fifth century BC. Meton[62] and Euctemon established star calendars and estimated the length of the seasons. New research was added by Eudoxus of Cnidus and Callippus of Cyzicus in the fourth century BC. Later, Archimedes constructed planetaria and calculated the distance of the earth from the moon and sun. Eratosthenes measured the circumference of the world with remarkable precision. Above all, Hipparchus,[63] arguably the greatest astronomer of antiquity, studied the movements of the planets, calculated, with a fair degree of accuracy, the distance of the moon from the earth, made a catalogue of 1,000 stars, and discovered a nova[64] and the precession of the equinoxes. In the practical field he invented measuring instruments such as the astrolabe.

Astronomy, as Le Bœuffle[65] writes, was never reduced to mathematical formulae, or abstract speculation by the Romans, as the *Natural History* confirms, nor was it a privilege reserved for some experts, but was in reach of everyone—countrymen, sailors, and scholars alike a readership intended by Pliny

---

[62] An Athenian astronomer who introduced the cycle of nineteen years by which he adjusted the course of the sun and moon. See Le Bœuffle, *Le Gel*, 6.

[63] Born in Nicaea, Bithynia (fl. 160–145 BC), Hipparchus lived in Rhodes and Alexandria. Cf. *NH* 2. 53.

[64] O. Pedersen, 'Some Astronomical Topics in Pliny', in French–Greenaway, *Roman Science*, 191 ff. Pliny (2. 95) writes: 'Hipparchus...discovered a 'nova' that came into existence in his lifetime. He was led to wonder with regard to this star, because of the change in brightness with which it flared up, whether this happened quite frequently and whether the stars we think are fixed actually move.' ('Hipparchus...novam stellam in aevo suo genitam deprehendit, eiusque motu qua fulsit ad dubitationem est adductus anne hoc saepius fieret moverenturque et eae quas putamus adfixas.')

[65] *Le Ciel*, 11.

himself. Similarly, the star calendars of antiquity served a practical purpose both in agriculture[66] and in navigation.[67] The rising and setting of stars were also considered useful in weather-forecasting. The majority of Stoics believed that the stars exerted a direct influence on the weather, although such theories were abandoned by Panaetius. Cicero (Mur. 36) writes, somewhat equivocally: 'For as storms are often stirred up by some particular constellation, they are often occasioned, unexpectedly, for no certain reason but because of some obscure cause' ('nam, ut tempestates saepe certo aliquo caeli signo commoventur, saepe improviso nulla ex certa ratione, obscura aliqua ex causa concitantur').

For recording the passage of historical time, however, ancient peoples employed a variety of different systems, complicated by the fact that the calendar year began at different times in the seasonal year in different places. Some intercalation was thus necessary.

These problems, however, are outside the scope of the present survey and it suffices to observe that Pliny employs the relatively straightforward expedient of dating events by reference to individual magistrates' terms of office and an overall chronology based on the elapse of years since the foundation of Rome (AUC), namely 753 BC.[68]

---

[66] Ibid. 16 ff. cf. Verg. G. 1. 1–12 and 252 ff. Similarly, Germanicus' Latin translation of the *Phaenomena* of Aratus (12–14) stresses the importance of stars both for farmers and sailors. So Pliny (18. 206) sums up: 'And it must be confessed that these matters chiefly depend on the weather—as, in fact, Virgil enjoins first, before all else, to learn the winds and the habits of the stars and to observe them just in the same way as they are observed for navigation' ('et confitendum est caelo maxime constare ea quippe Vergilio iubente praedisci ventos ante omnia ac siderum mores, neque aliter quam navigantibus servari.')

[67] Le Bœuffle, *Le Ciel*, 25 ff.: 'De nombreux détails évoqués précédemment s'appliquaient déjà aux marins, à ceux d'Ostie, d'Antium, ou de Misène, aussi bien qu'aux paysans du Latium: parapegmes, pronostics, météorologiques fournis par les astres, procédés destinés à évaluer l'heure pendant la nuit. Mais une autre utilisation de signes célestes leur était propre entre toutes, c'était l'orientation d'après les constellations polaires, qui, en quelque sorte, leur servaient par anticipation de boussole [mariner's compass] dans les traversées nocturnes.'

[68] See generally, E. J. Bickerman, *Chronology of the Ancient World*² (London, 1980), and Samuel, 'Calendars'.

19

Pliny and the Environment

It has often been claimed—but more often implied—that Pliny's moral diatribe against luxury is the result of an uncontrolled, spontaneous outburst, reflecting little more than the declamations against luxury he had learnt in the Schools.

But any in-depth examination of the *Natural History* shows that this is far from the truth. The idea that man, impelled by his desire for an extravagant lifestyle (*luxuria*) and blinded by greed (*avaritia*), behaves in an unnatural way, is central to Pliny's purpose

Man ruthlessly exploits our planet by misusing the gifts of Nature in all aspects of his life. 'In the *Natural History*', writes Wallace-Hadrill,[1] 'the function of science is to reveal the proper use of nature and so save mankind.'

In general, however, the situation in the first century AD was less complex, and the scale of the pollution was of a lower order, than we experience in the twentieth century. Pliny was not faced with problems such as the destruction of the Amazonian rain forests, with the attendant 'greenhouse effect', or with 'acid rain' resulting from the pollution of the atmosphere by an ever-increasing volume of chemical compounds discharged into the air. The land and sea were not continually poisoned by the discharge of toxic waste or effluents. So that, although, in a sense, Pliny may be considered an early environmentalist, it is important to bear in mind that the motives for his *indignatio*,[2] his impassioned outcry against luxury, differ

---

[1] A. Wallace-Hadrill, 'Pliny the Elder and Man's Unnatural History', *Greece and Rome*, NS 37 (1990), 80–96. See also D. S. Wallace-Hadrill, *The Greek Patristic View of Nature* (Manchester, 1968).

[2] So Juvenal, *Sat.* 1. 79: 'si natura negat facit indignatio versum.'

significantly from those of present-day 'Greens': his aims are not exclusively related to ecology but stem from an earnest desire not only to reverse the decay of Roman morality (in 9. 104 and 29. 27, Pliny refers to this as 'populatio morum' and 'lues morum') but also to effect a return to the strict code of behaviour under the early Roman Republic. In this he is a part of a tradition that began with Cato the Censor and Lucilius and which is seen even more strikingly later, in the Satires of Juvenal.

19.1 MINING AND QUARRYING

The worst example of the exploitation of our planet's resources, in Pliny's assessment, is mining, for which the sole motive is not the improvement of man's lot, but greed for metals, minerals, and precious stones.

Pliny, not inappropriately, begins book 33 as follows:

I shall now discuss metals and the natural resources we use to pay for goods. We search for these deep within the earth diligently and in a number of ways. In some places, we dig for riches, when our lifestyle requires gold, silver, electrum,[3] and copper, and, in others, for luxury, when gems and pigments for wall-painting and woodwork are required, and in yet other places, out of rash stupidity, when iron is sought, a metal even more welcome than gold amid the bloodshed of war. We follow up the lodes in the earth and live above mined areas, amazed that sometimes the earth gapes open, or begins to tremble, as if in truth it were not possible that our holy parent might express her indignation in this manner. We search for riches deep within the bowels of the earth, and where the spirits of the dead have their abode, as though the surface upon which we walk is not sufficiently bountiful and productive. Among all these things our least important goal is the discovery of remedies for diseases. Indeed, what fraction of mankind has medicine as its reason for digging?

---

[3] As electrum is included within a list of metals, Pliny must be referring to the naturally occurring mixture of gold and silver, which he defines (33. 80): 'All gold contains silver in various proportions. ... wherever the proportion of silver is one-fifth, the ore is called "electrum"; grains of this are formed in channelled gold' ('... ubicumque quinta argenti portio est, electrum vocatur; scobes hae reperiuntur in canaliensi'). All gold does *not* in fact contain silver—only auriferous ore found in placer deposits, the product of the weathering of parent lodes. This is the 'white gold' of Herodotus (1. 50. 1).

metalla nunc ipsaeque opes et rerum pretia dicentur, tellurem intus exquirente cura multiplici modo, quippe alibi divitiis foditur quaerente vita aurum, argentum, electrum, aes, alibi deliciis gemmas et parietum lignorumque pigmenta, alibi temeritati ferrum, auro etiam gratius inter bella caedesque. persequimur omnes eius fibras vivimusque super excavatam, mirantes dehiscere aliquando aut intremescere illam, ceu vero non hoc indignatione sacrae parentis exprimi possit. imus in viscera et in sede manium opes quaerimus, tamquam parum benigna fertilique qua calcatur; et inter haec minimum remediorum gratia scrutamus, quoto cuique fodiendi causa medicina est?

To compound his crimes in burrowing beneath the earth, man appears to gloat over his assault on Nature, as Pliny (33. 73) vividly describes:[4] 'The mountain splits open and caves in over a wide area with a rending sound which is beyond the power of men's imagination and is accompanied by an incredible shock wave. The miners, like conquerors,[5] gaze on the collapse of Nature. Nor is there any gold at that point; indeed, they did not know for sure that any gold existed when they began to dig. The mere hope of obtaining what they coveted was reason enough for facing such dangers.' 'mons fractus cadit ab sese longe fragore qui concipi humana mente non possit, aeque et flatu incredibili. spectant victors ruinam naturae. nec tamen adhuc aurum est nec sciere esse, cum foderent, tantaque ad pericula et impendia satis causae fuit sperare quod cuperent.')

The emperor Gaius Caligula is described (33. 79) as 'avidissimus auri', and such greed (colourfully referred to as 'auri fames'[6]) is an ongoing vice of mankind. Thus Lucretius (*DRN*

---

[4] Beagon, *Roman Nature*, 41 ff., makes an interesting observation on this mining operation: 'The mountains which protect *Nature* from man and man from himself have been destroyed, yet Pliny cannot avoid his optimistic tendency to admire the human achievement, even in an activity which violently rejects *naturae providentia*. It comes across ... as he follows the process step by step, up to the final collapse of the rock. Moreover Pliny's last thought on the matter adds an interesting pragmatic touch to his otherwise idealistic criticism. ... he implies that the trouble taken to obtain the substance is disproportionate when set against its moral worth.'

[5] Ibid.: 'These *victores* are to be contrasted with the heroes who have conquered in *NH* 2. 54. 'Their probing of the heavens, unlike these men's probing of the earth, was righteous because it freed many by conquering his irrational fear.'

[6] *NH* 33. 72: 'auri fames durissima'. Cf. Verg. *Aen.* 3. 57; Curtius 5. 1. 4.

5. 1113–14) had linked materialism with gold: 'Later came the invention of property and the discovery of gold which speedily robbed the strong and good-looking of their pre-eminence' ('posterius res inventast aurumque repertum|quod facile et validis et pulchris dempsit honorem').

The pursuit of luxury leads to extravagant gestures and immoderate behaviour. Lollia Paulina's emeralds and pearls,[7] Titus Petronius'[8] myrrhine ware, the vast amount of silver plate of Roman dinner-tables,[9] Antonia's gift of earrings to her favourite lamprey,[10] and the immoderate amount of incense at Poppaea's funeral[11] are just a few of the excesses symptomatic of the age which affected Nature.

Horace (*Odes* 3. 24. 47 ff.) also observes:

If we are truly penitent of our crimes, let us dispatch our gems and precious stones and our useless gold, the root of the worst evils, into the nearby sea.

> vel nos in mare proximum
> gemmas et lapides, aurum et inutile,
> summi materiem mali,
> mittamus, scelerum is vere paenitet!

The Greeks did not normally mine soft minerals, since such an operation carried with it the ever-present risk of roof collapse. Theophrastus (*Lap.* 52) discussing the ochre[12] mines of Cappadocia, writes: 'The risk of suffocation, they say, is a serious matter for miners, since this can happen to them quickly

---

[7] *NH* 9. 117. At her betrothal dinner Lollia was covered with alternating emeralds and pearls, which glittered all over her head, hair, ears, neck, and fingers, to the value 40,000,000 sesterces.

[8] *NH* 37. 20; cf. Tac. *Ann.* 16. 18: 'He was admitted into the small circle of Nero's intimates, as Arbiter of Taste. To the blasé emperor nothing was smart and elegant unless Petronius had given his seal of approval.' ('inter paucos familiarum Neroni adsumptus est, elegantiae arbiter, dum nihil amoenum et molle adfluentia putat nisi quod ei Petronius adprobavisset').

[9] *NH* 33. 141. This amounted to more than the silver owned by the whole of Carthage.

[10] *NH* 9. 172. Pliny adds: 'This over-the-top gesture by Antonia, the wife of Drusus, made people extremely eager to visit Bauli (Bacoli) where she had her country house.'

[11] *NH* 12. 83. Nero is alleged to have burned more than the annual output of incense of Arabia at the obsequies of his consort Poppaea.

[12] Healy, *Mining*, 137.

and takes a very short time.' Inadequate timbering[13] or other supports in galleries, and indeed shafts cut in soft, dangerous ground, added to the hazards. Statius (*Theb*. 6. 880–5) vividly likens the burial of a miner under a falling roof to the sudden crushing of a wrestler by his opponent: 'Just as when the Iberian miner burrows beneath a hill, and leaves far behind the living day, then, if the suspended ground has rocked and the tunnelled earth crashed down with sudden roar, overwhelmed by the fallen mountain he lies within, nor ever does his crushed and utterly broken corpse deliver up the indignant soul to its own skies.'

Skeletons of more than fifty men were discovered in some very old workings near Iconium in Asia Minor:[14] they had evidently been entombed in an underground chamber by the sudden collapse of the gallery leading to the open air.

Quarrying[15] is equally destructive of the environment, although Pliny subscribes to the long-held belief that marble regenerates itself.[16] Marble, however, is quarried not for practical purposes so much as for use in decoration. This is a particular affront to Nature since (*NH* 25. 7) 'she created mountains for her own benefit as a framework to hold together the inner part of the earth, and, at the same time, to make it possible for her to subdue violent rivers, and to break the force of heavy seas—and to hold in check her least settled elements with the hardest material of which she is made. ... We quarry these mountains and drag them off for no reason than for luxury. ... headlands are laid open to the sea and nature is flattened.' ('montes natura sibi fecerat ut quasdam compages telluribus visceribus densandis, simul fluminum impetus domandos fluctusque frangendos ac minime quietas partes coercendas durissima sui materia. ... caedimus hos trahimusque nulla alia quam deliciarum causa. ... promuntaria aperiuntur mari, et rerum natura agitur in planum').

---

[13] Ibid. 81. Greek miners were unused to the need for roof-shoring. Galleries at Laurium for example, were cut in schist and limestone following naturally solid strata. This fact, together with the small size of the cross-section of the galleries, meant that they did not require any timber supports.
[14] Healy, *Mining*, 137.
[15] See further J. F. Healy, 'Mines and Quarries'. in Grant–Kitzinger, *Mediterranean Civilization*, ii. 779–93. [16] See *NH* 36. 125.

Horace (*Odes* 3. 1. 33 ff.) had criticized land reclamation:

The fish feel their waters contract as piles are driven into the deep; at this point many a contractor with his slaves, and the owner who is bored with the land, shoot down building rubble.

> contracta pisces aequora sentiunt
> iactis in altum molibus; huc frequens
> caementa demittit redemptor
> cum famulis dominusque terrae
> fastidiosus.

Similarly, wells may provide unseen hazards, as Pliny (31. 49)[17] explains: 'When wells have been sunk deep, the diggers are killed if they meet with sulphurous or aluminous fumes.[18] A test for this danger is to let down a lighted lamp and see if it goes out.'[19] ('depressis puteis sulpurata vel aluminosa occurrentia putearios necant. experimentum huius periculi est demissa ardens lucerna si extinguitur.') He did not, of course, realize that the lamp was extinguished because of the lack of oxygen.

More subtle harm is done by the release into the atmosphere of gases ('odores'), or fumes especially as a result of mining, or metallurgical operations.

The Greeks were already well aware of the environmental consequences of silver-mining and of the deadly fumes caused by the refining of argentiferous galena. Xenophon (*Mem.* 3. 6. 12) vividly describes the devastation of the region round about Laurium, and Strabo (3. 2. 8) explains that furnaces in Turdetania were built with tall chimneys because of their emissions. Pliny (33. 98) also observes that the exahalations from silver-mines are dangerous to all animals, especially to dogs.[20] He refers to poisoning by salivation: 'odor ex argenti

---

[17] Cf. also Vitr. *De Arch.* 8. 6. 13.

[18] When air and dampness come into contact with sulphur in rock, oxidation produces dangerous sulphur dioxide gas (SO_2) and the equally poisonous hydrogen sulphide (H_2S).

[19] This method of testing the air was developed in the form of the Davy safety lamp used in coal-mines to detect methane gas (CH_4).

[20] Zehnacker, *NH 33*, p. 192 ad loc. The gases would sink to ground level, causing danger to animals of low height or small size.

fodinis inimicus omnibus animalibus, sed maxime canibus' (here 'odor' signifies 'fumes': these are likely to have been of hydrogen sulphide—carbon monoxide and carbon dioxide are both odourless).

Lucretius (*DRN* 6. 808–17), without realizing the fact provides evidence of a further kind of danger encountered in mines.

When miners are following veins of gold and silver, groping with their picks in the bowels of the earth, what fumes[21] are emitted from the pits of Scapte Hyle? what malignant breath is exhaled by goldmines! How it acts upon men's features and complexions! Have you not seen or heard how speedily men die and how their vital forces fail when they are driven by dire necessity to endure such work? All these gases, then, are given off by the earth and blown out into the open, into the unconfined spaces of the air.

> denique ubi argenti venas aurique sequuntur,
> terrai penitus scrutantes abdita ferro,
> qualis exspiret Scaptensula subter odores?
> quidve mali fit ut exhalent aurata metalla!
> quas hominum reddunt facies qualisque colores!
> nonne vides audisve perire in tempore parvo
> quam soleant et quam vitai copia desit,
> quos opere in tali cohibet vis magna necessis?
> hos igitur tellus omnis exaestuat aestus
> exspiratque foras in apertum promptaque caeli.

Rickard[22] and others[23] have correctly attributed the ill health and sickly appearance of the miners not to fumes, but to infestation with hookworm[24] (*Ancylostoma duodenale*) which causes iron-deficiency (anaemia), sometimes leading to death.

---

[21] The identification of the fumes—other than as foul air—poses a problem.
[22] T. Rickard, *Man and Metals* (London, 1932), i. 495 ff. esp. 496.
[23] Healy, *Mining*, 93.
[24] See G. S. Martin, 'Miner's Hookworm at the Centenillo Mines, Spain', *Transactions of the Institute of Mining and Metallurgy*, 31 (1922), 558. Cf. also 'Hookworm infestation (ancylostomiasis)' in the *BMA Complete Family Health Encyclopedia*, ed. Dr T. Smith (London, 1990), 540 f.

19.2 'Industrial' Hazards

The danger involved in polishing cinnabar[25] (33. 122) made it necessary for workmen to wear protective face-masks[26] to prevent inhalation of the dust, which is very pernicious, while allowing them to see over the mask: 'qui minium in officinis poliunt, faciem laxis vesicis inligant, ne in respirando pernicialem pulverem trahant et tamen super illas spectent.'

Processes involving lead are equally hazardous. Pliny (34. 167) states:

> For medical purposes lead is melted in earthern vessels, a layer of finely powdered sulphur being put underneath it; on this, thin plates are laid and covered with sulphur and stirred up with an iron spit. While it is being melted, the breathing passages should be protected during the operation, otherwise the noxious and deadly vapour of the lead furnace is inhaled: it is hurtful to dogs, acting with special rapidity, but the vapour of all metals is so to flies and gnats, owing to which those annoyances are not found in mines.

> coquitur ad medicinae usus patinis fictilibus substrato sulpure minuto, lamnis impositis tenuibus opertisque sulpure, veru ferreo mixtis. cum coquatur, munienda in eo opere foramina spiritus convenit, alioqui plumbi fornacium halitus noxius sentitur et pestilens; nocet canibus ocissime, omnium vero metallorum muscis et culicibus, quam ob rem non sunt ea taedia in metallis.

No attempt appears to have been made to outlaw such dangerous and potentially lethal practices: there were no active 'Friends of the Earth' to champion the cause of the environment—which makes Pliny's contribution so significant—nor was there a Health and Safety Executive, or factory inspectors, to enforce even simple regulations.

Pliny (2. 155) reminds his readers, here and elsewhere on a number of occasions, that, in spite of all this, long-suffering Nature continues to provide her many benefits for mankind: 'But earth is kind, gentle, and indulgent, ever a servant of mortal men, producing under our compulsion, or of her own accord, what scents and savours, what juices, what surfaces to

---

[25] Currently, miners in cinnabar mines are only allowed to work four days in any one month.

[26] The loose masks of bladder skin ('laxis vesicis') were clearly worn like modern surgeons' masks.

touch, what colours! How honestly she repays the interest loaned her! What produce she fosters for our benefit!' 'at haec [terra] benigna, mitis, indulgens, ususque mortalium semper ancilla, quae coacta generat, quae sponte fundit quos odores, saporesque, quos sucos. quod tactus, quos colores! quam bone fide creditum faenus reddit! quae nostra causa alit!') This theme is developed in depth in book 22.

The recurrent criticism, throughout the *Natural History*, of man's misuse of Nature, confirms the genuine concern of Pliny for Roman society's failure in its 'stewardship' of planet Earth. He was not, however, an 'eco-warrior', in the modern sense of the term. Pliny did not dig tunnels, or sit in tree-houses, to prevent building works[27] both of which he would have considered a further affront against Nature!

---

[27] Unlike protesters in the recent campaigns against motorway extensions and the second-runway project at Manchester Airport.

20

The *Natural History* in the Middle Ages and After

The *Natural History*[1] became an influential book in the all-embracing field of natural science from the moment of its publication c. AD 77. Already cited, epitomized, and even plagiarized by late Roman and patristic authors, this survey of scientific knowledge was to survive the breakdown of Roman political and educational institutions. Although the transmission of such a long, detailed, and specialized text presented great problems, it was too valuable a work to be lost, or to be replaced adequately by derivative works. The *Natural History* was conceived on a larger scale than other works belonging to the encyclopaedic genre which the Romans had learned to imitate from the Greeks—for example, the lost nine volumes of Marcus Terentius Varro (*Libri IX Disciplinae*).[2] As Pedersen[3] observes, it has been praised as an inexhaustible source of information relating to the scientific knowledge and social conditions of the early Roman imperial period.

---

[1] This chapter is not intended as a comprehensive treatment of the influence of the *Natural History* on later encyclopaedists, but as a brief overview of an important theme that has already been the subject of a number of special studies by B. S. Eastwood, R. French, and others (see below). See generally G. Holmes (ed.), *The Oxford Illustrated History of Medieval Europe* (Oxford, 1992); J. Strayer *et al.*, *Dictionary of the Middle Ages*, 7 vols., (New York, 1982–6); Marjorie Chibnall, 'Pliny's Natural History and the Middle Ages', in T. A. Dorey (ed.), *Empire and Aftermath: Silver Latin*, II (London, 1975), 57–78. On science, see, for example, G. S. Sarton, *The Appreciation of Ancient and Medieval Science during the Renaissance, 1450–1600* (Philadelphia, 1955); B. S. Eastwood, 'Plinian Astronomy in the Middle Ages and the Renaissance', in French–Greenaway, *Roman Science*, 197–251; R. French, 'Pliny and Renaissance Medicine', in French–Greenaway, Roman Science, 252–81.

[2] Varro was an important encyclopaedist (116–27 BC) whose works provided a rich source of material for his successors. From his prolific output, however, only the *De re rustica* and the *De lingua Latina* survive.

[3] O. Pedersen, 'Some Astronomical Topics in Pliny', in French–Greenaway, *Roman Science*, 162.

The *Natural History*—Middle Ages and After 381

As early as the third century AD, Solinus[4] assembled his *Collectanea rerum memorabilium*, as an account of the regions of the world with the fables and marvels traditionally attributed to them. Although the material is taken from the *Natural History*, the compilation is different and does not include any of Pliny's occasional scepticism. The *Collectanea* was very popular in Europe and presented a world of people with huge feet or dogs' heads, legendary animals like the griffin and mantichore, and magic stones.

Books on the healing qualities of precious stones were common in Egypt and Syria and were used by Pliny, Dioscorides, and Galen. The lore transmitted by Byzantine, Arab, and Jewish sources came back to the West, in translation, in the twelfth and thirteenth centuries, when Marbod of Rennes[5] wrote his treatise *Liber lapidum* (*On Precious Stones*) based on Solinus and Isidorus.[6]

Sometime after AD 370. Symmachus[7] sent a copy of the *Natural History* to Ausonius[8] (*Symm. Letters* 1. 24). In the

---

[4] Solinus wrote a geographical compendium containing a brief sketch of the world, as known to the ancients, diversified by historical notices and observations on the origin, habits, religious rites, and social conditions of the nations enumerated. It displays little knowledge, or judgement.

[5] Marbod was Bishop of Rennes (*c*.1035–1123). His *Liber lapidum*, was a poem in hexameters on the properties and supposed virtues of precious stones—some of which were mythical. In addition, he composed hymns, love poems, and epistles and wrote on the lives of saints.

[6] Isidorus Hispalensis, Bishop of Seville (570–636), is one of the most important links between the learning of antiquity and that of the Middle Ages. His works include: *Chronica maiora*; *Historia Gothorum*; *Liber rotarum*; *De rerum natura*; *Differentiae*, *Quaestiones in vetus Testamentum*; and *Etymologiae*, or *Origines*. Isidorus does not often mention his sources, but his information is derived from a wide variety of authorities (including Pliny and Suetonius). See further J. Fontaine, *Isidore de Seville et la culture classique dans l'Espagne Wisigotique* (Paris, 1959).

[7] Quintus Aurelius Symmachus (*c*.340–*c*.402) was a distinguished scholar and statesman. Trained by a Gallic rhetor, he became the greatest orator of his time. In 369 Symmachus was sent on a deputation to Gaul where he became a friend of Ausonius. He wrote the greater part of his *Letters*—ten books survive—between 391 and 402; these were addressed to leading men of the day and their arrangement imitates that found in the *Letters* of the Younger Pliny. Some fragments of his *Orations* also survive.

[8] Decimus Magnus Ausonius (*c*.310–*c*.395), a grammarian, rhetor, and poet, born at Burdigala (Bordeaux), who wrote poems in a variety of metres, including hexameters, elegiacs, and hendecasyllabics. His *Mosella* is a rhetorically fashioned *laudatio*. See H. G. Evelyn White (ed. and trans.), *Ausonius: Poems and Letters*, 2 vols. (Loeb: London and Cambridge, Mass., 1919–21), and S. Prete, *Richerche sulla storia del testo di Ausonio*, (Temi e Testi, 7; 1960).

fourth century learned familiarity with Plinian material was impressive and Ammianus Marcellinus[9] (27. 3. 4) relates a curious, but telling, incident involving Symmachus' father. He was driven from Rome for an alleged remark to the effect that he would rather use his wine to make concrete than sell it at a discounted price to the people. This has rightly been taken as a specific reference to a passage in the *Natural History* (36. 181) where lime is slaked with wine ('glaeba vino restinguitur'). Ausonius' *Mosella* likewise contains various allusions that seem to have been derived from the *Natural History*.

Roman authors who helped to transmit the more academic classical knowledge include Macrobius,[10] Martianus Capella,[11] Calcidius,[12] and Isidorus.[13] Macrobius' *Commentary on the Dream of Scipio* (*Commentarii in Somnium Scipionis*)[14] and Capella's *On the Marriage of Philology and Mercury* (*De nuptiis Philologiae et Mercurii*)[15] both include cosmography that

---

[9] Ammianus Marcellinus, a Greek by birth, lived in Syrian Antioch in the 4th cent. AD. His history of the Roman empire, of which eighteen books survive, covers the period AD 353–78.

[10] See A. Cameron, 'The Date and Identity of Macrobius', *Journal of Roman Studies*, 56 (1966), 25 ff. Macrobius' works include: *De differentiis et societatibus Graeci Latinique verbi*; *commentarii in Somnium Scipionis*—a Neoplatonist exposition of this work; and seven books of *Saturnalia*.

[11] Martianus Capella, between 410 and 439, composed a didactic treatise addressed to his son in a mixture of prose and verse inherited from Menippean satire (see further below, n. 15). Capella was not so much a scholar as a compiler. For his sources see further W. H. Stahl *et al.*, '*Martianus Capella and the Seven Liberal Arts*, 2 vols. (New York, 1971), i. 171–202; cf. also W. H. Stahl, 'To a Better Understanding of Martianus Capella', *Speculum*, 40 (1965), 102–15. Capella's textbook on astronomy, however, with its strong debt to Pliny, continued to be an essential basis for scientific knowledge from the 9th to the 12th cent. and was copied and often referred to thereafter.

[12] Calcidius (Chalcidius), in the late 4th cent., produced a translation and commentary on Plato's *Timaeus* (to 53c only), using earlier Neoplatonic and Peripatetic exegesis, especially the scholarship of Adrastus, Gaius, and Porphyry: see Margaret T. Gibson, 'The Study of the Timaeus in the XIth and XIIth Centuries', *Pensamiento, Revista de investigacion y informacion filosofica*, 15 (1969), 183–94.   [13] See above, n. 6.

[14] W. H. Stahl, *Macrobius' Commentary on the Dream of Scipio* (USA [*sic*], 1952).

[15] J. A. Willis, *De nuptiis Philologiae* (Teubner: Leipzig, 1983). The first part of Capella's work describes the ascent to heaven of Philologia, accompanied by her handmaidens, the seven Liberal Arts, to be married to Mercury, the god of Eloquence. The rest is a compendium of elementary and superficial information about the Arts, and, as such, appealed to the later Carolingians and the 12th-cent. school of Chartres, especially on account of its cosmological content.

helped to determine medieval ideas on the world. Like Pliny, the authors were convinced that the world was a sphere and agreed on the zones (*climata*)[16] into which it was divided—ideas originally derived from Plato and the Pythagoreans. Of ancient scholars, only the Jews maintained a flat-earth theory.

Throughout the early Middle Ages, in both the East and West, scholars and writers aimed to emulate the literary canons and educational norms inherited from antiquity. Holmes,[17] however, adds a caveat, stating that to see the cultural history of this period in terms of the fragile survival of a classical tradition is in some ways misleading. This view overlooks its limited but significant original achievements, not to mention an immense oral tradition now virtually lost to us, but does reflect the priorities of early medieval intellectuals.

Isidorus (AD 570–636) wrote the *Origines* and *De rerum natura*, works which embraced the arts and physical sciences: the latter relied heavily on Pliny's astronomical and meteorological material.

Bede's[18] opinion of Isidorus was not high. He therefore relied much more on Pliny for astronomical doctrine, although adapting and correcting his work on tides. Bede also refers to the evidence of *Natural History* 37, when discussing the various gems mentioned in the Book of Revelation. Bede helped to popularize the *Natural History* rather than to ensure its actual survival.

Manuscripts of the *Natural History* multiplied and in the ninth century were to be found at Corbie, St Denis, Lorches,

---

[16] See Pliny *NH*, 6. 212.   [17] See, generally, Holmes, *Medieval Europe*.

[18] Bede is alleged to have written his *De rerum natura* to prevent his students from reading the 'lies' of Isidorus. See P. Meyvaert, 'Bede the Scholar', in G. Bonner (ed.), *Famulus Christi* (London, 1976), 49–69, esp. 58–60; see also C. W. Jones, 'Bede's Place in Medieval Schools', ibid. 261–85. B. S. Eastwood, 'Plinian Astronomy', 197 ff., esp. 201, discusses Bede's use of Pliny, which has been the focus of disagreement regarding the availability of the *Natural History* in Britain in the 8th cent. This question is complicated by the clear availability of intermediates like Isidorus, whose quotations from Pliny can easily set out false scents. For example, Bede cannot be shown to have used *NH* 18 directly, despite superficial similarities of part of that book with Bede's *De rerum natura* 26. See further Charles W. Jones (ed.), *Opera de Temporibus* (Medieval Academy of America; Cambridge, Mass., 1943), 359. *Opera* (Corpus Christianorum Series Latina, 123A; Turnhout, 1975), 222 f., shows in his source notes the Isidorean bases for this Plinian material.

Reichenau, and Monte Cassino. Some of its books became an established part of monastic culture, being used for astronomy and medicine and as a source for illustrations for biblical commentaries—even for sermons.

During the eleventh and twelfth centuries, schools increased in number and this led to a revival of the study of little-known classical texts through the medium of Latin translations from Greek and Arabic sources.[19] Avicenna,[20] the Arabian translator of Aristotle and writer on minerals that he classifies as 'stones, sulphur, minerals, metals, and salts', recognized the sulphide group of ore-minerals. And Pliny was read at Chartres, where John of Salisbury[21] knew of the *Natural History*, frequently citing the work as a source for the study of natural philosophy.

Pliny's astronomy enjoyed its highest reputation in the early Middle Ages, since its heliodynamic theme gave it a sense of coherence. However, the work of William of Conches,[22]

---

[19] See Holmes, *Medieval Europe*, 48.

[20] Avicenna (Ibn Sinā), 980–1037, was born in Bukhara, Iran. A Persian physician, he was the most famous and influential of the philosopher-scientists of Islam. Avicenna was particularly noted for his contributions in the fields of Aristotelean philosophy and medicine. He composed the *Kitab al-shifa* (*Book of Healing*), a vast philosophical and scientific encyclopaedia, probably the longest work of its kind ever written by one man. It treats of logic, the natural sciences, including psychology, the quadrivium, and metaphysics, but there is no real exposition of ethics, or of politics. Avicenna owes a great deal to Aristotle, but is also indebted to other Greek influences and to Neoplatonism. See M. Achema and H. Masse, *Le Livre de science*, 2 vols. (Paris, 1955–8). Also, Holmes, *Medieval Europe*, 50, observes that Islamic civilization owed its richness and cosmopolitanism to its take-over of Hellenistic and Persian traditions as well as to the prosperity and order of Muslim society; dazzling heights were attained in art and literature at the courts of the Abbasids, in Baghdad.

[21] John of Salisbury (1115–80) became Bishop of Chartres, which was probably his birthplace. Henry II regarded him as a champion of ecclesiastical independence which influenced his two works, the *Policraticus* and *Metalogicon* (both dated 1159). He unfavourably contrasted the way of life followed by courtiers and administrators with an ideal practice derived from Latin poets and from classical and patristic writers. He returned to England in 1170.

[22] William of Conches (c.1110–54) was a scholastic philosopher and leading member of the school of Chartres. He was a realist, whose ideas leaned towards pantheism, and consequently gave an atomistic explanation of nature, regarding the four elements (air, water, fire, and earth) as combinations of homogeneous, individual atoms. He wrote explanations of Plato's *Timaeus*, and Boethius' *De consolatione philosophiae*. In addition he composed two original works: *Philosophia mundi* (*Philosophy of the World*) and

c.1128, with its elemental theory, eventually replaced the *Natural History*.

At Oxford, Robert Crichlade produced the *Defloratio Historiae Naturalis Plinii Secundi* (1141) an abridgement in nine books that omitted material no longer considered relevant. In addition, by careful deletions and replacement of incompatible sections by one of his own entitled *De errore ipsius Plinii*, he was able to give a Christian interpretation to Stoic beliefs.

In the Middle Ages, encyclopaedias were popular sources of instruction. Thomas of Cantimpré[23] acknowledges his debt to Aristotle, Pliny, and Solinus. Vincent of Beauvais[24] produced the most comprehensive medieval encylopaedia, his *Speculum naturale* (thirty-two books) recording the sum total of thirteenth-century knowledge of natural history derived from many sources. He also wrote the *Speculum mundi* (in four parts), c.1244, which included natural phenomena, such as earthquakes, volcanoes, whirlwinds, thunder, and lightning. Vincent drew on Aristotle's *De caelo* and *Meteorologica*, Seneca's *Quaestiones naturales,* and Pliny's *Natural History*, as well as the encyclopaedias of Isidorus and William of Conches. Botany was in its infancy, and so Vincent in his account of this

---

Pragmaticon philosophiae (*The Business of Philosophy*). William is also thought by some to have been the author of the *Summa moralium philosophorum*. See further E. Jeaunneau, *Glossae super Platonem* (Paris, 1965); Th. Silverstein, 'Guillaume de Conches and the Elements', *Medieval Studies*, 26 (1964), 364; and G. Maurach, *Wilhelm von Conches' Philosophia*, 2 vols. (Pretoria, 1980), i. 53–4 and ii. 33–5.

[23] Thomas of Cantimpré wrote a medieval encyclopaedia, the *De natura rerum* (c.1228–44). His aim was that of St Augustine: to unite in a single volume the whole of human knowledge concerning the nature of things, particularly the nature of animals, with a view to using it as an introduction to theology.

[24] Vincent de Beauvais (c.1190–1264) was the author of the *Speculum maius*, probably the greatest European encyclopaedia up to the 18th cent. It was in three parts: (1) historical, (2) natural, and (3) doctrinal. A fourth part, the *Speculum morale*, was added in the 14th cent. by an unknown author. Perhaps the most notable aspect of Vincent's encyclopaedia is his familiarity with Graeco-Roman classical scholarship and his respect for the classics—especially the works of Aristotle, Cicero, and Hippocrates. The nature of his work was an indication of the disappearing hostility to antiquity after the 12th-cent. renaissance of learning. Vincent was extremely influential in his own day (cf. Chaucer's debt) and was well known to the humanist scholars of the Italian Renaissance.

subject was further indebted to Pliny as well as to Dioscorides,[25] Palladius,[26] Rhazes,[27] and Avicenna. The *Natural History*, in spite of the spread of Arab medicine, and the translations of Galen,[28] remained none the less important as a source for encyclopaedists. Pliny is quoted by Vincent regarding the curative properties of minerals, plants, and animal products.

Bartholomew of England[29] also had an extensive firsthand knowledge of Pliny; he wrote his *De proprietatibus rerum* for beginners and non-specialists and this became a popular textbook.

---

[25] Dioscorides, of Anazarba, in Cilicia, a Greek physician who probably lived in the 2nd cent. AD. His *Materia medica* became a text book for many centuries.

[26] Rutilius Palladius lived in the 4th cent. AD and was an expert on climate and soil variations. Almost one third of his work consists of passages repeated from Columella. Cf. K. D. White, *Roman Farming* (London, 1970), 30 f.

[27] Rhazes, or to give him his full name, Abu Bakr Muhhamed ibn Zakariya Ar-Razi (c.865–923/32), was born in Persia (Iran). Rhazes was a celebrated alchemist and Muslim philosopher who is also considered to have been the greatest physician in the Islamic world. He viewed himself as the Islamic version of Socrates, in philosophy, and of Hippocrates, in medicine. Ar-Razi's two most significant medical works were the *Kitab al-Mansuri* and the *Kitab al-Hawi*. In the latter, he surveyed Greek, Syrian, and early Arabic medicine, as well as some Indian medical knowledge. Throughout his works he added his own considered judgement and medical experience by way of commentary.

[28] Galen, Claudius Galenus (AD 130–c.200), born at Pergamum, was, next to Hippocrates, the most celebrated of ancient physicians. He wrote a great many works on medical and philosophical subjects. See J. Scarborough, 'Medicine', in Grant–Kitzinger, *Mediterranean Civilization*, ii. 1237 ff. and bibliography (1246 ff.). Galen practised in his native city and in Rome where he attended the emperors Marcus Aurelius and Lucius Verus. His earliest treatise, *On Anatomical Procedures*, meticulously demonstrates dissecting techniques gained from numerous investigations of the Barbary ape (similar to the rhesus monkey) that modern researchers can duplicate and verify. Ancient medicine survived until Vesalius challenged Galen's anatomy in the 16th cent. Harvey overturned his concept of vascular function in the early 17th cent. and the surgical procedures of Hippocrates, Celsus, and other classical authorities were abandoned after the discovery of reliable anaesthesia in the 19th. It was not, however until the 1880s that many drugs were replaced by laboratory products.

[29] Bartholomew of England (*fl. c.*1220–40) was a lecturer in divinity who became a Franciscan *c.* 1225. As an encyclopaedist, he is famous for his *De proprietatibus rerum* (*On the Properties of Things*). Although primarily interested in Scripture and Theology, in the nineteen volumes of his encyclopaedia he covered all the customary knowledge of his time and was the first writer to make conveniently available the views of Greek, Jewish, and Arabic scholars on medical and scientific subjects. The immense popularity

In 1262 Albertus Magnus[30] wrote his *Natural History*, five books of which were devoted to minerals, but only those which bestowed supernatural powers on their owners and were considered valuable.

The popularity of Pliny's astronomy may be judged from the fact that nine commentaries primarily, or solely, on book 2, were published between c.1480 and 1556.

Georgius Agricola[31], born in the mining region of the Erzgebirge, south of Leipzig, was a keen observer of minerals and a careful recorder, who formulated the first reasonable theory of ore-genesis.

With the loss of many Greek texts, Pliny's *Natural History* became a substitute for a general education, a role envisaged for it in his Preface. In the Middle Ages, many of the larger monastic libraries possessed copies, which, together with numerous versions, ensured Pliny an important place in European literature.

The Italian humanists, Petrarch,[32] Boccaccio,[33] Colluccio Salutati,[34] and Niccolo de Niccoli,[35] were all well versed in Pliny the Elder, for, in addition to the scientific information

---

of his work is shown by the very large number of manuscript copies of it found in European libraries. Translated into English by John Trevisa and printed c.1455, it was highly influential in Tudor England.

[30] Albertus Magnus (c.1200–80) was a student of the liberal arts of the University of Padua. In the Introduction to his *Physica*, he undertakes to make intelligible to the Latins all branches of natural science, logic, rhetoric, mathematics, astronomy, ethics, economics, politics, and metaphysics. His works represent the entire body of European knowledge of his time, not only in theology, but also in philosophy and the natural sciences. Among his disciples were St Thomas Aquinas who arrived in Paris in 1245. See further F. J. Kovach and R. W. Shohan (eds.), *Albert the Great: Commemorative Essays* (1980).

[31] Georgius Agricola (1494–1555) is often referred to as 'the Father of Mineralogy'. Of his seven geological books, the *De natura fossilium* (Basel, 1546) contains his major contributions to mineralogy and this has been called the first textbook in this field. He also wrote the *De re metallica*, or *On Mining*.

[32] Petrarch (1304–74), Italian scholar, poet, and humanist, whose love of classical literature led him to become a stubborn advocate of the idea of the continuity between classical culture and the Christian message. An interesting example of the reinterpretation of 'pagan' legend is seen in the acceptance of the classical story of the birth of Aphrodite in the Renaissance painting of Botticelli's *Birth of Venus* (see I. Chilvers, *Art and Artists* (Oxford, 1990), 56). Petrarch, by making a synthesis of two seemingly conflicting ideals (classical and Christian)—regarding the former as the rich

their predecessors had used, they found in the *Natural History* 'new' materials which appealed to their interests in classical civilization.

In the eyes of Petrarch, at least, the authors of the ancient world were 'living men'. The transmission of the text of the *Natural History* over the centuries had caused many problems, however. An interesting light is thrown on the state of Pliny's text by Petrarch, who complained about a codex of the *Natural History* purchased in Mantua on 6 July 1350. 'What', he wrote, 'would Cicero or Livy, or the other great men of the past, Pliny above all, think if they could return to life and read their own books?' He answered his question with the surmise that they would scarcely recognize these corrupt and barbarous texts as theirs. Accordingly, Petrarch began work on emending and annotating the text of the *Natural History* in an attempt to sort out the disorder that had been introduced into a number of manuscripts of the ninth to the eleventh centuries.

By the end of the fifteenth century the whole of the *Natural History* had come to be greatly admired by scholars. It was one of the earliest Latin texts to be printed; the first edition

---

promise and the latter, as it were, the fulfilment—can claim to be the founder and great representative of the movement known as European humanism. Petrarch rejected the sterile argumentation and endless dialectical subtleties to which medieval scholasticism had become prey and turned back for values and illumination to the moral weight of the classical world. See further Aldo Scaglione, *Francis Petrarch Six Centuries Later* (1975).

[33] Boccaccio (1313–75) was influenced by Petrarch. His most notable contribution to literature was probably the *Decameron*. Ostensibly a work that is no more than a collection of a hundred tales about love, but, when subjected to the interpretative scrutiny that Boccaccio himself recommends, one which takes on a far more serious tone, reminding us that Boccaccio structured his work on Dante's spiritual epic, *La Divina Commedia*.

[34] Colluccio Salutati (1331–1406), chancellor of Florence, was Petrarch's successor as the leader of the humanist movement. He acquired many manuscripts and built up a splendid library.

[35] Niccolo di Niccoli (c.1364–1437), a wealthy Florentine humanist whose collection of art objects and library of manuscripts of classical works helped to shape a taste for the antique in 15th-cent. Italy. Associated with Cosimo de' Medici, his chief service to classical literature consisted in his copying and collecting ancient manuscripts, correcting the texts, introducing chapters, and making tables of contents. Many of the most valuable MSS in the Laurentian Library in Florence are by his hand, among them those of Lucretius and of twelve comedies of Plautus.

appeared in Venice in 1469, offering, however, a distinctly imperfect text.

The most eminent early scholars to work on the recension of the text were Hermolaos Barbarus,[36] who wrote his *Castigationes Plinianae* in 1492–3, claiming to have corrected some five thousand textual erros, and Beatus Rhenanus,[37] who corrected the text for Erasmus' edition published in 1525.

In 1490 Leoniceno inaugurated a famous controversy on the errors of Pliny the Elder.[38] In that year he sent to Politian a critique of Ibn Sinā [Avicenna], in which he noted in passing that Pliny seemed to have confused the two herbs ivy and *cistus* because of the similarity of their Greek names (χίσσος and κισθός); Politian commended Leoniceno's castigation of Ibn Sinā but politely challenged his criticism of Pliny.

Leoniceno responded with a tract, *De Plinii et aliorum in medicina erroribus* (1492), in which he not only defended his original point but charged Pliny with many other errors stemming from verbal confusion. This work provoked an indirect response from Hermolaos Barbarus in support of Pliny and a direct attack on Leoniceno by the Ferrarese lawyer Pandolfo Collenuccio. Others joined in the fray on both sides, with Leoniceno himself contributing three additional tracts in 1493, 1503, and 1507, the latter in response to Alessandro Benedetti. (Leoniceno's tracts were apparently circulated in manuscript before being printed.) All four tracts were

---

[36] Hermolaos Barbarus provided the first serious recension of the text of the *Natural History*. His *Castigationes Plinianae* modified after Leoniceno's attack on Pliny, met some of the former's criticisms.

[37] Beatus Rhenanus (1485–1547) a German humanist, writer, and advocate of Christian reform, whose editorial work helped to preserve a wealth of classical literature. He was influenced by Tacitus' study of German history and culture. In 1531 he wrote his *Rerum Germanicarum libri tres*, the first extensive commentary on the origin and cultural achievements of Germanic peoples.

[38] See G. C. Gillespie (ed.), *Dictionary of Scientific Biography*, vii (New York, 1981), 248 ff. See further A. Castiglioni, 'The School of Ferrara and the Controversy on Pliny', in E. Underwood (ed.), *Science, Medicine, and History* (Oxford, 1953), i. 269–79: French, 'Pliny and Renaissance Medicine', esp. 263–8. Pliny (26. 15), like his fellow countrymen, regarded the Greeks in general as a very superficial race ('levissima gens') and (3. 42) a conceited people ('in gloriam sui effusissimum'). Among many other negative qualities he observes (2. 248) their folly ('vanitas') and (36. 91) their irresponsible invention ('fabulositas').

published in 1509 (Ferrara) in the *De Plinii et plurium aliorum medicorum in medicina erroribus*, and constituted the first great challenge to Pliny's scholarship.

Leoniceno was a fervent Hellenist and so, not surprisingly, Pliny fell short of his ideal in a number of ways. Pliny was not Greek and he did not write in Greek, although he was able to read Greek sources, on which much of his *Natural History* was based. Leoniceno attacked Pliny for having no adequate scientific method, unlike the older and more authoritative Theophrastus and Dioscorides. In particular, he asserted that Pliny had no special knowledge of philosophy and medicine— essential components of Leoniceno's Hellenist outlook. Leoniceno, a physician by profession, could hardly have been sympathetic to Pliny in the light of his scathing criticisms of Greek doctors and medicine.[39] He similarly attacked Alessandro Benedetti, Pliny's 'medical patron'.

In spite of his obvious shortcomings, however, Pliny was not without his defenders. Collenuccio appointed himself Pliny's 'patron', or advocate:[40] 'Even in cases of confessed crime,' he

---

[39] See, generally, *NH* 29, where Pliny gives doctors a bad press, drawing attention to their greed (7 ff., 20, and 22), gimmicks (ibid. 11), and immunity from punishment (ibid. 18). He also says (ibid. 11): 'The art of medicine changes daily and is constantly given a new look. We are swept along by the hot air of Greek intellectuals. It is well known that those who are successful speakers have the power of life and death over us, just as if thousands of people do not exist without doctors or medicine. The Romans existed for more than six hundred years!' ('mutatur ars cottidie totiens interpolis, et ingeniorum Graeciae flatu impellimur, palamque est, ut quisque inter istos loquendo polleat, imperatorem ilico vitae nostrae necisque fieri, ceu vero non milia gentium sine medicis degant nec tamen sine medicina, sicut populos Romanus ultra sexcentesimum annum.') Pliny (29. 27) sums up his basically hostile attitude: 'Assuredly there is no greater reason for the decay of morals than medicine. In this we have daily proof that Cato was a prophet and oracle when he said that one needed only to skim through the literature of Greek intellectuals rather than make a detailed study of it.' ('ita est profecto lues morum, nec aliunde maior quam e medicina, vatem prorsus cottidie facit Catonem et oraculum: satis esse ingenia Graecorum inspicere non perdiscere.')

[40] Collenuccio (*fl.* late 15th cent.) defended Pliny in his *Pliniana defensio Pandulfi Collenucci Pisaurensis iurisconsulti adversus Nicolai Leoniceni accusationem* (Ferrara—no firm date known). The Hellenists commonly regarded themselves as drinking from the 'pure founts' of Greek wisdom and eloquence, while the barbarians wallowed in 'turbid waters', phrases used, for example, by Erasmus and other scholars. See further French, 'Pliny and Renaissance Medicine', 265 n. 26.

The *Natural History*—Middle Ages and After 391

asserted, 'the defendant escapes punishment if he is sufficiently useful to, or a great ornament to, society.' All who defended Pliny saw him as such an 'ornament of Latin culture' ('res latina—a term reflecting a reaction against the *graecitas* of the Hellenists). The clearest proof of the widespread diffusion and popularity of the *Natural History* is the survival of more than two hundred complete, or partial, manuscript copies of the text.

The *Natural History* received a new lease of life in 1601 when Philemon Holland[41] published his translation and it has continued to influence literature down to the present day. It is a tribute in itself that Pliny's monumental work has outlived the compilations of Varro and Verrius Flaccus and, indeed, is far better known than most medieval encyclopaedias.

New Leoniceni have, in our time, like the legendary Phoenix, risen from the ashes and Pliny again deserves an advocate. It is, therefore, worth repeating (cf. p. ix above) the perceptive judgement of Zehnacker,[42] echoed by other scholars: 'Mais il est symptomatique de constater que nos progrès dans la connaissance de la technologie antique nous amènent toujours à apprécier positivement l'*Histoire Naturelle*. Pline résiste à l'épreuve des faits; mieux: il en sort grandi.'

Judged by any criteria, the *Natural History* has made a significant contribution to our knowledge of science (or natural sciences) and technology in the early period of the Roman empire.

As for its author himself, the last word, perhaps, should go to his nephew, writing to the historian Tacitus (Letters 6. 16. 3): 'The fortunate man is he to whom the gods have granted the power either to do something which is worth recording, or to write what is worth reading, and most fortunate of all is the man who can do both. Such a man was my uncle, as his own books and yours will prove.'

---

[41] Philemon Holland (1552–1637) had an accurate and profound knowledge of Greek and Latin and is well known for his translations of a number of classical authors, including Xenophon and Livy among others, but is, possibly, best known for his translation of the *Natural History*. This appears to have been an important source for William Shakespeare's allusions.

[42] *NH 33*, pp. 32 f.

Salve, parens rerum omnium Natura, teque nobis Quiritium solis celebratam esse numeris tuis fave. (*NH* 37. 205)

Farewell, Nature, mother of all things: look with favour on me, for I alone of Rome's citizens have praised you in all your aspects.

SELECT BIBLIOGRAPHY

The bibliography includes only publications which are relevant to science, the natural sciences, or technology, together with treatments of topics which have a bearing on Pliny's career and literary achievement: it is by no means exhaustive. See the entries for R. Calcoen, J. Goodwin, W. Sarjeant, and M. Whitrow for further reference works relating to science and technology.

AIANO, R., 'The Roman Iron and Steel Industry at the Time of the Empire' (MA diss., Aberystwyth, 1975).
AITCHISON, L. A., *History of Metals* (London, 1960).
ALFÖLDY, G. (trans. A. Birley), *History of the Provinces of the Roman Empire: Noricum* (London and Boston, 1974).
ALLAN, J. C. 'Considerations on the Antiquity of Mining in the Iberian Peninsula', *Royal Anthropological Institute, Occasional Paper*, 27 (London, 1970).
AMSTUTZ, G. C., *Glossary of Mining Geology* (Stuttgart, 1971).
ASTIN, A. E., *Cato the Censor* (Oxford, 1978).
BAILEY, see below, Shackleton Bailey, D. R.
BAILEY, K. C., *The Elder Pliny's Chapters on Chemical Subjects*, i and ii (London, 1929–32).
BANDINELLI, R. B. (trans. P. Green), *Rome the Centre of Power: Roman Art to AD 200* (London, 1970).
BARNES, J. W. et al., Geology and Ore Deposits of the Sizma-Ladik Mercury District, Turkey (Cento Rep., 1969).
BARTON, I. M. (ed.), *Roman Public Buildings* (Exeter Studies in History, 20; Exeter, 1989).
BATEMAN, A. M., *Economic Minerals*[2] (London, 1950).
BATTEY, M. H., *Mineralogy for Students*[2] (London, 1981).
BAUER, M., *Precious Stones*, 2 vols. (New York, 1968).
BAYSAL, H. H., CEYLAN, A., et al., *Guide to Denizli: Pammukale* (museum guide [n.d., c.1996]).
BEAGON, MARY, *Roman Nature: The Thought of the Elder Pliny* (Oxford, 1991).
BEARE, W., *The Roman Stage* (London, 1968).
BENNETT, C. E. (trans.), *Frontinus: Aqueducts* (Loeb: London, 1925).

BESSONE, L., 'Sulla morte di Plinio il Vecchio', *Rivista di Studi classici*, 17 (1969), 166–79.
BIANCHI, E., 'Teratologia e Geographia', *Acme*, 34 (1981), 227–49.
BICKERMAN, F. J., *Chronology of the Ancient World*[2] (London, 1980).
BIRD, D. G., 'The Roman Gold Mines of North-West Spain', *Bonner Jahrbücher*, 172 (1972), 36–64.
BIRKELAND, P. W., and LARSON E. E. (eds.), *Putnam's Geology*[5] (Oxford, 1989).
BLANCO, A., and LUZON, J. M., 'Pre-Roman Silver Mines at Rio Tinto', *Antiquity*, 43 (1969), 124–31.
BLÁZQUEZ, J. M., 'Explotaciones mineras en Hispania durante la Republica y el Alto Imperio Romano: Problemas economicos, sociales, y tecnicos', *Annuario de Historia social y economica*, 2 (1969), 9 ff.
BLÜMNER, H., *Technologie und Terminologie der Gewerbe und Kunste bei Griechen und Römern*, 4 vols. (Leipzig, 1875–87).
BOARDMAN, J., GRIFFIN, J., and MURRAY, O., *The Oxford Atlas of the Classical World* (Oxford, 1986).
BODSON, L., 'Aspects of Pliny's Zoology', in R. French and F. Greenaway (eds.), *Science in the Early Roman Empire: Pliny the Elder, his Sources and Influence* (London and Sydney, 1986), 98–110.
BŒUFFLE, A. LE, *Le Ciel des Romains* (Paris, 1989).
BOLGAR, R. R. *The Classical Heritage and its Beneficiaries* (Cambridge, 1973).
BONNIEC, H. LE, *Bibliographie de l'Histoire Naturelle de Pline l'Ancien*, (Collection d'études Latines, 1; Paris, 1946). (*Série scientifique*, xxi.)
BOREN, H. C., *Roman Society: A Social Economic and Cultural History* (London, 1977).
BRADY, J. E., HUMISTON, G. E., and HEIKKINEN, H., *General Chemistry: principles and structure*[3] (New York, 1985).
BRAGG, W. L., and CLARINGBULL, G. F., *Crystal Structures of Minerals* (London, 1965).
BRANNIGAN, K., 'Metal Objects and Metal Technology of the Cycladic Cultures', in J. Thimme (ed.), *Art and Culture of the Cyclades* (Karslruhe, 1977), 117–22.
BROTHERS, A. J., 'Buildings for Entertainment', in I. M. Barton (ed.), *Roman Public Buildings* (Exeter Studies in History, 20; Exeter, 1989), 97–125.
BRUÈRE, R. T., 'Pliny the Elder and Virgil', *Classical Philology*, 51 (1956), 228–46.
BUECHEL W., 'Zur Geschichte der Physik', *Philosophia Naturalis*, 22 (1985), 427–39.

Select Bibliography

BURKE, J. G., 'Bursting Boilers and the Federal Power', *Technology and Culture*, 7 (1966), 1–23.

BURNS, M. A., 'Pliny's Ideal Roman', *Classical Journal*, 69 (1964), 233–8.

CALCOEN, R., *Notes bibliographiques d'histoire des sciences* (Brussels, 1970).

CALEY, E. R., *Orichalcum and Related Alloys: Origin, Composition, and Manufacture, with Special Reference to the Coinage of the Roman Empire* (American Numismatic Society, *Notes and Monographs*, 151; New York, 1964).

—— and RICHARDS, J. F. (trans. and comm.), *Theophrastus on Stones* (Columbus, Oh., 1956).

CAMERON, AVERIL, 'The Date and Identity of Macrobius', *Journal of Roman Studies*, 56 (1966), 25–38.

CAMPBELL, D. J., *Natural History*[2] (Aberdeen University Studies, 118; Aberdeen, 1936).

CASPAR, J. W., *Roman Religion as Seen in Pliny's Natural History* (Chicago, 1934).

CASSON, L., *Travels in the Ancient World* (London, 1974).

——(text, trans., and comm.), *The Periplus Maris Erythraei* (Princeton, 1989).

CHARLES, J. A., 'Early Aresenical Bronzes—A Metallurgical View', *American Journal of Archaeology*, 71 (1967), 21–6.

CHARLESWORTH, M. P., *Trade Routes and Commerce of the Roman Empire* (Cambridge, 1926).

CHERNISS, H., and HELMBOLD, W. (trans.), *Plutarch: Moralia*, 15 vols (Loeb: London and Cambridge, Mass., 1968).

CHIBNALL, M., 'Pliny's Natural History and the Middle Ages', in T. A. Dorey (ed.), *Empire and Aftermath: Silver Latin*, ii (London and Boston, 1975), 57–78.

CHILVER, G. E. F., *Cisalpine Gaul: Social and Economic History from 49 BC to the Death of Trajan* (Oxford, 1941).

CHILVERS, I., *Art and Artists* (Oxford, 1990).

COGHLAN, H. H., 'Prehistoric Copper and Some Experiments in Smelting', *Transactions of the Newcomen Society*, 20 (1939), 49–65.

COLEMAN, M., and WALKER, S., 'Stable Isotope Identifications of Greek and Turkish Marbles', *Archeaometry*, 21/1 (1979), 107–12.

COLLINGWOOD, R. G., and RICHMOND, I. A., *The Archaeology of Roman Britain* (Harmondsworth, 1969).

COLLOCOTT, T. C., *(Chambers) Dictionary of Science and Technology* (London, 1971).

COLLS, D., DOMERGUE, C., LAUBENHEIMER, F., and LIOU, B., 'Les Lingots d'étain de l'épave, Port Vendres,' *II, Gallia*, 33 (1975), 69–94.

CONFORTO, L., FELICI, M., MONNA, D., SERVA, L., and TADDEUCCI, A., 'A Preliminary Evaluation of Chemical Data (Trace Elements) from Classical Marble Quarries in the Mediterranean,' *Archaeometry*, 17/2 (1975), 201–13.

CONOPHAGOS, C. E., *Le Laurium antique et la technique grecque de la production de l'argent* (Athens, 1980).

COPE, L. H., 'The Complete Analysis of a Gold Aureus by Chemical and Mass Spectrometric Techniques', in E. T. Hall and D. M. Metcalf (eds.), *Methods of Chemical and Metallurgical Investigation of Ancient Coinage* (Royal Numismatic Society, Special Publication, 8; London, 1972), 307–13.

—— 'The Metallurgical Analyses of Roman Imperial Silver and Aes Coinages', ibid. 3–47.

CORCORAN, T. H. (trans.), *Seneca: Naturales Quaestiones*, 2 vols (Loeb: London and Cambridge, Mass., 1971–2).

Cordingley R. A. (rev.), *Banister Fletcher: A History of Architecture on the Comparative Method*[17] (London, 1961).

CORNELL, T., and MATTHEWS, J., *Atlas of the Roman World* (Oxford, 1982).

CRADDOCK, P. T., 'The Golden Horses of San Marco', *Journal of Archaeological Science*, 4 (1977), 103–23.

—— 'Early Zinc Production in India', *Mining Magazine* (Jan. 1985), 42–52.

—— LANG, JANET, and PAINTER, K. S., 'Roman Horse-Trappings from Fremington Hagg, Reath, Yorkshire, NR', *British Museum Quarterly*, 37 (1973), 9–17.

CROSSLAND, M. P., *Historical Studies in the Language of Chemistry* (London, 1962).

CUNNINGHAM, A., 'Getting the Game Right: Some Plain Words on the Identity and Invention of Science', *Studies in the History and Philosophy of Science*, 19/3 (1988), 365–89.

CURTIS, R. I., *'Garum and Salsamenta: Production and Commerce in Materia Medica'* (Studies in Ancient Medicine; Leiden, 1990).

DAVIES, O., *Roman Mines in Europe* (Oxford, 1935).

DELLA CASA, A., 'Plinio grammatico', in *Atti del Convegno di Como (1979)—Plinio il Vecchio sotto il profilo storico e letterario* (Como, 1982), 109–15.

DESANGES, J., 'Les Sources de Pline dans sa description de la Trogodytique et de l'Éthiopie', *Helmantica*, 37 (1986), 277–92.

DICKS, D. R., *The Geographical Fragments of Hipparchus* (London, 1960).

DICKS, D. R., *Early Greek Astronomy to Aristotle* (London, 1970).

DIHLE, A., 'Plinius und die geographische Wissenschaft in der römischen Kaiserzeit', in *Atti del Convegno di Como*

(1979)—*Technologia, economica e società nel mondo romano* (Como, 1980), 121–37.
DOMERGUE, C., 'Introduction à l'étude des mines d'or du nord-ouest de la peninsule ibérique dans l'Antiquité', *Legio VII Gemina, Coloquio internacional* (León, 1970), 255–86.
—— 'Apropos de Pline, *Naturalis Historia*, 33. 70–8', *Archivo Español de Arte y Arqueologia*, 45–7 (1972–4), 499–548.
—— *La Mine antique d'Aljustrel (Portugal) et les tables de bronze de Vipasca* (Paris, 1983).
—— 'Les Techniques minières antiques de *Le De Re Metallica* d'Agricola', in C. Domergue (ed.), *Mineria y metalurgia en las antiguas civilizaciones mediterraneas y europeas*, ii (Madrid, 1989), 76–95.
—— and HÉRAIL, G., *Mines d'or romaines d'Espagne: Le District de la Valduerna (León)*, serie B, vol. 4 (Toulouse, 1978).
DRACHMANN, A. G., *Ktesibios, Philon and Heron: A Study of Ancient Pneumatics* (Acta historica scientarum naturalium et medicinalium, 4; Copenhagen, 1948).
—— *The Mechanical Technology of Greek and Roman Antiquity* (Copenhagen, 1963).
DRINKWATER, J. E., *Roman Gaul* (London, 1985).
DWORAKOWSKA, ANGELINA, *Quarries in Ancient Greece*, trans. Krystyna Kozlowska (Warsaw, 1975).
EARL, B., 'The Melting of Tin', *Journal of the Historical Metallurgical Society*, 20/1 (1986), 17–32.
EASTWOOD, B. S., 'Plinian Astronomy in the Middle Ages and the Renaissance', in R. French and F. Greenaway (eds.), *Science in the Early Roman Empire: Pliny the Elder, his Sources and Influence* (London and Sydney, 1986), 197–251.
EDELSTEIN, L., *The Meaning of Stoicism* (Martin Classical Lectures, 21; Cambridge, Mass., 1966).
—— and KIDD, I. G., *Posidonius*, i: *The Fragments* (Cambridge, 1972).
EDWARDS, A. B., *Textures of the Ore-minerals and their Significance* (Melbourne, 1954).
EICHHOLZ, D. E. (trans.), *Pliny the Elder, Natural History*, x: *Books 36–37* (Loeb: London and Cambridge, Mass., 1962).
ELLIS, W. S., 'Glass: Capturing the Dance of Light', *National Geographic Magazine*, 184/6 (1993), 37–69.
ELUÈRE, Christiane, 'A Prehistoric Touchstone from France', *Gold Bulletin*, 19/2 (1986), 58–61.
ERASMUS, C. S., 'Gold Particle Analysis by Radiochemical Neutron Activation Methods', *Annual Report to the Chamber of Mines Research Organisation* (1977), 25–33.

FARNSWORTH, M. S., SMITH, C. S., and RODDA, J. L., 'Metallographic Examination of a Sample of Metallic Zinc from Ancient Athens', *Hesperia*, suppl. 8 (1949), 126–9.
FLEURY, P., *La Mécanique de VITRUVE* (Caen, 1993).
FORBES, R. J., *Studies in Ancient Technology*, 9 vols. (Leiden, 1966–93).
FORTENBAUGH, W. W., HUBY, P. M., SHARPLES, R. W., and GATES, D., *Theophrastus of Eresus: Sources for his Life, Writings, Thought and Influence*, 2 vols (Leiden, New York, and Cologne, 1992).
FRANKE, P. R., and HIRMER, M., *Die griechische Münzen* (Munich, 1964).
FRENCH, R., 'Pliny and Renaissance Medicine', in R. French and F. Greenaway (eds.), *Science in the Early Roman Empire: Pliny the Elder, his Sources and Influence* (London and Sydney, 1986), 252–81.
—— *Ancient Natural History: Histories of Nature* (London and New York, 1994).
—— and GREENAWAY, F. (eds.), *Science in the Early Roman Empire: Pliny the Elder, his Sources and Influence* (London and Sydney, 1986).
FURLEY, D., *The Greek Cosmologists*, ii (Cambridge, 1987).
GALE, N. H., and STOS-GALE, Z. A., 'Some Aspects of Early Cycladic Copper Metallurgy', in C. Domergue (ed.), *Mineria y metalurgia en las antiguas civilizaciones mediterraneas y europeas*, i (Madrid, 1989), 21–36.
GEILMANN, G., and WEIBKE, F., 'Chemische und metallographische Untersuchung eines Spiegels aus der Römerzeit', *Nachrichten von der Gesellschaft der Wissenschaften zu Gottingen*, 1 (1935), 103–8.
GERMANN, K., HOLZMAN, G., and WINKLER, F. J., 'Determination of Marble Provenance Limits of Isotopic Analysis', *Archaeometry*, 22/1 (1980), 99–106.
GIANNINI, A. (ed.), *Paradoxographorum Graecorum Reliquiae* (Milan, 1965).
GIBSON, MARGARET T., 'The Study of the Timaeus in the XIth and XIIth centuries', *Pensiamento, Revista de investigacion y informacion filosofica*, 15 (1969), 183–94.
GILLESPIE, G. C. (ed.), *Dictionary of Scientific Biography*, vii (New York, 1981).
GOLLANCZ, H., *Adelard of Bath: Questiones Naturales* (London, 1920).
GOODWIN, J., *Current Bibliography in the History of Technology* (London, 1980).

GOTTSCHALK, H. B., 'Strato of Lampsacus: Some Texts', *Proceedings of the Leeds Philosophical and Literary Society: Literary and Historical Section*, 11/6 (1965), 95–182.

GRANT, M., KITZINGER, R. (eds.), *Civilization of the Ancient Mediterranean: Greece and Rome*, 3 vols. (New York, 1988).

GREEN, P. M., 'Prolegomena to the Study of Magic and Superstition in the Natural History of Pliny the Elder, with Special Reference to Book XXX and its Sources' (diss., Cambridge, 1952).

——(trans.), *Juvenal: The Sixteen Satires* (Harmondsworth, 1967).

GREENAWAY, F., 'Chemical Tests in Pliny', in R. French and F. Greenaway (eds.), *Science in the Early Roman Empire: Pliny the Elder, his Sources and Influence* (London and Sydney, 1986), 147–61.

GREENE, MOTT T., *Natural Knowledge in Preclassical Antiquity* (Baltimore, 1991).

GRIFFIN, M. T., *Seneca: A Philosopher in Politics* (London, 1976).

GRIFFITH, S. V., *Alluvial Prospecting and Mining*² (Oxford, New York, and Paris, 1961).

GRIMAL, P., 'Encyclopédies antiques', *Cahiers d'histoire mondiale*, 9 (1965), 459–82.

GRISÉ, Y., 'L'Illustre Mort de Pline le Naturaliste', *Revue des Études Latines*, 58 (1980), 338–43.

GRMEK, M. D., 'Les Circonstances de la mort de Pline: Commentaire médical d'une lettre destinée aux historiens', *Helmantica*, 37 (1986), 25–43.

GRÜNINGER, G., *Untersuchungen zur Personlichkeit des alteren Plinius: Die Bedeutung wissenschaftlicher Arbeit in seinem Denken* (diss., Freiburg, 1976).

GUDGER, E. W., '*Pliny's Historia Naturalis*: The Most Popular Natural History Ever Published', *Isis*, 6 (1924), 269–81.

GUNTHER, R. (ed.), *Dioscorides: Materia Medica* (Oxford, 1934).

GUTHRIE, W. K. C., *The Greeks and their Gods* (London, 1956).

——*A History of Greek Philosophy* (Cambridge, 1969).

——*The Sophists* (Cambridge, 1971).

HALBAUER, D. K., 'A Review of some Aspects of the Geochemistry and Mineralogy of the Witwatersrand Gold Deposits', in *Proceedings of the Twelfth Congress of the Institute of Mining and Metallurgy* (Johannesburg, 1982), 957–64.

HALL, E. T., and METCALF, D. M. (eds.), *Methods of Chemical and Metallurgical Investigation of Ancient Coinage* (Royal Numismatic Society, Special Publication, 8; London, 1972).

HALLEUX, R., 'Le Problème des métaux dans la science antique', *Bibliothèque de la Faculté de Philosophie et Lettres de l'Université de Liège*, fasc. 209 (Liège, 1974).

HALLEUX, R., 'Les Deux Métallurgies du plomb-argentifère dans *l'Histoire Naturelle de Pline*', *Revue de Philologie*, 49 (1975), 72–88.
HANFMANN, G. M. A., *Letters from Sardis* (Cambridge, Mass., 1972).
HARTLEY, H., *Studies in the History of Chemistry* (Oxford, 1971).
HAÜY, RENÉ-JUST, *Traité de minéralogie* (Paris, 1801).
HAYWOOD, R. M., 'The Strange Death of Pliny the Elder', *Classical Weekly*, 46 (1952), 1–3.
HEALY, J. F., 'The Electrum Coinage of Mytilene' (diss., Cambridge, 1955).
—— 'Notes on the Monetary Union between Mytilene and Phokaia', *Journal of Hellenic Studies*, 77/2 (1957), 267–8.
—— 'The Composition of Mytilenean Electrum', *Actes du Congrès international de Numismatique*, ii (Paris, 1957), 529–36.
—— 'A New Light on the Unique Stater of Mytilene', *American Numismatic Society, Museum Notes*, 8 (1958), 1–10.
—— 'The Use of Sicilian and Magna Graecian [Coin] Types in White Gold and Electrum Series of Asia Minor and the Islands', in *Atti del Congresso internazionale di Numismatica, Roma, Sept. 1961* (Rome, 1965), 37–44.
—— 'Artists, Engravers and Style in Greek Coinage', *Bulletin of the John Rylands Library*, 54/1 (Manchester, 1971), 149–72.
—— (with A. S. Darling), 'Micro-probe Analysis and the Study of Greek Gold-Silver-Copper Alloys', *Nature*, 231 (1971), no. 5303, 444–5.
—— 'Greek Refining Techniques and the Composition of Gold-Silver Alloys', *Revue belge de Numismatique*, 120 (1974), 19–33.
—— *Mining and Metallurgy in the Greek and Roman World* (London, 1978).
—— 'Mining and Processing of Gold Ores in the Ancient World', *Journal of Metals* (American Institute of Metallurgical Engineers), 31/8 (1979), 11–16.
—— 'Greek White Gold and Electrum Coin Series', in D. M. Metcalf (ed.), *Metallurgy in Numismatics*, i (Royal Numismatic Society, Special Publication, 24; London, 1980), 194–215.
—— 'Problems in Mineralogy and Metallurgy in Pliny the Elder's Natural History', in *Atti del Convegno di Como—Technologia, economica e società nel mondo romano* (Como, 1980), 163–201.
—— 'Pliny the Elder and Ancient Mineralogy', *Interdisciplinary Science Reviews*, 6/2 (1981), 166–80.
—— 'Pliny on Mineralogy and Metals', in R. French and F. Greenaway (eds.), *Science in the Early Roman Empire: Pliny the Elder, his Sources and Influence* (London and Sydney, 1986), 111–46.

HEALY, J. F., *Sylloge Numorum Graecorum, vii. Manchester University Museum: The Raby and Güterbock Collections* (Oxford, 1986).
—— 'Mines and Quarries', in M. Grant and R. Kitzinger (eds.), *Civilization of the Ancient Mediterranean: Greece and Rome*, ii (New York, 1988), 779–93.
—— 'The Language and Style of Pliny the Elder', in *Filologia e Forme Letterarie: Studi offerti a Francesco della Corte*, iv (Urbino, 1988), 1–24.
—— 'Greek and Roman Gold Sources: The Literary and Scientific Evidence' in C. Domergue (ed.), *Mineria y metalurgia en las antiguas civilizaciones mediterraneas y europeas*, ii (Madrid, 1989), 9–20.
—— 'The Gold Staters of Lampsakos: A Preliminary Investigation', in *Proceeding of the Tenth International Congress of Numismatics, 1986* (London, 1990), 45–50.
—— (trans. and comm.) *Pliny the Elder, Natural History: A Selection* (London, 1991).
—— *Miniere e metallurgia nel mondo greco e romano* (Rome, 1993) = revised and enlarged edition (with colour plates), of *Mining and Metallurgy in the Greek and Roman World* (London, 1978).
—— 'Mint Practice at Mytilene: The Evidence for the Use of Hubs', in M. M. Archibald and M. R. Cowell (eds.), *Metallurgy in Numismatics*, 3 (Royal Numismatic Society, Special Publication, 24; London, 1993), 7–18.
HEILMEYR, WOLF-DIETER, 'Titus vor Jerusalem', *Mitteilungen des deutschen Archäologischen Instituts, Römische Abteilung*, 82 (1975), 299–314.
HÉRAIL, G., and GARCIA, L. C. P., 'Intérêt archéologique d'une étude géomorpho-gîtologique: Les gisements d'or alluvial du nord ouest de l'Espagne', in C. Domergue (ed.), *Mineria y metalurgia en las antiguas civilizaciones mediterraneas y europeas*, ii (Madrid, 1989), 21–31.
HEURTIBISE, M., MONTOLOY, F., and LUBKOWITZ, J. A., 'Rapid Determination of Matrix Compounds by Neutron Activation Analysis. The Analysis of Gold Alloys', *Analytical Chemistry*, 5/1 (1973), 47–52.
HIGGINS, R., *Greek and Roman Jewellery* (London, 1961, and Berkeley and Los Angeles, 1980).
HINE, H. M. (ed. and comm.), *Seneca: Naturales Quaestiones*, ii (Salem, 1984).
HODGE, A. T., 'Aqueducts', in I. M. Barton (ed.), *Roman Public Buildings* (Exeter Studies in History, 20; Exeter, 1989), 127–49.
HOLLAND, PHILEMON, *Translation of Pliny's Natural History* (London, 1601).

HOLMES, A., *Principles of Physical Geology*² (London, 1978).
HOLMES, G. (ed.), *The Oxford Illustrated History of Medieval Europe* (Oxford, 1988).
HOLMYRAD, E. J., 'Kitab al-Shifa', *Nature*, 117 (1926), 305 [=Avicenna-Ibn Sinā].
HOOVER, H. C., *Georgius Agricola: De Re Metallica* (New York, 1950).
HOPPER, R. J. 'The Attic Silver Mines in the Fourth Century BC', *Annual of the British School at Athens*, 48 (1953), 200 ff.
—— 'The Laurion Mines: A Reconsideration', ibid. 63 (1968), 293 ff.
—— *Trade and Industry in Classical Greece* (London, 1979).
HOWE, N. P., 'In Defence of the Encyclopaedic Mode: on Pliny's Preface to the *Natural History*', *Latomus*, 44 (1985), 561–76.
HRADECKY, K., *Die Strichprobe der Edelmetalle* (Vienna, 1930).
HUNT, L. B., 'The Oldest Metallurgical Handbook', *Gold Bulletin*, 9/1 (1976), 24–30.
HURLBUT, C. S. (ed.), *Dana's Manual of Mineralogy*¹⁸ (London, 1971).
HUTTER, L. A., *Wasser und Wasseruntersuchungen*, v (Frankfurt a. M. and Arau, 1994).
IMMEHRWAHR, Sara A., 'The Use of Tin on Mycenean Vases', *Hesperia*, 35 (1966), 381–96.
ISAGER, G., *Pliny on Art and Society: The Elder Pliny's Chapters on the History of Art* (London, 1991.)
—— 'Plinio il Vecchio e le meraviglie di Roma: Mirabilia in terris et Romae miracula nel XXXVI libro della Naturalis Historia', *Analecta Romana Instituti Danici*, 15 (1986), 37–50.
JAEGER, W. (trans. R. Robinson), *Aristotle: Fundamentals of the History of his Development* (Oxford, 1962).
JONES, C. W. (ed.), *Opera de Temporibus* (Medieval Academy of America; Cambridge, Mass., 1943).
—— 'Bede's Place in Medieval Schools', in G. Bonner (ed.), *Famulus Christi* (London, 1976), 261–85.
JONES, J. E., 'The Laurion Silver Mines: A Review of Present Researches and Results', *Greece and Rome*, 2nd ser. 49/2 (1982), 171–83.
JONES R. F. J., and Bird, D. G., 'Roman Gold-Mining in North West Spain, II', *Journal of Roman Studies*, 62 (1972), 59–74.
KALCYK, H., *Untersuchungen zum attischen Silberbergbau Gebietsstruktur: Geschichte und Technik* (Europäische Hochschulschriften, Reihe III. Geschichte und ihre Hilfswissenschaften, 160; Frankfurt a. M. and Berne, 1982).
KARTUNEN, K., 'The Country of Fabulous Beasts and Naked Philosophers: India in Classical and Medieval Literature', *Arctos*, 21 (1987), 43–52.

KAYE, G. W. C., and LABY, T. H., *Tables of Physical and Chemical Constants*[16] (London, 1995).
KELLERMANN, R., and TREUE, W., *Die Kulturgeschichte de Schraube*[2] (Munich, 1962).
KIRK, G. S., RAVEN, J. E., and SCHOLFIELD, M., *The Presocratic Philosophers*[2] (Cambridge, 1983).
KIRSCH, J. H. (trans. K. A. Jones), *Applied Mineralogy for Engineers, Technologists and Students* (London, 1968).
KÖNIG, R., HOPP, J., and GLOCKNER, W., *Plinius: Naturkunde, Buch 31* (Zurich, 1994).
—— and WINKLER, G., *Plinius der Ältere: Leben und Werk eines antiken Naturforscher* (Munich, 1979).
KOVACH, F. J., and SHOHAN, R. W. (eds.), *Albert the Great: Commemorative Essays* (1986).
KOZLOWSKA, KRYSTYNA, see above, DWORAKOWSKA, ANGELINA, *Quarries in Ancient Greece* (Warsaw, 1975).
KRINOV, E. L., *Principles of Meterorites* (Oxford, 1960).
LANDELS, J. G. *Engineering in the Ancient World* (London, 1978).
—— 'Engineering', in M. Grant and R. Kitzinger (eds.), *Civilization of the Ancient Mediterranean: Greece and Rome*, i (New York, 1988), 323–52.
Larousse Dictionary of Scientists (New York, 1994).
LATHAM, R. E. (trans.), *Lucretius: On the Nature of the Universe* (London, 1997).
LAUFFER, S., 'Bergmannische Kunst der Antiken Welt', in H. Winckelmann (ed.), *Der Bergbau in der Kunst* (Essen, 1958).
LAZZARINI, L., MOSCHINI, G., and STIEVANO, B. M., 'A Contribution to the Identification of Italian, Greek and Anatolian Marbles through a Petrological Study and Evaluation of the Ca/Sr Ratio', *Archaeometry* 22/2 (1980), 173–82.
LEONICENO, N., *De Plinii et plurium aliorum medicorum in medicina erroribus* (Ferrara, 1509).
LEWIS, P. R., and JONES, G. D. B., 'Roman Gold-mining in North-west Spain', *Journal of Roman Studies*, 60 (1970), 169–85.
LINDBERG, D. C., *The Beginnings of Western Science: The European Scientific Tradition in Philosophical, Religious and Institutional Contexts, 600 BC–AD 1450* (Chicago, 1992).
—— and WESTERMAN, R. S., *Reappraisals of the Scientific Revolution* (Cambridge, 1990).
LING, R., *Roman Painting* (Cambridge, 1991).
LLOYD, G. E. R., *Aristotle: the Growth and Structure of his Thought* (Cambridge, 1968).
—— *Early Greek Science: Thales to Aristotle* (London, 1970).
—— *Greek Science after Aristotle* (London, 1973).

LLOYD, G. E. R., *Magic, Reason and Experience: Studies in the Origin and Development of Greek Science* (Cambridge, 1979).
—— *Science, Folklore and Ideology* (Cambridge, 1983).
—— *Methods and Problems in Greek Science* (Cambridge, 1991).
LOCHER, A., 'Plinius der Ältere uber das Eisen', *Archäologisches Eisenhüttenwesen*, 51/12 (1980), 487–92.
—— 'The Structure of the *Natural History*', in R. French and F. Greenaway (eds.), *Science in the Early Roman Empire: Pliny the Elder, his Sources and Influence* (London and Sydney, 1986), 20–9.
LOEWENTHAL, A. L., and HARDEN, D. B., 'Vasa Murrina', *Journal of Roman Studies*, 39 (1949), 31–7.
LUCAS, A., *Ancient Egyptian Materials and Industries*[4] (London, 1962).
LUTZ, G. J., 'Photon Activation Analysis—A Review', *Analytical Chemistry*, 43/1 (1979), 93–103.
MCDONALD, D. M., and HUNT, L. B., *A History of Platinum and its Allied Metals* (Johnson Matthey: London, 1982).
MCLEISH, A., *Geological Science* (London, 1986).
MAIURI, A., *Roman Painting* (London, 1975).
MANFRA, L., MASI, U., and TURI, B., 'Carbon and Oxygen Isotope Ratios of Marbles from Some Ancient Quarries of Western Anatolia and their Archaeological Significance', *Archaeometry*, 17/2 (1975), 215–21.
MARÉCHAL, J. R., *Zur Frühgeschichte der Metallurgie* (Lammersdorf über Aachen, 1962).
MARSDEN, E. W., *Greek and Roman Artillery*, i: *Historical Development;* ii: *Technical Treatises* (Oxford, 1969–71).
MARTIN, R., 'La Mort étrange de Pline l'Ancien, ou l'art de la déformation historique chez Pline le Jeune', *Vita Latina*, 73 (1979), 13–21.
MAXWELL-STUART, P., 'Studies in the Career of Pliny the Elder and the Composition of his *Naturalis Historia*' (diss., St. Andrews, 1996).
MEIJER, F., *A History of Seafaring in the Classical World* (London, 1986).
MELVILLE, R. (trans.), *Lucretius: On the Nature of the Universe* (Oxford, 1988).
METTE, H. J., 'Enkyklios Paideia', *Gymnasium*, 67 (1960), 300–7.
MEYVAERT, P., 'Bede the Scholar', in G. Bonner (ed.), *Famulus Christi* (London, 1976), 49–69.
MICHEL, A., 'L'Esthétique de Pline l'Ancien', *Helmantica*, 38 (1987), 55–67.
MILLER, J. I., *The Spice Trade of the Roman Empire, 29 BC–AD 641* (Oxford, 1969).

MOMMESEN, TH. (ed.), *C. Iulii Solini Collectanea Rerum Memorabilium* (Berlin, 1985).

MOORE, D. T., and ODDY, W. A., 'Touchstones: Some Aspects of their Nomenclature, Petrography, and Provenance', *Journal of Archaeological Science*, 12 (1985), 59–80.

MÜNZER, F., *Beiträge zur Quellenkritik der Naturgeschichte des Plinius* (Berlin, 1897).

MUSSCHE, H. F., and CONOPHAGOS, C. E., 'Ore-washing Establishments and Furnaces at Megala Pevka and Demoliaki', in *Thorikos VI—1969* (Ghent, 1973), 60–72.

NAUERT, C. G., 'Humanists, Scientists, and Pliny: Changing Approaches to a Classical Author', *American Historical Review*, 84 (1979), 71–85.

NEWMAN, E. G. V., 'The Gold Metallurgy of Isaac Newton', *Gold Bulletin*, 8/3 (1975), 90–5.

NICHOLSON, E. D., 'The Ancient Craft of Gold Beating', *Gold Bulletin*, 12/4 (1979), 161–6.

NIXON, I. G., 'The Volcanic Eruption of Thera and its Effect on the Mycenaean and Minoan Civilizations', *Journal of Archaeological Science*, 12 (1985), 9–24.

NORDEN, E., *Die Antike Kunstprosa von VI. Jhdt. v. Chr. bis in die Zeit der Renaissance*, 2 vols. (Leipzig, 1898; repr. Berlin, 1909).

NOTTON, J. F., 'Ancient Gold Refining', *Gold Bulletin*, 7/2 (1974), 50–6.

NUTTING, J., and NUTTALL, J. L., 'The Malleability of Gold: An Explanation of its Unique Mode of Deformation', *Gold Bulletin*, 10/1 (1977), 2–8.

NUTTON, V., 'The Perils of Patriotism: Pliny and Roman Medicine', in R. French and F. Greenaway (eds.), *Early Roman Science: Pliny the Elder, his Sources and Influence* (London and Sydney, 1986), 30–58.

ODDY, W. A., and HUGHES, J. M., 'A Reappraisal of the Specific Gravity Method for the Analysis of Gold Alloys', *Archaeometry*, 12/1 (1970), 1–11.

——and SCHWEIZER, F., A Comparative Analysis of Some Gold Coins', in E. T. Hall and D. M. Metcalf (eds.), *Methods of Chemical and Metallurgical Investigation of Ancient Coinage* (Royal Numismatic Society, Special Publication, 8; London, 1972), 171–82.

OGDEN, J., 'Platinum Group Metal Inclusions in Ancient Gold Artefacts', *Journal of the Historical Metallurgical Society*, 11 (London, 1977), 53–72.

OLIVIERA, F. DE, *Les Idées politiques et morales de Pline l'Ancien* (Coimbra, 1992).

OLLERENSHAW, A. E., *Blue John Cavern and Blue John Mine: Castleton via Sheffield* (Scarborough, undated).
—— HARRISON, R. J., and HARRISON, D., *The History of Blue John Stone: Methods of Mining and Working, Ancient and Modern*[2] (Scarborough, undated).
ÖNNEFORS, A., *Pliniana: Plinii Maioris Naturalis Historia, Studia Grammatica, Semantica, Critica* (Uppsala, 1956).
OROZ RETA, J., 'Présence de Pline dans les Étymologies de Saint Isidore de Séville', *Helmantica* (1987), 259–306.
PARK, C. F., and MCDIARMID, R. A., *Ore Deposits*[3] (London, 1975).
PARTINGTON, J. R., *General and Inorganic Chemistry for University Students*[4] (London, 1966).
PÁSZTHORY, E., 'Investigations of the Early Electrum Coins of the Alyattes Type', in D. M. Metcalf, *Metallurgy in Numismatics*, i (Royal Numismatic Society, Special Publication, 24; London, 1980), 151–6.
PEDERSEN, O., 'Some Astronomical Topics in Pliny', in R. French and F. Greenaway (eds.), *Science in the Early Roman Empire: Pliny the Elder, his Sources and Influence* (London and Sydney, 1986), 162–96.
PENHALLURICK, R. D., *Tin in Antiquity: Its Mining and Trade throughout the Ancient World with Particular Reference to Cornwall* (London, 1986).
PETROCHILIS, N. K., *Roman Attitudes to the Greeks* (Athens, 1974).
PHILIPS, E. D., *Greek Medicine* (London, 1969).
PRETORIUS, D. A., 'The Nature of the Witwatersrand Gold–Uranium Deposits', *Economic Geology Research Unit, University of Witwatersrand Information Circular*, no. 86 (Witwatersrand, 1977).
PYDDOKE, E., *The Scientist and Archaeology* (London, 1963).
RADL, A., *Der Magnetstein in der Antike: Quellen und Zusammenhänge* (Boethius: Texte und Abhandlungen zur Geschichte der exacten Wissenschaften, 19; Stuttgart, 1988).
RAMIN, J., *Le Périple d'Hannon* (BAR Suppl. ser. 3; Oxford, 1976).
—— *La Technique minière et métallurgique des anciens* (Collection Latomus, 153; Brussels, 1977).
—— *Mythologie et géographie* (Paris, 1979).
READ H. H., and WATSON, J., *Introduction to Geology*[2] (London, 1968).
Report by the Department of Industry, Technology, and Regional Development, Science and Technology Policy Branch (Canberra, 1993).
REYNOLDS, JOYCE, 'The Elder Pliny and his Times', in R. French and F. Greenaway (eds.), *Science in the Early Roman Empire: Pliny*

the Elder, his Sources and Influence (London and Sydney, 1986), 1–10.
RICKARD, T., *Man and Metals*, 2 vols. (London, 1932).
ROBERTSON, I., *Blue Guide to Cyprus* (London, 1981).
ROSENFELD, A., *The Inorganic Raw Materials of Antiquity* (London, 1965).
ROSTOVTZEFF, M. (rev. P. M. Fraser), *The Social and Economic History of the Roman Empire*[2] (Oxford, 1957).
ROSUMEK, P., 'Die Epitheta metallischer Stoffe im Lateinischen', in (ed.) C. Domergue, *Mineria y metalurgia en las antiguas civilizaciones mediterraneas y europeas,* ii (Madrid, 1989), 63–7.
ROTTLÄNDER, R. C. A., 'Glasherstellung bei Plinius dem Alteren', *Glastechnische Berichte*, 52 (1979), 265–70.
—— 'The Pliny Translation Group of Germany', in R. French and F. Greenaway (eds.), *Science in the Early Roman Empire: Pliny the Elder, his Sources and Influence* (London and Sydney, 1986), 11–19.
ROUSE, W. H. (trans.), and SMITH, M. F. (ed.), *Lucretius: De Rerum Natura* (Loeb: London and Cambridge, Mass., 1975).
RUSSELL, D. A., 'The Arts of Prose: The Early Empire', in J. Boardman, J. Griffin, and O. Murray (eds.), *The Oxford History of the Classical World* (Oxford, 1986), 653–76.
RUSSO, E., *Éléments de bibliographie de l'histoire des sciences et des techniques*[2] (Paris, 1969).
RUTLAND L. W., '"Fortuna sola invocatur": Pliny's Statement', *Classical Bulletin*, 56 (1979), 28–31.
SABBAH, G., 'Présence de la N.H. chez les auteurs de l'Antiquité tardive: L'exemple d'Ammien Marcellin, de Symmaque et d'Ausone', *Helmantica*, 38 (1987), 203–21.
—— and MUDRY, P., La Médicine de Celse: Aspects historiques, scientifiques et littéraires (St Étienne and Lyons, 1994).
SAINT-DENIS, E. DE (ed.), *Pline l'Ancien: Histoire Naturelle, Livre 37* (Budé: Paris, 1972).
SALLMANN, K., 'Plinius der Ältere 1938–1970', *Lustrum*, 18 (1975) 5–299; 345–52.
SAMBURSKY, S., *The Physical World of Late Antiquity* (London, 1962).
—— 'Nuances et jeux de lumière dans l'Histoire Naturelle de Pline l'Ancien', *Revue de Philologie*, 45 (1971), 218–39.
SAMUEL, A. E., 'Calendars and Time-Telling', in M. Grant and R. Kitzinger (eds.), *Civilization of the Ancient Mediterranean: Greece and Rome,* i (New York, 1988), 389–95.
SANDBACH, F. M., *The Stoics* (London, 1975).

SARJEANT, W., *Geologists and the History of Geology: An International Bibliography from the Origins to 1978*, 5 vols. (London, 1980).
SCARBOROUGH, J., *Roman Medicine* (London, 1969).
—— 'Pharmacy in Pliny's Natural History: Some Observations on Substances and Sources', in R. French and F. Greenaway (eds.), *Science in the Early Roman Empire: Pliny the Elder, his Sources and Influence* (London and Sydney, 1986).
—— 'Medicine', in M. Grant and R. Kitzinger (eds.), *Civilization of the Ancient Mediterranean: Greece and Rome*, iii (New York, 1988), 1227–48.
—— and NUTTON, V., 'The Preface of Dioscorides' Materia Medica: Introduction, Translation and Commentary', *Transactions and Studies of the College of Physicians of Philadelphia*, 4/3 (1982), 187–227.
SCHILLING, R., 'La Place de Pline l'Ancien dans la littérature technique', *Revue de Philologie*, 52 (1978), 272–83.
SCHMIDT, G., *Heronis Alexandri opera quae supersunt omnia*, i: *Pneumatica et Automata* (Teubner: Leipzig, 1899–1914).
SCHOFF, W. H., *The Periplus of the Erythraean Sea* (London, 1912).
SCHRODINGER, E., *Nature and the Greeks* (Oxford, 1954).
SCONOCCHIA, S., 'La Structure de la *NH* dans la tradition scientifique et encyclopédique romaine', *Helmantica*, 38 (1987), 307–16.
SELTMAN, C. T., *Masterpieces of Greek Coinage* (Oxford, 1949).
—— *Greek Coins*[2] (London, 1955).
SERBAT, G. F., 'Il y a Grecs et Grecs! Quel sens donner au prétendu antihellénisme de Pline?', *Helmentica*, 38 (1987), 272–82.
SHACKLETON BAILEY, D. R., *Martial: Epigrams*, 3 vols. (Loeb: London and Cambridge, Mass., 1997).
SHAW, B. D., 'The Elder Pliny's African Geography', *Historia*, 30 (1981), 424–71.
SHEPHERD, R., 'Hannibal the Rockbreaker', *Minerals Industry International*, 1088 (Sept. 1992), 39–47.
SHERWIN WHITE, A. N., 'Pliny the Man and his Letters', *Greece and Rome*, NS 16 (1969), 76–90.
—— *Fifty Letters of Pliny*[2] (Oxford, 1969).
SKYDSGAARD, J. E., *Varro the Scholar* (Copenhagen, 1968).
SMITH, C. S., *A History of Metallography* (New York and London, 1965).
STAHL, W. H., *Roman Science: Origins, Development and Influence to the Later Middle Ages* (Maddison, 1962).
—— *Martianus Capella and the Seven Liberal Arts*, 2 vols. (New York, 1971).

STEINER, G., 'The Scepticism of the Elder Pliny', *Classical Weekly*, 48 (1955), 137–43.
STENICO, A., *Roman and Etruscan Painting* (London, 1963).
STILES, W. S., 'The Physics of the Dazzled Eye', *Proceedings of the Royal Society*, 104 B (London, 1929), 322–51.
STRAYER, J. et al., *Dictionary of the Middle Ages*, 7 vols. (New York, 1982–6).
STRONG, D., *A History of Roman Painting* (Harmondsworth, 1976).
SYME, R., *Tactius*, 2 vols. (Oxford, 1979).
—— 'Pliny the Procurator', in A. R. Birley, (ed.), *The Roman Papers of Sir Ronald Syme*, ii (Oxford, 1979), 742–73.
—— 'The Consular Friends of Pliny the Elder', in A. R. Birley, (ed.), *The Roman Papers of Sir Ronald Syme*, vii (Oxford, 1991), 496–511.
SZABÓ, Z., 'Az aranyfinometásról' [gold-refining], *Múzeumi Mútárgyvédelem*, 2 (1975), 105–19.
TARN, W. W., and GRIFFITH, G. T., *Hellenistic Civilization*[3] (London, 1966).
TAUBE, E., 'Mining Terms of Obscure Origin', *Science Monthly*, 58 (1944), 45.
THÉRASSE, J., 'Croyances et Crédulités des Romains, d'après Pline l'Ancien et les écrivains latins', in J. B. Caron, M. Fortin, and G. Mahoney (eds.), *Mélanges Lebel* (Quebec, 1980), 283–319.
THOMAS, K., *Man and the Natural World: Changing Attitudes in England, 1500–1800* (London, 1984).
TYLECOTE, R. F., *Metallurgy in Archaeology: A Prehistory of Metallurgy in the British Isles* (London, 1962).
—— *The Early History of Metallurgy in Europe* (London and New York, 1987).
VALLENCE, J., 'Theophrastus and the Study of the Intractable Scientific Method, *De Lapidibus* and *De Igne*', in W. W. Fortenbaugh and R. W. Sharples (eds.), *Theophrastean Studies* (Oxford, 1988), 25–40.
VITTORI, O., 'Pliny the Elder and Gliding', *Endeavour*, 3/3 (1973), 128–31.
—— 'Pliny the Elder on Gliding: A New Interpretation of his Comments', *Gold Bulletin*, 12/1 (1979), 35–9.
—— and MESTITZ, ANNA, 'Artistic Purpose of Some Features of Corrosion on the Golden Horses of Venice', *Burlington Magazine*, 117 (1975), 132–9.
WALLACE-HADRILL, A., *Suetonius: The Scholar and his Caesars* (New Haven, 1984).
WALLACE-HADRILL, A., 'Pliny the Elder and Man's Unnatural History', *Greece and Rome*, NS 37 (1990), 80–96.

WALLACE-HADRILL, D. S., *The Greek Patristic View of Nature* (Manchester, 1968).
WALTER, H., *Die 'Collectanea rerum memorabilium' des C. Iulius Solinus: Ihre Enstehung und die Echtheit ihrer Zweifassung* (Wiesbaden, 1969).
WARDMAN, A., *Rome's Debt to Greece* (London, 1976).
WARMINGTON, E. H., *The Commerce Between the Roman Empire and India*^{2} (Cambridge, 1974).
—— and CARY, M., *The Ancient Explorers*^{2} (London, 1963).
WATT, W. S. 'Notes on Pliny, *Naturalis Historia*, 33–37', *Classical Quarterly*, 38/1 (1988), 206–14.
WEHRLI, F., *Die Schule des Artistoteles, v: Straton von Lampsakos*^{2} (Basle and Stuttgart, 1969).
WELLMANN, M., 'Dioskuridis (12)', *RE* 5, pt. 1 (Stuttgart, 1903), cols. 1131–42.
—— (ed.), *Pedanii Dioscuridis Anazarbei: De Materia Medica*, 3 vols. (Berlin, 1906–14).
—— 'Beiträge zur Quellen analyse des Alteren Plinius', *Hermes*, 59 (1924), 129–56.
WENDER, DOROTHEA S., *Hesiod: Theogony, Works and Days* and *Theognis: Elegies* (London, 1979).
WERTIME, T. A., and MUHLY. J. D. (eds.), *The Coming of the Age of Iron* (Newhaven, 1980).
WEST, M. L. (ed.), *Hesiod: Theogony* (Oxford, 1966).
—— 'Anaxagoras and the Meteorite of 467 BC, *Journal of the British Astronomical Association*, 70/8 (1966), 368–9.
—— 'The Cosmology of "Hippocrates"' *De Hebdomadibus*', *Classical Quarterly*, 21/2 (1971), 365–88.
—— (trans.) *Hesiod: Theogony, Works and Days* (Oxford, 1988).
—— *Hesiod: Works and Days*^{2} (Oxford, 1997).
WHITE, K. D., 'Latifundia', *Bulletin of the Institute of Classical Studies*, 14 (1967), 62–79.
—— 'The Economics of the Gallo-Roman Harvesting Machines', in *Hommages à Marcel Renard*, iii (Collection *Latomus*, 102. Brussels, 1969), 807–9.
—— *Roman Farming* (London, 1970).
—— *Greek and Roman Technology* (London, 1984).
WHITROW, M., *Isis Cumulative Bibliography: A Bibiliography of the History of Science Formed from Isis Critical Bibliographies, 1–90 (1913–65)* =i–ii; and *Isis*, 1966– .
WHITTEN, D. G. A., and BROOKS, J. R. V., *A Dictionary of Geology* (London, 1974).
WIGHTMAN, E. M., *Gallia Belgica* (London, 1983).

WILLIAMS, A. R., 'The Production of Saltpetre in the Middle Ages', *Ambix (Journal of the Society for the History of Alchemy and Chemistry)*, 22 (1975), 125–33.

WILLIAMS, G., *Change and Decline: Roman Literature in the Early Empire* (Berkeley and Los Angeles, 1978).

WILLIS, J. (ed.), *De Nuptiis Philologiae et Mercurii* (Teubner: Leipzig, 1983).

WINSPEAR, A. D., *Lucretius and Scientific Thought* (Montreal, 1963).

WOLVERTON, R. E., 'The Encomium on Cicero in Pliny the Elder', *Classical, Medieval and Renaissance Studies in Honour of B. L. Ullmann*, i (Rome, 1964), 159–64.

YOUNG, W. J., and WHITMORE, F. E., *Application of the Laser Microprobe and Electron Microprobe in the Analysis of Platiniridium Inclusions in Gold* (Boston, 1973).

ZEHNACKER, H., *Pline l'Ancien: Histoire Naturelle, Livre 33* (Budé: Paris, 1983).

—— *Pline l'Ancien: Histoire Naturelle, Livre 3* (Budé: Paris, 1998).

—— 'Pline l'Ancien, lecteur d'Ovide et de Seneque', in H. Zehnacker and G. Hentz (eds.), *Hommages à R. Schilling* (Paris, 1983), 437–46.

ZIRKLE, C., 'The Death of C. Plinius Secundus AD 23–79', *Isis*, 58 (1967), 553–9.

INDEX LOCORUM

PLINY THE ELDER
Natural History, Preface
 1: 38, 40
 6: 40, 78
 12: 39
 13: 10 n. 39, 38, 83
 14: 37, 71, 87 n. 98
 15: 39
 17: 10, 36, 77, 78
 18: 26, 107, 108
 19: 25 n. 76
 20: 34
 21 ff.: 78
 21: 28
 28: 35
Natural History, Book 2
 10: 108, 109
 50: 20
 54: 9 n. 37
 66: 97 n. 291
 82: 265 n. 13
 89 ff.: 86 n. 65
 89: 86 n. 53
 96: 86 n. 58, 87 n. 97
 101: 152 f.
 102: 109
 117 f.: 27
 118: 29 n. 86, 97 n. 289
 139: 96 and n. 241
 140 f.: 110
 141: 110
 147 f.: 257
 147: 40, 110, 327
 148: 159
 149: 258
 150: 20, 327
 155: 379
 160: 96 n. 240
 164: 150
 181: 360
 182 f.: 366
 186: 361
 187: 363
 188: 107, 208, 362
 208: 95 n. 217, 107 f.
 235: 255, 256
 239: 150
Natural History, Book 3
 4: 55
 6: 8 n. 34
 28: 8
 30: 277
 31 ff.: 20
 60: 97 n. 283
 93 f.; 246
 99 ff.: 117 n. 10
 112: 120 n. 24
 117: 93 n. 183, 278, n. 33
 122: 22
 123: 93 n. 181
 136: 51
Natural History, Book 4
 43: 278 n. 34
 64: 45 n. 26
 66: 51
 70: 45 n. 26
 95: 45 n. 25, 94 n. 198
 98: 96 n. 266

Index Locorum 413

105: 18
115: 278 n. 32
120: 94 n. 207
147: 228 n. 140
Natural History, Book 5
  1 f.: 12
  3: 12
  4: 30, 65
  7: 64
  8: 11, 53
  9 f.: 54
  11: 48
  12: 65
  14: 11, 13, 49
  16: 11 n. 46
  22: 15
  25: 13
  33: 13
  34: 15, 120, 266 n. 14
  35 ff: 266 n. 14
  36: 266 n. 14
  37: 17
  38: 14
  44 ff.: 16 f.
  46: 66
  47: 45 n. 22
  51: 16
  53: 93 and n. 189
  57: 16
  72: 255 f.
  110 f.: 295 n. 89
  115: 14
  118: 259 n. 89
  119: 278 n. 35
  129: 45 n. 22
  145: 118 n. 12, 209 n. 64
  146: 118 n. 13
  156: 88
Natural History, Book 6
  18: 45 n. 22
  40: 118 n. 17
  43: 118 n. 17
  60 f.: 278 n. 36
  61: 50
  64: 45 n. 26
  70: 45 n. 26
  104: 93 n. 193
  127: 94
  143: 96 n. 238
  147: 120 n. 23
  160: 50 f.
  163: 45 n. 22
  197: 45
  198 ff.: 45
  200: 45 n. 25
  212: 183 n. 16, 383 n. 16
  213: 95 n. 226
  214: 95 nn. 226, 227
  215: 95 nn. 226 and 227, 295 n. 89
Natural History, Book 7
  5: 95 n. 218
  8: 111
  9: 65
  10: 65, 278 n. 31
  11: 66
  12: 65
  13: 67
  14: 67
  15: 66
  16: 64, 67
  17: 65
  21: 64
  23 f.: 65, 66
  24: 66
  25: 65
  30: 66
  31: 66
  33: 67
  35: 56, 67
  36: 51
  39: 67
  45: 6
  48: 67
  49: 67
  69 f.: 240 n. 179

PLINY THE ELDER – (cont.):
  Natural History, Book 7– (cont.):
    80: 4, 86, 87 n. 100
    81: 67
    85: 149, 151
    86: 110, 159
    87: 67
    91 ff.: 67
    124: 89 n. 133
    125: 45 n. 26, 162
    129: 45 n. 23
    130: 28
    133–46: 28
    147: 95 n. 219
    155: 45 n. 25
    159: 295 n. 89
    160: 96 n. 272
    174: 111
    188: 28 f., 74
    190: 96 n. 251
    191 ff.: 350 f.
    202: 162
    212: 362
    213: 364
  Natural History, Book 8
    37: 67
    39: 94 n. 199
    42: 64
    44: 73
    69: 93 n. 187, 93 n. 194
    70: 93 n. 179
    75: 67
    80 ff.: 67
    82: 30
    114: 14
    130: 86 n. 63
    162: 34
    169: 96 n. 239
    174: 15, 93 n. 191
    191 f.: 18
    195: 96 n. 250
    217: 11
    219: 86 n. 68
  Natural History, Book 9
    8: 55
    9 f.: 54, 67
    11: 55, 56, 67 f.
    26: 15, 95 n. 220
    29: 21
    65: 96 n. 255
    90 ff.: 53
    92 f.: 68
    94: 68
    103: 97 n. 286
    104: 29
    106 ff.: 29
    110: 280 n. 42
    113: 87 n. 72
    114: 86 n. 66
    115: 47, 63
    117: 4, 374 n. 7
    120 f.: 130 f.
    125 ff.: 280 n. 43, 259
    125: 138, 280 n. 43
    127: 29, 138, 141 n. 18
    131 f.: 139
    133 f.: 139
    136: 141 n. 18
    137 ff.: 87 n. 70, 267 n. 19
    140: 139
    143: 96 n. 244
    148: 87 n. 94
    150: 86 n. 55
    168 f.: 29
    168: 160 n. 34
    172: 374 n. 10
  Natural History, Book 10
    3: 87 n. 84
    5: 68
    8: 87 n. 96
    37: 96 n. 258
    53: 97 n. 281
    54: 94 n. 203
    56: 87 n. 92

133 ff.: 29
144: 86 n. 56
179: 86 n. 68
Natural History, Book 11
  6: 64
  8: 112
  27: 95
  41: 77, 85 n. 44, 96 n. 270
  77: 85 n. 44
  111: 278 n. 30
  121: 93 n. 178
  124: 93 n. 188
  130: 87 n. 92
  156: 95 n. 231
  212: 97 n. 277
  255: 97 n. 290
  261: 97 n. 292
  280: 95 n. 224
Natural History, Book 12
  9: 10 n. 41, 52, 68
  10 ff.: 68
  10: 95 n. 223
  39: 93 n. 192
  40: 96 n. 274
  48: 95 n. 224
  66: 228 n. 142
  83: 374 n. 11
  103: 86 n. 54
  132: 87 n. 81
Natural History, Book 13
  47: 86 n. 62
  74 f.: 359
  74: 358
  77: 358, 359
  81: 359
  82: 359
  83: 4, 57
  84: 46
  88: 10 n. 42, 52
  104: 93 n. 190
  119: 63, 68 f.
  121: 97 n. 278

Natural History, Book 14
  3 f.: 76 f.
  5: 77, 95 n. 221
  17: 96 n. 242
  54: 52
  56: 35, 81 n. 11
  92: 234
  130: 133
Natural History, Book 15
  51: 18
  59 f.: 96 n. 271
  84: 96 n. 267
  88: 325 n. 182
  91: 55
  121: 97 n. 293
Natural History, Book 16
  5 ff.: 57
  15: 89 n. 133
  21: 97 n. 294
  33: 21 f.
  49: 95 n. 234
  51: 87 n. 79
  52: 257
  56: 257
  96: 96 n. 248
  107: 97 n. 280
  114: 97 n. 278
  116: 97 n. 295
  120: 87 n. 75
  142: 89
  158: 18
  167: 96 n. 240
  232: 139 n. 10
  233: 97 n. 285
  245 ff.: 21, 87 n. 75
Natural History, Book 17
  21 ff.: 68
  41: 13
  42: 87 n. 77
  44: 22 93 n. 177, 211
  45: 210
  46: 93 n. 180
  106: 95 n. 230

PLINY THE ELDER – (cont.):
 Natural History, Book 17 –
 (cont.):
  123: 96 n. 253
  189: 97 n. 296
  208: 95 n. 232
  227: 95 n. 237
  236: 95 n. 229
  243: 68
 Natural History, Book 18
  6: 89
  30: 15
  75: 55
  86: 96 n. 269
  108: 96 n. 260
  114: 247
  172: 94 n. 208
  188: 14
  190: 14
  273: 352
  296: 19, 85 n. 44, 349
  306: 13, 87 n. 87
  317: 165
 Natural History, Book 19
  3 ff.: 21
  10: 10
  14: 87 n. 95
  17: 257 n. 30
  19 f.: 197
  21: 96 n. 250
  26 ff.: 10
  30: 162
  46 f.: 140
  86: 96
  110: 96 n. 257
  121: 97 n. 280
  139: 96 n. 265
 Natural History, Book 20
  1: 109
  4: 89 n. 142
  95: 89 n. 142
  99: 97 n. 287
  100: 87 n. 90

  193: 89 n. 143
  229: 89 n. 142
  264: 89 n. 138
 Natural History, Book 21
  45: 87 n. 70
  56: 95
  123: 86 n. 59
  156: 95 n. 236
 Natural History, Book 22
  2 ff.: 140 f.
  2: 93 n. 182, 140 f.
  3 f.: 141
  4: 138
  80: 308
  92: 86 n. 57
  105: 95 n. 236
  137: 96 n. 268
  105: 95
 Natural History, Book 23
  54: 131
  57: 131
  61: 89 n. 141
  97: 86 n. 59
  114: 96 n. 262
 Natural History, Book 24
  56: 23 n. 73, 32 n. 23
  94: 140 n. 13
  115: 94 n. 212
  120: 97 n. 294
  123: 87 n. 90
  145: 89 n. 143
 Natural History, Book 25
  21: 94 n. 201
  30: 95 n. 235
  60: 87 n. 90
  94: 97 n. 279, 140
  107: 95 n. 235
  161: 96 nn. 249, 274
  167: 96 n. 245
  184: 93 n. 185
 Natural History, Book 26
  15: 111, 389 n. 38
  16: 160 n. 34

21: 89 n. 140
121: 90 n. 144
Natural History, Book 27
　52: 136 n. 67
　71: 84 n. 27
　119 87 n. 93
　105: 96 n. 243
Natural History, Book 28
　4: 30
　10: 95 n. 273, 110
　33: 96 n. 273
　35: 96 n. 239
　63: 96 n. 261
　178: 239 n. 175
　183: 97 n. 296
　187: 84 n. 27
　191: 94 n. 200
　257: 96 n. 264
Natural History, Book 29
　4: 89 n. 135
　5: 89 nn. 136 f.
　7 ff.: 390 n. 39
　11: 390 n. 39
　14: 30
　17: 89 n. 134
　20: 390 n. 39
　21: 96 n. 247
　22: 390 n. 39
　24: 89 n. 138
　27: 390 n. 39
　65: 110
　67: 87 n. 88
　96: 87 n. 92
　100: 87 n. 78
　124: 188
Natural History, Book 30
　75: 96 nn. 259, 268
　98: 89 n. 135
　137: 111
Natural History, Book 31
　9: 4, 64
　12: 19, 123 f.
　20: 124

26 f.: 124
28: 124
29 f.: 124 f., 209
43: 97 n. 282
44: 107 f., 168
46: 168
49: 247, 376
67: 116
70: 121
71: 85 n. 37
73 ff.: 117 f.
74: 118
75: 118
77: 118 f.
78: 120 and n. 21
79: 120, 188
81: 15, 119, 121
82: 121
86: 121
90 f.: 134
93 ff.: 87 n. 73
106: 199 n. 25
110: 141
113: 134, 198
114: 134
123: 87 n. 94
Natural History, Book 32
　4: 96 n. 252
　15: 96 n. 249
　22: 97 n. 294
　32: 96 n. 255
　37: 89 n. 143
　60: 84 n. 27
　62 f.: 52
　116: 87 n. 88
　152: 87 n. 94
Natural History, Book 33
　1: 29, 84 n. 20, 86 n. 46
　1 f.: 373
　2: 84 n. 19
　4: 189
　5: 228
　10: 86 n. 45

PLINY THE ELDER – (cont.):
Natural History, Book 33 –
(cont.):
   12: 128 n. 64
   23: 95 n. 222
   26: 97 n. 298
   29: 22
   32 ff.: 3
   40: 93 n. 186
   46: 93
   51: 161
   53: 87 n. 83
   59: 84 n. 31, 181, 183, 272, 298
   61 f.: 288 f.
   61: 181, 272
   62: 84 n. 18
   63: 57
   64 ff.: 43, 290
   65: 195
   66 ff.: 278 f.
   66: 161
   67: 91 n. 161, 279
   68: 55 n. 36, 84 n. 16, 208, 280
   69: 92 n. 169
   70 ff.: 281 n. 46
   70: 91 n. 159
   71: 85 n. 37, 131, 133
   72: 9, 373 n. 6
   73: 373
   74 ff.: 282
   74: 85 n. 35, 169
   75: 9, 85 n. 34, 92, 282
   76: 282 f.
   77 f.: 283
   77: 91 n. 161, 92 n. 163
   79: 181, 235, 236, 373
   80: 283, 286, 372 n. 3
   81: 136
   82: 87 n. 74
   84: 84 n. 21, 126, 127, 286, 354
   85: 84 n. 23, 189, 215
   86 ff.: 84 n. 23
   86: 189, 215
   88: 141
   89: 195
   92: 89 n. 139
   95 ff.: 58
   95: 179, 188, 299 f., 322
   96: 57, 299
   97: 169
   98: 375 f.: 97, 169, 377
   99: 22, 286 f., 341
   100: 291
   101: 135, 181, 339
   102: 86 n. 59, 87 n. 86, 340 n. 231
   103 ff.; 246, 340
   106 ff.: 58, 92 n. 174, 323
   106: 321
   108: 322, 326
   109: 126
   110: 325
   111: 59, 65
   113 f.: 215 f.
   114: 217
   115: 216
   116: 316 n. 91
   118: 59, 216 n. 91, 218
   119 ff.: 135, 142
   119: 342
   122: 130
   123: 129, 129 n. 46, 342
   125: 291
   126: 224, 296
   127: 296 n. 91
   128 ff.: 144–6
   128: 150 and n. 13
   129: 145, 146 n. 7
   130: 287
   131: 127, 128, 137
   141: 374 n. 9
   143: 5
   152: 57, 87 n. 82, 143

157: 96 n. 256
158: 96 n. 275
160: 224
194 ff.: 356
Natural History, Book 34
  2 ff.: 311, 391
  2: 203, 213, 301, 311
  3: 22
  5: 273
  6 ff.: 307
  6: 139 n. 11
  8: 308
  9 f.: 58, 307
  30 ff.: 60
  39: 86 n. 64
  41: 86 n. 64
  46: 273
  55: 86 n. 69
  63: 289
  83: 47 n. 37
  86: 47 n. 38
  94 ff.: 272 f., 302 f., 308
  94: 137, 296 n. 95
  96: 22, 308
  97 ff.: 308 f.
  97: 273
  98 ff.: 313
  100 ff.: 312
  100: 204
  101: 204
  108 ff.: 86, 87 n. 91, 204
  112: 135, 136
  117: 90 n. 153, 213
  119: 312 n. 135
  121: 214
  123 ff.: 122, 136 n. 67
  138: 326
  142: 223, 327, 329
  143: 328
  144: 273, 334, 336
  145: 96 n. 246, 328
  146: 181, 257, 334
  148: 158

156 ff.: 315, 316, 323
156: 273, 317
157: 183, 212, 271 n. 2, 315, 316
158 ff.: 301 n. 109, 323
159: 57, 188
160: 319
162: 319
164: 178, 273, 324
165: 273 f., 324 f.
167: 378
172: 181
173: 58, 92 n. 174, 322
175: 261
177: 236, 261 f.
178: 235
183: 212
Natural History, Book 35
  3: 208 n. 62
  5: 87 n. 11, 76
  6: 87 n. 89
  7: 57
  11: 57, 46 n. 33
  29 ff.: 259
  30: 235, 260
  31 ff.: 6, 259
  33: 259
  36: 261
  37: 260, 261
  38: 261
  39: 262
  40: 262
  41: 136 n. 67, 260
  44: 259
  46: 139, 259
  47: 260
  51: 6
  74: 97 n. 288
  106: 87 n. 99
  128: 86 n. 64
  134: 86
  139: 139
  149: 90 n. 155

PLINY THE ELDER – (cont.):
Natural History, Book 35 –
(cont.):
  150: 194 f.
  151: 87 n. 85
  152: 87 n. 71
  164: 52
  168: 86 n. 60
  169: 17
  174: 247
  175: 248
  176 ff.: 249
  176: 94
  177: 249
  179: 256
  183 ff.: 193
  184: 193 f.
  185: 194 n. 9
  186: 194
  192: 84 n. 32, 185
  194: 88 n. 107, 214, 257
  195 ff.: 88 n. 106
  198: 219 f., 248
Natural History, Book 36
  6: 208 n. 62
  7: 208 n. 62
  14: 59, 69, 188, 206, 208 n. 62, 227
  18: 208 n. 62
  27: 27
  30: 90 n. 149
  44: 207, 208 n. 62, 350
  46: 207, 208 n. 62
  47: 207, 208 n. 62
  48: 208 n. 62
  49: 208 n. 62
  51: 16, 208, 221
  52: 237
  53: 237, 239
  55: 208 n. 62
  56: 88 n. 127
  57: 88 n. 120
  58: 214 n. 83
  59: 218 n. 97, 221 n. 105, 237, 265
  61: 218 n. 97
  63: 88 n. 125
  64 ff.: 43
  66: 163, 228 n. 142
  67: 48, 167
  70 ff.: 351
  71 f.: 37 f.
  72: 365
  79: 43 n. 10
  81 ff.: 351
  82: 69
  83 ff.: 351
  84: 69
  86: 207, 208 n. 62
  87 ff.: 208 n. 62
  88: 90 n. 149
  91: 111, 389 n. 38
  95 ff.: 351
  96 f.: 69, 164
  99 f.: 69, 160
  100: 90 nn. 150, 151
  101 ff.: 69, 351
  105: 351
  114: 208 n. 62
  116: 351
  117 f.: 166
  121 ff.: 351
  123: 169
  124 f.: 351
  125: 58, 69, 110, 205
  126 ff.: 69
  126: 155
  127 ff.: 185, 188, 296 n. 91
  127: 156 f., 228
  128: 228 n. 141
  129: 88 n. 102, 157
  130: 158
  131: 52, 59, 69, 125, 131, 180, 240
  132: 180, 208 n. 62, 237
  133: 125

135: 59, 208 n. 62
136: 224, 264
139: 196, 269
141 f.: 181
141: 84 n. 30
144 f.: 222
144: 181, 188
146 f.: 223
146: 59 n. 71, 88 n. 102, 189
147: 88 n. 103, 185, 214 n. 83
148: 17, 185, 269
154 f.: 60, 87 n. 80, 238
155 ff.: 239
156: 239 n. 176
157: 88 nn. 118 and 125, 185, 214 n. 83
158: 208 n. 62
159 f.: 19, 225
160 ff. 235 n. 162
160: 17, 225
161: 58, 226
162: 226
163: 84 n. 25, 236 f.
164 f.: 19, 269
165: 93 n. 184
166: 17, 236 f.
171: 90 nn. 146 f.
172: 90 n. 148
174: 60
181: 382
182 f.: 211
184: 90 n. 154
186: 90 n. 153
191 ff.: 56, 347, 352 f.
192: 353 f.
195: 270
198: 45, n. 23, 230
199: 149
Natural History, Book 37
1: 85 n. 43, 229, 230, 231, 263
18 ff.: 85 n. 43
18: 96 n. 247, 229, 231
20: 232, 374 n. 8
21: 177, 218 n. 97, 229 f., 231
23 ff.: 59, 177, 180, 187, 191 n. 3, 220
23: 59, 88 n. 105, 180
24: 60
27: 60, 221
28: 88 n. 110
30: 61, 86 n. 67
31 f.: 250, 251 n. 6
31: 61
33: 251
34: 60 n. 79, 251 n. 7
35: 59
36: 94 n. 197
37: 94 n. 210, 153
38 ff.: 252
39: 94 n. 205, 96 n. 272
40: 94 n. 209
42 f.: 252
42: 84 n. 17, 94 n. 204, 252
45: 253
46: 61, 252
47: 253
48: 153
50: 6
54 ff.: 29
55 ff.: 88 n. 111
55: 192, 344
56 ff.: 84 n. 25, 191
56: 187
57: 84 n. 25, 180, 188, 191
58: 188, 192
60 f.: 84 n. 26, 181, 192
61 ff.: 60
62 ff.: 241
62: 84 n. 18
64: 147, 242
65 ff.: 243
66: 84 n. 26

PLINY THE ELDER – (cont.):
Natural History, Book 37 –
(cont.):
  69: 61
  70 ff.: 244
  71: 179
  72: 181
  73: 88 n. 129, 208
    n. 62, 244
  74 f.: 245
  75: 60
  76: 202, 241
  78: 202 f.
  79: 61 n. 85, 140,
    140 n. 12, 187
  80 ff.: 264
  83: 140
  84: 88 n. 123
  85 60, 263
  86: 60, 264
  87: 60
  89: 265
  90: 265
  92 ff.: 265
  92: 88 n. 109
  94 f.: 60, 88 n. 121
  95 f.: 17
  96: 266
  97: 60, 66
  99: 254, 266
  103: 154, 227, 266
  104: 154, 266 n. 14
  105: 265
  106: 60, 84 n. 28
  107 ff.: 266
  107: 88 n. 128, 94 n. 21
  108: 60, 266
  110: 218 n. 97, 267
  113: 267
  114: 88 n. 122, 267
  115: 84 n. 26, 88 n. 112
  117: 88 n. 108
  118: 89 n. 131
  119 ff.: 267
  121: 84 n. 24, 264 n. 6, 267
  122: 268
  123: 264, 88 n. 124
  125: 268
  126: 268
  127 ff.: 185
  129: 264
  131: 88 n. 115, 218 n. 97, 268
  132: 88 n. 116, 218 n. 97, 268
  134 f.: 88 n. 117
  134: 218 n. 97
  135: 59
  136: 88 n. 119, 148, 224, 269
  139 ff.: 60
  139: 269
  143: 88 n. 113
  146: 60, 196
  148: 188, 269
  149: 60
  164: 269
  169: 88 n. 102, 188
  173 87 n. 71
  181: 88 n. 126
  182 ff.: 211
  186: 88 n. 103, 188
  187: 188
  188: 88 n. 104, 188
  189: 88 n. 118, 181
  192: 88 n. 114
  197 ff.: 141 n. 19
  200: 161
  201: 32
  205: 392

AMMIANUS MARCELLINUS
  27. 3. 4: 382

AUSONIUS
  Mosella
    362–4: 350

Index Locorum

CAESAR
  Bellum Civile
    2. 11: 164
    2. 89: 15 n. 53
    3. 40: 164
  Bellum Gallicum
    3. 21. 3: 22
    5. 12: 317
    5. 14: 140 n. 14
    6. 13 ff.: 21
    6. 27: 94 n. 198
CATO
  de Re Rustica
    39. 1: 246 n. 210
CATULLUS
  1. 1–2: 60 n. 238
  22. 8: 60 n. 238
CELSUS
  2. 8: 308
  2. 18: 130 n. 53
  2. 21: 130 n. 53
  3. 4: 89
  4. 5: 247 n. 219
  5. 5: 235 n. 160
  5. 18: 89
  6. 6. 5: 87
  16. 142: 89
  18. 6: 89
CICERO
  ad Atticum
    1. 14. 6: 86
    4. 5. 1: 86
    12. 45. 2: 86
    15. 14. 4: 86
    16. 5. 3: 86
  ad Familiares
    8. 1. 4: 91
    16. 18. 2: 363 n. 53
  ad Quintum Fratrem
    2. 14. 3: 86
  de Finibus
    2. 32. 101: 87
    4. 12. 30: 84 n. 18
  de Legibus
    1. 1. 2: 97 n. 276
  Natura Deorum
    2. 2. 5: 56 n. 55
    2. 5. 14: 63 n. 1
    2. 9. 24: 109
    2. 84: 108
  de Officiis
    2. 7. 25: 265 n. 13
  de Oratore
    1. 14. 62: 89
  de Partitione Oratoria
    64: 39 n. 9
  Tusculans
    1. 37. 90: 56 n. 55
  Philippics
    8. 27: 18 n. 58
  pro Murena
    36: 370
  pro Plancio
    66: 26
  Actio in Verrem
    2. 28: 248 n. 225
    2. 2. 50: 90
COLUMELLA
  1. 1. 12: 21 n. 69
  1. 3. 52: 180 n. 26
  8. 5. 11: 246 n. 210
  10. 403: 97 n. 278
CURTIUS
  5. 1. 4: 373 n. 6

HORACE
  Odes
    3. 1. 33 ff.: 376
    3. 8. 3: 265 n. 13
    3. 24. 47 ff.: 374
    4. 14. 46: 94 n. 206
  Satires
    2. 7. 110: 84 n. 18

ISIDORUS
  Origines
    16. 18. 2: 92 n. 167
    19. 31: 92

JUVENAL
  Satires
    1. 6: 288
    1. 79: 371 n. 2
    3. 194 ff.: 4
    3. 232 f.: 26 n. 79
    4. 21: 180 n. 26, 226, n. 133
    6. 155 f.: 232 n. 152
    6. 156: 228 n. 143
    10. 153: 131 n. 58
    13. 152: 289

LIVY
    2. 10. 8: 95 n. 222
    4. 11: 248 n. 225
    21. 5. 8: 278 n. 32
    21. 37. 2–3: 132, 280 n. 44
    28. 11. 9: 95 n. 222
    29. 31: 15 n. 52, 94 n. 195
    39. 12. 12: 248 n. 220

LUCAN
  Bellum Civile
    2. 89: 15 n. 53, 94 n. 195
    3. 256: 94 n. 206
    3. 261: 94 n. 206
    7. 160: 246 n. 213
    7. 755: 278 n. 32
    8. 370: 94 n. 206

LUCIUS PISO
  Annals
    1: 110

LUCRETIUS
  de Rerum Natura
    1. 24 f.: 38
    1. 63 ff.: 40 n. 12
    1. 86: 95 n. 216
    1. 136–9: 82
    1. 146–58: 75 f.
    1. 315: 95 n. 214
    1. 830–3: 82
    1. 887: 248 n. 225
    1. 926 f.: 106 n. 3
    1. 945 ff.: 75
    1. 947: 40 n. 11
    2. 368: 97 n. 277
    3. 378 ff.: 289 n. 83
    3. 751: 97 n. 277
    3. 765: 97 n. 297
    4. 1–25: 106 n. 3
    4. 54 ff.: 142 n. 1
    4. 269 ff.: 143 f.
    4. 292 ff.: 145
    4. 302 ff.: 146
    4. 311 ff.: 146–7
    4. 415: 95 n. 214
    4. 436: 147 f.
    4. 724 ff.: 288
    5. 419: 94 n. 213
    5. 460–5: 177
    5. 805: 95: n. 215
    5. 939: 97 n. 276
    5. 982: 95 n. 215
    5. 1113 f.: 374
    5. 1199: 95 n. 215
    5. 1238: 95 n. 215
    6. 219 ff.: 246 n. 212
    6. 221: 248 n. 220
    6. 553: 248 n. 225
    6. 747 f.: 246 n. 215
    6. 802: 265 n. 13
    6. 808–17: 377
    6. 908: 228 n. 140
    6. 910–16: 156
    6. 1002: 156
    6. 1042–55: 157

MANILIUS
  Astronomicon
    4. 926: 190

MARTIAL
  Epigrams
    1. 4. 42: 249 n. 229
    1. 41. 4: 249 n. 228
    1. 50. 15: 278 n. 32
    5. 11. 1: 245 n. 201
    8. 14: 180 n. 26, 226 n. 133

8. 33. 1 f.: 289
8. 62. 1: 287
9. 60. 20: 245 n. 201
10. 3. 3: 249
10. 10. 8: 15 n. 52
10. 80. 1: 228 n. 142, 229 n. 144
10. 96. 3: 278 n. 32
12. 57. 9: 14, 92, 249 n. 228
13. 97: 93
14. 58: 198 n. 20
14. 113: 231 n. 147, 234

OVID
Amores
1. 9. 37: 95 n. 216
1. 15. 34: 278 n. 32
3. 15. 17: 97 n. 277
Fasti
2. 315: 238 n. 173
4. 217: 95 n. 218
4. 739: 248 n. 220
Metamorphoses
2. 251: 278 n. 32
3. 159: 238 n. 173
3. 374: 240 n. 221
7. 107 f.: 133
7. 701 f. : 97 n. 277
8. 561. 238 n. 173
14. 9: 97 n. 279
14. 602: 97 n. 277
14. 791: 248 n. 222
15. 351: 248 n. 222
15. 393: 68 n. 19

PALLADIUS
7. 4: 19 n. 63
PERSIUS
Satires
2. 25. 3: 246 n. 210
5. 126: 84 n. 18

PLAUTUS
Poenulus
86: 94 n. 196
1269: 96 n. 254
Pseudolus
740 f.: 234
Rudens
2. 6. 48: 265 n. 13
PLINY THE YOUNGER
Letters
1. 14. 4: 3
2. 17. 1: 26
2. 17. 4: 180, 226 n. 133
2. 17. 21: 180
3. 3. 5: 6, 32
3. 5. 1–6: 32, 33 n. 12, 34
3. 5. 4: 34
3. 5. 6: 38
3. 5. 7 ff.: 25
3. 5. 9: 7 n. 26, 53 f.
3. 5. 17: 8
6. 16. 3: 391
6. 16. 4–20: 3 n. 12
6. 16. 4 ff.: 64 n. 12
6. 16. 4: 23 n. 72
6. 16. 19: 23 n. 74
7. 27. 13: 87 n. 82
9. 33. 4 ff.: 15 n. 55, 20 n. 67
PROPERTIUS
2. 2. 21: 216 n. 91
4. 5. 26: 231, 232 n. 151

QUINTILIAN
Institutio Oratoria
1. 10. 1: 36
3. 1. 21: 35
10. 1. 75: 218 n. 98
11. 3. 143: 35

SALLUST
Iugurtha
18. 8: 15 n. 51

SALLUST
  Iugurtha – (*cont.*):
    18. 12: 94
    46. 5: 15 n. 51
SENECA
  Apocolyntosis
    2. 2: 366
  de Brevitate Vitae
    13. 3. 9: 70
    14. 1. 1: 70
  de Otio
    5. 8. 1: 29 n. 85
  de Providentia
    4. 9: 226 n. 133
  de Vita Beata:
    7: 96 n. 264
  Epistulae
    24. 14: 87
    24. 19: 367 n. 58
    42. 1: 68 n. 19
    66: 96 n. 264
    86. 11: 180 n. 26,
      226 n. 133
    90. 25: 180 n. 26
    119. 3. 1: 233 n. 154
  Quaestiones Naturales
    1. 5. 5: 145
    1. 5. 14: 145
    1. 15. 8: 145 n. 6
    1. 17. 8: 145 n. 6
    2. 54. 1: 58
    3. 15. 5: 247
    3. 25. 10: 59, 177
    4. 13. 7: 226 n. 133
    3. 25. 10: 177
SILIUS ITALICUS
  Punica
    3. 642: 130 n. 54
    4. 349: 97 n. 283
STATIUS
  Silvae
    2. 4. 36: 68 n. 19
SUETONIUS
  Caesar (Iulius)
    24: 93 n. 178
    28: 1 n. 5
  Caligula
    18: 216 n. 91
  Domitian
    14: 236 n. 167
  Vespasian
    4: 7 n. 25
SYMMACHUS
  Letters
    1. 24: 381

TACITUS
  Agricola
    42: 95 n. 225
  Annals
    6. 28: 68 n. 19
    12. 27: 5
    13. 31: 35
    16. 5: 3
    16. 18: 374 n. 8
  Germania
    6. 11, 14, 18, 24: 94 n. 202
    31: 86 n. 46
    45: 94 n. 204
  Histories
    2. 101. 1: 35
TIBULLUS
  2. 3. 43–5: 205

VARRO
  de Lingua Latina
    9. 66: 130 n. 51
  de Re Rustica
    1. 7. 7 ff.: 21 n. 69,
      121 n. 26
    1. 7. 7: 97 n. 278
    1. 7. 8: 121 n. 25
    3. 2. 4: 90
    3. 5. 17: 363 n. 53
    39. 1: 246 n. 210
VERGIL
  Aeneid
    1. 52 ff.: 246 n. 216

1. 167: 248 n. 220
1. 421: 15 n. 54, 94 n. 196
3. 57: 94, 373 n. 6
4. 259: 15 n. 54, 94 n. 196
4. 261: 245 n. 201
4. 262: 138 n. 7
5. 214: 238 n. 173
6. 236 ff.: 246 n. 214
8. 77: 97 n. 277
10. 174: 328 n. 196
Eclogues
   4. 45: 262
   10. 27: 216 n. 91
Georgics
   1. 1–12: 370 n. 66
   1. 94: 248 n. 225
   1. 252 ff.: 370 n. 66
   3. 40: 15 n. 52
   3. 340: 94 n. 196
   4. 44: 238 n. 173
   4. 119: 97 n. 278

VITRUVIUS
de Architectura
   1. 6. 4: 368
   1. 6. 6: 363 n. 51
   2. 8. 7: 90
   3. 5. 17: 363 n. 54
   5. 12: 86
   7. 7. 5: 235 n. 160
   7. 8. 1: 217 n. 93
   7. 8. 5: 286
   7. 9. 4: 217 n. 93, 248 n. 225
   7. 9. 5: 135 n. 65
   7. 14. 6: 140 n. 14
   8. 1: 167
   8. 3: 281 n. 44
   8. 3. 1: 131, 133
   8. 3. 104: 209 n. 65
   8. 4–5: 167
   8. 6. 13: 376 n. 47
   8. 9. 1: 290
   8. 16. 13: 376 n. 17
   9. 8. 1 ff.: 365, 367 n. 59
   9. 8. 4–15: 363 n. 53
   10. 1. 1: 162 n. 43
   10. 8: 86
   10. 11: 86
   15. 12: 86

AESCHYLUS
Persae
   578: 196

ANTIGONUS OF CARYSTUS
Mirabilia
   120: 266 n. 15

APOLLONIUS RHODIUS
Argonautica
   4. 123–6: 12 n. 48, 276 n. 21

ARATUS
Phaenomena
   12–14: 370 n. 66

ARISTOPHANES
Ecclesiazusae
   378: 216 n. 91
Lysistrata
   470: 240 n. 181
Wasps
   93: 367 n. 58
   857: 367 n. 58

ARISTOTLE
Athenaion Politeia
   67. 2: 367 n. 58
frag. 495: 196
Historia Animalium
   $552^b 10$: 213 n. 76
   $606^b 20$: 11 n. 43, 64
Metaphysics
   $1078^a 16$: 161 n. 41
Meteorologica
   2. 3. 42: 120 n. 24
   3. $378^{a–b}$: 175, 215, 341 n. 233
   3. 383^b 10 ff.: 175 f.
   4. $383^{a–b}$: 334
   4. 386^b: 272

de Mirabilibus auscultationibus
    834^b25: 337 n. 220
    842^b 15: 254 n. 20
de Mundo
    359^b23: 175 n. 11
de Partibus Animalium
    639^a1 : 71
    640^a2: 38 n. 8
    651^a26: 340 n. 230
Physics
    198^a 22: 38 n. 8
Poetics
    76^a 24: 162 n. 44
Politics
    1337^b15: 36
    1451^a8: 367 n. 58
Problemata
    966^b28: 235 n. 160

BACCHYLIDES
    fr. 10: 296 n. 92

CALLIMACHUS
    Hymn to Artemis
    49: 175 n. 11

DEMOSTHENES
    18. 87: 162 n. 44
DIO CASSIUS
    36. 18: 131 n. 58
DIODORUS SICULUS
    1. 72. 1: 254 n. 20
    2. 52. 1–4: 58, 59, 177
    3. 12. 1–13: 280 n. 44
    3. 14. 1–4: 284 n. 52
    5. 13. 1–2: 332
    5. 22. 2: 212 n. 74, 317
    5. 27. 2: 182
    5. 38. 4: 314 n. 144, 316
    19. 98: 255 n. 27
DIOGENES LAERTIUS
    1: 363 n. 51
    2. 8: 174 n. 11
    3. 67: 282 n. 47

DIONYSIUS HALICARNASSUS
    3. 67: 282 n. 47
DIONYSIUS PERIEGETES
    Periplus Maris Erythraei
    49: 232 n. 151, 233
DIOSCORIDES
    1. 72. 1: 254 n. 20
    1. 73: 256 n. 28
    5. 84: 246, 339, 339 n. 226, 340
    5. 87: 322
    5. 95: 128
    5. 104: 215, 235 n. 160
    5. 106: 194 n. 11
    5. 115: 196 n. 15
    5. 127: 194 n. 11
    5. 138: 196
    5. 156: 88 n. 106

EURIPIDES
    frag. 567A: 155 n. 25
    Hippolytus
    741: 61 n. 90

HERODOTUS
    1. 1. 1: 38 n. 4
    1. 50. 1: 372 n. 3
    1. 50. 2: 183
    1. 67. 3: 326 n. 186
    1. 86: 91 n. 160
    1. 179: 254 n. 20
    2. 44: 244 n. 200
    2. 109: 362 n. 46
    2. 118f.: 38 n. 5
    2. 125: 162 n. 44
    2. 168: 87
    3. 57. 2: 277 n. 25
    3. 83: 162 n. 44
    3. 102–5: 278 n. 30
    3. 115: 314 n. 144
    3. 116: 278, n. 31
    4. 13: 278 n. 31
    4. 191: 216 n. 91
    4. 195: 256 n. 28

4. 217: 278 n. 31
5. 62: 237
5. 101: 275
6. 85: 92
6. 119: 254 n. 20
7. 69: 216 n. 91
8. 57: 162 n. 44
HESIOD
Shield
  122: 337 n. 220
  142: 250 n. 1
Theogony
  144: 63 n. 3
  722 ff.: 170
  862 ff.: 326 n. 185
  924: 102 n. 9
Works and Days
  109 ff.: 102 n. 10, 347 n. 1
HOMER
Iliad
  4. 277: 256 n. 28
  7. 473: 326 n. 185
  12. 380: 204 n. 33
  14. 415: 246 n. 211
  18. 468: 331
  18. 565: 59
  18. 574: 59
  18. 613: 59
  23. 826: 331
Odyssey
  1. 63 ff.: 63 n. 3
  1. 70: 63 n. 3
  1. 106 ff.: 63 n. 3
  1. 184: 326 n. 185
  1. 397: 63 n. 3
  1. 416: 63 n. 3
  4. 73: 250 n. 1
  9. 371 ff.: 63 n. 3
  9. 499: 204 n. 53
  12. 104 ff.: 63 n. 4
  12. 417: 246 n. 211
  12. 441: 63 n. 4
  15. 460: 250 n. 1
  21. 178: 340 n. 230

22. 481: 249 n. 226
23. 327: 63 n. 4
Homeric Hymn
  6. 9: 337 n. 220
Oxyhynchus Papyrus
  985: 162 n. 44

IBYCUS
apud POxy
  1790: 337 n. 220

PAUSANIAS
  5. 10. 2 : 237
  8. 37. 4: 145
PINDAR
Pythians
  10. 67: 294
Placita Philosophorum
  1. 7. 19: 109
PLATO
Cratylus
  425d: 162 n. 44
Critias
  112a: 89 n. 130
Hippias minor
  368b ff.: 36
Ion 533d–e: 155
  535d–e: 228 n. 140
Phaedrus
  96a: 38 n. 6
  229a: 68 n. 21
Timaeus
  59b–c: 184
  60b–c6: 175
  80c: 228 n. 140
PLUTARCH
Moralia
  434a: 213 n. 76
Quaestiones Platonicae
  7. 7: 153
Marcellus
  19: 363 n. 51
On Curiosity
  517 f.: 70

POLYBIUS
1. 48. 2: 162 n. 44

STEISICHORUS
88: 337 n. 220
STRABO
2. 2. 8: 285 f.
2. 5. 24: 363 n. 51
3. 2. 2: 218 n. 100
3. 2. 8: 92, 276, 277 n. 29, 285, 300, 376
3. 2. 10: 182
5. 1. 91: 250 n. 2
5. 2. 6: 178, 332
6. 2. 8: 175 n. 11
7. 5. 8: 255 n. 27
10. 5. 1 88 n. 106
11. 2. 19: 276 n. 19
12. 3. 40: 200, 201, 261
13. 1. 23: 275
13. 4. 14: 209 n. 65
14. 1. 35: 204 n. 33
14. 4. 2: 118 n. 14
14. 4. 14: 209 n. 65
15. 2. 4: 235 n. 160
15. 2. 10: 317
16. 3. 3: 120 n. 23
17. 2. 5: 340 n. 230

THEOCRITUS
Idylls
12. 36: 296 n. 92
16. 10: 254 n. 20
THEOGNIS
417: 294 n. 86
447: 196450: 294 n. 86
1105: 294 n. 86
THEOPHRASTUS
de Causis Plantarum
2. 17. 1: 21
Historia Plantarum
3. 16. 1: 21
9. 18. 2: 61 n. 87
9. 20. 2: 340 n. 230

de Igne
46: 240 n. 181
de Lapidibus
1: 176
4: 296 n. 91
6: 205
7: 237
9: 204 n. 33, 335
16: 254 n. 19, 266 n. 15
19: 265
24: 244
25 f.: 244 n. 194
26: 215 n. 84
29: 228 n. 140
33: 60, 265
35: 243 nn. 190 and 192
40: 235 n. 160
41: 157, 353 n. 32
42: 225 n. 129
45 ff.: 294
46 f.: 224
49: 310
51: 200
52: 200, 374
58 f.: 128 n. 42, 215
60: 128, 341, 342 n. 240
64 f.: 211 f.
69: 211 f.
THUCYDIDES
2. 13: 92 n. 168
2. 76: 91 n. 160
2. 77: 249 n. 227
4. 100: 249 n. 227
6. 2: 63 n. 3
6. 91: 216 n. 89
7. 27: 216 n. 89, 257 n. 31

XENOPHON
Anabasis
1. 2. 6: 209 n. 64
de Vectigalibus
3. 6. 12: 301
Memorabilia
1. 1. 11: 38, 104
3. 6. 12: 376

GENERAL INDEX

Abu Bakr Muhammed ibn
  Zakyriya al Razi (Rhazes)
  386 n. 27
Abydos 327
Academy (Athens) 68
acetic acid (vinegar) 130 f., 132
acetone 136
Achates (r.) 269
acids, uses of 130
  see also fire-setting, solvents
acier 337 n. 218
acies 337
acorn-bearing trees 21
acoustics 159 ff
Act of Union 298
adamas (ἄδαμας) 181, 190, 191, 192, 344
adeps, see stear (στέαρ, τό)
Adepsus (Euboea) 124, 209
Aedemon 48
Aegean passim
Aeginetan bronze 307, 308
Aegospotamoi (r.) 258
Aegraei 50
Aelius Gallus 50
Aemilius Paulus 159
Aeolian Islands (Lipari) 246, 247
Aeolus, king of the Winds 246
aes 303, 307, 308, 309, 311
aes coronarium 137
Aeschylus 250
Aethalia (Elba) 332
Aethiopian language, words from 93

aetiology 73
Africa 13, 55, 225, 360
Africa, circuit of 53
African language, words from 93
Africanus (the Elder) 263
Agamemnon, death mask of 289
agaric 21
agate 232 nn. 151–2, 269, 354
Agatharcides 67
  account of gold refining 284, 285
Aglaosthenes 45
agogae (ἀγωγαί) 92, 282
Agora, zinc found in 337
Agricola 32
agriculture 370 n. 66
Agrigento 121
Agrigentum 121, 256
Agrippa, Marcus Vipsanius 42
Agrippina 56
aides-mémoire 60 n. 75
airports, see Gatwick, Manchester
alabastrites (ἀλαβαστρίτης) 237
alamandine 265
alarm clock, clepsydra as 368 n. 60
Albertus Magnus 387, 387 n. 30
Albucrara (Gallaecia) mine 283
Alderley Edge (Cheshire), copper mines 201
Alessandro Benedetti 389
Alexander the Great 51, 72, 289, 360
Alexander Polyhistor 43, 47, 63

432 General Index

Alexandria 366, 369 n. 63
alkaline earth, lime 354
alkanet (anchusae radix)
    253 n. 15
allochromatic minerals 179
allotropes:
    lead 321
    sulphur 246
alloys, see brass, bronze,
    electrum, pewter, silver
alluvial deposits:
    gold, 275
    tin 314, 315
alluvial gold, trace elements
    (PGE) in 283
Alps 19, 220
alumen 193 ff.
    astringent properties of 193
    cleansing properties of
        alum 195
    occurrences of, Africa,
        Armenia, Egypt, Lipara,
        Macedonia, Melos,
        Pontus, Sardinia, Spain,
        Strongyle 193 f.
    uses of 194 f.
alumina 354
aluminous fumes in wells 376
alunogen (alum) 193 n. 6
alutiae 183, 212, 315
amalgam 128, 292
amalgamation 286
Amantes 120
Amazonian rain forests 371
Ambassador's house 362
amber 59, 250, 257
    exported to Panonia 253
amber:
    ants, gnats, lizards in 252
    kinds of 253
    pyroelectric properties of 153
    Sicilian and Romanian 253
amethyst 264
amethystus 264

amiantus ($\dot{\alpha}\mu\dot{\iota}\alpha\nu\tau\sigma$), asbestos 196
Amisus 328
Ammaeensian (Mts.) 60
Ammianus Marcellinus 382
Ammon, Oracle of 120
Ammoniac salt 120
amorphous iron hydroxides
    328 n. 197
ampelitis ('vine' earth) 257
amphitheatre 165, 253
anaemia 377
anaesthesia, discovery of
    in nineteenth century
    386 n. 28
Anatolia 129, 343
Anaxagoras 174, 327
    prediction of meteorite 258
Anaximander 363, 369
Anaximenes (of Miletus) 363
ancylostoma duodenale 377
Andeira 337
Andromeda 56, 67
Andronicus Cyrrhestes
    (Andronikos of Kyrrhos)
    367
animal feats 51
animal kingdom 67
    remedies from 111
Annaeus Seneca L. (the Younger
    Seneca) 28, 44 n. 17, 59,
    177, 385
    disapproval of interest in
        mirabilia 70
    portable sundials 366
Annaeus Seneca M.,
    Quaestiones naturales 39
    on dials 366
annalists 46
annealing furnaces, glass
    354 n. 33, 355
Antaeus 12
anthracites 17, 185, 197, 254
Anthropophagi (Cannibals) 65
Antigonus Gonatas, king 73 n. 3

General Index 433

Antignotus (of Carystus) 47
antimony:
  metallic 135
  ores of 201, 246, 300, 338
  refining of 339 f
  sources 338
  trisulphide 339
  uses of 340
antiquarianism 73
antiseptic, vinegar used as 133
Antium 370 n. 67
Antonia 374
Antonine Column 365
Antonius, M. 130
ants:
  as miners in India 278
  models of 149
Anubis, stained portrait of 137
Anuvio (near Castleton) 233
apatite 181
aphronitrum 134, 198
Apion 245
Apollonius Rhodius 250
apologia 100 n. 5
apostrophe, of Cicero 98
apsyctos (ἄψυκτος) 269
ἄπυρον, epithet of sulphur
    (θεῖον τό) 247
aqueducts 168 f. 281, 351
Aquilius, stomach the best
    time-piece 364
Ara Pacis 365
Arabia 50, 220, 237
Arabic language, words from 93
Arabs 50
Aratus 73, 370 n. 66
Arcadia 124, 196
ἀρχή ἡ basic element 103
Archelaus, king of Cappadocia
    252
Archimedes 166, 183, 369
Archimedean screw, *see* coclea
architect, Utopian education of
    348

architraves (*epistylia*), temple of
    Diana 69
archive rooms 57
Arellius Fuscus 57
argentariae 261
argentarium plumbum (EX
    ARGENT) 85
argentiferous galena 182
argentum vivum 129, 341
argillaceous lias, fire-setting
    tests with 133
Argo, anchor of 69
argyritis (ἀργυρῖτις) 321, 325
Arii 118
Arimaspi 65
Arimphaean language, words
    from 93
Aristander 68
Aristides 45
Aristophanes, *Acharnians,
    Knights, Wasps* 58
Aristotle 36, 64, 66, 71, 104,
    112, 162, 175, 176, 215,
    277, 310, 384, 385
  and the animal kingdom 72 f.
  Meteorologica 58
  μηχανικά τά Mechanics 162
Arles 5
Armenia 260
armenium (azurite) 259
  see also pigments
Armorica (Gallia Aquitana) 18
arrhenicum 235
Arrian 44 n. 12
arrugia 91
arsenic 200, 201, 306
  arsenic sulphide 181, 236
  association with 200
  and bronze alloys in
    Aegean region 310
  compounds used as pigments
    201
  in early Bronze Age 201
  solubility in copper 306

arsenopyrite 200
Arsinoe, temple of 158
Arsinoe, wife of Ptolemy II 267
Artemidorus 43, 66
Artemis temple, the first 162
asafoetida, see silphium
ἀσβέστινον τό 197
asbestos, ἄσβεστος ὁ 196
Asia 360
Asia Minor 337
Asinius Pollio 218 n. 98
Aspendus 118
asphalt, ἄσφαλτος ἡ 254 f.
assay, fire test 136
Assay Master (Birmingham), touchstone, description of 297 n. 98
assimilations from foreign languages (other than Greek and Spanish) 93 f.
see also separate entries for African, Arabic, Arimphaean, Carthaginian, Egyptian, German, Indian, Persian, Phoenician, Raetian, Scythian, Syrian, Trogodyte
Assos, stone of 125
Assos (Troad) 239
asteria (star stone) 268
astringent, alum used as 194
astronomy, hellenistic times in 362
Astyra (Troad) 275
Aswan, granite from 208
Ateius Capito 234
Athena, born from the head of Zeus 102
Atlas (Mts.) 12, 48, 49
atomic absorption analysis (AA) 295 n. 88
ἄτομοι οἱ *rerum primordia* 82

atramentum, see shoemaker's black
attritus 161
Aufidius Bassus 33, 34
Augustus, emperor 4, 11, 42, 350, 365
Augustus' sundial 364 f.
Aulus Cornelius Celsus *passim*
Aulus Gellius 364
aurariae 261
aurea domus, Golden House of Nero 4
Aurelius, emperor 386 n. 28
auri nodus (χρυσοῦ ὄζος) 192
aurichalcum 301
auripigmentum 235
Ausonius 381, 350, 382
autographs, Cicero, Augustus, Virgil 57
avaritia 371
Avernus (Lake) 246
Avicenna 384, 386
Ayios Minas 207 n. 60
azurite (*armenium*) 140, 179, 260

Babylon 121, 255, 256
Babylonians, day measurement of 161
Bacchylides 296 n. 92
Bactrian emerald 242
Baebalo mine 169
Baebius Macer 8, 33
Baetica 55
Baeton 51
Bagradas (r.) snake in 67
Balbus 108
Balearic Islands 10, 92
baling 169
Baltic amber (succinite) 253
balux (*baluca*) 92, 283
Bambolus (r.), river Non 54
Bandinelli, R. B. 365 n. 56
bar copper 302

barium titinate monohydrate 154
barley water 55
barrack room language 83
Bartholomew of England 386
basalt, (Lydian stone) 293
basanite (βασανίτης λίθος) 293, 294, 294 n. 86
basanos, see touchstone
βάσανος cf. πειρῶντι δὲ καὶ χρυσὸς ἐν βασάνῳ πρέπει καὶ νόος ὀρθός 294
basic element (ὑποκείμενον τό) 103
Bath Spa, waters at 19
Bauer, M. 232
Bauli (Bacoli) 374 n. 10
Beagon, Mary 9 n. 37, 29, 373 nn. 4–5
Beatus Rhenanus 389
Bede 383
Beilby layer 161 n. 40
bellows 331
benzine 248
Berenice, queen 224, 266
Berenice, city of Cave-dwellers 366
Bergamum (Bergamo) 203, 301
beryl (*beryllus*) 202, 203, 241, 264
  hexagonal crystal system 241
  hexagonal shape 203
Bessi, from Paeonia 277
bête noire, Pliny the Elder viewed as 80
Beyet, J., *Recherches sur le monde* 39
Biblical commentaries 384
bi-metallic (Greek) coins 284
binary alloy 296
binoculars 151
births, unusual, or multiple, see *mirabilia*
Biscaya, black lead found in 323
bitumen 121, 254, 255
bituminous quartzite 294

Bituriges 21
Bizerte (Hippo Diarrhytus) 15 n. 55
black lead 273
black pigments, carbon based 260
Black Sea 202, 328
bladder skin, protective masks of 378 n. 26
bladder stones 133
Blemmyae 16
bleu jaune 230 n. 146
block and tackle 163
bloom, iron 330, 331
bloomery hearths 332
blowing-pipe, in glass manufacture 355
Blue John:
  preparation and working of 230, 235
  properties of 229 f.
  sources of 228 ff.
Boccacio 387
Bocchus, see Cornelius Bocchus
bog iron 328, 331
boilers 349
boiling point, *see separate entries for metals*
Bologna 225
Bolos of Mende 61
bombril, refined gold, term for 298
book-roll, smoothing edges of 60
borax, as a flux 298 n. 100
bordered robes, purple in 309
bornite 179
botryitis (cf. βότρυς) 204, 312
Botticelli, *Birth of Venus* 387 n. 32
boussole, mariner's compass 370 n. 67
Boyle's Law 101

brass 310 ff.
  cast 313
  cementation method of
    production 312, 314
  modern 313
  physical properties of 313
  Roman coins series of 313
  vertical crucible used in
    manufacture 314
  wrought 313
bright lead (plumbum
  argentarium) 325
Brindisi (Brundisium), mirrors,
  composition of 287
Brinell hardness *see individual
  metals*
Britain:
  source of stibnite 245
  white marl used in 210
bronze 305 ff.
  antimony 309
  Brinell hardness of 306
  cold-worked 306
  bronze substrate, in gilding 292
Buffon, G. L. Leclerc, comte de
  111
building, Greek words used
  for 90
Bull Beef vein, fluorspar 230
buoyancy 167
Byzacium 13
Byzantine sources 381

Cadiz 67, 120
Cadiz (Gulf) 67
cadmea 203, 213, 301, 311, 312,
  338
Cadurci 21
Caesar, *see* Julius Caesar
calamine 204, 311
calaverite 275 n. 12
Calcidius 382
calcite, calcium carbonate 204 ff.

calcium bicarbonate, travertines
  (petrified 'waterfalls') of
  209 f.
calcium carbonate (calcite) 125,
  181, 204, 240
calcium fluoride 233
caldera, Santorini 239 n. 177
calendars 368 ff.
  and time-telling 360 n. 42
Caleti 21
Caligula, emperor 48, 68, 235,
  373
callaina (cf. κάλλαινος) 267
Callias (Kallias) 216 f.
Callimachus 64
Callippus of Cyzicus 369
Callistratus 60
Callixenus 48 n. 42, 167
calor 161
cambrics 10
Campania 203, 311
Campanian towns 3
  see also individual names
Campus Martius 364
Campus Raudius 159
Camulodunum (Colchester) 361
canalicium 208
canaliense 208
Canarii 49
cannibals (Anthropophagi) 65
Cape Sunium 179
Cappadocia 118, 119, 223,
  225, 374
  siri found in 13
carbon, used in carburization
  334
carbon/oxygen, isotopic ratios
  of 206 n. 58
carboniferous rock, fire-setting
  tests with 133
carbunculus (ruby) 17, 265 f.
carburization of iron 333, 336,
  337

General Index

carcer 363
career, see Plinius Secundus C.
careening ships' bottoms 257
Caria 154, 227
Carmania 218, 231, 267, 269
carnelian 265
Carteia 53, 68
Carthage 17, 53, 67, 151
Carthaginian electrum coins 184
Carthaginian language, words from 94
Caspian Gates 118
Cassandria 327
Cassiopeia 56 n. 56
Cassiterides 314, 316
cassiterite ($\kappa\alpha\sigma\sigma\iota\tau\epsilon\rho o s\acute{o}$) 183, 212, 315
Cassius Dio 131 n. 58
Cassius Hemina 46
Castor 289
Castor and Pollux 159
catapult 348
  see also 'machine gun' catapult
Catina (Sicily) 363
Cato the Censor (the Elder) 47, 47 n. 36, 60, 93, 371
Cato the Elder, on lime 60
Cato the Younger 31
Catullus, see Valerius Catullus
Caucasus 267
  antimony as native metal in 338
caulking, ships' timbers 18
Cave-dwellers (Trogodytae) 16, 266, 366
Cayster plain (near Ephesus) 217
Cayster (r.) 295
celestial bodies, motions of 362
Celsus, medical writer see Cornelius Celsus
Celtic language, words from 22, 93

cementation, gold refining by 126, 284 f.
cementite structure of iron 333
cenchros, possibly diamond described by Manilius 191
census figures, Spain 8
Cerasus (r.) 223
Cerbani 50
cerussa (psimithium) 259, 261
cerussite 201, 222
chalcanthon ($\chi\acute{\alpha}\lambda\kappa\alpha\nu\theta o\nu\ \tau\acute{o}$), flower of copper 122
chalcedony 265
chalcitis ($\chi\alpha\lambda\kappa\hat{\iota}\tau\iota s$), chalcopyrite 122, 179, 213, 301
chalcocite 213 n. 80, 304
chalcopyrite, see *chalcitis*
chalk 14, 128
Chalybean iron, difficulty of smelting 328
Chalybes 328
Champion, W. 314
Chandragupta, King 44 n. 12
'channelled' gold 208
chaplet copper 127, 272
charcoal fire 286
charcoal burning 265
Chares 61, 250
charge, of ore and charcoal 304
Chartres 384
Chatromitae 50
Chattii 5
Chauci 5
chemistry 115 ff.
  acids (solvents) 130 f.
  analytical techniques of 116
  chemical reactions 126 ff.
  chemical technology, of Theophrastus 176
  discriminatory tests 116, 134 ff.
  distillation 128 f.
  dyes and dyeing 138 ff.
  efflorescence 125 f., 198, 260

chemistry – (cont.):
  evaporation and precipitation 117 ff.
  filtration 121
  fire-setting 130 ff., 280, 280 n. 44
  hydrometallurgy 122, 314
  reliance on observation 136
  water, properties of 123 ff.
Cheng Ssu-hsiao 179 n. 19
Chersiphron 162
chert 214, 294
Chia terra candicans 214
Chilvers, I. 387 n. 32
China clay (?) 214
Chios, marble from 205, 207, 214
'chloran' 244
chromite 327
chrysitis (χρυσῖτις) 321
chrysocolla (modern mineral) 189, 215
chrysocolla (ancient mineral) χρυσόκολλα 189, 215
chrysolithus (χρυσόλιθος) 268
chrysotile, fibrous form of serpentine 196, 196 n. 16
χρυσοῦ ὄζος, knot of gold 344 n. 25
Cicero, *see* Tullius M.
Cicero, Greek words in *Letters* of 86
Cilbi 216
Cilbian plain 217
Cimbri 159
Cimolus 219
  earth of 248
cinnabar 59, 128, 129, 215, 217, 341, 343
  colourant, used as 215
  dressing of 217
  miniariae, mines 130
  polishing, industrial hazard in 130, 378
  refining and smelting of 218
  sources of 215 ff.
  see also industrial hazard (in polishing), pigments
Circeii 52
Circus Maximus 37, 226
cire perdue 274
civil engineering 168
civil service 2
classical and Christian ideals 387 n. 32
Claudius, emperor 2, 48, 56, 263
cleavage, in minerals 180
Cleopatra 130
clepsydra (water clock) 367, 367 n. 58, 368
Clitarchus 45
Cloaca Maxima 351
clock 348
clod sulphur (*glaeba*) 248
cloth of gold 57
coal 254
coal mines, methane gas in 376 n. 19
coal tar, distillation of 255
Cocanicus (Lake) 117
cochineal beetle (*kermes*) 139, 141, 216
coclea 169, 348
codex, Natural History of 388
Coghlan, H. H., experiments in smelting copper 304
coins, imperial Roman 309
cold-mercury gilding 292
Collenucio, *see* Pandolfo Collenuccio
Colluccio Salutati 387
'*colligae*' 198
color 184
colorimetry 298
Colossae (Phrygia) 209
colour, minerals of 178
Columella, *see* Junius Columella

Como Cathedral, statue of
    Pliny 24
concave mirrors 146
concerraneum (v.1. *concerronem, congerronem*)
condensers, in mercury
    production 343
condensing, zinc 338
conduits, for water supply
    168
cones (*tubuli*) 322
conrivatio 281
Constantine the Great,
    emperor 207
contubernalis, Pliny and
    Titus 7
convex mirrors 146
copper 301 ff.
    added to glass mix 353
    alloys of 305 ff.; *see also* brass,
        bronze
    antimony and silver 307
    arsenate 201
    bun ingots 305
    cadmea 203, 213, 301, 311,
        312, 338
    chalcanthon (flower of)
    chalcitis, native 305
    in early Bronze Age 306
    hydrometallurgy 122, 314
    Marian 301
    melting point of 304
    mining 22
    misy, *see* pyrites 196
    ores of *passim*
    physical properties of 301 ff.
    pyrites (*misy*) 126, 179, 196
    quenching preparatory to
        gilding 195
    Roman ingots of 305
    smelting, modern experiments
        in 303 f.
    sources 301

substrate 291
types of 307 ff.
coracles 315
coral (*Gorgonia*, or Gorgon's
    head) 269
Corbie, MSS at, in 9th cent.
    383
Corduba copper 301
Corinth 124
corn grinding 350
Cornelius Bocchus, on lead 57,
    60, 92, 220, 323
Cornelius Celsus A. 36, 48,
    48 n. 41, 61, 107
Cornelius Tacitus 3, 35, 94, 391
Cornelius Valerianus 68
Cornwall:
    buried alluvial tin ores
        316 f.
    tin slags 318
corrugi 85
corundum 181
Coryphas, oysters from 52
Cosimo de Medici 388 n. 35
cosmography 382
counterfeit gemstones 140
cranes 163
Crates (of Pergamum) 66, 67
Crawford Vase 233
creta Cimolia 219
creta Eretria, *see* pigments
Crete 119, 261
critical judgement, *see* Plinius
    Secundus C. 18
Croceae (near Gythion), marble
    quarries 205
crocodiles 16
    but see 16 n. 56 (alligators)
Croesus, refineries at Sardis
    284 f.
cross-cuts 169
crowns, given to actors 137
crucible, covered 303

crudaria 92
'cry' (tin) 273
cryptocrystalline constituents (limonite) 328 n. 197
crystal:
   mode of growth 123, 187, 188
   static electricity produced by 152
   systems 186; Pliny's understanding of 264; see also diamond, emerald, iris, quartz
crystallography 186 ff.
crystallum 220 f.
Ctesias 44, 65, 251
Ctesibius 162
   water-clocks 367 n. 59
cucumber seed, 'knot of gold', size of 192, 344
   see also platinum
cuniculus, see mine adit
cupellation, use of alum in 196
Curio 351
   revolving theatre 165 f., 353
cushions 18
cuttlefish 53
Cuvier, G. L. 111 f.
Cyclades 239
Cycladic early metallurgy 309
Cyclops 65
Cynics 33
Cyprian *analcime* 192
Cyprus 211, 220, 225
   amiantus, mines in Troodos Mts. 196
   copper 311
   and Perhaebia 211
   pyrites, best, found in 214
Cyrenaica 13 n. 49
Cyrene 13
Cyzicus 159

Dacia 140
Dalmatia 279

Damlatas (s. Turkey), caves 124 n. 32
damp course, temple of Diana 351
Dardanelles (Straits) 68
Davy Safety lamp 376 n. 19
day of equinox 366
Dead Sea:
   bitumen found in 255
   salinity of 117
death, no life after 28
deep lead (buried) placers, tin 316 f.
Defloratio Historiae Plinii Secundi 385
dehydrated gypsum (lime) 210
Delos 51, 307
Delphi 43 n. 11, 184
Democritus 60, 61, 351
Demostratus 60, 251, 263
densatio 121
density (specific gravity):
   definition of 182
   awareness of, by Greeks 183
desalination 121
diamantiferous sandstone 178
diamond (*adamas*) 190 ff.
   cleavage of 191
   crystal system of 191
   heat effect on 192
   six varieties of 192
   types of 191 f.
Diana, temple at Ephesus 69, 163, 351
di-electric crystals 153
Dihle, A. 45 n. 18
Diodorus Siculus 57, 58, 59, 182, 316, 317, 332
Diognetus 51
Dionysius Periegetes 43, 232 n. 151, 233, 245
diorite 227
Dioscorides 47, 136, 215, 246, 340, 381, 386

Dipileza 179
discriminatory tests 116, 134 ff.
distillation, of mercury 287, 343
Dog star (Sirius), rising of 138, 250
dogs, toxic fumes affect 376
dolomite 260
dolphin 20
Domitian, emperor 31, 32, 365
Domitius Corbulo 5
Domitius Piso 36, 78
Domus Aurea 4, 350
Don (r.) 202
Dracaena, Pterocarpus 216 n. 91
Drach, Cuevas del (Majorca) 124 n. 32
dragon's blood (*sanies draconis*) 216 n. 91
drainage-wheels 169
Dream of Scipio, Commentary on 382
dream sequence, ghost of Drusus 34
Druids 21
dry exhalation 175
ductility 271
duties (*negotia, officia*) 26
Duvius Avitus 7
dyes and dyeing 138
earring, pearl, of Cleopatra 130
Earth:
   crust, composition of 327
   curvature of 151
   despoliation of 372 f.
   flat disc 369
   measurement of circumference by Eratosthenes 369
   oblate spheroid 171
   rotation of 368
earthenware pipes 169
earthquakes, on Delos 51
Earth Sciences 173 ff.
echoes 159 f.
eco-warrior 379

edging ores (*stricturae*) 336
education, Greek 36 n. 1
efflorescence 125 f., 198, 260
   white gold in Lusitania 276
Egnatius Calvinus 53
egula 248
Egypt 119, 164, 239, 244, 366
Egyptian language, words from 94
Elba, iron from 332
Elbe (r.) 178, 332
electricity 152 ff.
   current and frictional 152
   static 152 f.
Electrides, islands 250
Electronprobe micro-analysis (EPMA) 295 n. 88
electrum:
   amber, electrostatic properties of 136, 250
   nativum, alloy 136, 139, 184
   coin series 194
   and silver interface, characterization of 205
elegantiae arbiter, Petronius as 374 n. 8
elements, four basic 103, 108
ἐλευθέραι ἐπιστῆμαι αἱ 104
embaenetarii trierum piscensium 91
embaeneticam facere 91
emerald 147, 181, 241, 242, 243, 263, 264
Emerald (Mt.) 244
emperor's council 7
emporitica (cf. ἐμποριτικά), brown paper 359
encaustic painting 90
ἐγκύκλιος παιδεία, encyclopaedia 36, 37, 74
engineering, conservatism in 350
environment, Pliny and 371
Ephesus 52, 69, 162, 216
Ephorus 43, 45
Epicurus 74

ἐπιστήμη (ἡ) 38
epistylia, architraves 69
Eporedia 22
eporedias (Celtic) 22
eques (Knight), *passim*
equestrian order (*ordo equester*) 1, 3
equinox, day of 366
equinoxes, precession of 370
Eratosthenes 44, 369
Eretrian earth 181, 185, 259
Eridanus (r.) 250
erythrodanum, dye 140
esparto 10
Etesian winds 242
Ethiopia 45, 158, 218
Etna (Mt.), summit of 151
Euboea, coloured marble from 205
Euctemon 369
Eudoxus (Cnidus) 45, 66, 73, 369
eunuch apple (*spadonia*) 18
Euripides 155, 250
Eurymenae 124, 209
eutectic temperature 320
pewter 320
EX ARG EX ARGENT inscriptions 324
Exekias 20 n. 67
exotic foods 29
eye make-up 340
'eyeballs', glass 356, 357
eye-salves 185
exotic foods 29

Fabius Dossenus 234
Fabius Pictor 47 n. 36
Fabius Quintilianus, M. 36
Fabius Vestalis 363
face-masks 378
factory inspectors 378
Facundus Novius, mathematician 364

Falernian wine 234
fantasies, *see mirabilia*
fayalite slag 335
ferro-electric barium titinate 154
ferrous sulphate (*atramentum*) 260
fervor 161
Field of Research Classifications 113 f.
fig-tree 68
fire-gilding, hot mercury 290
fire-resisting stone (πυρίμαχος λίθος) 335
fire-setting 130 ff., 280, 280 n. 44
First Contact (Attica) silver ore from 179
First edition (Venice), of *Natural History* 388 f.
First Punic War 45 n. 29, 362
First Triumvirate 1 n. 4
Five Ages of Mankind (Hesiod) 102
flame-throwers 255 n. 24
flat irons, water tasting of 19
flat tables and helicoidal washeries 169, 182
Flavians 7, 27
opposition to 32
flax and linen 197
fleeces:
absorption of sea-spray by 121
cushions for stuffing 18
Fleury, P. 79, 79 n. 2
flint 214
Floating Bodies (Archimedes) 167
flos (efflorescence) 126
flos salis 134
fluoride 232
fluorspar (*myrrha*) 177, 181, 230, 232
fluxes 335
foam, salt 118

General Index

formaceus 17
formation of stones 58
'formigas', of miners 278 n. 30
Fortuna 28
Fortune, statue and temple (Praeneste) 27, 288, 289
fossil pitch (maltha) 255
fossil resin 152
fossiles 175
foundation of Rome (AUC) 370
fractaria 85
freaks, see *mirabilia*
Fremington Hagg (Yorkshire) 137
French, R. 72 n. 1, 80 n. 4, 100, 105
French words 86
friabilis 191
'Friend(s) of the Earth', Greens 29, 378
Frisii 7
fritting, glass 354 n. 33
frontes 238
fuller's earth 213, 219
fumite bomb 249
furnace chimneys (Laurium) 300
fusibility, see *separate entries for metals*
fusion, glass

gabbro 227
Gades 54
Gadir 94
Gaetulian coast 138
gagates (jet) 181
Gaius Caesar 59
Gaius Caligula, emperor 373
Gaius Cassius 31
Gaius Epidius 68
Gaius Gracchus 57
Gaius Memmius 40 n. 11
Galen 381, 386, 386 n. 28
galena (lead sulphide) 181, 221

gall nuts, infusion of 136
Gallaecia, black lead *not* found in 323
galleries and shafts (mines) 375
Gallia:
    Aquitana (Armorica) 18
    Belgica 18, 19, 225
    Bracata 20
    Celtica (Lyonese) 18
    Comata 18, 18 n. 58
    Narbonensis 14, 18
    togata 18 n. 58
    Transpadana 22
Gallic harvester (*pecten*) 349
Gamphasantes 16
Ganga (r.) 278
Garamantes 15
Garden of the Hesperides 12
garimperos, miners in Brazil 278 n. 30
garnet 265, 268
Garonne (r.) 18
Gatwick airport 158 n. 31
Gaul and Germany, provinces of 121
gears 348, 362
Gellius, Aulus 46 n. 33
gems and precious stones 173, 263 ff.
Georgius Agricola, from Erzgebirge 387
Ger (r.) 49
German language, words from 94
Germanicus 5, 370 n. 66
Germany 5, 42, 57, 203, 301, 311
Gerra (Arabia) 120
Gilbert, William 152, 158
gilding 290 ff.
glaeba (sulphur) 247, 382
glaesum (glass) 252
glass:
    additions to melt 353, 354
    blowing 353, 355

444 General Index

glass – (*cont.*):
  colour change on cooling 355 n. 36
  coloured by malachite and azurite 354
  dark lumps of 355 n. 36
  discovery of 352
  'flexible' 357
  formers 354
  furnaces 353
  fusion of 357
  globules, magnification by 356
  jewellery 354
  manufacture of 221, 351
  marbles, as children's toys 357
  myrrhine ware, painted to resemble 230
  properties of 357
  technology 352 ff.
gnomon (γνώμων), of sundial 365
Goat-Pans 16, 64
Goat river (Aegospotami) 327
gold 274 ff.
  absolute purity of 299
  alluvial deposits 275 ff.
  alum used in purification of 196
  amalgamation 286 f.
  analysis (chemical and non-destructive), 295, 295 n. 88
  beater 289
  cementation 283
  coating 287
  colorimetry 298
  dust from Pactolus (r.) 285
  gilding 290 ff.
  imagery, in mining 9
  knot of (χρυσοῦ ὄζος and *auri nodus*) 344
  leaf 288, 290, 291, 292
  malleability of 288
  mines and mining 277 ff., 283

  mining: techniques 277 ff.; terms 91 ff.
  nugget of 92
  numismatic evidence for refining 284
  physical properties of 288 f.
  Praenestian leaf 288
  processing 58
  production in Asturia, Gallaecia, Lusitania 283
  prospectors 279
  purified 284
  Quaestorian leaf 289
  refiners and refining 284, 285 ff.
  refining: Egyptian 284; Pliny's description of 286; Strabo's account of 285 f.
  sources 277
  tellurides 275
  trace elements (PGE) 277, 283 ff.
  Trial of 1710 (Sir Isaac Newton) 298
  washeries (χρυσιοπλύσια) 276
  'white' (λευκὸς χρυσός) 284
gold–bronze interface 292
Golden Fleece 276
Golden Horses of St Mark's Basilica, Venice 291
Golden House (*Domus Aurea*) 4, 237
Goodyear, F. 98
Gorgonia 269
Gorgon's head 56 n. 56
Goriano (Mt.) 266 n. 14
gorse 263
gossan 223
Gough's caves, Cheddar Gorge (Avon) 124 n. 32
Graecinus, Julius 69 n. 23
graecitas 391
Graecostasis 363

General Index

Grande Corniche (French Riviera) 51 n. 45
granite (syenite) 205
gravitas, weight/density 183
gravity 179
  'feed' boiler 304
  flow 168
  separation 169; of cinnabar 217; of silver 169;
  see also tin
greed (*avaritia*) 371
Greek:
  and Latin words 10, 91 ff.
  medicine, scathing criticism of 390 n. 39
  mining terms 91 ff.
  orators 58
  science, terms for 100
  theatre, acoustics of 159
  words 86
  words Latinized 90
Greeks, conceit and imagination of 270
Greene, Mott T. 103 nn. 16–17
greenhouse effect 371
'Greens' 372
griffins 278
groove cam 348
gum 359
gypsum 60, 181, 211

Hades, entrance to 246
Hadrian, emperor 92
haematite 179, 181, 222
Hagia Sophia 207 n. 61
Halicarnassus 208
Halleux R. 92
Halley's comet 258 n. 34
Hannibal, crossing of the Alps 131
  watchtowers of 360
Hanno 11, 45

hardness (metals), Brinell (B), Vickers (HV) 273
Harley Manuscript, in British Library 24
Harmodius and Aristogeiton, the *Tyrannicides* 47 n. 38
Harvey, Sir William 386 n. 28
Haüy, René-Just, physicist and mineralogist 186 n. 55
Health and Safety Executive 378
Heathcote, J. M., BDS 232 n. 148
Hebrus (r.) 278
Hecataeus 43 n. 9
helicoidal washeries 170, 182
heliodynamic theme 384
Heliopolis 365
helix, mechanical 348
hellenistic period, widening of research in 73
hellenistic silver plate 137
Helvidius Priscus 31
Heraclean stone 155, 224, 296
Herculaneum 3, 350
Hercules 12
Hercules' temple (Tyre) 244
Hermes Psychopompus 137 n. 1
Hermolaos Barbarus 389
Hermus (r.) 275, 295
Hero 160, 348
Herodotus 11, 16, 43, 183, 275, 326
Hesiod 43 n. 9, 76, 102, 104, 170
Hesperides, Garden of 12
hexagonal system, emarald 187, 202
Hierapolis (Pamukkale) 209
hieroglyphs, on Egyptian obelisks 37
Hiero's crown 183 f.
Hippias of Elis 36, 104
Hippo Diarrhytus (Bizerte) 15 n. 55

hippocentaur, preserved in
    honey 56, 67
Hippocrates 386
Hispania Tarraconensis 7 ff., 22
historia (ἱστορία) 38, 83
ἱστορίαι, αἱ τῶν περὶ τὰς πράξεις
    γραφόντων 38 n. 2
Hither Spain 10
hollow balls, used in filtration
    121
Holmes G., 383
Homer 43, 63, 104, 149, 216,
    288, 331
  knowledge of iron 326
Homeric Hymn 337 n. 220
Homeritae 50
'Homes of the Muses' 238
homocentric spheres 45 n. 21
homoeomeria (ὁμοιομερεία) 82
hookworm infestation 377
Horace, see Horatius Flaccus
Horatius Flaccus, Q. 90, 374, 376
horizon, from observer 171
horologion, of Andronicus
    Cyrrhestes 367
horse-trapping (*phalera*),
    metallographic analysis of
    138
hours of daylight, at Meroe,
    Alexandria, Britain 361
houses built of salt, in Babylon,
    Utica, Crete 15
human blood and marrow,
    use of 30
hybrid language (*lingua franca*),
    of miners 10
Hyderabad 190
hydrargyrus 129
hydraulicking 121
hydrocarbons 248, 254 ff.
hydrochloric acid 297
hydrogen sulphide 376 n. 18
hydrology 167 ff.
hydrometallurgy 122, 314

hydrostatics 166, 167
Hymettus (Mt.) marble quarries
    205
Hypasis (r.) 366
hypocaust (ὑποκαύστος ὁ) 160
ὑποκείμενον τό underlying
    substance 103
hysgine (cf. ὕσγινον) 139

iaspis (ἴασπις) 265
Iberia, tin found in 316
ibis 53
Ibycus 337 n. 220
Iconium (Asia Minor) 375
iconography 347
Ictis, St Michael's Mt., 314 n. 142
Ida (Mt.) 228
idochromatic minerals 179
Ilissus (r.) 68 n. 21
imagery 12, 174, 281 f.,
indestructibility of matter,
    Law of 75 n. 11
India 43, 44, 154, 190, 202, 220,
    227, 268
Indian:
  diamond 187
  glass 353
  language, words from 94
indicators, of presence of water
    168
indignation (*indignatio*) 98
indigo 139
Indus and Ganga (r.) 43 n. 12
industrial:
  diamonds 192
  hazards 378
  operations (*cura*) 86
influences on Pliny's style, see
    Cato the Elder, Columella,
    the Younger Seneca, Varro,
    Virgil
ingots 184 see also copper, lead,
    silver, tin
insulae, tenement blocks 4

General Index

intelligible principle (Stoic) 109
intercalation 370
Invalides, Les 207 n. 60
Ion 155
Ionian school, earliest
    philosophers 102 ff.
iris ($ἶρις$) rock crystal 269
iron 326 ff.
    acies (steel) 337
    amorphous hydroxide 223
    bloom 330, 331 f.
    bloomeries 331 f.
    by-product of gold-refining
        329
    carbide 333
    carbonate ores 331
    carburization 334 ff.
    cementation, produced by 329
    chromite 327
    Egypt, objects found in 329
    filings 157
    hardness of 333
    melting point of 330
    meteoric and meteorites
        257 ff., 327
    ores 328; iron content of 330;
        oxide 328; see also bog
        iron, haematite, limonite,
        magnetite, siderite,
        sphaerosiderite, spathic
    preliminary roasting of 330 f.
    properties of 273
    pyrites association with 224
    reduction 329 f.; difficulties in
        328
    refining 335
    rejection of 158; see also
        magnetism
    roasting 330 f.
    shaft furnaces 332 f.
    slag 330
    smelting, accounts by
        [Aristotle], Diodorus
        Siculus, Strabo 337
    steel, see acies
    telluric 327
    treatments of (thermal and
        mechanical) 333 ff.
    wrought ($εἰργασμένος$ $σίδηρος$)
        335
ironstone (siderites) 156, 192, 228
Isidorus (Hispalensis), Bishop of
    Seville
Isigonus and Nymphodorus
    63, 65
Isis, temple of 16
isostructural compounds
    (minerals) 182
isotopic ratios (marble) 206,
    206 nn. 57–8
Isselmeer 7
Istria 52
Italian Alps 349
Italian Confederacy 1 n. 3
Iulianus, unknown Roman
    knight 253
ivory (or shell), paper smoothed
    by 358

Jaffa 68
Jason 276
jasper (iaspis) 265
jet propulsion 160
jewellery, Late Roman glass 137
Jewish, flat earth theory 383
Jhelum (r.) 44 n. 15
John of Salisbury 384
Joppa (Jaffa) 56
Juba II, King of Mauretania,
    11, 44, 59, 60, 78, 218, 220,
    237, 244, 265
Julius Caesar C. 21, 53, 317
Julius Graecinus 69 n. 23
Junius Moderatus Columella 2,
    93, 107
Junius Rusticus 31, 32 n. 7
Jupiter, emerald obelisk in
    temple 241

Jurassic limestone, fire-setting tests with 133
Justinian, emperor 207
Juvenal 4, 288, 371

κατακεκαυμένα τa 176
κέγκρος ὁ, diamond 191
keen eyesight 151
Kemel Ataturk 207 n. 61
Kennedy, J. A. 352 n. 1
kermes (coccum) cochineal 139, 141, 216
kinetic energy 161
κλέψυδρα ἡ, see clepsydra
Knaresborouh (N. York.), petrifying waters at 124 n. 32
knuckle bones, tin in form of 318
Kosmos (κόσμος ὁ) 82
κρύσταλλος ὁ
see also crystallus 220
La Tène 336
La Turbie 51
labyrinth (Egyptian) 69, 207, 245, 351
Ladik (Anatolia) 129, 343
Laenas 367
Lake Nemi 293
lamprey 374
Landels, J. G. 349
language and style 79 ff.
lapis lazuli 244, 267
lapis Lydius 224
lapis specularis (selenite) 17, 119, 177, 180, 224, 237
Larcius Licinus 8, 53
Latera 20
latifundia 19
latitude, daylight varies with 361
Laurentine Library MSS 388 n. 35
Laurentine villa 26

laurex 92
Laurium 222, 376
  litharge found at 322
Laws of Physics 170
lead 57, 201, 272, 320 ff.
  acetate (cerussa) 261
  allotropic forms 321
  argentiferous galena 221, 300, 322
  'bright lead' (plumbum argentarium) 325
  desilvering of 324
  EX ARG EX ARGENT 324
  liquation 248
  litharge 126 f., 320, 321
  mining 324
  ores of, see argentiferous galena, cerussite
  orthorhombic 321
  Pattinson process 323 f.
  physical properties of passim
  refining 321 f.
  sources 324
  tubuli, cones of litharge 322
  two metallurgies of 322 n. 173, 323
  uses of 324, 325
leisure (otium) 24, 25 n. 77
  Greek ideal of 26
Leoniceno 80, 389
Leptis Magna 13
Lesbos 74 n. 8
Les Invalides 207 n. 60
Letters, Cicero, Greek words in 86
leucochrysus (λευκόχρυσος) 268
Leucogaei (Campania), hills of 247
λευκότατον 310
levers 164 f.
Lex Vatinia 1
Lex Vipascensis 92
liberal arts 77 f.

General Index

Licinius Mucianus 10, 51, 52, 59, 64, 68, 69, 240
lifting devices 163 f.
lignite 266, 331
Liguria 250, 254
Lilybaeum (Cape Boeo) 67, 151
lime (calcium hydroxide) 210
   slaked 210, 382
limestone 204, 240, 335
   schist, effect of acetic acid on 132
λίμνη ἀσφαλτίτης, Dead Sea 255 n. 27
limonite 330
Lipari Islands 246
literary 'subset', technical vocabulary constitutes 83
lithium sulphate monohydrate 154
lithostrata 90
lithotripsy 133
Littré, E. 111
Livia, wife of the emperor Augustus 221
Livy 15, 131, 388
Lollia Paulina, jewellery of 4, 374
longevity 66
longitude 360
Lorches 383
Lower Silurian 218
Lucan, *see* Annaeus Lucanus
Lucilius 371
Lucius Aelius 234
Lucius Ateius Captio 234
Lucius Lucullus 52
Lucius Papirius Cursor, sundial set up by 363
Lucius Piso 46, 110
Lucius Verus, emperor 386 n. 28
Lucretius Carus, T. 28, 38, 58, 74 ff., 94, 147 f., 155, 170, 177, 274, 288, 373 f., 377

Lucretius, manuscripts of 388 n. 35
Lucullus 255
ludibria 65 n. 14
Luna, marble from 206
Lusitania 276, 315, 316
   Ammaeensian Mts. 220
Luxor 288
luxury (*luxuria*) 29, 371, 372
lychnis (tourmaline) 226, 265
lychnites (λυχνίτης) 59, 206, 207, 227
Lycia 52, 68
Lycus (r.), modern Aksur
lyddite 214
Lydia 240
Lydian stone (*lapis Lydia*) 224, 296, 298
lyncurium 251

Macedonia 124
machina 85, 162
machinalis scientia 162
'machine gun' catapult 348
Machlyes (μαχλός), bi-sexual 66
Macrobius 382
Madder 140
Maenian column 362
mafic minerals 294
magalia 15
Magi 196
Maglev 158
Magnes 156, 328
Magnesia 155, 228
magnesium carbonate 260
magnetism 69, 155 ff., 185
magnetic properties, metals 274
magnetite, ferrimagnet 69, 185
   in gravels in Nubia 329
magydaris 140
Mahandi (r.) 190
Mahon 133 n. 61
malachite 179, 215, 244, 260, 303

μαλάττεται, iron grows soft 335
maltha, fossil pitch 255
man, first in the order of
    Creation 65, 108
Manchester airport, second-
    runway project 379 n. 27
Manilius 190, 191
Manius Valerius Messala 363
mantichora 67
Mantua 388
manuscripts (ninth century) 383 f.
mapalia 15
marble (calcite) 204 ff.
    carbon/oxygen (isotopic ratios)
        206
    Cycladic sculptures 204
    geochemical data 206
    identification of 205
    minor and trace elements in
        205 f.
    petrology of 205
    Roman sources 205
    Spilia, quarries at 204
    trace elements in 206
    varieties, of: Aphrodisias 206;
        Chian 205; Croceae
        (Gythion) 205; Docimean
        (Phrygian) 206; Doliana
        (Tegea) 205; Ephesian
        206; Hymettan 205;
        Lacedaemonian 204 f.;
        Luna 206; Numidian
        208; Naxos 204; Parian
        (Mt. Marpessa) 204;
        Pentelic 204 f.; porphry
        205; Proconnesian
        (Marmara) 206, 207;
        Synnadic 208 n. 62;
        Taenerian 208
        n. 62; Thasian 208 n. 62;
        Thebes (Egypt), possibly
        mottled red granite
        (syennite) 205; Tiberian
        208 n. 62

veneer 208
Marbod (of Rennes) 381
marcasite 213
Marcus Agrippa 42, 42 n. 4
Marcus Aurelius 386 n. 28
Marcus Brutus 31
Marcus Crassus, killed at
    Carrhae 327
Marcus Porcius Cato,
    see Cato the Elder
Marcus Scaurus 56, 67
Maréchal, J. R. 306
mariner's compass (boussole)
    370 n. 67
marl:
    red (*agaunum*) 22
    white 210, 211
marmor, *see* marble (calcite)
Maronean wine 52
Marriage of Philology and
    Mercury 382
Martial, *see* Valerius Martialis M.
Martianus Capella 382
Massalia 361
mass people-transporter 158
mast, ship's 172
matches 249, 249 n. 229
matrix, quartz 279, 283
Mattiacum, hot springs of 124
Mauretania 11, 48
Mauryan Empire (India) 44 n.
    12
Mausoleum of Augustus 365
Mausolus, king of Caria, palace
    of 208
Maxwell-Stuart, P. 7, 8
μηχανή ἡ 162
μηχάνημα τό 162
mechanical clocks 362
mechanical deformation,
    of metals 272
mechanical properties, *see entries
    for individual metals*
mechanical reaper (*pecten*) 19, 85

General Index

mechanics 161 f.
Medea 255
Median stones 179, 243
medicamentum (mordant) 141, 194
medieval sources 381
Megasthenes 44, 65
Meleager 251, 252
melinum, white marl 259, 260
Melos 247
melting and evaporation 160, 161
melting point, see *individual metals*
Menander 60
Menderes Massif 294, 295 n. 89
Mendips 254
Meninx (North Africa) 138
mercury (*hydrargyrus*) 128, 129, 215, 291, 340 ff.
   argentum vivum 341
   Callias 216 f.
   distillation of 343
   industrial hazard, in polishing 378, 378 n. 26
   miniariae, mines 342
   ores of 215 ff., 341;
     see *also* cinnabar
   physical properties of 340 f.
   prills 219
   processing of 342 f.
   separation of 128
   Sisapo 341 f.
   sources of, see cinnabar
   Theophrastus' account of 341
Meroe (island) 344
metallurgical processes 347
metals, see antimony, copper, gold, iron, lead, mercury, platinum, silver, tin, zinc
   mechanical properties of, see ductility, fusibility, hardness (Brinell and Vickers), melting and boiling points
   metamorphic rocks, associations with 227

meteorites 257 ff.
methane gas 376 n. 19
Meton 369
Metrodorus 60, 61
mica, see selenite
Middle Ages:
   the *Natural History* in 380 ff.
   Plinian astronomy in 384
Mieza (Macedonia) 124
Milesian Monists 103
militia equestris 2, 5
miltos ($\mu\acute{\iota}\lambda\tau o \varsigma \ \acute{\eta}$), cinnabar used as a pigment 216
Minaei 50
mine adit (*cuniculus*) 11
minerals 190 ff.
   see *also separate entries*
minerals characteristics of 178 ff.
   allochromatic 179
   Aristotle, theory of formation of 175 f.
   atomic structure of 174
   cleavage 180
   colour 178 f.
   crystal systems 186 ff.
   crystallography 186
   density (SG) 182 ff.
   hardness (*duritia*), Mohs scale 180 f.
   idiochromatic 179
   Lucretius, Posidonius, Seneca, on formation of 178
   magnetism 185 f.
   mirabilia 69
   nomenclature of 87 f., 188 f.
   origin, early theories of 174 ff.
   piezoelectric and pyroelectric properties 185
   regeneration of 177 f.
   streak characteristics of 184 f.
   tenacity 181
   Theophrastus' theory of formation of 176

miners' disease, hookworm 377
miners, suffocation of 374
miniaturization 149
mining, *see individual metals*
minium 201
mirabilia 11, 61, 63 ff.
  animal kingdom from 67 f.
  births, multiple (superfetation), premature, stillborn, unusual 67
    in Ethiopia 64
    freaks 65 f.
  hellenistic interest in 63
  in Homer 63, 63 nn. 3–4
  in human behaviour 65
  inanimate objects
    in India 64
  legendary creatures 63, 63 nn. 3–4
  man, height, keen sight, longevity 66 f.
  in mineralogy 69
  paradoxographers 64 n. 8
  Plutarch, disapproval of interest in 70 seen by Mucianus 64 n. 10
  structures, man-made 69
  'sub-science' represented by 63
  synonyms, *miracula, prodigia, portenta* 63
  trees 68 f.
  unusual physical attributes 66; powers possessed by certain races 67
  X-rated 70
  in zoology 65 f.
mirabilis, mirum 63
mirrors, *see* concave, convex, parabolic
Misenum 22
mistletoe 21
misy (copper pyrites) 213

Mithraeum 350
Mithridates, defeat by Pompey 228
Mithridates Eupator 251
Mohs scale 181, 217, 232
mollitia 181
molybdaena (μολύβδαινα) 322
molybditis (μολυβδῖτις) 321
Monists, Milesian natural scientists 103, 104
Monocoli 66
monolith, sundial 364
monstra 67
Montebras (Creuse), tin sources 317
moon 119, 119 n. 19
moonstone 268, 269
mordant (*medicamentum*) 141, 194, 195
Moselle (r.) 350
Mossynoeci 310
motorway extensions 379 n. 27
moufflon 10
Mucianus, *see* Licinius Mucianus
μύδρος διάπυρος, red hot mass of stone 174
mullet 20
multinational workforce, in mines of Hispania Tarraconensis 91
munus debitum 4
Münzer, F., 7
murales machinae 162
Murano (Venice) 355 n. 36
murex 141
Myanmar (Burma) 252 n. 13
Mycenean vases, sheet tin applied to 319
Myrmecides, miniature model-maker 149
myrrha (μύρρα, ἡ) 228 ff.
  see also Blue John, fluorspar

myrrhine ladle 232
Mytilene, electrum coins of 184

Nabataeans 267 n. 20
napalm 255 n. 24
naptha 248, 255
Narbonne (Gallia Narbonensis) 20, 21
Naturae historiae 33
Natural History 36 ff.
  readership of 40
  scope of 38
  sources for 42 ff.:
  archives 57; *commentarii* 53 f.; earth sciences, minerals and pigments, authorities on 58; encyclopaedists 46 f.; explorers 54; historians 46; local authorities 55; knights 54 f.; monographs 47 f.; military commanders 48 ff.; mining and metallurgy 57 f.; objects and creatures brought to Rome 56 f.; personal observation 56; precious stones, authorities on 60 f.; provincial governors 51 ff; special commissions 54; specific minerals 59 f.; Theophrastus, Vitruvius, Celsus 61; topography, authorities on 42 ff.; unnamed 55 f.
  style of 97 f.
natural sciences, beginnings of 102 ff.
natural scientists 103, 177
naturalis 39
Nature 27, 28, 40
  degradation of 112
nauplius, cuttlefish 68

navigation 370 n. 67
Naxos, marble quarries 205
Neapolis (Campania) 247
Near Eastern civilizations 134
Nearchus 44
Necron, island (Red Sea) 220
Nemausus 20
Nemi (Lake) 293
neologisms 95
Nepos 45, 45 n. 28
Nereids and Tritons, seen near Lisbon 67
Nero, emperor 3, 4, 6, 32, 237, 242, 279, 289, 350
Nero Claudius Drusus 34
Neronian conspiracy 6
Neutron Activation Analysis 295 n. 88
Newton, Sir Isaac 298
Nicaea 51
Nicander 73, 228, 250
niccolite 179
Nicias 98
Nicolo de Niccoli 387
niello 137 f.
  horse-trapping inlaid with 137 f.
  spectrographic analysis of 137
nila, sanskrit for dark blue 268
Nile (r.) 268
Nile water 359
Nilios 267
Nilotic paintings 66 n. 16
nitric acid 128
nitrum 199
  see also *aphronitrum*
Nomads 15, 50
non-destructive analysis 295 n. 88
  see also Atomic Absorption (AA), colorimetry, Electron-probe

non-destructive analysis –
  (*cont.*):
    Microanalysis (EPMA),
    Proton Induced X-Ray
      Emission spectroscopy,
    X-ray Photoelectron
      Spectroscopy (XPS),
    X-ray Fluorescence (XRF)
Norden, E., Pliny's style 80
Noricum, iron at 328
North Africa 11, 42
northern Afghanistan 267
nouns formed from passive
  participle 96
Nova 369
Nova Carthago 10
Novum Comum 1
Nubia 329
nucleus ferri 85, 336
nuggets (gold) 283
Numa, King 110
Numidia 15
Nymphodorus 63 f.

obelisk 163, 351
obrussa (cf. χρύσιον ὄβρυζον) 92
'obsian' 140
oceans, salinity of 117
ochre 175, 223, 259
ochre mines (Cappadocia) 200,
  374
octahedral system, diamond 187
octopus 68
  see also polypus
odeon, acoustical properties of
  159
Odysseus 63
officia 26
oil, boiling 249
Olisipo (Lisbon) 54
olivenite 201
Onesicritus 4, 366
ὀνυχίνη λιθεία καὶ μουρρίνη 233

onyx marble 265, 354
oolitic limonite 223, 328
opal (*opalus*) 264, 354
Ophiogenes 67
Oplontis 3
ordo equester, *see* equestrian order
ore, scraping from 296
ores, *see entries for individual
  metals*
ore-washing 169 f.
ὀρείχαλκος, 'brass' 337
Orga 14
organ 348
organic minerals:
  bitumens, waxes, resins 254
  oil, natural gas 254
Oromenus 118
orpiment (trisulphide of arsenic)
  200, 235, 260, 310
orthoclase 181
Orthosia 154, 274
ὀρύσσειν 91
Ostia 15 n. 55
  Antium, Misenum 370 n. 67
Othrys (Mt.) 201
otium 24, 25 n. 77
Ovidius Naso. P. 133
Oxyrhynchus papyri 162 n. 44
oyster:
  British 52
  purple 141

Pactolus (r.) 224, 275, 278
Padus (r.) 278
padus, pine-tree 22
paedaros 264
Paetus Thrasea 31
pagan legend 387 n. 32
πάλα ἡ, Spanish and Greek
  mining terms *palacra*,
  palacrana, *palacurna*,
  palaga 92
Palatine Hill 4

General Index

Paliputra (India) 44 n. 12
Palladius 386
Panaetius 26 n. 80, 370
Pandolfo Collenuccio, Ferrarese lawyer 389
Pangaeus (Mt.) 277
panning 276
Pano Amiandos (near Troodos Mts., Cyprus) 197
paper 358 ff.
  ancient manufacture of 358, 359
  types of: Augustan (hieratic) 359; *emporitica* (brown paper) 359; 'woven' 358
Papirius Fabianus 58, 69, 107, 110, 177, 205
papyrus 154, 358
paradoxographers 64 n. 8
paraetonium 259, 261
Paroikia (Paros) 207 n. 60
Paros, marble quarries 204, 205, 206
Parthia 231, 255
passernices 19
paste, used in paper manufacture 359
Patalene 268
patrii sermonis egestas 81
Pattinson process, lead refining 323
pax romana 27, 323
pearl, dissolving of 130
pearls 29, 280
  see also unio
pecten, part of harvesting device 19, 349
Pedersen, O. 369 n. 6, 380
pegmatites 212
Peleus, son of 331
Peloponnese, topography of 360
Pelusium 120
pendent rings 155, 156

pensilis balnearum usus (hot-air baths) 160 n. 34
πεπυρωμένα τά 176
περὶ φύσεως ἱστορία ἡ 38
peridot 60
periodite 227
periphrases, philosophical terms for 94 f.
Perperena (Mysia) 124, 209
Perseus, King 159
Persian language, words from 94
Petra 267
Petrarch 387, 387 n. 32, 388
petrification (petrifaction) 124, 240
petrified trees 209
petroleum 285
petrology 69
Petronius 232
petroti 356
pewter, alloy of tin and lead 320
PGE (platinum group elements) 277, 284
  high melting points of 344 f.
  insolubility of 343 f.
Phaethon 250
phalera, see horse-trapping
'pharantis' 268
Pharos (Alexandria) 351
Phausia 124
phengites (φεγγίτης, ὁ) 236
phialai, silver 137
Philemon 60, 61
Philemon Holland 391
Philip II (Macedon), stater of 343
Philippi, gold mines 192
Philippus 364
Philon (of Byzantium) 58, 267, 348
Philonides 360
φιλοσοφία *passim*
Philoxenus 250

Phoenician language, words
    from 94
Phoenicians 315
Phoenix 68, 167, 391
Phokaia, electrum coins of 184
phosphorus:
    in Lake iron 328
    in wrought iron 337
Photon Activation Analysis 295
    n. 88
phreatomagnetic explosion 103
    n. 16
Phrygia 118
phyllites 194
φυσικά τά 142
Physics 142 ff.
    acoustics 159 ff.: echo,
        phenomenon of 159;
        in Greek and Roman
        theatre 159
    earth's curvature 150 f.
    electricity 152 ff.: piezo- 154;
        Plutarch, phenomenon
        related to hot exhalations
        by 153; pyro- 153; static
        152 f.; St Elmo's Fire
        152 f.
    gravity 170
    heat 160 f.: friction (*calor,
        fervor*) 161; friction,
        sawing by 161; Hero's
        device 160; hypocaust
        160; jet propulsion 160;
        steam 160
    hydrology 167 ff.: aqueducts
        168 f.; gravity separation,
        used in 169 f.; mines in
        169
    hydrostatics, water
        resources, management
        of 167 f.: Archimedes
        166 f.; Hiero's crown
        167
    magnetism 155 ff.: Lucretius
        and Pliny on 156 f;
        excitation of iron filings by
        157; Maglev 158, 158 n.
        31; polarity 155, 157 n. 28
    mechanics 161 ff.: levers
        164 f.; lifting devices
        163 f.; *machina*, Vitruvius'
        definition 162, 162 n. 43;
        machinalis scientia 162;
        machinis aedificationum
        162; *murales machinae*
        162; revolving theatre, of
        Curio 165 f.
    optics 142 ff.: distorted
        images 145 n. 6; emeralds
        concentrate the vision
        147; horizon, from
        observer 171 f.;
        kaleidoscope 148;
        magnification 149 f.;
        miniaturized exhibits, at
        St Ouen 149; mirrors
        143, 146, 146 n. 8;
        multiple images 145;
        optometrists 143 n. 3;
        parabolic mirrors 146,
        see also mirrors *passim*;
        periscope 146 n. 9; prism
        148; reflection, Lucretius
        on 142 ff.; Pliny's account
        of 144 f.; refraction:
        definition of 147;
        rainbow stone 148;
        submerged parts of ships
        appear bent 147 f.; reversal
        of images in mirrors 145;
φύσις ή 38, *passim*
pigments 259 ff.
    see also armenium (azurite)
        atramentum, creta
        Eretria, melinum,
        orpiment, psimithium

General Index

(*cerussa*), sandaraca (realgar), sandyx
Pindar 98, 294
Pinus succinifera 253
piscina 85
Pisonian Conspiracy 6
πίσσα ἡ 256 n. 28
pissasphalton 256
pitch (*pix*) 248, 257
Pitch Lake (San Fernando, Jamaica) 254 n. 21
pivot, revolving 165
placers 275 ff.
    exploited by the Bessi in Macedonia 277
placitis (πλακῖτις) 204, 312
plaited reed (sand)bags 163
plane-tree 51, 68
planets 369
    motions of earth, moon, sun 368
plaster of Paris 211, 212
Plataea 288
platinum 192, 343 f.
    native 343; Egyptians, used by 343; gold stater of Philip II in 343;
    see also knot of gold
Plato 72, 153, 175, 274, 383
plaumoratum 94
Plautus 234
Pleiads, rising of 351
PLINIO PRAEF EQ 5, 138
Plinius Caecilius Secundus C. (the Younger Pliny) 3, 6, 7, 8, 10, 24, 107, 391
Plinius Secundus C. (Pliny the Elder) *passim*
    attitude to research 71
    career, chronology of 3 ff.
    character 23 ff.
    critical judgement exercised by 62, 109 ff.

death, manner of 23
duovir 7
early years 1 ff.
equestrian order, member of 1
eyewitness 9, 10, 13, 17, 20, 56
Flavians, connections 7, 27
'Friend(s) of the Earth', Green(s) 29, 378
literary influences on 81
literary movements, interest in 4
militia equestris 5 ff.
Misenum 22
praefectus: alae 2; *classis Misenensis* 2 n. 10, 22; *vigilum* 2, 22
procuratorship 2, 5 n. 17. 7 ff.; Africa, Hispania Tarraconensis 81
religion, attitude to 27
research 71 ff.
rhetoric, influence of 97 f.
Rome, time spent in 3 f.
status (*eques*), preoccupation with 2 f.
Vesuvius, death in the eruption of 23
Pliny Translation Group 85, 325 n. 181
Plistonices, the Cantankerous One 245
plumbum:
    album 317
    argentarium 325
Plutarch 70, 153
Plymouth, possible source of tin 314 n. 145
πνεῦμα νοερὸν καὶ πυρῶδες οὐκ ἔχον μόρφην 109
pneumatic devices 348
Po (r.) 250
polarity 155

pollution, atmospheric at
    Laurium 301
Polybius 45
Polyphemus 63, 63 n. 3
polypus (octopus) 53, 55
Pompeii 3, 66, 350
    and Herculaneum contrasted 3
Pompeius Flaccus 55
Pompeius Magnus, Cn. 228.
Pompeius Paulinus 5
Pompeius Strabo 1
Pompey, see Pompeius
    Magnus Cn.
Pomponianus, son of Pomponius
    Secundus 23
Pomponius Mela 107
Pomponius Secundus 4, 6
Pont du Gard (Nîmes) 168
Poppaea, incense burned at
    funeral 374
porphyry-copper 213
portable sundial (*vasum
    horoscopicum*) 366
portico, known as the
    "Seven Voices" 160
Porticus Vipsania 42 n. 4
Portugal 60
Port Vendres 318
Poseidon 56 n. 56, 104
Posidonius 58, 59, 177
potassium nitrate 134, 199
Praetorian Guard 2
Praxibulus 216
Precambrian, or Palaeozoic,
    massif 294
precession of the equinoxes 370
precious stones (gems),
    see separate entries
President's House 69
Presocratics 174
primordia rerum 94
Prison, The 362
process (discriminatory) tests 116

procuratorships 7 ff.
    see also Africa, Hispania
        Tarraconensis,
        Gallia Belgica, Gallia
        Narbonensis
Prometheus, myth of 348
Propertius 231, 233
prostitute, *see* 'The Clepsydra'
Proton Induced X-ray Emission
    spectroscopy 295 n. 88
Psammeticus II 37 n. 3
Psemetnepserphreus, king
    37 n. 3
ψευδάργυρος ὁ (zinc) 337
psimithium, *see* pigments
Psylli 67
Ptolemais 366
Ptolemy, King 48, 158, 267
publicani 200
Publius Pomponius Secundus 81
Puerto de Santa Maria 120
pumex 238
pumice 60, 238
    medicinal compounds 238
    shutes, at Skala Fira for
        loading ships with 239
    on Santorini 238 n. 177
    sources: Aeolian Islands,
        Melos, Nisyros 238
    used in Suez canal, 239
    uses of 238
    see also pumex
Pump Room (Bath) 19
pure water (ὕδωρ καθαρόν) 177
purple-fish 280
purpurae insania 29
Puteoli, lead processed at 321
Pygmies 66
πυκνότατος 184
Pyramids 69, 351
πυρίμαχος λίθος 335
pyrope (πυρώπης) 265
pyroclastic rocks 238

Pyrrhus, king of Epirus 45 n. 29
Pythagoras, officer of Ptolemy 220
Pythagoreans 383
Pytheas of Massalia, Greek explorer 361

quadrivium 37, 384 n. 20
quarantine-station, Mao estuary island (Minorca) 133 n. 61
quarrying 375
quartz (κρύσταλλος ὁ) 181, 187, 203
quenching (metals) 333
quicksilver:
  artificial 291, 342
  mercury 129, 291, 341
Quintilian, *see* Fabius Quintilianus
Quirinus, temple of 363
Quirites 392

rabbit 10
Raetian language, words from 94
rainbow stone (*iris*) 148
Rameses II, King 163
ramentum, sulphuratum 249
reagents, acids 297
realgar (arsenic sulphide) 175, 200, 201, 259, 261, 262, 310
  sources: Cappadocia, Carmania, Paphlagonia, Mysia 201
Reate 46
Recherches sur le monde, Beyet 39
Record Books 152
recrystallization temperature (bronze) 306
red lead 342
red ochre 134, 216, 262
red wine 133
reduction, *see individual metals*
reduction, shaft furnaces in 333

regeneration of metals and stones 178
Reichenau 384
rerum natura 39
reservoirs, water for hydraulicking 281
resin 61, 152
  American 233
  Pliny's account of indebted to Theophrastus 61
reumenen (cf. ῥεομένην) 322, 322 n. 172
réveil-matin 368 n. 60
Reynolds, Joyce 8, 33 n. 12
Rhazes 386
rhetoric, influence of 98
res Latina 391
Rhodes 124, 178, 369 n. 63
rhodochrosite 179
rhodonite 179
Rhône (r.) 250
Rickard, T. A, 377 n. 22
Ridley's Stripper 19 n. 61
Rio Tinto (Huelva) 122
roasting, *see individual metals*
Robert of Crichlade 385
rock-crystal 59, 60, 191, 224, 354
Rodari, Tomasso 24
Roman:
  Agora (Athens) 167
  coins, argentiferous bronze series 309
  empire 42
  ships 293
  silver plate 137
  theatre 159
Rome:
  Ambassador's house 362
  Antonine Column 365
  Ara Pacis 365
  Augustus' sundial 364 f.
  Campus Martius 364

Rome: – (*cont.*):
   Circus Maximus 37, 226
   Cloaca Maxima 351
   conditions in 26
   foundation date of 370
   great fire in 350
   Mausoleum of Augustus 365
   Mithraeum 350
   obelisk 163, 351
   Palatine Hill 4
   Porticus Vipsania 42 n. 4
   President's House 69
   Rostra 362
   San Clemente 350
   Senate House (Curia) 362
   roof tiles (*tegulae*), 'Peacock style' 19
rostra, sundial on column 363
Rottländer, R. C. A., 325, 356 n. 39
Rubellius Plautus 31, 32
rubies 354
ruddle, red ochre 175, 259
Ruteni 21
Ruwar (r.), tributary of Moselle 350

Sabaei 50
sail cloth, weaving of 21
St Elmo's Fire 152 f.
sal nitrum 358
salinity:
   of Dead Sea 119
   of oceans 117
salivation, poisoning by 376
Sallustian and Livian mines 22
Sallustius Crispus, C. 15
salt (sodium chloride) 134
   cementation process in gold-refining 285
   evaporation and precipitation 117 ff.
   evaporation, terms for 117

   natural 117, 118
   pans 119
   precipitation at night 119, 119 n. 19
   spontaneous production of 118
salutatio 7, 24
Samian earth 214
Samland (Peninsula) 253
Samos 260
Samosata (Commagene) 255
Samothracian rings, excitation of 157
Sandaracurgium (Mt.) 200
sandyx 259, 262
Santorini 239
'sapenos' 268
sapphire (*hyacinthus*) 268, 354
Saqqara and Thebes, gold-beating at 288
Saramatae 140
sarcophagus stone 52, 59, 70, 125, 173
sarda 264
Sardes, reconstruction of gold-processing at 285
Sardinia, earth of 248
sardonyx 263, 264
Sarpedon 52
Satyrs 16, 64
Satyrus 167, 250
'saw', used in cutting marble 161
'scaffolding' 163
Scapte Hyle, fumes from mines 377
schistos ($\sigma\chi\iota\sigma\tau os$), alum 194
   see also alumen, haematite
sciothericon ($\sigma\kappa\iota o\theta\eta\rho\iota\kappa o\nu$, τo) 363 n. 51
   see also $\sigma\kappa\iota o\theta\eta\rho\iota\kappa o\iota\ \gamma\nu\omega\mu o\nu\epsilon s$
Scipio Aemilianus 54
Scipio Nasica 367
sclerometric hardness 180
Scylla and Charybdis 63

General Index

Scyros 124, 209
scytanum, used in dyeing 141
Scythia 251, 278
Scythian language, words from 94
Scythians 85, 118
sea monsters 67 f., 141
sea-level 172
Second and Third Punic War 46 n. 31
secondary silica 294
'secondios' 268
Segobriga (Hither Spain) 225
segullum (segullo) 91, 279
Seianus, shrine of 237
selenite (σεληνίτης λίθος) 17, 119, 177, 180, 237
  mining and quarrying of 225, 226
  see also *lapis specularis*
selenium 275
semantics, exercise in 101
Senate House (Curia) 362
Seneca, see Annaeus
Septimus Severus, emperor, debasement of coinage by 309
Serapis, statue of 245
sered, Persian term for yellowish-red 264
sermons 384
serpentine 260
Serra Pelada (Brazil), alluvial gold at 298
Sesothis 37 n. 3
Seti I 37 n. 3
'Seven Voices', colonnade (ἑπτάφωνος στοά) 160
sewerage, at Rome 351
sex change 51
Sextus Niger, on lead 58, 92, 323
shadow tables 362
shadows, theory of 363

shaft furnaces, development of 332
  early Central European type 332
sheepskins, used to collect alluvial gold 275
Shepherd, R. on fire-setting 132 n. 59 *passim*
shode (tin), vein ore 316
shoemaker's black (*atramentum*) 122
Sicily 225, 244, 269
Sicyon, journey time to Elis from 360
siderite (σιδηρῖτις), meteorite composed of 258
Sidon 256, 352
Sierra Leone 45 n. 19
sieves, for ore washing 182
Silenus 69
  image in marble block 206
silex 133
silica 329 f., 354
siliceous schist, *see* lapis Lydius
silicon, part of earth's crust 327
silicon dioxide (quartz) 186
sillimanite crucible 285
silphium (σίλφιον τό) 13, 13 n. 49
silver 299 ff.
  argentiferous galena 221 f., 299, 300; *see also* galena
  chloride crystals, electrum–silver interface 285
  furnaces at Laurium 300 f.; trachyte, refractory material 300 f.
  gravity separation 169
  Laurium 222
  melting point 317
  overlay 138
  plate, Roman 137, 374
  prospecting for 179

silver – (cont.):
    Slave Revolt 300
    sources 300
    Turdetania, furnaces at 301
    washeries (flat table, helicoidal) 169 f.
Silver Latin 79
    debt to rhetoric 97
Sinan, architect of Suleiman the Magnificent 207 n. 61
single-lever press 164, 164 n. 51
Siphnos 338
Sirens 63, 63 n. 2
sirus (σίρος ὁ) 13, 15
Sisapo (Almaden) 218
    cinnabar mines at 341
Skala Fira (Santorini) 239
σκαλαυθρῖτις ἡ (*vericulum*), spit 322 n. 171
skeletons, in mine at Iconium 375
slag hearth, tin 318
Slave Revolt 300
smaragdus (emerald) 241 ff.
smelting temperature, in metallurgical operations 160
smithing temperature 160
smithsonite, see cadmea
Soanes 276
soapstone 225 n. 129
society, materialism of 78
'socos' 268
Socrates' disenchantment with natural science 104
soda:
    add to glass mix 353
    foam of 134, 198
    traders in 352
sodium carbonate 126
Solinus, *Collectanea* 381
σόλον αὐτοχόωνον, cast iron 331
solubility and saturation 116
solutes 123

solvents 130
Sophists 36, 104
Sophocles 251
Sotacus 59, 157, 185, 223
southern Attica 218
spadonia (σπάδων) 18
Spanish mining terms 91 f.
Sparta 138
Spathic iron ore 328
specific heat 160
spectrum, colours of 148
specular iron ore 222
Speculum Mundi 385
spelt, grain (*alicia*) 14
sphaerosiderite 328
sphalerite (zinc sulphide) 179
Spilia, marble quarries 204
spinel, red 265
springs, rivers, wells 167
Sri Lanka 43, 66
Stabiae 3
stagnum 319
stalactites 123 f.
stalagmites 124 f.
stannite 212
    associated with cassiterite, chalcopyrite, tetrahedrite 315
Star Catalogue, Eudoxus 73
star clock 362
starlings 68 n. 20
Statius Papinius, P. 375
steam engine 348
steam pressures 349
steam turbine, efficiency of 349
stear (στέαρ τό), fat, paste 340, 340 n. 230
steatite 225 n. 129
Steisichorus 98
sterilis materia 39
stibnite 181, 245 f., 339
    stimi, stibi, alabastrum, larbasis 134 f.
Stoicism 26, 28, 370

στόμωμα τό hardened iron 335
Stonehenge 163 n. 48
'stoneware' 356, 357
Strabo (of Amaseia) 43, 58, 178, 182, 200, 275, 285, 317, 332, 337
Strabo, unknown except for keen-sightedness 67
calculations of his claim 171 f.
Strapfeet 16
Straton 58, 176
straws, attracted by amber 154
streak, *see minerals ores passim*
streak (*sucus*) 84, 181, 184 f.
plate 185
strigiles, grains, or nuggets of gold 84
stringere aciem 336
stucco 211
style, Pliny's 97
styptic earth (στυπτηριώδης γῆ) 286
sualiternicum 251
subdialia 90
substrates, bronze, copper 291
succinic acid 253
succinite 253
sucinum, distillation from pine tree 252
sucus, see streak
Sudines 61, 265
Suetonius Paulinus, C. 11, 12, 49
Suetonius Tranquillus, C. 7
Suez Canal, stabilization of 239
sulfurata fila 249 n. 229
Sulmona 328
sulphides, produced by hard-boiled egg 128
sulphur 127, 152, 175, 186, 246 ff.
aetherium 246
allotropic forms of 247
dioxide, bleaching agent 248, 340, 376 n. 18

egula 248
epithets of 248
glaeba (clod) 248
lampwick as 249
native (*vivum*) 246
physical characteristics of 246
rhombic 247
sources of 246 f.
sulphuratum ramentum 249
thunder and thunderbolts, associated with 246
uses in medicine, religious ceremonies, siege warfare 249
sulphurous fumes, in wells 376
sun and moon, course of, adjusted 369 n. 62
sundial:
art of making 363
of Augustus 364
calibration of original 363 f.
Greeks obtained from Babylonians 362
history of 365 n. 57
readings, change of 366
types of 365
Sunium (Cape) 179
sun's rays, concentration of 149
superfetation, *see mirabilia* (births)
Suplja Stena (Belgrade), cinnabar extracted at early date 215
surgeons' masks 378 n. 26
surgical procedures 386 n. 28
Sverdlovsk 202
Sybaris 159
sylvanite, associations with 275
symbolum 86
Syme, R. 5 n.17, 8
Symmachus 381
Syracuse, Papyrus Institute 359 n. 41

Syria 211, 256, 260
Syrian colour 259
Syrian language, words from 94
Syrtes 13

Tacape 13
Tacitus, *see* Cornelius Tacitus
Tagus (r.) 278
tailings 282
Takovaya (r.) 242 n. 189
talc 157, 181
talking trees, *see mirabilia*
'tall stories' (*fabulosa*) 65
talo 91 n. 161
talutium 91, 279
Taprobane 66
tar pits 254
Tarentine Lake and Lake Cocanicus 117
Tarifa 55
Tartarus 170
tasconium, argillaceous earth for crucibles 92
Taygetus (Mt.) 244
technical literature, place of *Natural History* in 106 ff.
technical terms, shared with other Roman authors 95
technical vocabulary, *see* language and style 79 ff.
technical vocabulary, specific 83
technology 347 ff.
teeth, damage to 232
hardness of (Mohs scale) 232
telescope 151
telluric iron 327, 328
tellurides 275
tenement blocks (*insulae*) 4
tenor 279

Terentius Varro, M. 21, 36, 46, 57, 59, 93, 110, 151, 207, 363, 380, 391
terminations (-itis, -ites), of minerals 188
terminologies, two for lead 92
Tertiary 214
test paper (cf. litmus) 136
tetrahedrite 315
Thales (of Miletus) 103 f., 155
'The Clepsydra', prostitute's nickname 367 n. 59
Thebes (Egypt) 205
θεῖον τό (sulphur) 249 n. 227
Theocrestus 61
Theocritus 254 n. 20, 296 n. 92
Theodotus (of Smyrna) 261
Theognis (Megara) 294, 294 n. 86
Theomenes 251
Theophrastus 58, 59, 61, 124, 128, 157, 173, 176, 179, 189, 200, 205, 211 f., 215 n. 84, 217, 224, 247, 251, 265, 270, 294, 295, 310, 335, 341, 374
Thera, eruption of volcano 103, 103 n. 16
thermal springs 209
Thesprotia, *anthracites* found in 254
Thomas of Cantimpré, debt to Aristotle, Pliny, Solinus 385
Thoricus, silver mines at 243
Thracian shield 145
Thucydides 58
Thule, island of 361
Tiberius, emperor 1, 31, 36, 54, 55
Tiberius Gracchus, documents in the hand of 57
Tibullus, on marble in Rome 205

General Index

Tigellinus 32 n. 6
Timaeus 16, 45, 45 n. 29, 61
Timagenes (of Alexandria) 59, 218
timbering, in mineral quarries and mines 375
time and measurement of 351, 360 ff.
time-measuring instruments 362 ff.
  see also hour glass, portable sundial (vasum horoscopicum), shadow tables, star clock, sundial, water-clock (clepsydra)
Timochares 158
Timosthenes 45
tin 314 ff.
  alluvial (*alutiae*), tin-streaming 316
  deep lead, buried placers 316 f.
  dipping, in Gallic provinces 319
  igneous rocks and pegmatites, found in 315
  ingots 318; analyses of 318 n. 155; Port Vendres, found at 318; in Thames (r.) 318
  lead (*plumbum candidum*), confusion with 317
  mineral associations with 315
  mining technique 316 f.
  ores 315; *see also* cassiterite
  Pytheas' visit to the miners of Belericum 314 n. 142
  pewter, alloy with lead 320
  in quartz matrix (at Vaulry) 319
  shode, vein ore 316
  slag hearths 318
  smelting of 317 f; archaeological evidence for 318
  sources 317; *see also* cassiterite
  stick 319
  streaming 183, 316
  tinning 319 f.
  uses of 319 f.
  vein ore 314
  weight of 273
  wiping hot bronze with 317
'Tin Islands' 314
tin–lead alloys (pewter) 320
titanium 329
Titans and Zeus 102
Titus, emperor 26, 40
  contubernalis 7
  Preface of *Natural History* addressed to 106
Titus Petronius 374
Tmolus (Mt.) 244, 295
Tomasso Rodari 24
tooth powder (*dentifricia*) 239
topaz 154, 181
Topazos 265
topazin, to seek 94
Topkapi dagger, emeralds in 147
Topkapi Sarayi (Istanbul) 147 n. 10
tortoise shells, staining of 139
touch-needles 297
touchstone (*Lapis Lydius*) 84, 214, 224, 293 ff.
  terms for 294
tourmaline 154, 226
Tower of the Winds (Athens), sundial, water-clock, weather vane 367, 368
  reservoir attached to 368
toxic fumes, from Vesuvius eruption 23 n. 74
trachyte, Laurium furnaces 300
tragic poets, Aeschylus, Sophocles, Euripides 61
Transpadane Gaul 1
travertines 210
Treak Cliff (Derbyshire) 230

Trebius Niger 52
trees, unknown species 49
Treffiw, springs at 19
'trenched' gold 279
'tres partes', meaning of 356
  see also glass
trichitis (alum) 194
trilitha (τρίλιθα), at Stonehenge 163 n. 48
Trinidad (Jamaica), pitch lake 254
Triton 54, 67, 368
trivium 37
Troas 74 n. 8
Trogodytae (Cave-dwellers) 16, 266
Trogodyte language, words from 94
Trophy of the Alps, see La Turbie
tubuli, lead 322
Tullius Cicero, M. 28, 38, 86, 91, 98, 149, 151, 370, 388
Tullus Hostilius 110
Tungri 122
Tunis 151
tunnel, mine adit 11
Turbie, see La Turbie
turbistum, used in dyeing 141
Turdetania 276, 300
  furnace chimneys of 376
turpentine, from pine trees 257
Turranius Gracilis 55
tuyères 169
Twelve Tables 362
Tympaea 211
Typhoeus 102
Tyre 138
Tyrian purple 140, 141, 259

Umbrella feet 66
Umbrians, measurement of day 361
unio, specimen pearl 130
Universe:
  nature of 104
  origin of 103
unusual earth, copper/arsenic 310
Upper Cretaceous, chert occurring in 214
Urals 202
urium, ouros (cf. ὄρος τό) 92
  see also Greek mining terms
Utica 15, 119
Utopian education, architect of 348

vade-mecum 41
Valerius Catullus, C. 60, 238
Valerius Martialis, M. 231, 234, 249, 287, 289
Valerius Messala, M. 363
vanitas, Greek 30, 61, 250
Varro, see Terentius Varro
vasa horoscopica 366
vascular function 386 n. 28
Vaulry, tin found at 319
Velitrae 68
veneers 161
verdigris (copper sulphate) 135
Verrius Flaccus 46, 59, 391
Verus, see Lucius Verus
Vesalius 386 n. 28
Vespasian, emperor 5, 14, 24, 26, 32
Vesuvius 3, 23, 64
Vetters and Schaaber, nucleus ferri 336
Vincent of Beauvais 385, 385 n. 24
vinegar (acetum) 130 ff., 261
  and alum, in fire-gilding 290
  used in paper making, 359
  salt, in lead production 325
Virgilius Maro, P. 15, 93, 262 340
viscera 373
vitae lepos 26
Vitellius, emperor 55
viticulture 76

Vitruvius, Pollio M. 47, 47 n. 39, 61, 133, 162, 167, 217, 286
Vittori. O. 292
vivum, as epithet, *see amiantus*, mercury, sulphur
Vocontii (Gaul)_ 200, 327
Volcanic Explosivity Index (VEI) 103 n. 16
volumen 238
Vulcano 246

Wales, copper ingots from 305
Wallace-Hadrill, A. 98, 371 n. 1
wall-painting, pigments for 372
warfare, pitch and sulphur used in 249
washeries, gold ($\chi\rho\upsilon\sigma\iota\sigma\pi\lambda\upsilon\sigma\iota\alpha$ $\tau\acute{\alpha}$) 276
water:
 conduits 282
 properties of 123
 tests for the presence of 168
water-clocks 367 n. 59
waterfalls, petrified 210
 see also travertines
Waterford Crystal 354
watermen, in mines 169
watermill 350
wedges, used to split marble 206
werewolves 67
wheeled plough 349
whetstone (basanites) 185
'white gold' coinage 298
white marl 210, 211
wicker containers, *see* coracles
William of Conches 384
William Gilbert 152, 158
William Shakespeare 391 n. 41
window glass (*specularia*) 180, 26
wine presses 164
woad (*vitrum*), body paint used by ancient Britons 140 f.
wolves, (Egyptian) ants as big as 278 n. 30

Women Councillors 263
wrestler, imagery of 375
wrought iron 337
 [Aristotle] on 335

Xanten (Vetera) 5, 137
Xenocrates 47, 60, 61, 251
Xenophon 376, 391
 Cyropaedia 44 n. 15
 Memorabilia 104
 Poroi 58
Xenophon (of Lampsacus) 45
Xenophon (of Sicyon) 47, 60, 61, 251
X-ray Fluorescence (XRF) 295 n. 88
X-ray photoelectron spectroscopy (XPS) 295 n. 88

Younger Pliny , *see* Plinius Caecilius Secundus C.

Zacynthus 256
Zehnacker H. 137 n. 1, 236, 391
 fluorspar, colour changes in 232 n. 150
Zenothemis 60
Zeus:
 anger of 104
 Athena born from the head of 104
zinc ($\psi\epsilon\upsilon\delta\acute{\alpha}\rho\gamma\upsilon\rho\sigma\varsigma$ $\acute{o}$) 337 f.
 Andeira, ore from 337
 cadmea 203 f., 338
 distillation 337
 melting point of 338
 ores, *see cadmea*
 oxide 337
 refining and retorting, modern production by 338
 vapour pressure 338
zopissa 257
zura, wild thorn 94